Dedicated to
Lord Jagannath,
the Lord of the Universe

CONTENTS

FOREWORD

Today's science is tomorrow's technology. Bringing the intriguing features of science, especially physics, to the use of engineers has been an important aspect of scientific books in the present arena: Engineering Physics. Engineering students have worked their way to an understanding of applied physics with mathematics, the most fundamental of sciences – using various available textbooks in the market. Bringing the concepts of physics to the basic understanding level of the budding engineers has been always a challenge for physicist, which I believe the present authors have taken up.

A good physics teacher always explains the concepts through examples making a smooth transition from the basic understanding of a student to higher learning. What makes me happy in this book by Dr Sadasiva Biswal and Dr (Mrs) Manjusa Jena is the methodical and coherent presentation of concepts through plenty of examples and solved problems. The take home points/messages are crystal clear and mentioned as separate notes in the book. This will make the learners engaged and encourage them for further studies. At the end, the task should not be clearing an academic course, rather should be to encourage the next generation engineers to be innovative and contribute in nation building. Through this, I intend to encourage the learners to have a basic understanding of the concepts, which not only in the long run, would make an engineer/scientist self-consistent, but also help in the next generation training. The book has been highly benefitted through a more than 3-decades long teaching experience of Dr Biswal, which would be quite helpful for the learners to digest the concepts easily. The scholastic approach of Dr Jena in putting all steps in a mathematical derivation in a lucid way and clearly deriving the physical interpretations is noteworthy. These give a different value to the current book and I am sure the students would benefit from this book.

I sincerely hope, this book would bring differently flavoured insights in looking into the conceptual problems in physics, thereby bringing higher adaptability. I wish all the best for the authors and the readers of this book and would like to see the book a grand success.

Indore
1st January 2019

Dr Raghunath Sahoo
Associate Professor of Physics,
Indian Institute of Technology Indore, India

PREFACE

This book *Engineering Physics* deals with physics courses prescribed for engineering students. It contains topics of mechanics, wave and oscillations, optics, electricity, magnetism and quantum mechanics, etc. The narrations are lucid and easily understandable. Mathematical portions are clear and explanatory. A few problems relating to each topic are solved and included at the end of the chapter. The subject matter has been discussed comprehensively keeping in view the examinations of universities. Some examination-oriented questions are given at the end of each chapter. Overall, the book is suitable for preparing for examinations.

We hope the book, in its present form, will serve the students better. However, the scope for further improvement is always there. So, any constructive criticisms, comments and suggestions for improvement of the present work would be highly appreciated.

In spite of all our efforts and care, there might be some printing mistakes which should be brought to our notice by the readers for correction in the further editions of the book.

Authors

ACKNOWLEDGEMENTS

In the preparation of this book, we have followed many valuable texts and also have drawn a few material from some standard works. For this, we are highly obliged to those authors and publishers.

We owe our indebtedness to our fellow workers who have encouraged us to write a book of this form.

We thank our students for their suggestions for the improvement of our work.

The contributions of Mrs Madhusmita Maharana, Associate professor of Physics, Suddhananda Engineering and Research Centre in preparation of the manuscript is highly praiseworthy. We are very much thankful to her.

Lastly, we are greatly indebted to the publisher for bringing out the book in its present form.

Dr Sadasiva Biswal
Dr (Mrs) Manjusa Jena

PART I

Chapter 1

Simple Harmonic Motion

Periodic Motion

A non-uniformly accelerated motion is the periodic motion in which the body repeats its path of motion after a fixed interval of time. The to and fro, or vibrating, motion of objects stretched or bent from their normal positions and then released is particularly important in mechanics. Such an object moves back and forth along a fixed path, repeating over and over a fixed series of motions and returning to each position and velocity after a definite period of time. This type of motion is produced by varying forces and hence the body experiences varying accelerations. The periodic motion is often called harmonic motion.

If a particle in periodic motion moves back and forth over the same path, we call the motion oscillatory or vibratory. The world is full of oscillatory motions. Some examples are the up and down motions which ensue when a weight hanging from a spring is pulled down and released, the vibration of the string or air columns of organ pipes and musical instruments, the vibration of a bridge or building under impact loads, and oscillation of the balance wheels of a watch or of a clock pendulum, atoms in molecules or in a solid lattice, and air molecules as a sound wave passes by.

Not only mechanical systems can oscillate. Radio waves, microwaves and visible light are oscillating magnetic and electric field vectors. Thus, a tuned circuit in a radio and a closed metal cavity in which microwave energy is introduced can oscillate electromagnetically. Simplest type of periodic motion in which body moves along a straight line is called simple harmonic motion.

Potential Well

In Fig. 1.1, a typical potential energy curve of a particle is shown, in which the point B corresponds to minimum potential energy U_0, which is the point of stable equilibrium. The dotted line representing total energy E cuts the potential energy curve at two points A and C. Such a region (as from A to C) is called 'potential well' with point B as its bottom. Further, for small values of displacement, the potential energy may be expressed as a power series in displacement s, as

$$U = U_0 + \left(\frac{\partial U}{\partial s}\right)_0 s + \frac{1}{2!}\left(\frac{\partial^2 U}{\partial s^2}\right)_0 s^2 + \frac{1}{3!}\left(\frac{\partial^3 U}{\partial s^3}\right)_0 s^3 + ... \qquad ...(1)$$

where s is the displacement, measured from point B, U_0 is the minimum potential energy at point B and $\left(\dfrac{\partial U}{\partial s}\right)_0$, $\left(\dfrac{\partial^2 U}{\partial s^2}\right)_0$, etc. are the values of differential coefficients of U. At the point B, they are constant.

Now by choosing the zero of the potential energy so that the energy of equilibrium configuration is zero, $U_0 = 0$.

Furthermore, when the particle is in equilibrium position, its potential energy must be minimum for $s = 0$, hence $\left(\dfrac{\partial U}{\partial s}\right) = 0$.

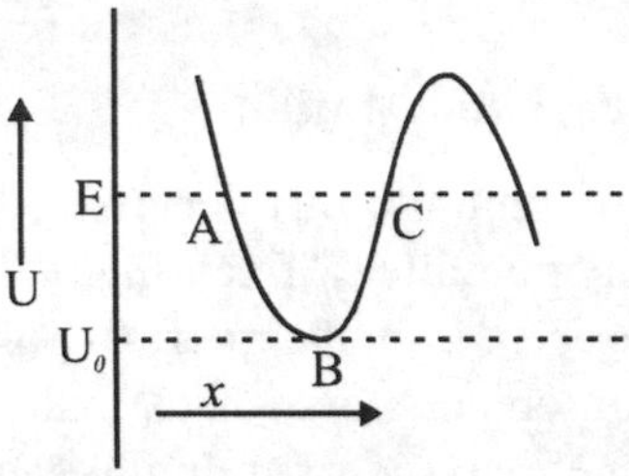

For sufficiently small amplitudes of vibration, the terms containing higher powers of s (i.e., s^3, s^4, etc.) can be neglected. Hence, equation (1) yields

Fig. 1.1

$$U = 0 + 0 + \frac{1}{2}\left(\frac{\partial^2 U}{\partial s^2}\right)_0 s^2 + 0 + \ldots$$

$$= \frac{1}{2}\left(\frac{\partial^2 U}{\partial s^2}\right)_0 s^2$$

or, $\qquad U = \dfrac{1}{2} ks^2 \qquad\qquad\qquad ...(2)$

where $\qquad k = \left(\dfrac{\partial^2 U}{\partial s^2}\right)_0$

Force acting on the particle is given by

$$F = -\frac{\partial U}{\partial s} = -\frac{\partial}{\partial s}\left(\frac{1}{2}ks^2\right)$$

$$= -\frac{1}{2}k\,(2s)$$

$$F = -ks \qquad\qquad\qquad ...(3)$$

This force F is a linear restoring force, having force constant k. In this region, the potential energy curve is parabolic. s is the displacement from the equilibrium position.

Simple Harmonic Motion

The type of vibrating motion in which the acceleration is proportional to the displacement and always directed towards the equilibrium position is called simple harmonic motion (SHM).

Mathematically the motion is given by

$$a \propto -s \qquad\qquad\qquad ...(1)$$

where a is the acceleration of the body and s is its displacement.

Hence, the acceleration is proportional to the displacement but always acts in the opposite direction.

Eqn. (1) can also be written as

$$a = - ks \qquad \qquad ...(2)$$

where k is known as the force constant.

Equation of Simple Harmonic Motion

From the definition of simple harmonic motion, we have

$$a \propto - s$$

or, $\qquad a = - ks$

where s is the displacement and a is the acceleration of the vibrating body.

If m is the mass of the body, then

$$ma = - ks$$

Since m is constant.

or, $\qquad a = - \dfrac{k}{m} s$

or, $\qquad a = - \omega^2 s \qquad \qquad ...(3)$

where $\qquad \omega^2 = \dfrac{k}{m}$

or, $\qquad \omega = \sqrt{\dfrac{k}{m}}$

Eqn. (3) is the equation of simple harmonic equation.

If s is the displacement, $\dfrac{ds}{dt}$ is the velocity and $\dfrac{d^2s}{dt^2}$ is the acceleration.

Putting this value of acceleration in eqn. (3), we get

$$\dfrac{d^2s}{dt^2} = - \omega^2 s$$

or, $\qquad \dfrac{d^2s}{dt^2} + \omega^2 s = 0 \qquad \qquad ...(4)$

This is the differential form of the equation of simple harmonic motion.

Solution of the Equation of SHM: The equation of simple harmonic motion is given by

$$\dfrac{d^2s}{dt^2} + \omega^2 s = 0 \qquad \qquad ...(1)$$

where s is the displacement,

t is the time, and

ω is the angular velocity.

Let the solution of the above eqn. be

$$s = A \sin \omega t \qquad \qquad ...(2)$$

Hence eqn. (2) must satisfy eqn. (1)

$$\frac{ds}{dt} = A\omega \cos \omega t$$

$$\frac{d^2 s}{dt^2} = - A\omega^2 \sin \omega t$$

Substituting this value of $\dfrac{d^2 s}{dt^2}$ in eqn. (1), we get

$$- A\omega^2 \sin \omega t + \omega^2 s = - \omega^2 s + \omega^2 s \quad \text{(Since } s = A \sin \omega t)$$
$$= 0$$

Thus, $s = A \sin \omega t$ is the solution of eqn. (1)

where s is the displacement of the vibrating body, and

A is the amplitude of motion.

Taking $s = B \cos \omega t$ as a solution, it will be proved that this will satisfy eqn. (1). So, $s = B \cos \omega t$ is also a solution of eqn. (1).

As there are 2 solutions of SHM, the sum and difference are also the solutions of SHM

i.e., $$s = A \sin \omega t \pm B \cos \omega t$$

The displacement of simple harmonic motion is given by

$$\boxed{s = A \sin \omega t} \qquad \qquad ...(3)$$

Velocity of Vibrating Body

$$V = \frac{ds}{dt} = \frac{d}{dt}(s)$$

$$= \frac{d}{dt}(A \sin \omega t)$$

$$= A\frac{d}{dt}(\sin \omega t)$$

i.e., $$V = A\omega \cos \omega t \qquad \qquad ...(4)$$

$$= A\omega\sqrt{1 - \sin^2 \omega t}$$

$$= A\omega\sqrt{1 - \frac{s^2}{A^2}}$$

$$= A\omega\sqrt{\frac{A^2 - s^2}{A^2}}$$

i.e., $$V = \frac{A\omega}{A}\sqrt{A^2 - s^2} = \omega\sqrt{A^2 - s^2} \qquad \qquad ...(5)$$

Maximum Velocity: At time $t = 0$, $\omega t = 0$, cos $\omega t = 1$ = maximum. Putting this value in eqn. (4), we get

$$V_{max} = A\omega \qquad \qquad ...(6)$$

Minimum Velocity: When

$$\omega t = \frac{\pi}{2}, \cos \omega t = 0 = \min^m$$

Putting this value in eqn. (4), we get

$$V_{min} = 0 \qquad \qquad ...(7)$$

Thus, the velocity of a particle executing simple harmonic motion has maximum value at the mean position and minimum value at the extreme positions.

Acceleration: From the definition of acceleration, we have

$$a = \frac{dv}{dt} = \frac{d}{dt}(v)$$

$$a = \frac{d}{dt}(A\omega \cos \omega t)$$

$$= A\omega \frac{d}{dt}(\cos \omega t)$$

$$= -A\omega^2 \sin \omega t \qquad \qquad ...(8)$$

$$= -\omega^2 s \text{ where } (A \sin \omega t = s) \qquad \qquad ...(9)$$

Maximum Acceleration: From eqn. (9) it is clear that when displacement is maximum, acceleration is maximum. But the displacement is maximum when

$$\sin \omega t = 1 = \sin \frac{\pi}{2}, \text{ i.e., } \omega t = \frac{\pi}{2}$$

Thus, the acceleration is maximum at the extreme positions.

Minimum Acceleration: Similarly from eqn. (9) it is seen that the acceleration is zero when s = 0, i.e. the acceleration is zero at the mean position. Again $s = 0 = A \sin \omega t$.

or, $\qquad \qquad \sin \omega t = 0 \qquad$ (Since A ≠ 0)

or, $\qquad \qquad t = 0$

i.e., the acceleration is zero at the mean position.

Conclusion: From the above discussion, it is clear that the acceleration is minimum where the velocity is maximum and the acceleration is maximum where the velocity is minimum.

The Projection of Uniform Circular Motion Upon a Diameter

When a body moves with uniform speed in a circle, the projection of this motion on a diameter is simple harmonic motion.

In Fig. 1.2, the body P is moving with uniform speed v_c and uniform angular speed ω in a circular path. The projection B moves up and down along the vertical diameter. Assume that the time is assigned a zero value when the body B passes through the equilibrium position C, moving upward. At any later time t the rotating body P will have moved through an angle $\theta = \omega t$. At this instant the projection B has a displacement.

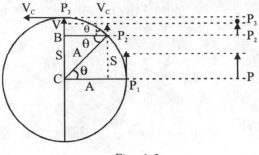

Fig. 1.2

$$s = A \sin \theta$$

Since $\qquad \omega = \dfrac{\theta}{t}, \theta = \omega t$

$$s = A \sin \omega t \qquad\qquad\qquad ...(1)$$

If a graph were made plotting various positions of s at different stages of SHM, the curve formed would be a sine curve (Fig. 1.3).

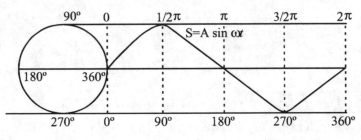

Fig. 1.3

The velocity v in the vibration is given by

$$v = \frac{ds}{dt} = \frac{d}{dt}(A \sin \omega t)$$

$$v = A\omega \cos \omega t$$

$$= Aw \cos \theta \qquad (\text{Since } \omega t = \theta)$$

$$= v_c \cos \theta \qquad\qquad\qquad ...(2)$$

where $\qquad v_c = A\omega$

$$= r\omega \quad (\text{Since } A = r)$$

r is the radius of the circle along which the body is moving.

Acceleration of B: The acceleration of B on the vertical diameter is given by

$$a = \frac{dv}{dt} = \frac{d}{dt}(v) = \frac{d}{dt}(A\omega \cos \omega t)$$

$$= A\omega \frac{d}{dt}(\cos \omega t)$$

$$= - A\omega^2 \sin \omega t$$

$$a = - a_c \sin \theta \quad \text{(where } \omega t = \theta \text{ and } A\omega^2 = a_c)$$

$$= - \omega^2 s \qquad \qquad ...(3)$$

$$\Rightarrow \qquad \qquad a \propto - s \qquad \qquad ...(4)$$

Thus, the acceleration of B is proportional to the displacement but is opposite in direction. Hence, the motion of B is simple harmonic motion. That means the projection of uniform circular motion upon a diameter is simple harmonic motion.

Amplitude: The amplitude of simple harmonic motion is defined as its maximum displacement from mean position.

The eqn. of SHM is given by

$$s = A \sin \omega t$$

When $\sin \omega t$ is maximum, the displacement becomes maximum. The maximum value of $\sin \omega t$ is equal to 1.

So $\qquad \qquad s = A$

where A is known as the amplitude of SHM

It is a scalar quantity. It is measured in m, cm and ft.

Time Period: The time period of a vibrating particle is defined as the time taken by the particle to complete one vibration. Mathematically is given by

$$T = \frac{2\pi}{\omega} \quad \left(\text{Since } \omega = \frac{2\pi}{T} \right)$$

From SHM, we have

$$a = - \omega^2 s$$

or, $\qquad \qquad \omega^2 = -\dfrac{a}{s}$

or, $\qquad \left(\dfrac{2\pi}{T}\right)^2 = -\dfrac{a}{s}$

or, $\qquad \dfrac{4\pi^2}{T^2} = -\dfrac{a}{s}$

or, $\qquad T^2 = -\dfrac{4\pi^2 s}{a}$

or, $\qquad T = 2\pi\sqrt{-\dfrac{s}{a}} \qquad \qquad ...(1)$

The period can be expressed in terms of the force constant of the spring or other agency that supplies the restoring force. The restoring force is given by

$$F = - ks$$

or, $\qquad ma = -ks$

or, $\qquad -\dfrac{s}{a} = \dfrac{m}{k}$

Substituting this value in eqn. (1) we get

$$T = 2\pi\sqrt{\dfrac{m}{k}} \qquad\qquad ...(2)$$

Time period is a scalar quantity. It is measured in seconds.

Frequency: The frequency of a vibrating body is defined as the number of oscillations made by the body in 1 sec. It is denoted by f. It is a scalar quantity. It is measured in cycles/sec or Hertz (Hz).

Relation Between Time Period and Frequency

From the definition of time period of SHM we know that T is the time taken for one complete oscillation. So,

in T sec the body completes 1 oscillation

in 1 sec the body completes $\left(\dfrac{1}{T}\right)$ oscillation.

From the definition of frequency we have

$$f = \dfrac{1}{T} \qquad\qquad ...(3)$$

where f is the frequency of SHM.

Phase: Phase of a particle is defined as its state as regards its position and direction of motion. The equation of SHM with the phase of the particle is given by

$$S = A \sin (\omega t \pm \phi) \qquad\qquad ...(4)$$

where ϕ is called the phase of the particle in SHM

Phase is a scalar quantity. It is measured in radians.

Energy in Simple Harmonic Motion

A particle executing simple harmonic motion vibrates to and fro about its mean position. Total energy 'E' of the particle, at any instant of time, is composed of two types of energies, viz.,

1. Kinetic energy, 2. Potential energy.

1. Kinetic energy (E_k): It is the energy possessed by the particle by virtue of its motion.

If 'v' is the instantaneous velocity of the particle, its kinetic energy 'E_k' is given by

$$E_k = \dfrac{1}{2}mv^2$$

$$= \dfrac{1}{2}m\left[\omega\sqrt{(A^2 - s^2)}\right]^2$$

where 's' is the instantaneous displacement and 'A' is its amplitude.

$$E_k = \frac{1}{2} m\omega^2 (A^2 - s^2) \qquad \qquad ...(1)$$

Since velocity of the particle is maximum in the mean position while it is at rest in the extreme position, its kinetic energy is *also maximum in mean position while it is zero at the extreme position.*

2. Potential energy (E_p): It is the energy possessed by the particle by virtue of its position.

As the particle moves away from the mean position, restoring force is always directed towards the mean position. So work has to be done to take it away. This work done in removing the particle to a position away from the mean position is called its potential energy. It can be calculated as follows:

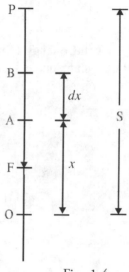

Let 'A' be the position of the particle of mass 'm' at any instant of time when its displacement is 'x'. Acceleration of the particle at A,

$$= -\omega^2 x$$

Force on the particle at A,

$$\left| \vec{F} \right| = -m\omega^2 x$$

Fig. 1.4

where negative sign indicates that it is directed towards the mean position.

If 'dW' is the work done in displacing it through a small distance 'dx' from A to B.

$$dW = \vec{F}.d\vec{x}$$

$$= F dx \cos 180° = -F dx$$

or, $\qquad dW = m\omega^2 x dx$

Work done 'W' in removing the particle from O to P can be obtained by integrating it between the limits o to s. This will be the potential energy 'E_p' of the particle at any instant.

$$E_p = W = \int_0^W dW$$

$$= \int_0^S m\omega^2 x \, dx$$

$$= m\omega^2 \int_0^S x \, dx$$

$$= m\omega^2 \left[\frac{x^2}{2}\right]_O^S$$

i.e.,
$$E_p = m\omega^2 \left(\frac{s^2}{2} - \frac{O}{2}\right)$$

i.e.,
$$E_p = \frac{1}{2} m\omega^2 s^2 \qquad \qquad ...(2)$$

Total energy (E): Total energy 'E' of the particle, at any instant of time, is the sum total of instantaneous kinetic and potential energies.

$$E = E_k + E_P$$

$$= \frac{1}{2} m\omega^2 (A^2 - s^2) + \frac{1}{2} m\omega^2 s^2$$

$$= \frac{1}{2} m\omega^2 (A^2 - s^2 + s^2)$$

i.e.,
$$\boxed{E = \frac{1}{2} m\omega^2 A^2} \qquad \qquad ...(3)$$

$$= \frac{1}{2} kA^2 \qquad \left(\because \omega^2 = \frac{k}{m}\right)$$

$\Rightarrow \qquad E \propto A^2$ (Since k is a constant)

From equation (3) it is clear that the instantaneous total energy is independent of the displacement of the particle, i.e. whatever the displacement may be, it is going to remain the same. In other words, total energy of a particle executing simple harmonic motion always remains constant. Kinetic energy 'E_k', potential energy 'E_p' and total energy 'E' plotted as a function of time are shown in Fig. 1.5. The curves indicate that 'E_k' and 'E_p' are complementary to each other. As the

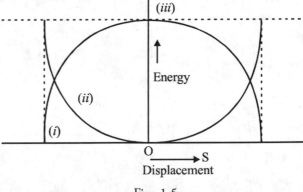

Fig. 1.5

particle moves away from mean position, kinetic energy gets transformed into potential energy till the transformation is completed at extreme position. While coming from extreme to mean position it gets converted from potential to kinetic energy. This conversion is in accordance with the laws of conservation of energy.

(i) Kinetic energy, (ii) Potential energy, (iii) Total energy.

(Energy versus displacement (i) Kinetic energy, (ii) Potential energy, (iii) Total energy.)

Simple Pendulum

An ideal 'simple pendulum' consists of a point mass suspended from a rigid support of infinite mass by a massless, inelastic, and perfectly flexible thread. In practice this is realised by taking a small bob of mass '*m*' suspended by a light inextensible thread of length 'L' from a fixed point.

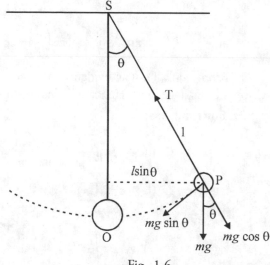

Fig. 1.6

When the bob is slightly drawn to a side and gently released, the bob will make a to and fro motion. The oscillations will be in a vertical plane about the equilibrium position 'O' (Fig. 1.6). The equation of motion of the bob can be derived in the following manner:

Let us consider the position P of the bob $\angle OSP = \theta$. The weight of the bob '*mg*' acts vertically downwards. This can be resolved into two components, '*mg* cos θ' along the thread and '*mg* sin θ' perpendicular to it. The tension T, acting along the thread, balances the component '*mg* cos θ'. Thus the component '*mg* sin θ' is the net force acting on the bob along the tangent to the arc of the circle which tends to restore the normal position of the bob.

Hence, the restoring force is '*mg* sin θ' which is directed towards the mean position, whereas the displacement of the bob is OP which is measured from the mean position. That means they oppose each other and the bob oscillates along the arc of a circle which can be approximated as a straight line provided the angular displacement 'θ' is small (within 4°). Now, restoring force,

$$F = mg \sin \theta \approx mg\, \theta \quad \text{(When } \theta \text{ is small)}$$

i.e.,

$$F = mg\, \frac{\text{Arc OP}}{\text{radius OS}}$$

$$= mg\left(\frac{-y}{l}\right)$$

[The negative sign is due to the fact that the displacement is in opposite direction to restoring force.]

Hence, acceleration, '*a*' is given by:

$$a = \frac{F}{m} = -\frac{mgy}{lm} = -\frac{gy}{l}$$

or, $a = -\left(\dfrac{g}{l}\right) y$

or, $a \propto -y \left[\text{Since}\left(\dfrac{g}{l}\right) = \text{constant}\right]$ ...(1)

From eqn. (1), it is evident that acceleration of the bob is directly proportional to the −ve of the displacement. Hence, the motion of the simple pendulum is simple harmonic.

Time period:

$$T = 2\pi\sqrt{-\dfrac{\text{displacement}}{\text{acceleration}}}$$

$$= 2\pi\sqrt{-\dfrac{y}{a}}$$

or, $T = 2\pi\sqrt{\dfrac{l}{g}}$ ...(2)

$$\left(\because -\dfrac{y}{a} = \dfrac{l}{g}\right)$$

Here, g = acceleration due to gravity
l = length of the pendulum

It is to be noted here that the time period of a simple pendulum is independent of the mass of the bob.

Second's Pendulum: A second's pendulum is that pendulum whose time period is 2 seconds.

For a second's pendulum,

$$T = 2\pi\sqrt{\dfrac{l}{g}}$$

$$2 = 2\pi\sqrt{\dfrac{l}{g}}$$

$\therefore$ $l = \dfrac{g}{\pi^2} = \dfrac{980}{\pi^2} = 99.25$ cm.

Therefore, the length of second's pendulum is 99.25 cm.

Laws of Simple Pendulum

Law of Isochronism: The oscillations of a pendulum are isochronous for small *amplitudes*. That means the pendulum takes equal time for each oscillation provided the amplitude of oscillation remains within 4°.

Law of Length: The time period of oscillation of a simple pendulum is directly proportional to square root of its length.

Mathematically,

$$T \propto \sqrt{l}$$

when 'g' is constant.

The length of a pendulum clock increases *in summer* due to which the time period increases and the clock loses time and *is said to run slower*. In winter, the length contracts due to which time period decreases and the clock gains time and is said to *run faster*.

Law of Acceleration: The time period of oscillation of a simple pendulum is inversely proportional to the square root of the acceleration due to gravity.

Mathematically,

$$T \propto \frac{1}{\sqrt{g}}$$

when 'l' is constant.

As the value of 'g' *decreases* (when the pendulum clock is taken from pole towards equator or from sea level to top of a mountain or bottom of a mine) *the time period increases* and the *clock loses time* and is said to *run slower*.

Again as the value of 'g' *increases* (when the pendulum clock is taken from equator to pole or from the top of a mountain or bottom of a mine to sea level) the *time period decreases* and the clock gains time and is said to *run faster*.

Law of Mass: *The time period* of oscillation of a simple pendulum *is independent of the mass* of material of the bob.

That means at a place as long as the length of the simple pendulum is kept constant the time period is same, whether one uses a brass bob or a copper bob, heavier bob or lighter bob, etc.

Torsion Pendulum

A torsion pendulum is a case of simple angular harmonic motion. It is achieved by suspending a heavy cylinder at the end of a thin rod. When this cylinder is twisted through a small angle θ about an axis through the rod and released, the cylinder will make angular oscillations which are harmonic in nature.

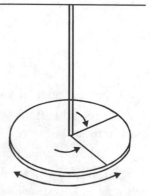

Theory: Here the rod supplies a restoring torque (L) proportion to the angle of twist which tries to bring the cylinder to the normal position.

Thus, $L = - c\theta$ ...(1)

where 'c' is the restoring torque per unit twist. Here, the negative sign indicates that L and θ are always in opposite directions.

Fig. 1.7

If the moment of inertia of the cylinder is I, then we have,

$$L = I.\alpha$$

where 'α' is the angular acceleration,

$$\alpha = \frac{d\omega}{dt}$$

where ω is the angular velocity

and

$$\omega = \frac{d\theta}{dt}$$

$$\therefore \quad \alpha = \frac{d^2\theta}{dt^2}$$

and hence

$$L = I\frac{d^2\theta}{dt^2}$$

Equation (1) then becomes

$$I\frac{d^2\theta}{dt^2} = -c\theta$$

or,

$$\frac{d^2\theta}{dt^2} = -\frac{c\theta}{I} \qquad \qquad ...(2)$$

or,

$$\alpha \propto -\theta$$

From eqn. (2) we find that the angular acceleration is directly proportional to the angular displacement (because c/I = constant) and is in opposite direction to it. So the torsion pendulum executes simple angular harmonic motion.

Time period:

$$T = 2\pi\sqrt{\frac{-\text{angular displacement}}{\text{angular acceleration}}}$$

$$= 2\pi\sqrt{-\frac{\theta}{\left(d^2\theta/dt^2\right)}}$$

or, $$T = 2\pi\sqrt{\frac{I}{c}} \qquad \qquad ...(3)$$

For the angular harmonic motion, I is the inertia factor and c is the spring factor.

Free vibration: When a body vibrates with its natural frequency, the vibration is called free vibration.

Example: vibration of air particles.

Damped vibration: The vibration in which the amplitude of vibration gradually decreased with time is known as damped vibration.

Example: the vibration of simple pendulum in air.

Forced vibration: When a body is forced to vibrate by an external agency whose frequency is different from the natural frequency of the body, the vibration of the body is called forced vibration. Here

$$n \neq f$$

where, n is the natural frequency of the body, and

f is the frequency of the external periodic force.

Example: The vibration of air particles in a sound box caused by the vibration of a tuning fork placed on the sound box. *The vibration of a hanging bridge caused by the marching of soldiers on it.*

Resonance: Resonance is the particular case of forced vibration. A body is said to be in resonance when its natural frequency of vibration becomes equal to the frequency of external periodic force, i.e.

$$n = f$$

where, n is the natural frequency of vibration of the body, and

f is the frequency of the external periodic force.

The amplitude of vibration becomes maximum at resonance.

Examples:

1. Resonance caused by radio circuits.
2. Resonance caused in sound box.
3. Resonance sets in hanging bridges due to marching of soldiers on it.
4. Resonance sets in musical instruments, like violin, harmonium, etc.
5. Resonance caused in air column, like in horns.

Various Other Methods of Solving the Equation of Simple Harmonic Motion

Method 1: In oscillatory motion, the restoring force is given by

$$F = - ks \qquad \qquad ...(1)$$

where, F is the restoring force,

k is the force constant, and

s is the displacement of the vibrating body.

From Newton's second law of motion, we know that

$$F = ma$$

so, $\qquad ma = - ks$

or, $\qquad a = -\dfrac{k}{m}s$

or, $\qquad \dfrac{d^2 s}{dt^2} = -\dfrac{k}{m}s$

or, $\qquad \dfrac{d^2 s}{dt^2} + \omega^2 s = 0 \qquad \qquad ...(2)$

where $\omega = \sqrt{\dfrac{k}{m}}$ = the angular frequency.

Now multiplying eqn. (2) by $2\dfrac{ds}{dt}$, and integrating with respect to t, we obtain

$$\int 2\frac{ds}{dt}\frac{d^2s}{dt^2}\,dt + \omega^2\int s\frac{ds}{dt}\,dt = 0$$

or, $\qquad \left(\dfrac{ds}{dt}\right)^2 = -\,\omega^2 s^2 + c$ ...(3)

where c is the constant of integration. When the displacement is maximum, the velocity is zero, i.e.,

when $\qquad s = \text{A}, \dfrac{ds}{dt} = 0$ ...(4)

The maximum displacement of the vibrating body from the mean position is called the amplitude of vibration.

Applying the condition (4) to (3), we have

$\qquad\qquad 0 = -\,\omega^2\text{A}^2 + c$

or, $\qquad c = \omega^2\text{A}^2,$ ...(5)

Substituting this value in (3), we obtain

$$\left(\frac{ds}{dt}\right)^2 = -\,\omega^2 s^2 + \omega^2\text{A}^2$$

$$= \omega^2(\text{A}^2 - s^2)$$

or, $\qquad \text{V}^2 = \omega^2(\text{A}^2 - s^2)$

or, $\qquad \text{V} = \omega\sqrt{\text{A}^2 - s^2}$ ...(6)

The eqn. (6) gives the velocity of the harmonic oscillator at any position.

Now, $\qquad \dfrac{ds}{dt} = \omega\sqrt{\text{A}^2 - s^2}$

or, $\qquad \dfrac{ds}{\sqrt{\text{A}^2 - s^2}} = \omega dt$

Integrating this equation, we have

$$\sin^{-1}\left(\frac{s}{\text{A}}\right) = \omega t + \phi$$

where ϕ is the another constant of integration, which can be known from initial conditions.

Hence, $\qquad \dfrac{s}{A} = \sin(\omega t + \phi)$

or, $\qquad s = A \sin(\omega t + \phi)$ ...(7)

This is the required solution of the equation of simple harmonic motion. Here, $(\omega t + \phi)$ represents the total phase of the body at any time t. With change of time t, $\sin(\omega t + \phi)$ varies between $+1$ and -1 and corresponding displacement varies from $+A$ to $-A$. The phase constant ϕ is called the initial phase of the vibrating body. If we count time from the mean position ($s = 0$), then $\phi = 0$.

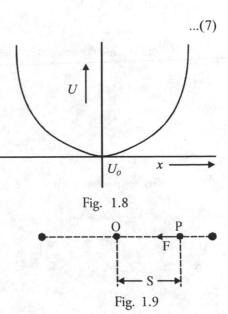

Fig. 1.8

Fig. 1.9

Method 2: Let us consider a particle P of mass m executing simple harmonic motion about an equilibrium position O.

The restoring force under the action of which the particle is oscillating, is given by

$$F = -ks$$ ...(1)

where s is the displacement of the particle from the mean position O, k is the force constant and F is the restoring force. During time t, the particle is displaced from O to P, i.e. its displacement is s.

According to Newton's second law, the force acting on the particle is equal to the product of the mass and the acceleration. Hence,

$$F = ma = m\dfrac{d^2s}{dt^2}$$ ...(2)

From (1) and (2), we have

$$m\dfrac{d^2s}{dt^2} = -ks$$

or, $\qquad \dfrac{d^2s}{dt^2} = -\dfrac{k}{m}s$

or, $\qquad \dfrac{d^2s}{dt^2} = -\omega^2 s \qquad \left(\text{taking } \dfrac{k}{m} = \omega^2\right)$

or, $\qquad \dfrac{d^2s}{dt^2} + \omega^2 s = 0$ ...(3)

This is the differential equation of a simple harmonic oscillator.

Let its solution be

$$s = ce^{\alpha t} \qquad \qquad ...(4)$$

where c and α are arbitrary constants. On differentiating (4) with respect to t, we obtain

$$\frac{ds}{dt} = c\alpha\, e^{\alpha t}$$

and $\qquad \dfrac{d^2 s}{dt^2} = c\alpha^2\, e^{\alpha t}$

Substituting the values of s and $\dfrac{d^2 s}{dt^2}$ in eqn. (3), we obtain

$$c\alpha^2\, e^{\alpha t} + \omega^2 c\, e^{\alpha t} = 0$$

or, $\qquad ce^{\alpha t}\,(\alpha^2 + \omega^2) = 0$

or, $\qquad \alpha^2 + \omega^2 = 0$

or, $\qquad \alpha^2 = -\omega^2 = i^2\omega^2$

or, $\qquad \alpha = \pm\, i\omega$

where, $\qquad i = \sqrt{-1},\ \alpha_1 = +\, i\omega$ and $\alpha_2 = -\, i\omega$

Hence, the general solution of eqn. (3) is

$$s = c_1 e^{\alpha_1 t} + c_2 e^{\alpha_2 t}$$
$$= c_1\, e^{+\, i\omega t} + c_2\, e^{-\, i\omega t} \qquad ...(5)$$

where c_1 and c_2 are the arbitrary constants. From eqn. (5),

$$s = c_1\,(\cos \omega t + i \sin \omega t) + c_2\,(\cos \omega t - i \sin \omega t)$$
$$= (c_1 + c_2)\cos \omega t + i\,(c_1 - c_2)\sin \omega t$$

Let us take

$$c_1 + c_2 = A \sin \phi$$

and $\qquad i(c_1 - c_2) = A \cos \phi$

where A and ϕ are new constants. Then,

$$s = A \sin \phi \cos \omega t + A \cos \phi \sin \omega t$$
$$= A\left[\sin \omega t \cos \phi + \cos \omega t \sin \phi\right]$$

or, $\qquad s = A \sin (\omega t + \phi) \qquad \qquad ...(6)$

This is the solution of the equation of SHM, where A is the amplitude of vibration and ϕ is the phase constant. The values of A and ϕ depend upon the commencement of the motion.

Method 3: Let us consider the case of a vibrating body represented in Fig. 1.10. At some instant of time, the displacement of the body from the mean position is s. The resultant force on it is simply the elastic restoring force F given by

$$F = - ks \qquad \qquad ...(1)$$

where k is the force constant.

From Newton's second law of motion,

$$F = ma$$

so, $$ma = - ks$$

or, $$m\frac{d^2s}{dt^2} = - ks$$

or, $$m\frac{d^2s}{dt^2} + ks = 0 \qquad \qquad ...(2)$$

This is a second order differential equation for the displacement s as a function of the time t.

Now, in order to solve equation (2), we can write

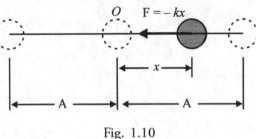

Equilibrium position

Fig. 1.10

$$\frac{d^2s}{dt^2} = \frac{d}{dt}\left(\frac{ds}{dt}\right) = \frac{dv}{dt}$$

$$= \frac{dv}{ds}\cdot\left(\frac{ds}{dt}\right) = \frac{dv}{ds}\,v$$

$$= v\frac{dv}{ds}$$

Substituting this value of $\frac{d^2s}{dt^2}$ in equation (2), we have

$$mv\frac{dv}{ds} + ks = 0 \qquad \qquad ...(3)$$

Instead of a second order differential equation for s as a function of t, we now have a first order equation for v as a function of s. This equation (3), however, can readily be integrated. After multiplying through by ds, we obtain

$$\int mv\,dv + \int ks\,ds = 0$$

which on integration yields

$$\frac{1}{2}mv^2 + \frac{1}{2}ks^2 = c_1, \qquad \qquad ...(4)$$

The first term on the left is the kinetic energy E_k of the body and the second is its elastic potential E_p. Equation (4) therefore states that the total energy of the system is constant, and the integration constant c_1 equals the total energy E. This is only true for a conservative system. Thus,

$$E_k + E_p = E \qquad \qquad ...(5)$$

In simple harmonic motion, the velocity is maximum at the mean position of the vibrating body and it is zero at the extreme positions. When velocity is maximum, kinetic energy is maximum and potential energy is zero. Similarly, when velocity is zero, the kinetic energy is zero and the potential energy is maximum. So, at the mean position of the vibrating body,

$$(E_k)_{max} = E \qquad (\because E_p = 0)$$

or, $\qquad \dfrac{1}{2} m v_{max}^2 = E$

or, $\qquad v_{max} = \pm \sqrt{\dfrac{2E}{m}} \qquad \qquad ...(6)$

The sign of v_{max} being positive or negative depending on the direction of motion.
On the other hand, at the extreme position of the motion,

$$(E_p)_{max} = E$$

or, $\qquad \dfrac{1}{2} k s_{max}^2 = E$

or, $\qquad s_{max} = \pm \sqrt{\dfrac{2E}{k}}$

or, $\qquad A = |s_{max}| = \sqrt{\dfrac{2E}{k}} \qquad \qquad ...(7)$

where A is a amplitude of motion, which is the maximum displacement from the mean position.

The velocity v at any displacement s is

$$v = \sqrt{\dfrac{2E - ks^2}{m}} \qquad \text{(from eqn. 4)}$$

$$= \sqrt{\dfrac{kA^2 - ks^2}{m}} \qquad \text{(from eqn. 6)}$$

$$= \sqrt{\dfrac{k}{m}} \left(\sqrt{A^2 - s^2} \right) \qquad \qquad ...(8)$$

We can now replace v with $\dfrac{ds}{dt}$.

so, $\dfrac{ds}{dt} = \sqrt{\dfrac{k}{m}}\left(\sqrt{A^2 - s^2}\right)$

or, $\dfrac{ds}{\sqrt{A^2 - s^2}} = \sqrt{\dfrac{k}{m}}\, dt$

Integrating both sides, we obtain

$$\sin^{-1}\left(\dfrac{s}{A}\right) = \sqrt{\dfrac{k}{m}}\, t + c_2 \qquad ...(9)$$

Let s_0 be the value of s at $t = 0$. The integration constant c_2 is then

$$c_2 = \sin^{-1}\left(\dfrac{s_0}{A}\right) \qquad ...(10)$$

That is, c_2 is the angle (in radians) whose sine equals $\dfrac{s_0}{A}$. Let us represent this angle by ϕ_0. So,

$$\sin \phi_0 = \dfrac{s_0}{A} \qquad ...(11)$$

(Note that this equation does not completely determine the angle ϕ_0, since $\sin(\pi - \phi) = \sin \phi$). Equation (9) can now be written as

$$\sin^{-1}\left(\dfrac{s}{A}\right) = \sqrt{\dfrac{k}{m}}\, t + \sin^{-1}\left(\dfrac{s_0}{A}\right)$$

$$= \sqrt{\dfrac{k}{m}}\, t + \phi_0$$

or, $\dfrac{s}{A} = \sin\left(\sqrt{\dfrac{k}{m}}\, t + \phi_0\right)$

or, $s = A\sin\left(\sqrt{\dfrac{k}{m}}\, t + \phi_0\right) \qquad ...(12)$

Thus, the displacement s is therefore a sinusoidal function of the time t. The term in parenthesis is an angle in radians. It is called the phase angle, or simply the phase of the motion. The angle ϕ_0 is the initial phase angle, and is also referred to as the **epoch angle**.

Systems Vibrating in SHM

There exist mechanical, electrical and magnetic systems which execute simple harmonic motions. When a body executes simple harmonic motion, two forces act on it, one is the **deflecting force** and the other is the **restoring force**.

(a) **Deflecting force:** The external force which deflects the body from its mean position. It can be represented as $m\dfrac{d^2x}{dt^2}$.

(b) **Restoring force:** It is the force which restores the body from its deflected position to its mean position. It is equal in magnitude and opposite in direction of the deflecting forces. It can be represented as $-kx$.

Example 1: Compound pendulum: Any rigid body which is so mounted that it can swing in a vertical plane about a horizontal axis passing through it is called a 'physical' or 'compound' pendulum.

Fig. 1.11 represents the vertical section of an irregular rigid body pivoted at a point S in it. In the equilibrium position, the centre of mass C of the body lies vertically below S. When the body is displaced to one side and then released, it oscillates in the vertical plane containing S and C about a horizontal axis perpendicular to this plane and passing through S. It is now acting as a physical pendulum.

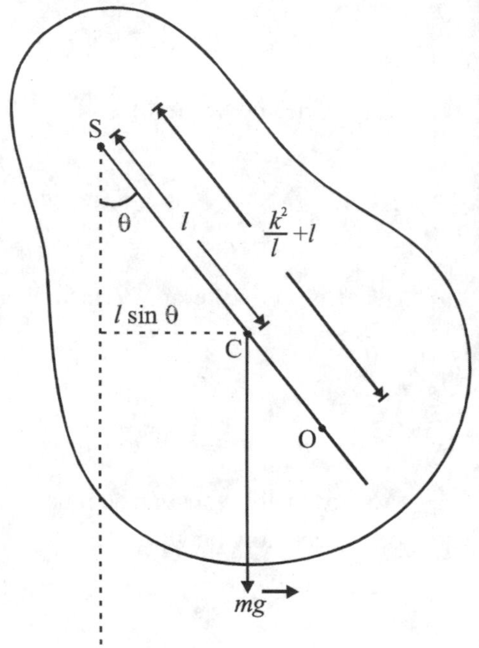

Let m be the mass of the body and l be the distance of C from S. Suppose at some instant during the oscillation, the body makes an angle θ with the vertical. At this instant, the moment of its weight mg (acting vertically at C) about the axis of oscillation through S is $mg\,(l\sin\theta)$. This is the restoring torque τ (say) at this instant tending to bring the body to its equilibrium position. Thus,

Fig. 1.11

$$\tau = -\,mgl\sin\theta$$

The negative sign is used because the torque acts in the direction of θ decreasing. If the angle θ is small, then $\sin\theta = \theta$ (in radian). Therefore,

$$\tau = -\,mgl\theta \qquad \qquad ...(1)$$

If I be the moment of inertia of the body about the axis through S, and $\dfrac{d^2\theta}{dt^2}$, the instantaneous angular acceleration. Then, the torque acting on the body at this instant is given by

$$\tau = I\frac{d^2\theta}{dt^2} \qquad \qquad ...(2)$$

From (1) and (2), we obtain

$$I \frac{d^2\theta}{dt^2} = - mgl\theta$$

or,

$$\frac{d^2\theta}{dt^2} = -\frac{mgl}{I}\theta = -\omega^2\theta$$

where,

$$\omega^2 = \frac{mgl}{I} \text{ (a constant)}$$

Hence,

$$\frac{d^2\theta}{dt^2} \propto -\theta$$

or,

$$\alpha \propto -\theta \qquad \qquad \qquad ...(3)$$

That is, the angular acceleration α is proportional to the angular displacement and always acts towards the mean position. The motion is, therefore, simple harmonic and its time period is given by

$$T = \frac{2\pi}{\omega} = 2\pi\sqrt{\frac{I}{mgl}} \qquad \qquad ...(4)$$

Let K be the radius of gyration of the pendulum about an axis passing through its centre of mass C and perpendicular to the plane of oscillation. Then, its moment of inertia about this axis will be mK^2. Therefore, the moment of inertia I about the axis through S will be

$$I = mK^2 + ml^2 \quad \text{(by theorem of parallel axes)}$$

$$\therefore \qquad T = 2\pi\sqrt{\frac{mK^2 + ml^2}{mgl}}$$

or,

$$T = 2\pi\sqrt{\frac{K^2 + l^2}{gl}}$$

$$= 2\pi\sqrt{\left(\frac{\dfrac{K^2}{l} + l}{g}\right)} \qquad \qquad ...(5)$$

This is the expression for the time period of oscillation of a compound pendulum.

Centre of oscillation: The period of oscillation of a physical pendulum is given by

$$T = 2\pi\sqrt{\left(\frac{\dfrac{K^2}{l} + l}{g}\right)}$$

Comparing it with the period of a simple pendulum

$$T = 2\pi\sqrt{\frac{L}{g}}$$

We observe that the period of a physical pendulum is the same as that of a simple pendulum of length $L = \left(\dfrac{K^2}{l} + l\right)$. Thus, $\left(\dfrac{K^2}{l} + l\right)$ is the 'length of the equivalent simple pendulum'.

Let us suppose that the whole mass of the physical pendulum is concentrated at a point O on SC produced, such that

$$SO = \left(\frac{K^2}{l} + l\right)$$

Thus, it would be like simple pendulum of the same period. The point O which is at a distance equal to the length of the equivalent simple pendulum from S, is called the 'centre of oscillation' corresponding to the centre of suspension S.

Interchangeability of the centres of suspension and oscillation: When the body is suspended from the centre of suspension S, its period of oscillation is given by

$$T = 2\pi \sqrt{\left(\frac{\dfrac{K^2}{l} + l}{g}\right)}$$

where l is the distance of S from C.

Let the body be now inverted and suspended from the centre of oscillation O. The distance of O from C is $\dfrac{K^2}{l}$. The new period of oscillation T′ is therefore obtained by substituting $\dfrac{K^2}{l}$ for l in the above expression, i.e.,

$$T' = 2\pi \sqrt{\frac{\left(\dfrac{K^2}{K^2/l}\right) + \left(\dfrac{K^2}{l}\right)}{g}}$$

$$= 2\pi \sqrt{\frac{l + \dfrac{K^2}{l}}{g}}$$

Hence, $\qquad T = T'$

That is, the period of oscillation about S and O are equal. Ostensibly, the centres of suspension and oscillation are interchangeable.

Example 2: Spring and mass system

Let us consider a massless spring of length l hanging vertically (Fig. 1.12a). When a block of mass m is attached to its lower end, its length extends say by x_0 (Fig. 1.12b). The spring due to its elasticity, exerts an (restoring) upward force F on the mass m. This force is, Hooke's law, given by

$$F = -kx_0$$

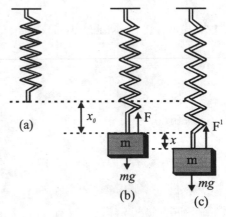

(a)

(b)   (c)

Fig. 1.12

where k is the force-constant of the spring and the extension x_0 is small. The positive force is taken in the downward direction. The other force acting on the mass is its weight, $+ mg$ (downward). Thus, the total force acting on m is $F + mg$. But the mass has no acceleration, and hence the total force, acting on it must be zero. That is,

$$F + mg = 0$$

$$\text{or, } -kx_0 + mg = 0$$

$$\text{or,} \qquad x_0 = \frac{mg}{k} \qquad\qquad ...(1)$$

This is the static equilibrium position of the block 'm'.

Now, suppose that, during the oscillation, the total extension in the spring is $x_0 + x$ (Fig. 1.12c). The (upward) force exerted on m by the spring is then

$$F' = -k(x_0 + x)$$

$$= -k\left(\frac{mg}{k} + x\right) \quad ...(\text{from eqn. 1})$$

$$= -mg - kx \qquad\qquad ...(2)$$

The other force acting on the mass m is its weight, $+ mg$ (downward). Therefore, the net force acting on the block is

$$F'' = F' + mg$$

$$= (-mg - kx) + mg \quad ...(\text{from eqn. 2})$$

$$= -kx \qquad\qquad ...(3)$$

This, by Newton's second law, must be m times the instantaneous acceleration $\dfrac{d^2x}{dt^2}$ of the body. That is

$$F'' = -kx = m\frac{d^2x}{dt^2}$$

$$\text{or} \qquad \frac{d^2x}{dt^2} = -\frac{k}{m}x = -\omega^2x \qquad\qquad ...(4)$$

where $\omega^2 = \dfrac{k}{m}$. This is the equation of the SHM. Thus, the block of mass m executes

SHM about the position given by $x_0 = \dfrac{mg}{k}$. The time period of oscillation of the block is

$$T = \frac{2\pi}{\omega} = 2\pi\sqrt{\frac{m}{k}} \qquad \qquad ...(5)$$

From eqn. (1), we have $k = \dfrac{mg}{x_0}$. Therefore,

$$T = 2\pi\sqrt{\frac{x_0}{g}} \qquad \qquad ...(6)$$

Equations (5) and (6), both can be used to determine the period of oscillation of the block-spring system. Eqn. (6) has the advantage that we can find the period of oscillation simply by measuring the extension x_0 of the spring, without knowing the mass of the block and the force constant of the spring.

Example 3: Oscillations of a mass fitted with two springs

Let us consider a body of mass 'm' fitted with two springs of force constants k_1 and k_2. The spring of force constant k_1 is tightly fixed with a rigid support as shown in the Fig. 1.13.

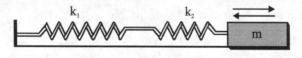

Fig. 1.13

When the mass m oscillates, both the springs are stretched (or compressed) simultaneously. Let x be the displacement of m from its equilibrium position at an instant and x_1, x_2 be the extensions (or contractions) in the length of the springs k_1, k_2 respectively at that instant. Then,

$$x = x_1 + x_2 \qquad \qquad ...(1)$$

The same elastic (restoring) force F is developed in each spring. By Hooke's law, we have

$$F = -k_1 x_1 = -k_2 x_2$$

so, $$x_1 = -\frac{F}{k_1} \text{ and } x_2 = -\frac{F}{k_2}$$

Substituting these values in eqn. (1), we obtain

$$x = -\frac{F}{k_1} - \frac{F}{k_2}$$

or, $$F = -\left(\frac{k_1 k_2}{k_1 + k_2}\right)x \qquad \qquad ...(2)$$

Thus, the force constant of the system is $\dfrac{k_1 k_2}{k_1 + k_2}$. Eqn. (2) can be written as

$$m \frac{d^2 x}{dt^2} = - \frac{k_1 k_2}{k_1 + k_2} x$$

or,

$$\frac{d^2 x}{dt^2} = - \frac{(k_1 k_2 / k_1 + k_2)}{m} x$$

or,

$$\frac{d^2 x}{dt^2} = - \omega^2 x \qquad \qquad ...(3)$$

where,

$$\omega^2 = \frac{(k_1 k_2 / k_1 + k_2)}{m}$$

or,

$$\omega = \frac{\sqrt{(k_1 k_2 / k_1 + k_2)}}{m}$$

Eqn. (3) is the equation of SHM. Hence, the mass executes simple harmonic motion. Such type of oscillations is called bifilar oscillations and the arrangement of a body connected with two springs as shown in Fig. 1.13 is known as bifilar arrangement.

Period of oscillation: The period of the bifilar oscillations is given by

$$T = \frac{2\pi}{\omega} = 2\pi \sqrt{\frac{(k_1 + k_2) m}{k_1 k_2}}$$

Frequency of oscillation: The frequency of oscillation is given by

$$n = \frac{1}{T} = \frac{\omega}{2\pi} = \frac{1}{2\pi} \sqrt{\frac{k_1 k_2}{(k_1 + k_2) m}}$$

Example 4: Oscillations of a mass fitted with two springs, one on either side of the spring

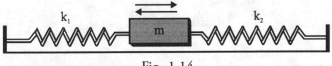

Fig. 1.14

When the mass m oscillates, then at any instant one spring is stretched and the other is compressed or vice-versa. Let x be the displacement of m from its equilibrium position at an instant. This will also be the extension in one and the compression in the other spring.

The restoring force developed in the two springs are:

$$F_1 = - k_1 x \text{ and } F_2 = - k_2 x$$

and act on the mass along the same direction. k_1 and k_2 are the force constants of the two springs. The total restoring force is

$$F = F_1 + F_2 = - (k_1 + k_2)\ x$$

Thus, the force constant of the system is $(k_1 + k_2)$, and the equation of oscillation is

$$m\frac{d^2x}{dt^2} = - (k_1 + k_2)\ x$$

or, $$\frac{d^2x}{dt^2} = -\frac{(k_1 + k_2)}{m}\ x$$

$$= - \omega^2 x \qquad \qquad ...(1)$$

Thus, the spring executes simple harmonic motion, where

$$\omega^2 = \frac{k_1 + k_2}{m}$$

or, $$\omega = \sqrt{\frac{k_1 + k_2}{m}} \qquad \qquad ...(2)$$

Period of oscillation: The period of oscillation of the spring is given by

$$T = \frac{2\pi}{\omega} = 2\pi\sqrt{\frac{m}{k_1 + k_2}} \qquad \qquad ...(3)$$

Frequency: The frequency of oscillation of the spring is given by

$$n = \frac{1}{T} = \frac{1}{2\pi}\sqrt{\frac{k_1 + k_2}{m}} \qquad \qquad ...(4)$$

Example 5: Motion of a floating cylinder

Let us consider a cylinder C in Fig. 1.15 of the material of density ρ, length l and radius r. Let this cylinder be floating in a liquid of density σ.

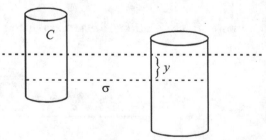

Fig. 1.15

Now, this cylinder is pressed such that it goes inside the liquid by a distance y, then released.

Deflecting force = (mass of the cylinder) × (acceleration)

$$= (\pi r^2 l\rho) \times \left(\frac{d^2y}{dt^2}\right) \qquad \qquad ...(1)$$

Restoring force = upward buoyant force

= weight of the displaced liquid

$$= - (\pi r^2 y \sigma)\ g \qquad \qquad ...(2)$$

Deflecting force = Restoring force

So, $\qquad (\pi r^2 l \rho) \times \dfrac{d^2 y}{dt^2} = - (\pi r^2 y \sigma)\ g$

or, $\qquad \dfrac{d^2 y}{dt^2} = - \left(\dfrac{g \sigma}{l \rho} \right) y$

$$= - \omega^2 y \qquad \qquad ...(3)$$

where, $\quad \omega^2 = \dfrac{gr}{l\rho}$ or $\omega = \sqrt{\dfrac{g\sigma}{l\rho}}$

Eqn. (3) shows that the motion of the cylinder is simple harmonic.

Time period of oscillation:

Here, $\qquad \omega = \sqrt{\dfrac{g\sigma}{l\rho}}$

or, $\qquad \dfrac{2\pi}{T} = \sqrt{\dfrac{gr}{l\rho}}$

or, $\qquad T = 2\pi \sqrt{\dfrac{l\rho}{g\sigma}} \qquad \qquad ...(4)$

Frequency: The frequency of oscillation is

$$n = \frac{1}{T} = \frac{1}{2\pi} \sqrt{\frac{g\sigma}{l\rho}} \qquad \qquad ...(5)$$

Example 6: Motion of a liquid in a U-tube

Let us consider a U-tube clamped vertically and a liquid of density ρ is filled in it, such that the liquid tube of length l levels at AA (Fig. 1.16a).

Now liquid level in one of the limbs of U-tube is pressed through a distance x and then released. Fig. 1.16b represents the displaced level of the liquid in the U-tube.

From the Fig. 1.16b, it is clear that the total displacement of the liquid is $2x$.

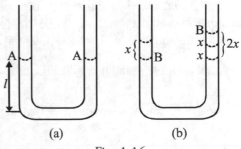

(a)   (b)

Fig. 1.16

Deflecting force = (mass of the liquid in the tube) × acceleration

$$= (\pi r^2 l \rho)\left(\frac{d^2x}{dt^2}\right) \qquad ...(1)$$

Restoring force $= - (\pi r^2)(2x)\rho g$

$$= - (2\pi r^2 \rho g)x \qquad ...(2)$$

Deflecting force = Restoring force

$$(\pi r^2 l \rho)\frac{d^2x}{dt^2} = - (2\pi r^2 \rho g)x$$

or,

$$\frac{d^2x}{dt^2} = -\left(\frac{2g}{l}\right)x$$

$$= - \omega^2 x \qquad ...(3)$$

where,

$$\omega^2 = \frac{2g}{l} \text{ or } \omega = \sqrt{\frac{2g}{l}}$$

Equation (3) shows that the motion of the liquid in the U-tube is simple harmonic.

Time period of oscillation: We have

$$\omega = \sqrt{\frac{2g}{l}}$$

or,

$$T = \frac{2\pi}{\omega} = 2\pi\sqrt{\frac{l}{2g}} \qquad ...(4)$$

Frequency of oscillation: Frequency of oscillation is

$$n = \frac{1}{T} = \frac{1}{2\pi}\sqrt{\frac{2g}{l}} \qquad ...(5)$$

Example 7: L-C circuit

Let us consider a capacitor of capacitance C, carrying a charge q_0, be discharged through a coil of inductance L and negligible resistance (Fig. 1.17).

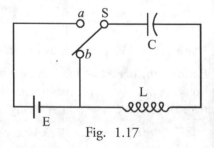

Fig. 1.17

At an instant t during the discharge, let q be the charge left on the capacitor and i be the discharging current through the inductance. Then the potential difference across the capacitor is $\frac{q}{C}$ and the induced e.m.f. in the inductance

is $L\left(\frac{di}{dt}\right)$ which acts opposite to that across the capacitor. Hence, the effective e.m.f. in the circuit is

$$E_{eff} = E_C + E_L, \text{ as per Kirchhoff's 2nd law}$$

where,

$$E_C = \frac{q}{C} \text{ and } E_L = - L\frac{di}{dt}$$

But, $E_{eff} = 0$ as there is no resistance in the circuit. So,

$$\frac{q}{C} - L\frac{di}{dt} = 0 \qquad \qquad ...(1)$$

This is the equation of a L-C circuit during discharge. This is a non-homogeneous equation. To make it homogeneous, let us put $i = -\frac{dq}{dt}$, where the negative sign enters as q is decreasing with time. Hence,

$$\frac{q}{C} - L\frac{d}{dt}\left(-\frac{dq}{dt}\right) = 0$$

or, $\qquad \frac{q}{C} + L\frac{d^2q}{dt^2} = 0$

or, $\qquad L\frac{d^2q}{dt^2} + \frac{q}{C} = 0$

or, $\qquad \frac{d^2q}{dt^2} + \frac{1}{LC}q = 0$

or, $\qquad \frac{d^2q}{dt^2} + \omega^2 q = 0 \qquad \qquad ...(2)$

where $\omega^2 = \frac{1}{LC}$. This is a linear differential equation of second order. Let the solution of this equation be $q = A\,e^{\alpha t}$

where A and α are arbitrary constants. Differentiating it with respect to t, we obtain

$$\frac{dq}{dt} = A\,\alpha e^{\alpha t}$$

and $\qquad \frac{d^2q}{dt^2} = A\,\alpha^2 e^{\alpha t}$

Substituting the values of $\frac{d^2q}{dt^2}$ and q in eqn. (2), we have

$$A\,\alpha^2 e^{\alpha t} + \omega^2 A\,e^{\alpha t} = 0$$

or, $\qquad \alpha^2 + \omega^2 = 0$

or, $\qquad \alpha^2 = -\omega^2$

or, $\qquad \alpha = \pm j\omega$

where, $j = \sqrt{(-1)}$. Thus, the eqn. (2) has two solutions which are

$$q = A\,e^{+j\omega t} \text{ and } q = A\,e^{-j\omega t}$$

As the equation is linear, any linear combination of these two solutions will also satisfy it. Hence, the most general solution of the eqn. (2) is

$$q = A_1 \, e^{+j\omega t} + A_2 \, e^{-j\omega t} \qquad \qquad ...(3)$$

where A_1 and A_2 are new arbitrary constants.

At $t = 0$; $q = q_0$, so that eqn. (3) yields

$$q_0 = A_1 + A_2 \qquad \qquad ...(4)$$

Differentiating eqn. (3) with respect to t, we have

$$i = \frac{dq}{dt} = A_1 \, (j\omega) \, e^{j\omega t} + A_2 \, (-j\omega) \, e^{-j\omega t} \qquad \qquad ...(5)$$

At $t = 0$, $i = 0$ when the discharging of the capacitor just starts. Then, from eqn. (5), we obtain

$$0 = j\omega A_1 - j\omega A_2$$

or, $\qquad \qquad j\omega A_1 = j\omega A_2$

or, $\qquad \qquad A_1 = A_2 \qquad \qquad ...(6)$

Now, from eqns. (4) and (6), we obtain

$$A_1 = A_2 = \frac{q_0}{2} \qquad \qquad ...(7)$$

Therefore, eqn. (3) becomes

$$q = \frac{q_0}{2} (e^{j\omega t} + e^{-j\omega t})$$

$$= q_0 \left(\frac{e^{j\omega t} + e^{-j\omega t}}{2} \right)$$

$\therefore \qquad \qquad q = q_0 \cos \omega t \qquad \qquad ...(8)$

This is the equation of the discharge of the capacitor. It shows that the discharge of the capacitor is oscillatory and simple harmonic.

The time period of oscillation is given by

$$T = \frac{2\pi}{\omega} = 2\pi\sqrt{LC} \qquad \qquad ...(9)$$

and the frequency of oscillation is

$$n = \frac{1}{T} = \frac{1}{2\pi\sqrt{LC}} \qquad \qquad ...(10)$$

Example 8: Vibrations of a magnet

Let us consider the oscillations of a magnet in the Earth's field of intensity H. During the course of its oscillations, it subtends an angle θ with the direction of H at a particular instant. If M be the moment of the magnet, the restoring couple acting on it is given by

Restoring Couple = MH sin θ

$$= MH \, \theta \qquad \qquad ...(1)$$

where θ is very small. This couple acts in opposite direction to the deflecting couple due to its

angular acceleration $\frac{d^2\theta}{dt^2}$, where θ is the angular displacement.

If I be the moment of inertia of the magnet, the deflecting couple is given by

$$\text{Deflecting Couple} = I\frac{d^2\theta}{dt^2} \qquad \qquad ...(2)$$

In the equilibrium condition of oscillation,
Deflecting Couple = Restoring Couple

$$\text{or,} \qquad I\frac{d^2\theta}{dt^2} = -MH\theta$$

$$\text{or,} \qquad \frac{d^2\theta}{dt^2} = -\frac{MH}{I}\theta$$

$$\text{or,} \qquad \frac{d^2\theta}{dt^2} = -\omega^2\theta$$

$$\text{or,} \qquad \frac{d^2\theta}{dt^2} + \omega^2\theta = 0 \qquad \qquad ...(3)$$

where $\omega^2 = \dfrac{MH}{I}$ or $\omega = \sqrt{\dfrac{MH}{I}}$. This equation (3) is the equation of SHM.

The time period of oscillation is given by

$$T = \frac{2\pi}{\omega} = 2\pi\sqrt{\frac{I}{MH}} \qquad \qquad ...(4)$$

The frequency of oscillation of the magnet is given by

$$n = \frac{1}{T} = \frac{1}{2\pi\sqrt{\dfrac{I}{MH}}} = \frac{1}{2\pi}\sqrt{\frac{MH}{I}} \qquad \qquad ...(5)$$

Note: Couple on a magnet placed in a uniform magnetic field:

Suppose a magnet NS of pole strength m and length $2l$ oscillates in the earth's uniform horizontal field (Fig. 1.18).

Let the axis of the magnet make an angle θ with the magnetic meridian at any instant of time and let H be the intensity of the uniform field.

Then the force acting on the north pole of the magnet is mH dynes in the direction TN and on the south pole, there is a force mH dynes in the direction PS. These two forces are equal and parallel, and hence constitute a couple.

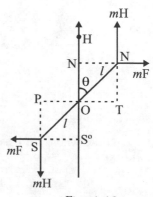

Fig. 1.18

The moment of the couple on NS = one of the forces × perpendicular distance between them = $mH \times PT$

$$= mH \times 2l \sin \theta$$
$$= (2ml) H \sin \theta$$
$$= MH \sin \theta$$

where, $2ml = M$, the moment of the magnet. Thus, the restoring couple acting on the magnet is MH sin θ.

Example 9: Oscillation of a mass suspended by a heavy spring

Let us consider a body of mass M suspended by a uniform spring of mass m and force constant k. If $m \ll M$, then the spring will stretch uniformly along its length. Let l be the length of the spring. Then, its mass per unit length is $\dfrac{m}{l}$. Let us consider an element of length dz at a distance z from the fixed end of the spring. The mass of this element is $\left(\dfrac{m}{l}\right) dz$.

Let x be the instantaneous displacement of the mass M from its position of rest. Since the displacement is proportional to the distance from the fixed end, the displacement of the element considered will be $\left(\dfrac{z}{l}\right) x$, and its instantaneous velocity is $\left(\dfrac{z}{l}\right) \dfrac{dx}{dt}$. Hence, the instantaneous kinetic energy of the element E'_k is given by

$$E'_k = \frac{1}{2}\left(\frac{m}{l} dz\right)\left(\frac{z}{l}\right)^2 \left(\frac{dx}{dt}\right)^2$$

$$= \frac{m}{2l^3}\left(\frac{dx}{dt}\right)^2 z^2\, dz \qquad \qquad ...(1)$$

If the spring is uniform, the total kinetic energy of the suspended mass and the spring is

$$(E_k) = \frac{1}{2}M\left(\frac{dx}{dt}\right)^2 + \frac{m}{2l^3}\left(\frac{dx}{dt}\right)^2 \int_0^l z^2\, dz$$

$$= \frac{1}{2}M\left(\frac{dx}{dt}\right)^2 + \frac{m}{2l^3}\left(\frac{dx}{dt}\right)^2 \frac{l^3}{3}$$

$$= \frac{1}{2}\left(M + \frac{m}{3}\right)\left(\frac{dx}{dt}\right)^2 \qquad \qquad ...(2)$$

Now, when the mass is at a distance x from its position of rest, the elastic potential energy of the system is

$$E_p = \frac{1}{2} kx^2 \qquad \qquad ...(3)$$

where k is the force constant of the spring. Therefore, the total energy of the system is

$$E = E_k + E_p$$

$$= \frac{1}{2}\left(M + \frac{m}{3}\right)\left(\frac{dx}{dt}\right)^2 + \frac{1}{2}kx^2 \qquad \qquad ...(4)$$

Since E remains constant, $\dfrac{dE}{dt} = 0$, that is

$$\frac{1}{2}\left(M + \frac{m}{3}\right)2\left(\frac{dx}{dt}\right)\frac{d^2x}{dt^2} + \frac{1}{2}k(2x)\frac{dx}{dt} = 0$$

or, $$\left(M + \frac{m}{3}\right)\frac{d^2x}{dt^2} + kx = 0$$

or, $$\frac{d^2x}{dt^2} + \frac{k}{\left(M + \dfrac{m}{3}\right)}x = 0$$

or, $$\frac{d^2x}{dt^2} + \omega^2 x = 0 \qquad \qquad ...(5)$$

This is the equation of a simple harmonic motion,

where $$\omega^2 = \frac{k}{\left(M + \dfrac{m}{3}\right)} \quad \text{or} \quad \omega = \sqrt{\frac{k}{\left(M + \dfrac{m}{3}\right)}}$$

Time period: The time period of oscillation of the body is given by

$$T = \frac{2\pi}{\omega} = 2\pi\sqrt{k \Big/ \left(M + \frac{m}{3}\right)}$$

$$= 2\pi\sqrt{\frac{\left(M + \dfrac{m}{3}\right)}{k}} \qquad \qquad ...(6)$$

The frequency is

$$n = \frac{1}{T} = \frac{1}{2\pi}\sqrt{\frac{k}{\left(M + \dfrac{m}{3}\right)}} \qquad \qquad ...(7)$$

If the spring be massless, the period will be

$$T = 2\pi\sqrt{\frac{M}{k}} \qquad \qquad ...(8)$$

Thus, when the mass of the spring is taken into account, the period is increased as per eqn. (6) and the frequency is decreased by eqn. (7).

Example 10: Oscillations of two masses connected by a spring

Let us consider two masses connected by a massless spring and free to vibrate along the axis of the spring. If the masses are slightly displaced from equilibrium in opposite directions, the spring extends (or contracts) and exerts a linear restoring (elastic) force on both the masses. Therefore, the masses, when released, vibrate about their equilibrium position, their centre of mass remaining at rest. Such a system is called a **'two-body oscillator'**.

Suppose two bodies of masses m_1 and m_2 are connected by a massless spring of force constant k (Fig. 1.19). Let x_1 and x_2 be the distances of m_1 and m_2 respectively from the origin O of an inertial frame of reference at an instant during the oscillation.

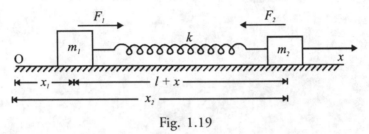

Fig. 1.19

Let l be the equilibrium length of the spring and x be the instantaneous change in length. Then, from the figure, we have

$$x_2 - x_1 = l + x \qquad \text{...(1)}$$

If x is positive, the spring is stretched and if x be is negative, the spring is compressed. Suppose in the figure, x is positive.

Now, the restoring force exerted by the spring on each mass has a magnitude kx. If x is positive, the mass m_1 is displaced to the left from its equilibrium position and the mass m_2 to the right. That is, the spring is tending to pull m_1 to the right (in positive direction) and m_2 to the left (in negative direction). Thus, the restoring force on m_1 is $F_1 = kx$, and on m_2 is $F_2 = -kx$. Hence, from Newton's second law of motion, we have

$$F_1 = kx = m_1 \frac{d^2 x_1}{dt^2} \qquad \text{...(2)}$$

and

$$F_2 = -kx = m_2 \frac{d^2 x_2}{dt^2} \qquad \text{...(3)}$$

Now, multiplying the equation (2) by m_2 and the equation (3) by m_1 and then subtracting (2) from (3) we obtain,

$$F_2 - F_1 = -k(m_1 + m_2)\, x$$

$$= m_1 m_2 \left(\frac{d^2 x_2}{dt^2} - \frac{d^2 x_1}{dt^2} \right)$$

or,

$$\frac{d^2 x_2}{dt^2} - \frac{d^2 x_1}{dt^2} = -k \left(\frac{m_1 + m_2}{m_1 m_2} \right) x$$

or, $\dfrac{d^2}{dt^2}(x_2 - x_1) = -k\left(\dfrac{m_1 + m_2}{m_1 m_2}\right)x$ ...(4)

From (1) and (4), we have

$$\dfrac{d^2}{dt^2}(x + l) = -k\left(\dfrac{m_1 + m_2}{m_1 m_2}\right)x \qquad ...(5)$$

As l is a constant, $\dfrac{d^2 l}{dt^2} = 0$

Hence, the eqn. (5) becomes

$$\dfrac{d^2 x}{dt^2} = -k\left(\dfrac{m_1 + m_2}{m_1 m_2}\right)x \qquad ...(6)$$

The quantity $\dfrac{m_1 m_2}{m_1 + m_2}$ is called the reduced mass of the two-body system and is denoted by 'μ'.

So, $$\mu = \dfrac{m_1 m_2}{m_1 + m_2} \qquad ...(7)$$

From eqns. (6) and (7), we obtain

$$\dfrac{d^2 x}{dt^2} = -\left(\dfrac{k}{\mu}\right)x$$

or, $$\dfrac{d^2 x}{dt^2} = -\omega^2 x \qquad ...(8)$$

where $$\dfrac{k}{\mu} = \omega^2 \text{ or } \omega = \sqrt{\dfrac{k}{\mu}}$$

Eqn. (8) is of the same form as the equation of a single-body simple harmonic oscillator, with the difference that here x is the relative displacement of the two bodies from their equilibrium separation. Hence, the relative motion of two masses m_1 and m_2 is just the same as the motion of a simple harmonic oscillator of mass $\mu = \dfrac{m_1 m_2}{m_1 + m_2}$.

The solution of the differential eqn. (8) is of the form

$$x = A \sin(\omega t + \phi) \qquad ...(9)$$

where A and ϕ are constants. A is the amplitude of vibration and ϕ is the phase angle.

Time Period of Oscillation: The time period of oscillation is given by

$$T = \dfrac{2\pi}{\omega} = 2\pi\sqrt{\dfrac{\mu}{k}} \qquad ...(10)$$

Frequency of Oscillation:

The frequency of oscillation is given by

$$n = \frac{1}{T} = \frac{1}{2\pi}\sqrt{\frac{k}{\mu}} \qquad \qquad ...(11)$$

Thus, the frequency of a two-body oscillator is exactly the same as the frequency of a single body oscillator of mass μ connected with one end of the same spring, the other end of which is fixed in an inertial frame of reference. In other words, in two-body oscillation, one body moves relative to the other as if the other body were fixed and the mass of the moving body were reduced to μ.

Example 11: Helmholtz resonator

A **'resonator'** is essentially a gas column having natural vibration frequency of its own. There are various types of them – the resonance box of a tuning-fork or an organ pipe open or closed, are all resonators. They will resound energetically to tuning-forks of appropriate pitches. Helmholtz, for his work on the analysis of complex sounds and other acoustical determinations, devised a series of special types of turnip-shaped resonators of glass or brass with varying periods of vibration.

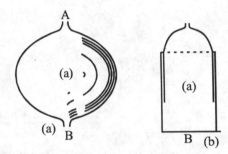

Fig. 1.20: Helmholtz resonators

They are known after him as Helmholtz's resonators. Two forms of Helmholtz's resonators in general use are shown in Fig. 1.20. They are either of globular form as shown in diagram (a) or cylindrical boxes as in diagram (b). They have a wide mouth A at one end to receive the exciting waves and a narrow neck B at the other to conduct the vibrations to the ear. The air cavity is almost completely enclosed with the result that only a small portion of the energy is radiated into the medium. The damping is, therefore, small and the tuning fairly sharp. For this reason, it is specially suitable for detecting sounds of a definite pitch and feeble intensity. If a compound note is sounded near such a resonator and if it contains the component to which the resonator can resound, it will 'pickout' and reinforce that component producing a loud sound. Such resonators are made in sets, the pitch of each being known and they can be used to detect the components of a complex note although these components may be too feeble to be detected by the ear alone. The cylindrical resonator is made in two pieces, one of which can slide inside the other, so that the volume of air inside can be varied at will and thus the same resonator can be turned to different frequencies.

Helmholtz carried out the analysis of a number of complex waves by using his resonators and applied it to the investigation of vowel sounds and musical quality.

Theory of Resonators: Let us now see how oscillations are caused in resonators and what is the frequency of such oscillations. Rayleigh, who developed the theory of these resonators, regards the air in the mouth as acting like a reciprocating piston alternately compressing and rarefying the air contained in the resonator. Further, it is assumed that the dimensions of the resonator are small as compared to the wavelength of sound in the free air and, therefore, at any instant, a condensation will be uniform throughout. The action of a resonator may be described as follows:

Let us suppose that the stream of air rushes in by the mouth A. It increases the pressure of the air inside. The increase of pressure checks the stream of air and causes an outward rush and the momentum, thus acquired, will carry the stream past its position of equilibrium and a deficit of pressure is thereby produced inside the resonator. Consequently, a stream flows in the reversed direction and thus periodic oscillations are set up. Comparing these oscillations with a mass attached to a spring, we can say that the air piston in the mouth is the mass and the air in the cavity is the spring. Let l and α denote the length and the sectional area of the mouth respectively and ρ denotes the normal density of the gas in the mouth. Then, the mass (M) of the air piston is $\rho l \alpha$. If the air in the cavity is made to resonate, the oscillations take place under adiabatic conditions. Now, for an adiabatic change in volume of the cavity, we have $pV^\gamma = a$ constant, where p and V are the pressure and volume of air in the cavity respectively. γ is the ratio between the specific heat at constant pressure (C_p) and the specific heat at constant volume (C_v) of air.

Differentiating $pV^\gamma =$ Constant, we obtain

$$\gamma\, pV^{\gamma-1}\, dV + V^\gamma\, dp = 0$$

or,

$$V^\gamma\, dp = -\gamma\, pV^{\gamma-1}\, dV$$

or,

$$dp = -\frac{\gamma pV^{\gamma-1}}{V^\gamma}\, dV$$

$$= -\frac{\gamma p}{V}\, dV$$

$$= -\gamma p\, \frac{dV}{V} \qquad\qquad ...(1)$$

Consider a layer of air in the neck moved through a distance y. Then the change in the volume of air $(dV) = \alpha y$ and the restoring force is

$$F = \alpha dp$$

or,

$$F = -\gamma p\, \frac{dV}{V}\, \alpha$$

$$= -\frac{\gamma p\alpha^2 y}{V} \qquad (\because dV = \alpha y) \qquad ...(2)$$

By Newton's second law of motion
$$F = Ma$$
$$= M\frac{d^2 y}{dt^2} \qquad \text{...(3)}$$
where M is the mass of the air piston.

From (2) and (3), we obtain
$$M\frac{d^2 y}{dt^2} = -\left(\frac{\gamma\, p\alpha^2}{V}\right) y$$

or,
$$\frac{d^2 y}{dt^2} = -\left(\frac{\gamma\, p\alpha^2}{MV}\right) y$$

or,
$$\frac{d^2 y}{dt^2} = -\omega^2 y \qquad \text{...(4)}$$

where
$$\omega^2 = \frac{\gamma\, p\alpha^2}{MV} \text{ or } \omega = \sqrt{\frac{\gamma\, p\alpha^2}{MV}}$$

Equation (4) represents the equation of a simple harmonic motion.

Time period of oscillation: The time period of oscillation is given by
$$T = \frac{2\pi}{\omega} = 2\pi\sqrt{\frac{MV}{\gamma\rho\alpha^2}} \qquad \text{...(5)}$$

or,
$$T = 2\pi\sqrt{\frac{\rho l \alpha V}{\gamma\, p\alpha^2}} \qquad (\because M = \rho l\alpha)$$

or,
$$T = 2\pi\sqrt{\frac{Vl}{v^2\alpha}} \qquad \left(\because v = \sqrt{\frac{rp}{\rho}}\right) \qquad \text{...(6)}$$

where, v is the velocity of sound in air.

Frequency of oscillation: The frequency of oscillation is
$$n = \frac{1}{T} = \frac{1}{2\pi}\sqrt{\frac{v^2\alpha}{Vl}} \qquad \text{...(7)}$$
$$= \frac{v}{2\pi}\sqrt{\frac{\alpha/l}{V}} \qquad \text{...(8)}$$

Thus, the frequency is directly proportional to the square root of the area α of the cross-section of the neck and inversely proportional to the square root of its length l.

The expression $\left(\dfrac{\alpha}{l}\right)$ has the dimensions of a length and is known as the conductivity of the neck. For a particular resonator, it follows from the above equation that
$$n^2 V = \text{Constant} \qquad \text{...(9)}$$

when a cork is rapidly withdrawn from a tightly stoppered bottle, a note is heard which is due to the adiabatic oscillation of the air in the bottle with the air in the neck serving as a piston.

Equation (9) can be verified experimentally by taking a cylindrical resonator with variable volume and forks of pitches varying from 256 to 512. A curve plotted with volume as ordinates and values of $\dfrac{1}{n^2}$ as abscissae should be a straight line through the origin if the theory is correct. But, in practice it will be found not to pass through origin although the relation is very nearly linear. The curve will obey closely the law

$$n^2(V + C) = \text{Constant} \qquad \qquad ...(10)$$

where the value of C is a correction applied to V. This may be regarded as **neck correction** and the ratio of C to the volume of the neck should be recorded.

Note: In the above calculation of time period of oscillation, we have assumed that the pressure is the same throughout the gas during the oscillations.

This is not true, since some time is required for the transmission of the pressure. The other assumption concerning the adiabatic character of the compression is very approximately true, especially as the neck is narrow and heat will not easily escape from the vessel or be transmitted to it. The piston will thus behave only roughly according to the formula and will have an approximate period of oscillation given by the above value.

Composition of SH motions at right angles to each other (Lissajous' Figures)

Two orthogonal simple harmonic vibrations can be compounded to give a resultant motion. The nature of the resultant motion depends upon the amplitudes, periods or frequencies and phase differences of the two constituent simple harmonic vibrations. The resultant motion can be traced in the form of a straight line, circle, ellipse, parabola, etc. Such geometrical representations of the resultant wave are termed as **Lissajous' Figures**.

We shall now consider some simple cases of the composition of two simple harmonic motions at right angles to each other. These cases can be considered in two ways:

(i) Graphical Method

(ii) Analytical Method

(i) Graphical Method

Case I: Composition of two rectangular SH motions of equal periods (or frequencies) with unequal amplitudes and same phase

Let one of component vibrations take place parallel to AB, the horizontal diameter of the lower reference circle and the other parallel to A'B', the vertical diameter of the upper reference circle (Fig. 1.21). Since the two vibrations are in the same phase, they pass their mean positions simultaneously. The reference circles have radii equal to the amplitudes a and b of the component vibrations. Let their circumferences be divided into equal number of parts, say 8. Each part will be travelled in the same time. We are compounding here the component of the motion of P along AB or XX′ and that of P′ along A′B′ or YY′. The positions zero on the respective circles coincide with the mean position O. This would show that the two SH motions have no phase difference between them initially. From each point 0, 1, 2, etc. on the circle AB draw ⊥ers on XOX′. Similarly, in case of A′B′ draw ⊥ers from 0′, 1′, 2′, etc. on YoY′. At the moment of start, both the ⊥ers intersect at 0, the position taken up by the vibrating point. In the positions 1, 1′

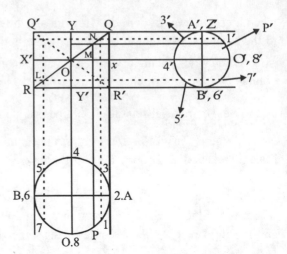

Fig. 1.21: Lissajous figure, periods 1:1, phase same

respectively, they intersect at N, in the positions 2, 2′ at Q and so on, till after a complete time period, the point O is reached again. Thus, positions Q, N, O, L, R are determined. By joining these points we get the **Lissajous figure**. In this case, the resultant path is a **straight line** QR, the diagonal of the rectangle of sides 2a and 2b. It is SH motion of the same time period and phase but amplitude

$$r = \sqrt{a^2 + b^2}$$

If the first vibration were to start from 4 (phase π) and the second from O′, then the path will be the other **diagonal** Q′R′ shown dotted in the figure.

Case II: Composition of two rectangular SH motions of equal periods (or frequencies) with unequal amplitude and phase $\dfrac{\pi}{4}$

This case has been exhibited in Fig. 1.22. Here the numerals in the reference circle AB are shifted forward by $\dfrac{\pi}{4}$, those in A′B′ remaining in the same positions as before (in case I). By drawing ⊥ers and proceeding exactly as in case I, the locus of the particle is found to be an **oblique ellipse** within the rectangle QQ′RR′.

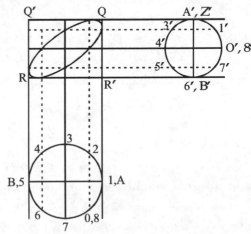

Fig. 1.22: Lissajous figure, periods 1:1 and phase $\dfrac{\pi}{4}$ (Oblique ellipse)

Case III: Composition of two rectangular SH motions of equal periods (or frequencies) with unequal amplitudes and phase $\frac{\pi}{2}$

In this case also the zero position in the circle A′B′ is kept the same while that in the circle AB is shifted forward by $\frac{\pi}{2}$. Proceeding as above, the resultant is found to be a **symmetrical ellipse** whose axes are XOX′ and YOY′ and which is inscribed within the same rectangle (Fig. 1.23). The curve will be a **circle** if the amplitudes are equal (Fig. 1.24).

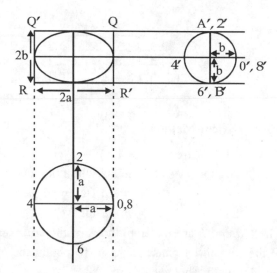

Fig. 1.23: Lissajous figure for frequencies 1:1, unequal amplitudes and phase $\frac{\pi}{2}$

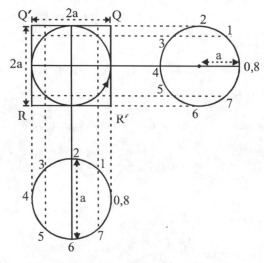

Fig. 1.24: Lissajous figure, periods (or frequencies) 1:1, phase $\frac{\pi}{2}$ and amplitudes equal (Circle)

The curves obtained by the above method for various values of the phase difference varying from 0 to π are shown in the Fig. 1.25. As the phase difference increases from π to 2π, the various figures are obtained in the reverse order, thus completing a cycle of changes.

Fig. 1.25: Lissajous figures with varying phases, periods 1:1
or frequencies 1:1

(ii) Analytical Method

(Periods 1:1 or frequencies 1:1):

Let the two rectangular vibrations be represented by

$$x = a \sin (\omega t + \phi) \qquad \qquad ...(1)$$

and $\qquad y = b \sin \omega t \qquad \qquad ...(2)$

where a and b are the amplitudes of the component vibrations and the first vibration is ahead of the second by phase angle ϕ. To obtain the equation of the path of the resultant vibration, we eliminate t between them.

Now, $\qquad x = a \sin (\omega t + \phi)$

or $\qquad \dfrac{x}{a} = \sin (\omega t + \phi)$

$$= \sin \omega t . \cos \phi + \cos \omega t \sin \phi \qquad \qquad ...(3)$$

From (2), we have

$$\sin \omega t = \frac{y}{b}$$

or $\qquad \cos \omega t = \sqrt{1 - \dfrac{y^2}{b^2}}$

Substituting the values of $\sin \omega t$ and $\cos \omega t$ in eqn. (3), we obtain

$$\frac{x}{a} = \frac{y}{b} \cos\phi + \sqrt{1 - \frac{y^2}{b^2}} \sin \phi$$

or, $\qquad \dfrac{x}{a} - \dfrac{y}{b} \cos\phi = \sqrt{1 - \dfrac{y^2}{b^2}} \sin \phi$

Squaring both sides, we have

$$\frac{x^2}{a^2} + \frac{y^2}{b^2} \cos^2 \phi - \frac{2xy}{ab} \cos \phi = \sin^2 \phi - \frac{y^2}{b^2} \sin^2 \phi$$

or, $\qquad \dfrac{x^2}{a^2} + \dfrac{y^2}{b^2} (\cos^2 \phi + \sin^2 \phi) - \dfrac{2xy}{ab} \cos \phi = \sin^2 \phi$

or, $\qquad \dfrac{x^2}{a^2} + \dfrac{y^2}{b^2} - \dfrac{2xy}{ab} \cos \phi = \sin^2 \phi \qquad \qquad ...(4)$

Which is the equation of an ellipse inclined to the axes of co-ordinates. This ellipse may be inscribed in a rectangle whose sides are $2a$ and $2b$.

Particular Cases:

(i) When $\phi = 0$, i.e. when there is no phase difference between the two motions, $\sin \phi = 0$ and $\cos \phi = 1$, so that the eqn. (4) becomes

$$\frac{x^2}{a^2} + \frac{y^2}{b^2} - \frac{2xy}{ab} = 0$$

or, $$\left(\frac{x}{a} - \frac{y}{b}\right)^2 = 0$$

or, $$\pm\left(\frac{x}{a} - \frac{y}{b}\right) = 0 \qquad \qquad ...(5)$$

Which represents a pair of coincident straight lines lying in the first and third quadrants represented by the diagonal shown in curve number 1 of Fig. 1.25.

It corresponds to the diagonal QOR of Fig. 1.21, such that it makes an angle $\tan^{-1}\left(\frac{b}{a}\right)$ with the axis XX′.

(ii) When $\phi = \frac{\pi}{4}$, eqn. (4) becomes

$$\frac{x^2}{a^2} + \frac{y^2}{b^2} - \frac{\sqrt{2}xy}{ab} = \frac{1}{2} \qquad \qquad ...(6)$$

Which is an oblique ellipse (Curve number 2 of Fig. 1.25).

(iii) When $\phi = \frac{\pi}{2}$, $\sin \phi = 1$ and $\cos \phi = 0$, then eqn. (4) becomes

$$\frac{x^2}{a^2} + \frac{y^2}{b^2} = 1 \qquad \qquad ...(7)$$

Which is the equation of the symmetrical ellipse shown in Fig. 1.25, curve number 3, whose major and minor axes coincide with the direction of the two given motions and whose semi-axes are a and b respectively.

(iv) When $\phi = \frac{\pi}{2}$ and $a = b$, then eqn. (4) becomes

$$\frac{x^2}{a^2} + \frac{y^2}{a^2} = 1,$$

or, $$x^2 + y^2 = a^2, \qquad \qquad ...(8)$$

Which is the equation of a circle. In this case, the particle describes a circle once in the same time as taken by any one of the component motions. A uniform circular motion may, therefore, be regarded as a combination of two equal or similar SH motions at right angles to each other and differing in phase by $\dfrac{\pi}{2}$.

(v) When $\phi = \pi$, $\sin \phi = 0$, $\cos \phi = -1$, then eqn. (4) becomes

$$\frac{x^2}{a^2} + \frac{y^2}{b^2} + \frac{2xy}{ab} = 0$$

or $\qquad \left(\dfrac{x}{a} + \dfrac{y}{b}\right)^2 = 0,$ \hfill ...(9)

which represents the diagonal shown in Fig. 1.25, curve number 5. It corresponds to the diagonal Q'OR' of Fig. 1.21.

Composition of two orthogonal simple harmonic vibrations of periods in the ratio of 1:2 (or frequencies in the ratio of 2:1)

(i) Graphical Method

Case I: Amplitudes unequal, phase same

Here we shall proceed in the same manner as for equal periods or equal frequencies. But, as the period of one of the component vibrations, say, along YOY' is double that along XOX', during the time the particle P' describes its journey once round the reference circle A'B', the particle P of which the frequency is double, describes two journeys. Divide the reference circle A'B' into 8 equal parts and the circle AB into 4 equal parts and put numerals as indicated in Fig. 1.26. Both the vibrations start from their central positions simultaneously, i.e. they are in same phase. As before to get the position of the vibrating point, draw ⊥ers on XOX' and YOY' from the corresponding points on the circles of reference and determine the points of intersections I, II, III, etc. By joining these points we obtain the resultant curve. This is a figure of '8'.

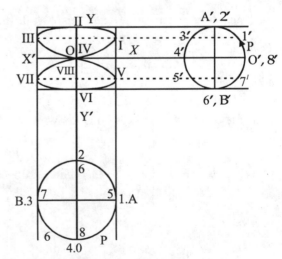

Fig. 1.26: Lissajous figure, periods 1:2 (or frequencies 2:1) and phase same

Case II: Amplitudes unequal and initial phase difference corresponding to a quarter of the smaller period

This case has been illustrated in the Fig. 1.27. The zero of the circle AB has been shifted to 1, while that of the circle A′B′ is in the same position. On proceeding as above, we find that the resultant path is a parabola (Fig. 1.27).

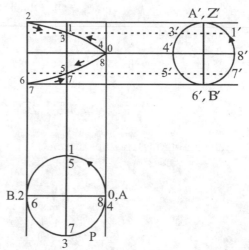

Fig. 1.27: Lissajous figure, periods 1:2 phase diff.: quarter of smaller period

Likewise the figures for different initial phases can be drawn by varying these phases from 0 to π of the smaller time period. Such curves are shown in (Fig. 1.28a). If the ratio of time periods is not exactly 1:2, the curve will change gradually forward and backward through the series of (Fig. 1.28a). If the periods is in the ratio of 2:3, then the curves will be like those of (Fig. 1.28b).

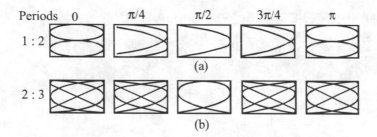

Fig. 1.28: Lissajous figures with varying period ratios and phases

(b) Analytical Treatment: (Periods 1:2 or frequencies 2:1)

Let the simple harmonic motions be represented by

$$x = a \sin (2\omega t + \phi) \qquad \qquad ...(10)$$

and

$$y = b \sin \omega t \qquad \qquad ...(11)$$

where ϕ is the phase angle by which the first motion is ahead of the second.

(i) If we take $\phi = 0$, eqn. (10) becomes

$$x = a \sin 2\omega t = 2a \sin \omega t . \cos \omega t$$

or, $\qquad \dfrac{x}{a} = 2 \sin \omega t . \cos \omega t \qquad\qquad\qquad ...(12)$

From eqn. (11), we have

$$\sin \omega t = \frac{y}{b}$$

So, $\qquad \cos \omega t = \sqrt{1 - \sin^2 \omega t} = \sqrt{1 - \dfrac{y^2}{b^2}}$

Substituting these values in eqn. (12), we obtain

$$\frac{x}{a} = 2 \frac{y}{b} \sqrt{1 - \frac{y^2}{b^2}}$$

or, $\qquad \dfrac{x^2}{a^2} = 4 \dfrac{y^2}{b^2} \sqrt{1 - \dfrac{y^2}{b^2}}$

$$= -4 \frac{y^2}{b^2} \left(\frac{y^2}{b^2} - 1 \right)$$

or, $\qquad \dfrac{x^2}{a^2} + \dfrac{4y^2}{b^2} \left(\dfrac{y^2}{b^2} - 1 \right) = 0 \qquad\qquad ...(13)$

which represents a figure of '8' as shown in Fig. 1.26.

(ii) If we take $\phi = \dfrac{\pi}{2}$, we obtain

$$\frac{x}{a} = \sin \left(2\omega t + \frac{\pi}{2} \right) \text{ from eqn. (10)}$$

$$= \cos 2\omega t$$

$$= 1 - 2 \sin^2 \omega t$$

$$= 1 - 2 \frac{y^2}{b^2} \qquad \left[\because \sin \omega t = \frac{y}{b} \right]$$

or, $\qquad \dfrac{2y^2}{b^2} = 1 - \dfrac{x}{a} = -\dfrac{x-a}{a}$

or, $\qquad y^2 = -\dfrac{b^2}{2a}(x - a) \qquad\qquad\qquad ...(14)$

which represents the equation of a parabola with vertex at $(a, 0)$ as in Fig. 1.27.

Varying initial phases from 0 to π, the Lissajous figures can be drawn (Fig. 1.28).

SOLVED EXAMPLES

Q.1. **A 5 N ball is fastened to the end of a flat spring. A force of 2 N is sufficient to pull the ball 6 cm to one side. Find the force constant and the period of vibration.**

Sol. W = 5 N, F = 2 N,

s = 6 cm = 6 × 10^{-2} m, k = ?, T = ?

$$k = \frac{F}{s} = \frac{2\,N}{6 \times 10^{-2}} = \frac{10^2}{3} = 33.3 \text{ N/m}$$

$$m = \frac{W}{g} = \frac{50\,N}{9.8 \text{ m/s}^2} = 0.51 \text{ kg}.$$

$$T = 2\pi \sqrt{\frac{m}{k}} = 2\pi \sqrt{\frac{0.51 \text{ kg}}{33.3 \text{ N/m}}} = 0.78 \text{ sec}.$$

Q.2. **A 2 kg body vibrates in simple harmonic motion with an amplitude of 3 cm and period of 5 sec. Find the speed, acceleration, the kinetic energy and the potential energy: (i) at the mid-point of the vibration, (ii) at the end of the path, (iii) at a point 2 cm from the mid-point, and (iv) Find the force on the body at the point 2 cm from the mid-point.**

Sol. m = 2 kg, A = 30 cm = 3 × 10^{-2} m,

T = 5 sec, v = ?, a = ?, E$_k$ = ?

(i) At the mid-point of the vibration

$$v_{max} = 2\pi \frac{1}{T} - 2 \times \pi \times \frac{3 \times 10^{-2}}{5}$$

$$= 3.142 \times \frac{6}{5} \times 10^{-2}$$

$$= \frac{18.852}{5} \times 10^{-2}$$

$$= 3.774 \times 10^{-2} = 3.8 \times 10^{-2} \text{ m/s}$$

In SHM the acceleration is given by

$$a = 0 - \omega^2 s$$

where s is the displacement from the mid-point.

When the body is at mid-point, s = 0

So, $\qquad a = 0$

The kinetic energy at the mid-point is given by

$$E_k = \frac{1}{2} m v_{max}^2 = \frac{1}{2} \times 2 \times (3.8 \times 10^{-2})^2$$

$$= 14.44 \times 10^{-4} = 1.444 \times 10^{-3} \text{ Joule}.$$

The potential energy at the mid-point is

$$E_p = \frac{1}{2}Ks^2 = 0 \qquad (\because x = 0)$$

(ii) At the end of the path

At the end point of path the velocity is given by

$$v = A\omega \cos \omega t$$

$$= A\omega^2 \sin \frac{\pi}{2} = A\omega^2$$

$$= A\left(\frac{2\pi}{T}\right)^2 = \frac{4\pi^2 A}{T^2} = \frac{4p^2 A^2}{T^2 A} = \left(\frac{2pA}{T}\right)^2 \times \frac{1}{A} = \frac{V_{max}^2}{A}$$

$$= \frac{(3.8 \times 10^{-2})^2}{3 \times 10^{-2}} = \frac{14.44 \times 10^{-4}}{3 \times 10^{-2}}$$

$$= 4.81 \times 10^{-2} \text{ m/sec.}$$

$$E_k = \frac{1}{2}mv^2 = 0 \qquad (\because v = 0)$$

$$E_p = \frac{1}{2}KA^2 \qquad (\because s = A \text{ here})$$

$$= \frac{1}{2}\omega^2 m \times A^2$$

$$= \frac{1}{2}(v_{max})^2 \times m = \frac{1}{2}mv_{max}^2$$

$$= \frac{1}{2} \times 2 \text{ kg} \times (3.8 \times 10^{-2})^2$$

$$= 14.44 \times 10^{-4} = 1.444 \times 10^{-3} \text{ Joule.}$$

(iii) At a point 2 cm from the mid-point

Velocity is given by

$$v = A\omega \cos \theta$$

$$= A\omega\sqrt{1 - \sin^2 \theta}$$

But, $\qquad s = A \sin \theta$

$$\sin \theta = \frac{s}{A}$$

Then, $\qquad v = A\omega\sqrt{1 - \frac{s^2}{A^2}}$

$$s = 2 \text{ cm} = 2 \times 10^{-2} \text{ m}$$

$$v = 3 \times 10^{-2} \text{ m} \times \frac{2\pi}{T} \times \sqrt{1 - \frac{4 \times 10^{-4}}{9 \times 10^{-4}}}$$

$$= 3 \times 10^{-2} \times \frac{2 \times 3.142}{5} \sqrt{\frac{9-4}{9}}$$

$$= \frac{6}{5} \times 3.142 \times 10^{-2} \sqrt{\frac{5}{9}}$$

$$= \frac{18.852}{5} \times 10^{-2} \times \frac{\sqrt{5}}{3}$$

$$= \frac{18.852}{15} \times 10^{-2} \times \sqrt{5} = 1.2568 \times 10^{-2} \times \sqrt{5}$$

$$= 1.2568 \times 2.2 \times 10^{-2}$$

$$= 2.8 \times 10^{-2} \text{ m/sec.}$$

The magnitude of acceleration

$$a = A\omega^2 \sin\theta$$

$$= \omega^2 \times s$$

$$= \left(\frac{2\pi}{T}\right)^2 \times 2 \times 10^{-2}$$

$$= \frac{4 \times 9.87 \times 2 \times 10^{-2}}{25} = \frac{8 \times 9.87 \times 10^{-2}}{25}$$

$$= \frac{78.96 \times 10^{-2}}{25} = 3.1584 \times 10^{-2}$$

$$= 3.2 \times 10^{-2} \text{ m/s}^2$$

The acceleration is always −ve to the displacement.

$$E_k = \frac{1}{2}mv^2$$

$$= \frac{1}{2} \times 2 \times (2.8 \times 10^{-2})^2$$

$$= 7.84 \times 10^{-4} \text{ Joule}$$

$$= 7.9 \times 10^{-4} \text{ Joule}$$

$$E_p = E - E_k$$

$$= 1.4 \times 10^{-3} \text{ J} - 7.9 \times 10^{-4} \text{ J}$$

$$= 14.4 \times 10^{-4} - 7.9 \times 10^{-4}$$

$$= (14.4 - 7.9) \, 10^{-4}$$

$$= 6.5 \times 10^{-4} \text{ J.}$$

(iv) Force on the body at the point 2 cm from the mid-point

$$s = 2 \text{ cm} = 2 \times 10^{-2} \text{ m}$$

$$F = Ks = m\omega^2 s$$

$$= m \times s \times \left(\frac{2\pi}{T}\right)^2$$

$$= 2 \times 2 \times 10^{-2} \times 4 \times 9.87 \times \frac{1}{25}$$

$$= \frac{16 \times 9.87}{25} \times 10^{-2} = \frac{157.92}{25} \times 10^{-2}$$

$$= 6.316 \times 10^{-2} \text{ N.}$$

The direction of force is –ve to the displacement.

Q.3. **A solid cylinder of mass 5 kg and radius 6 cm is suspended by a vertical wire as a torsion pendulum. The axis of the cylinder is along the line of the moment of the torsion of the wire. Find the moment of torsion of the wire if the period of vibration of the cylinder is 4 seconds.**

Sol. $\qquad m = 5 \text{ kg, R} = 6 \text{ cm} = 6 \times 10^{-2} \text{ m,}$

$$\text{T} = 4 \text{ sec}, c = ?$$

$$\text{I} = \frac{1}{2}m\text{R}^2$$

$$= \frac{1}{2} \times 5 \times 36 \times 10^{-4}$$

$$= 18 \times 5 \times 10^{-4} = 90 \times 10^{-4}$$

$$= 9 \times 10^{-3} \text{ kg/m}^2$$

The time period of vibration of the cylinder is given by

$$\text{T} = 2\pi \times \sqrt{\frac{\text{I}}{c}}$$

where c is called the moment of torsion of the wire.

$$\text{T}^2 = 4\pi^2 \frac{\text{I}}{c}$$

$$\Rightarrow \quad c = 4\pi^2 \frac{\text{I}}{\text{T}^2} = \frac{4 \times 9.87 \times 10^{-3} \times 9}{16}$$

$$= \frac{9}{4} \times 9.87 \times 10^{-3}$$

$$= \frac{88.83}{4} \times 10^{-3} = 22.27 \times 10^{-3}$$

$$= 2.2 \times 10^{-2} \text{ mN/radian.}$$

Q.4. A spring has a force constant of 100 N/m. It oscillates at a frequency of 40 cycles/min when an object is attached to it. What is the mass of the object?

Sol. $\qquad$ K= 100 N/m,

$\qquad f = 40$ cycles/min

$$= \frac{40}{60} \text{ cycle/sec} = \frac{2}{3} \text{ cycle/sec}$$

$\qquad m = ?, F = Ks$

We know,

$$\omega^2 = \frac{K}{m}$$

$$(2\pi f)^2 = \frac{K}{m}$$

$$\Rightarrow \qquad m = \frac{K}{4\pi^2 f^2} = \frac{100}{4 \times 9.87 \times \dfrac{4}{9}}$$

$$= \frac{900}{16 \times 9.87} = \frac{900}{157.92} = \frac{900}{158}$$

$$= 5.68 = 5.7 \text{ kg.}$$

Q.5. What is the force constant of a spring that is stretched 100 cm by a force of 50 N? What is the period of vibration of a 100 N body if it is suspended by this spring?

Sol. $\qquad$ K = 100 cm = 10 × 10^{-2} m = 10^{-1} m

$\qquad$ F = 50 N, W = 100 N, K = ?, g = 9.8 m/s^2

$$K = \frac{F}{s} = \frac{50 \text{ N}}{10^{-1}} = 500 \text{ N/m}$$

$$T = \frac{W}{g} = \frac{100 \text{ N}}{9.8 \text{ m/s}^2} = \frac{100}{9.8} \text{ kg}$$

$$T = 2\pi \sqrt{\frac{m}{K}}$$

$$= 2 \times 3.142 \times \sqrt{\frac{100}{9.8 \times 500}}$$

$$= 6.284 \times \sqrt{\frac{1}{9.8 \times 5}}$$

$$= 6.284 \times \sqrt{\frac{1}{49.0}}$$

$$= 6.284 = 0.897 \text{ sec.}$$

Q.6. **A 10 kg mass is hung on a spring. When an additional 1 kg mass is added the spring stretches 0.8 m. If the spring and the 16 kg mass is stretched 2 m and then released, what is the period of vibration?**

Sol. $m = 16$ kg, $s = 0.8$ m, T = ?

$$K = \frac{F}{s} = \frac{mg}{s}$$

$$= \frac{1\,kg \times 9.8\,m/s^2}{0.8} = \frac{9.8}{0.8}\,N/m$$

$$= \frac{49}{4} = 12.25\,N/m$$

$$T = 2\pi\sqrt{\frac{m}{K}} = 2\pi\sqrt{\frac{16}{12.25}}$$

$$= 2\pi \times 4\sqrt{\frac{\dfrac{1}{1225}}{100}}$$

$$= 8 \times \pi \times \frac{10}{35} = \frac{80}{35}\pi = \frac{16}{7}\pi$$

$$= \frac{16 \times 3.142}{7} = \frac{50.272}{7}$$

$$= 7.181 \approx 7.2\,sec.$$

Q.7. **A 4 kg body is caused to vibrate with SHM by means of a spring. If the amplitude is 30 cm and the time of a complete vibration is 0.6 sec., Find: (i) the maximum speed, (ii) the maximum kinetic energy, (iii) the minimum kinetic energy, and (iv) the force constant of the spring.**

Sol. A = 30 cm = 30×10^{-2} m,

$m = 4$ kg, T = 0.6 sec

(i) v_{max} = ?

$$v_{max} = A\omega = A\frac{2\pi}{T}$$

$$= \frac{3 \times 10^{-1} \times 2 \times 3.142}{0.6} = 3.142\,m/sec.$$

(ii) $E_{k_{max}}$ = ?

$$E_{k_{max}} = \frac{1}{2}mv^2_{max} = \frac{1}{2} \times 4 \times 9.87$$

$$= 2 \times 9.87 = 19.74\,Joule.$$

(iii) $\qquad E_{k_{max}} = 0$

(iv) The force constant $= k$

$$T = 2\pi\sqrt{\frac{m}{k}}$$

$$T^2 = 4\pi^2 \frac{m}{k}$$

$\Rightarrow \qquad k = \dfrac{4\pi^2 m}{T^2} = \dfrac{4 \times 9.87 \times 4}{36 \times 10^{-2}}$

$$= \frac{4}{9} \times 9.87 \times 10^2 = 4 \times 1.096 \times 10^2$$

$$= 4.384 \times 10^2 = 438.4 \text{ N/m}.$$

Q.8. What is the period of a vibrating object which has an acceleration of 8 m/sec² when its displacement is 1 m?

Sol. $\qquad a = 8 \text{ m/sec}^2,\ s = 1 \text{ m}$

$$T = 2\pi\sqrt{\frac{s}{a}}$$

$$= 2 \times 3.142 \times \sqrt{\frac{1}{8}}$$

$$= 6.284 \times \frac{1}{\sqrt{8}}$$

$$= \frac{6.284}{2\sqrt{2}} = \frac{3.142}{\sqrt{2}}$$

$$= \frac{3.142}{1.414} = 2.2 \text{ sec}.$$

Q.9. An 8 kg body performs SHM of amplitude 30 cm. The restoring force is 60 N when the displacement is 30 cm. Find: (i) the period, (ii) the acceleration when the displacement is 12 cm, (iii) the maximum speed, and (iv) the kinetic energy and potential energy when the displacement is 12 cm.

Sol. $\qquad m = 8 \text{ kg},$

$\qquad A = 30 \text{ cm} = 30 \times 10^{-2} \text{ m} = 3 \times 10^{-1} \text{ m}$

$\qquad F = 60 \text{ N},\ s = 30 \text{ cm} = 30 \times 10^{-2} \text{ m}$

(i) $\qquad T = ?$

$$K = \frac{F}{s} = \frac{60 \times 10}{3} = 200 \text{ N/m}$$

$$T = 2\pi\sqrt{\frac{m}{K}}$$

$$= 2 \times 3.142 \sqrt{\frac{8}{200}}$$

$$= 2 \times 3.142 \frac{2\sqrt{2}}{10\sqrt{2}}$$

$$= \frac{4 \times 3.142}{10} = 1.2568 \text{ sec} = 1.25 \text{ sec}.$$

(ii) $a = ?$, $s = 12$ cm $= 12 \times 10^{-2}$ m

$$a = -\omega^2 s = -\left(\frac{2\pi}{T}\right)^2 s$$

$$= -\frac{4\pi^2}{(1.25)^2} \times 12 \times 10^{-2}$$

$$= -\frac{4 \times 9.87 \times 12 \times 10^{-2}}{1.5625}$$

$$= -\frac{48 \times 9.87 \times 10^{-2}}{15625 \times 10^{-4}}$$

$$= -\frac{47376 \times 10^{-4}}{15625 \times 10^{-4}}$$

$$= -\frac{47376}{15625} = -3.03 \text{ m/sec}^2.$$

The –ve sign shows that the acceleration is opposite to the displacement.

(iii) $v_{max} = ?$

$$v_{max} = \omega A = \frac{2\pi}{T} \times A$$

$$= \frac{2 \times 3.142 \times 3 \times 10^{-1}}{1.25}$$

$$= \frac{6 \times 3.142}{1.25} \times 10^{-1}$$

$$= \frac{18.852}{1.25} \times 10^{-1}$$

$$= \frac{18852 \times 10^{-3} \times 10^{-1}}{125 \times 10^{-2}}$$

$$= \frac{18852}{125} \times 10^{-2} = 1508 \times 10^{-2}$$

$$= 0.15 \text{ m/sec}.$$

(iv) $\qquad E_k = ?, E_p = ?$

$\qquad s = 12 \text{ cm} = 12 \times 10^{-2} \text{ m}$

$$E_k = \frac{1}{2} K s^2$$

$$= \frac{1}{2} \times 20 \times 144 \times 10^{-4}$$

$$= 14400 \times 10^{-4} \text{ J} = 144 \times 10^{-2} \text{ J}$$

$$E = \frac{1}{2} m v_{max}^2$$

$$= \frac{1}{2} \times 8 \times 2.25$$

$$= 4 \times 2.25 = 9.00 \text{ J}$$

$$E_k = 9 - E_p = (9 - 144 \times 10^{-2})$$

$$= 9 - 1.44 = 7.56 \text{ J}.$$

Q.10. A body moves with SHM of an amplitude of 24 cm and a period of 1.2 sec. (i) Find: the speed of the object when it is at its mid position and when 24 cm away, (ii) the magnitude of the acceleration in each case.

Sol. $\qquad A = 24 \text{ cm} = 24 \times 10^{-2} \text{ m}, T = 1.2 \text{ sec}$

(i) $\qquad v_{max} = ?$

$\qquad v_{max} = \omega A$

$$= \frac{2\pi A}{T} = \frac{2 \times 3.142 \times 24 \times 10^{-2}}{1.2}$$

$$= \frac{6.284 \times 24 \times 10^{-2}}{12 \times 10^{-1}}$$

$$= 12.568 \times 10^{-1} = 1.2568 \text{ m/sec}$$

$$s = 24 \text{ cm} = 24 \times 10^{-2} \text{ m}$$

$$v_1 = A\omega \cos \theta$$

$$= A\omega \sqrt{1 - \sin^2 \theta}$$

$$= A\omega \sqrt{1 - \frac{s^2}{A^2}} \quad (\because s = A \sin \theta)$$

$$= \omega \sqrt{A^2 - s^2}$$

$$= \frac{2\pi}{T} \sqrt{(24 \times 10^{-2})^2 - (24 \times 10^{-2})^2} = 0$$

(ii) At the mid position, $s = 0$

$$A = -\omega^2 s = 0$$

$$s_1 = 24 \text{ cm} = 24 \times 10^{-2} \text{ m}$$

$$A_1 = -\omega^2 s$$

$$= -\left(\frac{2\pi}{T}\right)^2 \times 24 \times 10^{-2} = -\frac{4\pi^2}{T^2} \times 24 \times 10^{-2}$$

$$= -\frac{4 \times 9.87}{1.2 \times 1.2 \times 10^{-2}} \times 24 \times 10^{-2}$$

$$= -\frac{9.87}{3} \times 2 = -\frac{19.74}{3}$$

$$= -6.58 \text{ m/sec}^2.$$

Q.11. **A 4 kg mass as attached to the end of a flat spring which is pulled 0.08 m to one side by a force of 10 N. Find the force constant, the period of vibration and the frequency of vibration.**

Sol. $m = 4$ kg, $s = 0.08$ m, F = 10 N,

$K = ?, T = ?, f = ?$

$$K = \frac{F}{s} = \frac{10}{0.08} = \frac{1000}{8} = 125$$

$$T = 2\pi\sqrt{\frac{m}{K}} = 2 \times 3.142 \times \sqrt{\frac{4}{125}}$$

$$= 6.284 \times \frac{2}{5\sqrt{5}}$$

$$= \frac{12.568}{5\sqrt{5}} = \frac{12.568}{5 \times 2.2}$$

$$= \frac{12.568}{11.0} = 1.1425 \text{ sec}$$

$$f = \frac{1}{T} = \frac{1}{1.1425} = 0.87 \text{ cycle/sec}.$$

Q.12. **A body having simple harmonic motion of amplitude 5 cm has a speed of 50 cm/sec when its displacement is of 3 cm, what is its period?**

Sol. $A = 5$ cm $= 5 \times 10^{-2}$ m,

$v = 50$ cm/sec $= 50 \times 10^{-2}$ m/sec

$s = 3$ cm $= 3 \times 10^{-2}$ m

$$v = \omega\sqrt{A^2 - s^2} = \frac{2\pi}{T}\sqrt{A^2 - s^2}$$

$$T = \frac{2\pi}{v}\sqrt{A^2 - s^2}$$

$$= \frac{2 \times 3.142}{50 \times 10^{-2}} \sqrt{25 \times 10^{-4} - 9 \times 10^{-4}}$$

$$= \frac{2 \times 3.142}{50 \times 10^{-2}} \times 4 \times 10^{-2}$$

$$= \frac{4}{25} \times 3.142$$

$$= 0.16 \times 3.142$$

$$= .50272 \text{ sec.}$$

Q.13. A simple pendulum is used to determine the value of g. When the length of the pendulum is 98.45 cm the period is measured to be 1.99 sec. Find the value of 'g'.

Sol. $l = 98.45$ cm, $T = 1.99$ sec, $g = $?

$$T = 2\pi \sqrt{\frac{l}{g}}$$

$\Rightarrow \qquad T^2 = 4\pi^2 \frac{l}{g}$

$\Rightarrow \qquad g = 4\pi^2 \frac{l}{T^2}$

$$= 4 \times 9.87 \times \frac{98.45 \times 10^{-2}}{1.99 \times 1.99}$$

$$= \frac{39.48 \times 98.45 \times 10^{-2}}{3.9491}$$

$$= \frac{3886.8060}{3.9601} \times 10^{-2}$$

$$= \frac{38868060}{39601} \times 10^{-2}$$

$$= 989.5 \times 10^{-2} = 9.8 \text{ m/s}^2.$$

Q.14. A simple pendulum is 1 m long. What is its period?

Sol. $l = 1$ m, $T = $?, $g = 9.8$ m/s^2

$$T = 2\pi \sqrt{\frac{l}{g}} = 2 \times 3.142 \sqrt{\frac{1}{9.8}}$$

$$= 6.284 \times \frac{1}{3.13} = \frac{6284}{313} \times \frac{10^{-3}}{10^{-2}}$$

$$= 20.07 \times 10^{-1} = 2.007 \text{ sec.}$$

Q.15. **At a certain place a simple pendulum 100 cm long makes 250 complete vibrations in 8.38 min. What is the length of a simple second's pendulum at that place?**

Sol.
$$l = 100 \text{ cm} = 1 \text{ m}$$

$$f = \frac{250 \text{ vib}}{8.38 \text{ min}} = \frac{250}{8.36 \times 60}$$

$$= \frac{25}{8.36 \times 6} = \frac{25}{50.28} \text{ vib/sec}$$

$$T = \frac{1}{f} = \frac{50.28}{25} = 2.01 \text{ sec}$$

$$T = 2\pi\sqrt{\frac{l}{g}}$$

$$T^2 = 4\pi^2 \frac{l}{g}$$

$$\Rightarrow \qquad l = \frac{T^2 g}{4\pi^2} = \frac{2^2 \times 9.77}{4 \times 9.87}$$

$$= \frac{977}{987} = 0.989 \text{ m} = 98.7 \text{ cm}.$$

Q.16. **A watch has a balance wheel which moves with an angular acceleration of 4.1 rad/sec² when it is displaced 15° from its equilibrium position. What is its frequency?**

Sol.
$$\alpha = 42 \text{ rad/sec}^2, \ \theta = 15°, \ f = ?, \ T = ?$$

$$T = 2\pi\sqrt{\frac{s}{A}}, \text{ for linear motion}$$

For angular motion this becomes

$$T = 2\pi\sqrt{\frac{\theta}{\alpha}}$$

$$= 2 \times 3.142 \times \sqrt{\frac{\frac{\pi}{180} \times 15}{41}}$$

$$= 6.284 \times \sqrt{\frac{\pi}{12 \times 41}}$$

$$= 6.284 \times \frac{\sqrt{\pi}}{\sqrt{492}} = 6.284 \times \frac{1.77}{22.1}$$

$$= \frac{11.12268}{22.1} = 0.503 \text{ sec} \approx 0.5 \text{ sec}$$

$$f = \frac{1}{T} = \frac{1}{0.5} = \frac{10}{2} = 2 \text{ cycles/sec}.$$

Q.17. A torsion pendulum begins moving with an angular acceleration of 15 rad/sec² when displacement is 90°. What is the frequency of the pendulum?

Sol. $\alpha = 15$ rad/sec², $\theta = 90° = \dfrac{\pi}{2}$ rad

$$T = 2\pi\sqrt{\dfrac{\theta}{\alpha}} = 2\pi\sqrt{\dfrac{\pi/2}{15}}$$

$$= 2\pi\sqrt{\dfrac{\pi}{30}} = 2\pi\dfrac{1.77}{5.4}$$

$$= \dfrac{6.284 \times 1.77}{5.4}$$

$$= \dfrac{11.12268}{5.4} = 2.05 \text{ sec}$$

$$f = \dfrac{1}{T} = \dfrac{1}{2.05} = \dfrac{100}{205} = 0.4 \text{ cycle/sec.}$$

Q.18. A watch has a balance sheet which moves with an angular acceleration of 25 rad/sec². When it is displaced 45° from its equilibrium position, what is its frequency?

Sol. $\theta = 45° = \dfrac{\pi}{4}$ rad, $\alpha = 25$ rad/sec²

$$T = 2\pi\sqrt{\dfrac{\theta}{\alpha}} = 2\pi\sqrt{\dfrac{\pi}{4 \times 25}}$$

$$= \dfrac{2 \times 3.142 \times 1.77}{10} = \dfrac{6.284 \times 1.77}{10}$$

$$= \dfrac{11.12268}{10} = 1.112268 \text{ sec}$$

$$f = \dfrac{1}{T} = \dfrac{1}{1.11} = \dfrac{100}{111} = .90 \text{ cycle/sec.}$$

Q.19. A particle executing SHM completes 600 oscillations per minute and has an amplitude of 40 cm. Write down the equation of motion of particle and find the: (i) maximum velocity, (ii) the time period needed to move from mean position to a point 8 cm from it and (iii) acceleration when the displacement is 8 cm.

Sol. A= 10 cm

$f = 600$ vib/min

$$= \dfrac{600}{60} \text{ vib/sec} = 10 \text{ vib/sec}$$

$$T = \dfrac{1}{f} = \dfrac{1}{10} \text{ sec}$$

$$\omega = \frac{2\pi}{T} = \frac{2\pi}{1/10} = 20\,\pi \text{ vib/sec}$$

Equation of motion of the particle executing SHM is

$$s = A \sin \omega t$$
$$= 10 \sin 20\pi t$$

(i) Maximum velocity

$$v_{max} = A\omega = 10 \times 20\pi$$
$$= 200 \times \pi \text{ cm/sec} = 200\,\pi \text{ cm/sec}.$$

(ii) $$s = A \sin \omega t$$

$$\Rightarrow \quad \sin \omega t = \frac{s}{A} = \frac{8}{10} = 0.8$$

$$\Rightarrow \quad \omega t = \sin^{-1}(0.8) = 53.13° = 0.928 \text{ rad}$$

$$\Rightarrow \quad t = \frac{0.928 \text{ rad}}{\omega} = \frac{0.928}{20\pi} = 0.0148 \text{ sec}.$$

(iii) Acceleration, $a = -\omega^2 s = -400\pi^2 \times 8$
$$= -400 \times 9.87 \times 8$$
$$= -31608 \text{ cm/s}^2.$$

Q.20. A particle executing SHM has a maximum velocity of 81.4 cm/sec and maximum acceleration of 314 cm/sec². Find the frequency of oscillation and the amplitude.

Sol. Maximum velocity, $\quad v_{max} = A\omega = 31.4 \text{ cm/s}$

Maximum acceleration, $\quad a_{max} = A\omega^2 = 31.4 \text{ cm/s}^2$

$$\omega = \frac{A\omega^2}{A\omega} = \frac{314}{31.4} = 10 \text{ rad/s}$$

Hence, $$f = \frac{\omega}{2\pi} = \frac{10}{2\pi} \text{ vib/sec}$$

$$A = \frac{A\omega}{\omega} = \frac{v_{max}}{\omega} = \frac{31.4}{10} = 3.14 \text{ cm}.$$

Q.21. A particle executing SHM has a period of 0.05 sec. It passes through the mean position with a velocity of 3.14 cm/sec. Find the maximum displacement and maximum acceleration.

Sol. $\quad$ T = 0.05 sec, v = 3.14 cm/s

$$T = 0.05 \text{ sec} = \frac{5}{100} = \frac{1}{20} \text{ sec}$$

$$f = \frac{1}{T} = 20 \text{ vib/s}$$

$$\omega = 2\pi f = 2\pi \times 20 = 40\pi \text{ rad/s}$$

When the particle passes through the mean position the velocity is maximum.

$$\therefore \quad v_{max} = A\omega = 3.14 \text{ cm/sec} = \pi \text{ cm/sec}$$

Maximum displacement, i.e. amplitude is given by

$$A = \frac{v_{max}}{\omega} = \frac{\pi \text{ cm/sec}}{40\pi \text{ vib/sec}} = \frac{1}{40} = 0.025 \text{ cm.}$$

Maximum acceleration $= a = \omega^2 A$

$$= (40\pi)^2 \times \frac{1}{40}$$

$$= 3.95 \text{ cm/s}^2.$$

Q.22. **A simple harmonic motion is given by:**

$s = 0.5 \sin (1000t + 0.08)$

where *s* is in metre and t in sec. Find: (i) initial phase, (ii) amplitude, (iii) angular frequency, and (iv) time period.

Sol. $s = 0.5 \sin (1000t + 0.8)$

Comparing the equation with the standard form

$s = A \sin (\omega t + \phi)$, we have

(i) Initial phase = 0.80 rad/sec

(ii) Amplitude, A = 0.5 m

(iii) Angular frequency, ω = 1000 rad/sec

(iv) Time period, $T = \frac{2\pi}{\omega} = \frac{2 \times 22}{7 \times 1000} = 6.28$ millisec.

Q.23. **A spring elongates 2 cm for a load of 15 gm. If a body of mass 294 gm is attached to the spring and made to vibrate, find the force constant and the time period of vibration.**

Sol. $m = 15$ gm, $m_1 = 294$ gm

For constant, $k = \dfrac{\text{Weight of the load}}{\text{Elongation}}$

$= 15$ gm $\times 980$ cm/sec^2

$= 1350$ dyne/cm

Time period of vibration,

$$T = 2\pi\sqrt{\frac{m_1}{k}} = 2 \times 3.14\sqrt{\frac{294}{7350}} = 1.26 \text{ sec.}$$

Q.24. **A simple pendulum has a time period of 2 sec. The supporting fibre broke and was shortened by 2 cm. Find the change in period.**

Sol. $T = 2$ sec

Let, l = initial length of the simple pendulum

T = initial time period = $2\pi\sqrt{\dfrac{l}{g}}$

$$l = \frac{gT^2}{4\pi^2} = \frac{980 \times 4}{4 \times 3.14 \times 3.14} = 99.39 \text{ cm}$$

The new length $l' = 99.39 \text{ cm} - 2 \text{ cm} = 97.39 \text{ cm}$ and new time period is T'.

$$T' = 2\pi\sqrt{\frac{l'}{g}} = 2 \times 3.14\sqrt{\frac{97.39}{980}} = 1.95 \text{ sec}$$

Change in time period,

$$T - T' = 2 \text{ sec} - 1.95 \text{ sec} = 0.05 \text{ sec}.$$

Q.25. A pendulum clock shows a correct time. If the length increases by 0.1 per cent, find the error in the time per day.

Sol. The correct number of seconds per day is 86400. If the length increases, the time period will increase and consequently the frequency will decrease. If the incorrect number is given by the clock is $86400 - x$, then from eqn.

$$T = 2\pi\sqrt{\frac{l}{g}}, \text{ we have}$$

$$\frac{86400 - x}{86400} = \sqrt{\frac{l}{1 + 0.001l}} = (1 + 0.001)^{-\frac{1}{2}}$$

$$= 1 - \frac{1}{2} \times 0.001$$

$$\Rightarrow \qquad \frac{x}{86400} = \frac{0.001}{2}$$

$$\Rightarrow \qquad x = 43.$$

$\therefore$ The clock goes slower by 43 seconds each day.

Q.26. A torsion pendulum twists through an angle of 7.5 when it is acted upon by a torque of 10 m. The frequency of the pendulum is 90 vibrations per minute. What is the moment of inertia of the torsion pendulum?

Sol.

$$T = 2\pi\sqrt{\frac{I}{c}}$$

$$T^2 = 4\pi^2 \frac{I}{c}$$

$$\Rightarrow \qquad I = \frac{CT^2}{4\pi^2}$$

Here,

$$c = \frac{10 \text{ Nm}}{7.5} = \frac{10 \text{ Nm}}{\dfrac{\pi \times 7.5 \text{ rad}}{180}}$$

$$= \frac{10 \text{ Nm}}{\dfrac{\pi}{24} \text{ rad}} = \frac{240}{\pi} \text{ Nm/rad}$$

$$T = \frac{1}{f} = \frac{1}{\dfrac{90 \text{ vib}}{60 \text{ sec}}} = \frac{60}{90} \text{ sec} = \frac{2}{3} \text{ sec}$$

$$I = \frac{CT^2}{4\pi^2} = \frac{240}{\pi}\left(\frac{2}{3}\right)^2 \times \frac{1}{4\pi^2} \text{ kg m}^2$$

$$= \frac{80}{3\pi^3} \text{ kg m}^2 = 0.86 \text{ kg m}^2.$$

Q.27. **If a particle executes simple harmonic motion of period 8 s and amplitude 40 cm, find maximum velocity and acceleration.**

Sol. $T = 8$ sec, $A = 40$ cm

$$\omega = \frac{2\pi}{T} = \frac{2\pi}{8 \text{ sec}}$$

or, $\omega = \dfrac{\pi}{4}$ rad/sec

$$v_{max} = \pi\omega = 40 \times \frac{\pi}{4}$$

$$= 31.42 \text{ cm/sec}$$

$$a_{max} = \omega^2 A = \left(\frac{\pi}{4}\right)^2 \times 40$$

$$= \frac{\pi^2}{16} \times 40 = 2468 \text{ cm/sec}^2.$$

Q.28. **A particle is executing SHM with an amplitude of 15 cm and frequency 4 vib/s. Compute the following:**

(i) Maximum velocity of the particle.

(ii) Acceleration when displacement is 9 cm.

(iii) Time required to move from mean position 10 cm.

Sol. $A = 15$ cm, $f = 4$ vib/sec

$\omega = 2\pi f = 2\pi \times 4 = 8\pi$ rad/sec

(i) $v_{max} = A\omega = 15 \times 8\pi = 120\pi$ cm/sec

$= 377.04$ cm/sec.

(ii) $s = 9$ cm

$\therefore$ Acceleration $= \omega^2 s = (8\pi)^2 \times 9$

$= 5668.47$ cm/sec^2.

(iii) $\qquad s = 12$ cm

In case of SHM, $s = A \sin \omega t$

$\Rightarrow \qquad 12 = 15 \sin \omega t$

$\Rightarrow \qquad \sin \omega t = \dfrac{12}{15} = 0.75$

$\Rightarrow \qquad \omega t = \sin^{-1}(0.75) = 48.6° = 0.848$ rad

$\Rightarrow \qquad t = \dfrac{0.848}{\omega} = \dfrac{0.848}{8\pi} = 0.03375$ sec.

Q.29. A bob executes SHM of period 16 s. Two seconds after it passes through its centre of oscillation its velocity is found to be 5 cm/s. Find the amplitude.

Sol. Instantaneous velocity 'v' of a body executing SHM is,

$\qquad v = A\omega \cos \omega t$

Here, $\qquad$ T $= 16$ sec, v $= 5$ cm/sec,

$\qquad t = 2$ s^{-1}, $r = ?$

$$\omega = \frac{2\pi}{T} = \frac{2\pi}{16} = \frac{\pi}{8} \text{ rad/sec}$$

$\qquad v = A\omega \cos \omega t$

$\Rightarrow \qquad 5 = \dfrac{A\pi}{8} \cos\left(\dfrac{\pi}{8} \times 2\right)$

$\qquad\qquad = \dfrac{A\pi}{8} \cos\left(\dfrac{\pi}{4}\right)$

$\Rightarrow \qquad 5 = A\left(\dfrac{\pi}{8}\right) \times \dfrac{1}{\sqrt{2}}$

$\Rightarrow \qquad A = \dfrac{40\sqrt{2}}{\pi} = 18$ cm.

Q.30. A particle executing SHM has maximum velocity of 100 cm/s and a maximum acceleration of 157 cm/s². Calculate its time period.

Sol. $\qquad v_{max} = A\omega = 1000$ m/sec

$\qquad\qquad a_{max} = \omega^2 A = 157$ cm/sec²

$\qquad \dfrac{A\omega^2}{A\omega} = \dfrac{157}{100}$

$\Rightarrow \qquad \omega = 1.57$ rad/sec

$\therefore \qquad T = \dfrac{2\pi}{\omega} = \dfrac{2\pi}{1.57} = 4$ sec.

Q.31. A particle is moving with SHM in a straight line. When the distance of particle from the equilibrium position, have values 'x_1' and 'x_2' respectively, the corresponding values of velocities given by v_1 and v_2. Show that the period of oscillations is given by:

$$T = 2\pi \sqrt{\frac{x_2^2 - x_1^2}{t_1^2 - t_2^2}}$$

Sol. Velocity 'u' of the particle executing SHM is given by:

$$u = \omega\sqrt{A^2 - s^2}$$

or, $u^2 = \omega^2(A^2 - s^2)$

        At $s = x_1$

$$u_1^2 = \omega^2\left(A^2 - s_1^2\right)$$

$$u_1^2 = \omega^2 A^2 - \omega^2 s_1^2 \qquad\qquad ...(1)$$

        At $s = x_2$

$$u_2^2 = \omega^2 A^2 - \omega^2 s_2^2 \qquad\qquad ...(2)$$

Subtracting (1) from (2),

$$u_2^2 - u_1^2 = \omega^2 x_2^2 - \omega_2^2 x_1^2 = \omega^2\left(x_2^2 - x_1^2\right)$$

$$\therefore \quad \omega^2 = \frac{u_1^2 - u_2^2}{x_2^2 - x_1^2}$$

or, $\omega = \sqrt{\dfrac{u_1^2 - u_2^2}{x_2^2 - x_1^2}}$

Since $T = \dfrac{2\pi}{\omega}$

$$\therefore \quad T = \frac{2\pi}{\sqrt{\dfrac{u_1^2 - u_2^2}{x_2^2 - x_1^2}}}$$

$$\Rightarrow \quad T = 2\pi\sqrt{\frac{x_2^2 - x_1^2}{u_1^2 - u_2^2}}$$

Q.32. A body weighing 40 g has a velocity of 4 cm/sec after one second of its starting from the mean position. If the time period is 8 s, find its kinetic energy, potential energy and the total energy.

Sol. $m = 10$ gm, $v = 40$ m/sec, $T = 8$ sec

$$\omega = \frac{2\pi}{T} = \frac{2\pi}{8} = \frac{\pi}{4} \text{ rad/sec}$$

Instantaneous velocity of the body $= A\omega \cos \omega t$

At $t = 1$ sec

$v = A\omega \cos \omega t$

$\Rightarrow$ $4 = A \frac{\pi}{4} \cos\left(\frac{\pi}{4} \times 1\right)$

$\Rightarrow$ $4 = A \times \frac{\pi}{4} \times \frac{1}{\sqrt{2}}$

$\Rightarrow$ $A = \frac{16\sqrt{2}}{\pi} = 7.20$

$$E_k = \frac{1}{2}mv^2$$

$$= \frac{1}{2} \times 10 \times (7.20)^2 = 160 \text{ erg}.$$

Total energy of the body

$$= \frac{1}{2}m\omega^2 A^2 = \frac{1}{2} \times 10 \times \frac{\pi^2}{16} \times (7.2)^2$$

$$= 160 \text{ erg}.$$

$\therefore$ Potential energy

$$= \text{Total energy} - \text{Kinetic energy}$$

$$= 160 - 80 = 80 \text{ erg}.$$

Q.33. A particle executes SHM of period 8.5. After what time of its passing through the mean position will the energy be half kinetic and half potential?

Sol. $T = 8$ sec.

$$\omega = \frac{2\pi}{T}$$

$$= \frac{2\pi}{8} \text{ rad/sec} = \frac{\pi}{4} \text{ rad/sec}$$

$$E_k = \frac{1}{2}mv^2 = \frac{1}{2}m\omega^2(A^2 - s^2)$$

$$E_p = \frac{1}{2}m\omega^2 s^2$$

Since kinetic energy and potential energy are equal,

$$\frac{1}{2}m\omega^2(A^2 - s^2) = \frac{1}{2}m\omega^2 s^2$$

$$\Rightarrow \quad (A^2 - s^2) = s^2$$

$$\Rightarrow \quad 2s^2 = A^2$$

$$\Rightarrow \quad s^2 = \frac{A^2}{2}$$

$$\Rightarrow \quad s = \frac{1}{\sqrt{2}}A$$

Since, $s = A \sin \omega t$

$$\Rightarrow \quad \sin \omega t = \frac{1}{\sqrt{2}} = \sin \frac{\pi}{4}$$

$$\Rightarrow \quad \omega t = \frac{\pi}{4}$$

$$\Rightarrow \quad t = \frac{\pi/4}{\omega} = \frac{\pi}{4} \times \frac{4}{\pi} = 1 \text{ sec.}$$

QUESTIONS

1. Define simple harmonic motion. Deduce the equation of simple harmonic motion. Solve it to find the displacement of the vibrating particle.

2. Define amplitude, time period, frequency and phase of simple harmonic motion. Establish a relation between time period and frequency.

3. Form and solve the equation of simple harmonic motion. Find expressions for velocity and acceleration of the vibrating particle.

4. Prove that the projection of uniform circular motion upon a diameter is simple harmonic.

5. Compute the energy in simple harmonic motion.

6. Show that an oscillating simple pendulum executes simple harmonic motion. Hence find the expression for time period and frequency of the pendulum.

   What is a second's pendulum?

7. State and explain the laws of simple pendulum.

8. What is a torsion pendulum? Show that the torsion pendulum executes simple angular harmonic motion. Hence find the time period of the pendulum.

9. Define the following:
   (i) Free vibration
   (ii) Forced vibration
   (iii) Damped vibration
   (iv) Resonance

10. What are Lissajous' figures? Construct the figures when two orthogonal simple harmonic vibrations with same frequency are superposed. Consider phase differences from 0 to 2π.

11. What do you mean by Lissajous' figures?

   What type of figures will you obtain when you compound two simple harmonic motions having frequency in the ratio 2:1 and differing in phases from 0 to 2π.

12. Show graphically that the resultant motion of a particle acted upon by two SHMs of equal frequencies and amplitudes along two directions at right angles, but differing in phase by $\frac{\pi}{2}$, is circular.

13. Explain 'periodic motion' and 'simple harmonic motion'. Find mathematically the resultant of two simple harmonic motions at right angles to each other having same frequency and differing in phase by $\frac{\pi}{4}$.

Problems

1. Two simple harmonic vibrations P and Q of equal periods and differing in phase by $\frac{\pi}{2}$ are impressed on the same particle. The amplitudes of P and Q are 4 cm and 6 cm respectively. Find the amplitude of the resulting vibration and the phase difference from Q.
   (Ans. 7.211 cm; 0.5882 radian)

2. A particle executes a SHM of period 10 seconds and amplitude 5 cm. Calculate its maximum velocity and acceleration.
   (Ans. 3.142 cm/sec; 1.974 cm/sec²)

3. The path of a body executing SHM along a straight line is 4 cm long and its velocity, when passing through the centre of its path is 16 cm/sec. Calculate its time-period.
   (Ans. 0.7854 cm)

4. The maximum velocity of a particle undergoing a SHM is 8 ft/sec and its acceleration at 4 ft from the mean position is 16 ft/sec². Compute its amplitude and period of vibration.
   (Ans. 4 ft; 3.142 sec)

5. A particle executes SHM of period 16 sec. Two seconds after it passes the centre of oscillation, its velocity is found to be 4 ft per second. Find the amplitude.
   (Ans. 14.41 ft)

6. A particle describes SHM in a line 4 cm long. Its velocity, when passing through the centre of the line, is 12 cm per second. Find the period.
   (Ans. 1.047 sec)

7. A particle executing SHM has an acceleration of 0.5 cm/sec^2, when the displacement is 2 cm. Find its period.

   (**Ans.** 12.57 sec)

8. Calculate the time period of a circular disc of radius r, oscillating about an axis through a point, distant $\dfrac{r}{2}$ from its centre and perpendicular to its plane.

   (**Ans.** $2\pi\sqrt{\dfrac{3r}{2g}}$)

9. A particle moving in a straight line with SHM of period $\dfrac{2\pi}{\omega}$, about a fixed point O, has a velocity $\sqrt{3}\, b\omega$, when at a distance b from O. Show that its amplitude is $2b$ and it will cover the rest of its distance in time $\dfrac{\pi}{3\omega}$.

10. The total energy of a particle executing a SHM of period 2π sec is 10,240 ergs. $\dfrac{\pi}{4}$ seconds after the particle passes the mid-point of the swing, its displacement is $8\sqrt{2}$ cm. Calculate the amplitude of the motion and the mass of the particle.

    (**Ans.** 16 cm; 80 gm)

●●●

Chapter 2

Damped Harmonic Motion

Damped Harmonic Motion

The motion of the particle which we now attempt to consider is subject to: (i) a restoring force proportional to the displacement, and (ii) a resisting force varying as the velocity, such motions are called resisted damped oscillations. All vibrating bodies are subjected to these forces, for otherwise the system would be conservative and there would be no emission of energy, etc.

Let us suppose that the resisting force is given by $F = r\left(\dfrac{dy}{dt}\right)$ where r is a constant

depending on the dimensions of the body, the minus sign indicating that the force opposes the direction of y. The rate at which work would be done against friction, etc.

$$= F\left(\frac{dy}{dt}\right) = -\left(\frac{dy}{dt}\right)^2$$

This is the rate at which the total energy of the system (K + P) would be dissipated, therefore

$$\frac{d}{dt}(K+P) = -r\left(\frac{dy}{dt}\right)^2 \qquad \qquad ...(1)$$

where K is the KE and P is the PE.

Now KE and PE is given by

$$K = \frac{1}{2}mv^2 = \frac{1}{2}m\left(\frac{dy}{dt}\right)^2 \qquad \qquad ...(2)$$

$$P = \int_{0}^{y} \mu y \; dy = \frac{1}{2}\mu y^2 \qquad \qquad ...(3)$$

Substituting the values of K and P from (2) and (3), we have

or, $\quad m\left(\dfrac{dy}{dt}\right)\dfrac{d^2y}{dt^2}+\mu y\dfrac{dy}{dt}=-\left(\dfrac{dy}{dt}\right)^2$

or, $\quad m\dfrac{d^2y}{dt^2}+r\dfrac{dy}{dt}+\mu y=0$

or, $\quad \dfrac{d^2y}{dt^2}+\dfrac{r}{m}\dfrac{dy}{dt}+\dfrac{\mu}{m}y=0$

Let $\quad \dfrac{r}{m}=2k \ \text{ and }\ \dfrac{\mu}{m}=\omega^2$

$\therefore \quad \dfrac{d^2y}{dt^2}+2k\dfrac{dy}{dt}+\omega^2 y=0 \qquad\qquad \text{...(4)}$

This is the differential eqn. for damped harmonic motion which may be solved in the form, $y=e^{\alpha t}$. From this, we have

$$\frac{dy}{dt}=\alpha e^{\alpha t} \ \text{ and }\ \frac{d^2y}{dt^2}=\alpha^2 e^{\alpha t}$$

So that on substituting the values in equation, (4) we get

$$\alpha^2 e^{\alpha t}+2k\alpha e^{\alpha t}+\omega^2 e^{\alpha t}=0$$

or, $\quad \alpha^2+2k\alpha+\omega^2=0$

where $\quad \alpha=-k\pm\sqrt{k^2-\omega^2}$

Since α has two values, there are two particular solutions of eqn. (4),

$$y=A'\,e^{\left(-k+\sqrt{k^2-\omega^2}\right)t} \ \text{ and }\ y=B'\,e^{\left(-k+\sqrt{k^2-\omega^2}\right)t}$$

where A′ and B′ are any arbitrary constants.

Now when two solutions such as

$$y-A'\,e^{\left(-k+\sqrt{k^2-\omega^2}\right)t}=0 \ \text{ and }\ y-B'\,e^{\left(-k-\sqrt{k^2-\omega^2}\right)t}=0$$

has been found, it obviously follows that

$$2y-A'\,e^{\left(-k+\sqrt{k^2-\omega^2}\right)t}-B'\,e^{\left(-k-\sqrt{k^2-\omega^2}\right)t}=0$$

or, $\quad y=\dfrac{A'}{2}\,e^{\left(-k+\sqrt{k^2-\omega^2}\right)t}+\dfrac{B'}{2}\,e^{\left(-k-\sqrt{k^2-\omega^2}\right)t}$

writing A for $\dfrac{A'}{2}$ and B for $\dfrac{B'}{2}$, we get the general solution

$$y = A' e^{\left(-k+\sqrt{k^2-\omega^2}\right)t} + B' e^{\left(-k-\sqrt{k^2-\omega^2}\right)t} \qquad ...(5)$$

Here A and B are constants whose values can be determined from the critical displacement and velocity.

Let us define the boundary conditions. If the system starts with maximum displacement but no velocity, then

$$\text{and} \quad \left. \begin{array}{l} y = y_0 \\ \dfrac{dy}{dt} = 0 \end{array} \right\} \text{ where } t = 0$$

Hence $A + B = y_0$ $\qquad\qquad ...(6)$

and $\left(-k + \sqrt{k^2 - \omega^2}\right)A + \left(-k - \sqrt{k^2 - \omega^2}\right)B = 0$

or $-k(A + B) + \sqrt{k^2 - \omega^2}\,(A - B) = 0$ $\qquad\qquad ...(7)$

But $A + B = y_0$

$\therefore$ $-ky_0 + \sqrt{k^2 - \omega^2}\,(A - B) = 0$

or $A - B = \dfrac{y_0 k}{\sqrt{k^2 - \omega^2}}$ $\qquad\qquad ...(8)$

From (6) and (8)

$$A - B = \dfrac{y_0 k}{\sqrt{k^2 - \omega^2}}$$

$$2A = y_0 \left(1 + \dfrac{k}{\sqrt{k^2 - \omega^2}} \right)$$

$$2B = y_0 \left(1 - \dfrac{k}{\sqrt{k^2 - \omega^2}} \right)$$

or, $A = \dfrac{y_0}{2} \left(1 + \dfrac{k}{\sqrt{k^2 - \omega^2}} \right)$

$$B = \dfrac{y_0}{2} \left(1 - \dfrac{k}{\sqrt{k^2 - \omega^2}} \right)$$

Substituting the values of A and B in eqn. (5), we have

$$y = y_0 \left[\frac{1}{2}\left(1 + \frac{k}{\sqrt{k^2 - \omega^2}}\right) e^{\left(-k+\sqrt{k^2-\omega^2}\right)t} + \frac{1}{2}\left(1 - \frac{k}{\sqrt{k^2 - \omega^2}}\right) e^{\left(-k-\sqrt{k^2-\omega^2}\right)t} \right] \qquad ...(9)$$

Eqn. (9) indicates that the motion depends upon the relative magnitudes of k and ω. Three cases may arise:

Case I: Aperiodic motion

This will happen when there are large frictional forces that is, $k > \omega$ and thus $k^2 > \omega^2$, the roots of eqn. (9) will be real and negative.

The expression for y shows that the displacement after passing its first maximum decays asymptotically to zero, (Fig. 2.1a), the retarding force is so great that the particle does not vibrate, that is the value of y does not change sign but falls from its maximum value to zero in an infinite time. Such a motion is called *aperiodic* or *dead-beat*. This is realized in practice in a moving coil galvanometer with a low resistance across its terminals or in a pendulum set swinging in a very viscous fluid. This case is of little importance in acoustics.

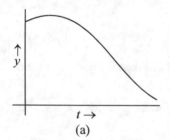

(a)

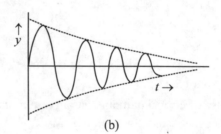

(b)

Fig. 2.1

Case II: Critically damped motion

When $k = \omega$ and so $k^2 = \omega^2$, eqn. (9) evidently breaks down, for in this case two of the coefficients becomes infinite. Returning to eqn. (5), let us see what from will it take when $\sqrt{k^2 - \omega^2}$ has not vanished but is reduced to a small quantity h.

Then $$y = Ae^{(-k+h)t} + Be^{(-k-h)t} = e^{-kt}\left[Ae^{ht} + Ae^{-ht}\right]$$

writing e^{ht} and e^{-ht} in the from of series

$$e^{ht} = 1 + ht + \frac{h^2t^2}{2!} + \frac{h^2t^2}{3!} + ...$$

and $$e^{-ht} = 1 - ht + \frac{h^2t^2}{2!} - \frac{h^2t^2}{3!} + ...$$

We have
$$y = e^{-ht}\left[A\left(1 + ht + \frac{h^2 t^2}{2!} + ...\right) + B\left(1 - ht + \frac{h^2 t^2}{2!} - ...\right)\right]$$

Now h being small, the terms containing h^2 and higher power of h may be neglected but the quantities Ah and Bh have unknown magnitudes since A and B are known.

Then,
$$y = e^{-kt}\left[(A + B) + (A - B)ht\right]$$

Denoting the two constants $(A + B)$ and $(A - B)h$ by G and H respectively, we have

$$y = e^{-kt}(G + Hr)$$

This is also a non-oscillatory motion and the case is known as that of 'critical damping', since there is just no oscillation. The displacement at first is by virtue of the factor $(G + Hr)$ but the exponential term becomes relatively important as t increases.

If now $y = y_0$ when $t = 0$, $\qquad G = y_0$

and $\qquad \dfrac{dy}{dt} = 0$ when $t = 0$, $\qquad H = ky_0$

$$y = y_0 e^{-kt}(1 + kt) \qquad\qquad\qquad ...(10)$$

The chain curve in (Fig. 2.1b) represents the equation in this case and the motion is said to be critically damped.

The case of critical damping has an important bearing on the problem of recording sound vibrations either mechanically or electrically.

Case III: Oscillatory motion

A very important case which is of particular interest in the study of sound arises when the frictional forces are small that is $k < \omega$ or $k^2 < \omega^2$. The roots of eqn. (9) then are imaginary since $(k^2 - \omega^2)$ is a negative quantity. Let $\sqrt{k^2 - \omega^2}$ be written as $\sqrt{-1}\sqrt{\omega^2 - k^2}$ or $i\beta$, where $i = \sqrt{-1}$ and $\beta = \sqrt{\omega^2 - k^2}$ so that the quantity under the root sign is again real.

Eqn. (9) then becomes

$$y_0 = y_0\left[\frac{1}{2}\left(1 + \frac{k}{i\beta}\right)e^{(-k+i\beta)t} + \frac{1}{2}\left(1 + \frac{k}{i\beta}\right)e^{(-k-i\beta)t}\right]$$

$$= y_0 e^{-kt}\left[\left(\frac{1}{2} + \frac{k}{2i\beta}\right)e^{+i\beta t} + \left(\frac{1}{2} + \frac{k}{2i\beta}\right)e^{-i\beta t}\right]$$

$$= y_0\, e^{-kt} \left[\frac{e^{i\beta t} + e^{-i\beta t}}{2} + \frac{k}{\beta}\, \frac{e^{i\beta t} + e^{-i\beta t}}{2i} \right]$$

Now, the exponential form of $\cos \beta t$ is $\dfrac{e^{i\beta t} + e^{-i\beta t}}{2}$ and that of $\sin \beta t$ is $\dfrac{e^{i\beta t} - e^{-i\beta t}}{2i}$.

Putting these values in our equation, we get

$$y = y_0\, e^{-kt} \left\{ \cos \beta t + \frac{k}{\beta} \sin \beta t \right\}$$

$$y = y_0\, \frac{e^{-kt}}{\beta} \left\{ \beta \cos \beta t + k \sin \beta t \right\}$$

Let ϕ' be angle such that

$$\beta = a' \sin \phi \quad \text{and} \quad k = a' \cos \phi'$$

$$\therefore \qquad y = y_0\, \frac{a'\, e^{-kt}}{\beta} \left\{ \sin \beta t \cos \phi' + \cos \beta t \sin \phi' \right\}$$

$$= y_0\, \frac{a'\, e^{-kt}}{\beta} \sin (\beta t + \phi')$$

or $\qquad y = a e^{-kt} \sin(\beta t + \phi') \qquad\qquad\qquad …(11)$

where $\qquad a = \dfrac{y_0 a'}{\beta}$ and $\beta = \sqrt{\omega^2 - k^2} = \sqrt{\dfrac{\mu}{m} - \dfrac{r^2}{4m^2}}$

This expression for y represents geometrically a simple harmonic curve of which the amplitude $a e^{-kt}$ diminishes exponentially to zero with increasing time and whose periodic time is given by $T' = \dfrac{2\pi}{\beta}$. Since the periodic term $\sin(\beta t + \phi)$ alternates between the limits ± 1, the space time displacement y of the curve lies between the curves $y = \pm a e^{-kt}$. The motion is represented by the continuous oscillatory curve of (Fig. 2.1b). The measure of the decay of the amplitude is the constant 'k' called damping coefficient. Note that the equation differs from the equation of the undamped motion by the introduction of the factor e^{-kt}.

A further consideration of equation (11) shows that if time is counted from the instant of zero displacement, that is $t = 0$ when $y = 0$, the equation gives

$$\sin \phi' = 0 \quad \text{or} \quad \phi' = 0$$

Hence under these initial conditions, the motion is given by

$$y = ae^{-kt} \sin \beta t \qquad \qquad ...(12)$$

To obtain successive amplitudes we equate $\dfrac{dy}{dt} = 0$, for the particle velocity is zero at maximum displacement.

Differentiating eqn. (12)

$$\frac{dy}{dt} = ae^{-kt} \beta \cos \beta t - ake^{-kt} \sin \beta t$$

whence

$$\tan \beta t = \frac{\beta}{k}$$

The values of βt remain unchanged if βt increases by $n\pi$ where n is an integer, 1, 2, 3, ..., etc. Thus,

$$\tan \beta t = \tan \beta = \left(1 + \frac{\pi}{\beta}\right) = \tan \beta \left(t + \frac{2\pi}{\beta}\right) = ...$$

$$\Rightarrow \tan \beta = \left(1 + \frac{n\pi}{\beta}\right)$$

From this it follows that the maximum amplitude occurs at intervals of $\dfrac{\pi}{\beta}$ counted from the instant when y has a maximum value. Now y is maximum when $\sin \beta t = 1$ or $\beta t = \dfrac{\pi}{2}$ or $t = \dfrac{\pi}{2\beta}$ measured from start and this maximum value is equal to $ae^{-k\pi/2\beta}$.

The successive maximum amplitudes occurring at times $\dfrac{3\pi}{2\beta}, \dfrac{5\pi}{2\beta}, \dfrac{7\pi}{2\beta}$ are $ae^{-3k\pi/2\beta}$, $ae^{-7k\pi/2\beta}$ respectively.

The ratio of the successive maximum amplitudes on opposite sides of the zero displacement is $e^{k\pi/\beta}$ and the logarithm of this ratio is known as the logarithmic decrement of the oscillations. Its value is given by

$$\lambda = \log_e e^{k\pi/\beta} = \frac{k\pi}{\beta} = k\frac{T'}{2}$$

Since $\dfrac{\pi}{\beta} = \dfrac{T'}{2}$, where T′ is the periodic time of the damped oscillation.

The value of k may be determined experimentally as follows:
If a_0 be the initial amplitude of vibration, then a_p is the amplitude after p vibration, i.e. after a time, $t = pT = \dfrac{p}{f}$, where f is the frequency, from eqn. (12) it is given by

$$a_p = a_0\, e^{-kt}$$

or, $$\frac{a_0}{a_p} = e^{kt} = e^{kp/f}$$

whence $$\log_e\left(\frac{a_0}{a_p}\right) = \frac{k_p}{f}$$

$$k = \frac{f}{p}\log_e\left(\frac{a_0}{a_p}\right) = 2.303\,\frac{f}{p}\log_{10}\left(\frac{a_0}{a_p}\right) \qquad\qquad ...(13)$$

So knowing f, p, and the ratio $\dfrac{a_0}{a_p}$, k may be calculated.

Energy of Damped Vibration

The displacement of an oscillatory damped system is given by

$$y = ae^{-kt}\cos(\beta t + \phi')$$

Differentiating $\quad \dot{y} = a\beta e^{-kt}\cos(\beta + \phi') - ake^{-kt}\sin(\beta t + \phi')$

Hence kinetic energy is $\mathrm{E}_{kin.} = \dfrac{1}{2}m\dot{y}^2$

$$= \frac{1}{2}ma^2 e^{-2kt}\left[\beta^2\cos^2(\beta t + \phi') - 2k\beta\cos(\beta t + \phi')\sin(\beta t + \psi') + k^2\sin^2(\beta t + \phi')\right]$$

and potential energy is

$$\mathrm{E}_{pot.} = \frac{1}{2}\mu^2 y^2 = \frac{1}{2}m(\beta^2 + k^2)y^2 \qquad \left[\because \frac{\mu}{m} = \omega^2 = \beta^2 + k^2 \text{ or } \mu = m(\beta^2 + k^2)\right]$$

$$= \frac{1}{2}m(\beta^2 + k^2)\left[a^2 e^{-2kt}\sin^2(\beta t + \phi')\right]$$

$$= \frac{1}{2}m\beta^2 a^2 e^{-2kt}\sin^2(\beta t + \phi') + \frac{1}{2}mk^2 a^2 e^{-kt}\sin^2(\beta t + \phi')$$

Hence the total energy, $\mathrm{E} = \mathrm{E}_{kin.} + \mathrm{E}_{pot.}$

$$= \frac{1}{2}m\beta^2 a^2 e^{-2kt} - mka^2 e^{-2kt}\sin(\beta t + \phi')\times\left[\beta\cos(\beta t + \phi') - k\sin(\beta t + \phi')\right]$$

$$= \frac{1}{2}m\beta^2 a^2 e^{-2kt}; \qquad \text{approx, if } k \text{ is small.}$$

Effect of Damping on Frequency

The existence of damping produces two effects: (i) the decrease in amplitude as explained above, and (ii) the increase of time-period. In the case of undamped oscillations, we have seen already that the periodic time,

$$T = \frac{2\pi}{\omega} \text{ and frequency } f = \frac{\omega}{2\pi}, \text{ whereas in the case of damped oscillations, } T' = 2\pi/\beta$$

and $f = \beta/2\pi$, where $\beta^2 = \omega^2 - k^2$.

which shows that the time period has become greater than what it would be in the absence of damping. In most cases occurring in practice, however, the difference between ω and $\sqrt{\omega^2 - k^2}$ is a very small quantity of the second order and may usually be neglected, i.e. the effect of damping on the period is usually negligible except in a few extreme cases where the vibrations are strongly damped. Experiment has shown that it lowers the frequency by 1 part in 5 millions, let us illustrate it by an example.

Example: Suppose $f = \frac{\beta}{2\pi} = 500 \text{ cycles/sec}$ and amplitude falls to $\left(\frac{1}{e}\right)$ of its original value in one second, then by eqn. (11)

$$ae^{-kt} = \frac{a}{e}$$

$$\text{or} \qquad ae^{-kt} = \frac{a}{e} \qquad\qquad \left[\because t = 1\sec\right]$$

and so $\qquad k = 1.$

Also $\qquad \beta^2 = f^2 4\pi^2 = (5\omega)^2\, 4\pi^2 = 10^6 \times \pi^2 = 10^7$

and $\qquad \omega^2 = \beta^2 + k^2 = 10^7 + 1$

Thus $\qquad \delta(\beta^2) = 1$

$\therefore \qquad 2\beta\delta\beta = 1$

and so $\qquad \dfrac{\delta\beta}{\beta} = \dfrac{1}{2\beta^2} = \dfrac{1}{2 \times 10^7}$

i.e., ω exceeds β by 1 part in 2×10^7 and so the effect of this damping on the frequency is negligible. But, on the other hand, in the case of an air column k is so large that vibrations die out after a few oscillations with appreciable increase in the periodic time.

Relaxation Time

The relaxation time is defined as the time taken for the total mechanical energy to decay to $\left(\frac{1}{e}\right)$ of its original value. The mechanical energy of damped harmonic oscillator is given by

$$E = \frac{1}{2} a^2 \mu \, e^{-2kt}, \text{ where '}a\text{' is the amplitude at } t = 0$$

Let $\qquad E = E_0$ when $t = 0$

$\therefore \qquad E_0 = \frac{1}{2} a^2 \mu$

Now $\qquad E = E_0 \, e^{-2kt}$ ...(1)

Let τ be the relaxation time, i.e. at $t = \tau$, $E = \dfrac{E_0}{2}$

Making this substitution in eqn. (1), we have obtain

$$\frac{E_0}{e} = E_0 \, e^{-2k\tau}$$

or, $\qquad e^{-1} = e^{-2k\tau} \qquad$ or, $\qquad -1 = -2k\tau$

or, $\qquad \tau = \dfrac{1}{2k}$ ...(2)

From eqns. (1) and (2), we have $\quad E = E_0 \, e^{-t/\tau}$ ...(3)

Quality Factor

The quality factor is defined as 2π times the ratio of the energy stored in the system to the energy lost per period.

$$Q = 2\pi \frac{\text{energy stored in system}}{\text{energy lost per period}} = 2\pi \frac{E}{PT}$$

where P is the power dissipated and T is periodic time.

$$Q = 2\pi \frac{E}{\left(\dfrac{E}{\tau}\right)T} = \frac{2\pi\tau}{T} = \omega\tau \qquad\qquad \left(\because P = \frac{E}{\tau} \right)$$

where $\quad \omega = \left(\dfrac{2\pi}{T} \right) = $ angular frequency.

So, it is clear that higher the value of Q, higher would be the value of relaxation time τ, i.e. lower damping.

Logarithmic Decrement

Logarithmic decrement measures the rate at which the amplitude dies away. The amplitude of damped harmonic oscillator is given by

$\text{Amplitude} = a e^{-kt}$

At $\qquad t = 0$, amplitude $a = a_0$

Let a_1, a_2, a_3, ... be the amplitudes at time t = T, 2T, 3T, ... respectively, where T is the time period of vibration. Then

$$a_1 = ae^{-kT}, \; a_2 = ae^{-2kT}, \; a_3 = ae^{-3kT}$$

Hence, $$\frac{a_0}{a_1} = \frac{a_1}{a_2} = \frac{a_2}{a_3} = ... = e^{kT} = e^{\lambda}$$ ...(1)

(where kT = λ)

λ is known as the logarithmic decrement.

Taking the log of eqn. (1), we obtain

$$\lambda = \log_e \frac{a_0}{a_1} = \log_e \frac{a_1}{a_2} = \log_e \frac{a_2}{a_3} = ...$$ ...(2)

Thus, logarithmic decrement is defined as the natural logarithm of the ratio between two successive maximum amplitudes which are separated by one period.

Moving Coil Galvanometer

The damping has effects on the moving parts of the moving coil galvanometer. As a result of which, the oscillation of a moving coil galvanometer is damped harmonic.

This galvanometer was first devised by Lord Kelvin and then modified for laboratory use by D'Arsonval.

Construction

The moving coil galvanometer consists of coil C, of large number of turns of fine copper wire wound over a conducting metallic hoop, suspended by means of a phosphor-bronze strip St (Fig. 2.2), in between cylindrically cut pole-pieces N and S of a powerful compound horseshoe magnet. The coil is placed symmetrically over a soft-iron cylindrical core I. A plane mirror M is attached to the suspension fibre and helps to record the deflection of the coil with a lamp and scale arrangement. The phosphor-bronze strip provides one connecting lead to the coil, the other being provided by another phosphor-bronze strip in the form of a spring (Sp) attached to the lower end of the coil.

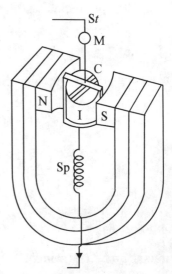

Fig. 2.2: Moving coil galvanometer

Principle

The current to be measured is fed into the coil and the coil experiences a torque which rotates it through a certain angle, thus twisting the suspension fibre that provides the restoring couple. In the equilibrium position of the coil,

Deflecting Couple = Restoring Couple

Theory: Let us consider a coil ABCD carrying a current *i* placed in between the pole-pieces of a powerful horse-shoe magnet of uniform magnetic field H, where the flux density B = μH. Now, the current *i* in the arms AB and CD of length *l* each is at right angles to the magnetic field and hence two forces (F and F) each equal to B*il* will act on these arms tending to rotate the coil in the clockwise direction, shown by arrows (Fig. 2.3), in accordance with the Fleming's Left Hand Rule. If φ be the inclination of the plane of the coil with the magnetic field, the perpendicular distance between these forces is *b* cos φ, where BC = *b* and the direction of the force B*il* is mutually perpendicular to both *i* and B. Hence, the deflecting couple is

$$B il \times b \cos \phi = B i (lb) \cos \phi$$

$$= BiA \cos \phi$$

where,   *lb* = A = area of the coil.

If the coil consists of N turns, then the current in each turn produces a similar turning moment and so the total deflecting couple is N*i*BA cos φ.

Now, if θ be the twist produced in the suspension fibre due to the rotation of the coil, then the restoring couple is Cθ, where C is the couple per unit twist in the fibre. In the equilibrium position, the deflecting couple is equal to the restoring couple, i.e.

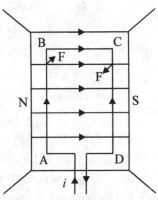

$$N i BA \cos \phi = C\theta$$

or $$i = \frac{C}{NBA} \cdot \frac{\theta}{\cos \phi}$$ ...(1)

and hence the current depends upon φ, i.e. on the initial position of the coil. If however, φ is zero, the current will become

$$i = \left(\frac{C}{NBA}\right)\theta = k\theta$$ ...(2)

Fig. 2.3: Principle of the moving coil galvanometer

where, $$k = \frac{C}{NBA} = \text{Constant}$$

So, $i \propto \theta$ ...(3)

The current is directly proportional to the deflection.

As the soft iron cylindrical core is symmetrically placed in-between the pole pieces cut cylindrically, the magnetic field between the poles N and S of a U-shaped magnet becomes *radial* in the air gap Fig. 2.4. This has the virtue of giving a field that is constant in magnitude and always parallel to the plane of the coil as the coil rotates.

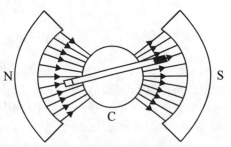

Fig. 2.4: Radial field with the help of a soft iron core C

The actual value of l and b are immaterial and only the area of the coil A determines the current. The shape of the coil may, therefore, be circular, the soft iron core spherical and the pole pieces cut spherically.

It is necessary that the magnet should retain its pole strength. The magnet is, therefore, a laminated compound magnet of U-shape, which can preserve its strength for a long time. Because of the strong field of the magnet, the earth's horizontal component of the magnet, its variation, or any stray magnetic field has very little influence on the deflection of the moving coil galvanometer and hence it does not require any setting in the magnetic meridian. For the same reason, the instrument is housed in an iron case for magnetic screening and protection from the effect of external magnetic field.

The current sensitivity of a moving coil galvanometer is given by

$$\frac{d\theta}{di} = \frac{\text{deflection}}{\text{current}} = \frac{\text{NBA}}{\text{C}}$$

It can be increased by increasing N, B and A, and by decreasing C, the torsional couple per unit twist. It is uniform for all values of the current and hence for all values of the deflection.

Moving-coil Ballistic Galvanometer

This is analogous to the suspended coil galvanometer. Let a current i flow through the coil, placed in a magnetic field of flux density B, then the force on each of the arms *ab* and *cd* (length l) is Bil and constitute a couple Fig. 2.5. The direction of these forces is given by Fleming's Left Hand Rule. There is no resultant force along the *bc* and *ad* which are parallel to the magnetic field. If the current acts for a short interval dt, the impulse of the force is B$ildt$ and the total impulse is

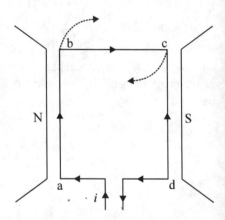

$$\int_0^T B ildt = Bl \int_0^T i\, dt = Blq$$

and the moment of this about the axis of suspension is

Fig. 2.5: Moving-coil ballistic galvanometer

$$Blq \times (bc) = Blqb = Bq(lb) = BqA$$

where, lb = A, the area of the coil and therefore, the moment of the impulse is BAq. If the coil contains N turns, then the effective area A is equal to N times the area of each turn. Now, if I is the moment of inertia of the moving system and ω is the angular velocity, we have

$$BAq = I\omega. \qquad\qquad ...(1)$$

The kinetic energy of the system is $\frac{1}{2}I\omega^2$ and is used up in twisting the suspension fibre.

If C is the restoring couple per unit twist, the couple for twist θ is $C\theta$ and work done for an additional twist dq is $c\theta d\theta$, so that the whole work done in twisting the suspension fibre is

$$\int_0^\theta C\theta \, d\theta = \frac{1}{2}C\theta^2$$

$\therefore \qquad \frac{1}{2}C\theta^2 = \frac{1}{2}I\omega^2$ \hfill ...(2)

or, $\qquad \omega^2 = \frac{C}{I}\theta^2$

or, $\qquad \left(\frac{BAq}{I}\right)^2 = \frac{C}{I}\theta^2$ \hfill (from eqn. 1)

or, $\qquad \frac{B^2A^2q^2}{I^2} = \frac{C}{I}\theta^2$

or, $\qquad q^2 = \frac{C^2I}{B^2A^2C}\theta^2$

or, $\qquad q^2 = \frac{C^2}{B^2A^2}\frac{I}{C}\theta^2$ \hfill ...(3)

Now, the period of vibration of a body of moment of inertia I, where C is the restoring couple per unit twist is given by

$$T = 2\pi\sqrt{\frac{I}{C}}$$

or, $\qquad \frac{T^2}{4\pi^2} = \frac{I}{C}$ \hfill ...(4)

Substituting this value of $\frac{I}{C}$ in eqn. (3), we obtain

$$q^2 = \frac{C^2}{B^2A^2} \times \frac{T^2}{4\pi^2} \times \theta^2$$

or, $\qquad q = \left(\frac{CT}{2\pi\,BA}\right)\theta$

$$= \frac{T}{2\pi} \cdot \frac{C}{BA} \cdot \theta = \frac{C}{BA} \cdot T \cdot \frac{\theta}{2\pi} \qquad \ldots(5)$$

or, $\qquad q = kT \dfrac{\theta}{2\pi}$ $\qquad\qquad\qquad\qquad\qquad\qquad$ ...(6)

where $k = \dfrac{C}{BA}$. This eqn. (6) is a relation between the charge flowing and the ballistic throw θ of the galvanometer, where k is the galvanometer constant.

Sensitivity of the Galvanometer

The current in a moving coil galvanometer is given by

$$i = \frac{C}{BA}\theta = k\theta$$

and the current sensitivity $= \dfrac{\theta}{i} = \dfrac{\text{deflection}}{\text{current}} = \dfrac{BA}{C}$

The charge sensitivity of the ballistic galvanometer

$$= \frac{\text{deflection}}{\text{current}} = \frac{2\pi}{T} \cdot \frac{BA}{C} \qquad \ldots(7)$$

Hence the charge sensitivity of a ballistic galvanometer is $\dfrac{2\pi}{T}$ times its current sensitivity.

Condition when a moving coil galvanometer is Dead-beat or Ballistic

Let us investigate the condition when a moving coil galvanometer can act as a dead-beat galvanometer or as a ballistic galvanometer. The resistance to the moving parts on open circuit is due to: (i) friction of air, (ii) viscosity of the suspension fibre, and (iii) induced current in any neighbouring mass of metal. The first two of these factors constitute mechanical damping, while the third factor accounts for electromagnetic damping. The couple due to damping varies as the angular velocity $\omega = \dfrac{d\theta}{dt}$ and may be taken as $p\dfrac{d\theta}{dt}$, where p may be called the coefficient of damping, it being the ratio of the retarding couple $p\dfrac{d\theta}{dt}$ on the moving system to the rate of change of displacement $\left(\dfrac{d\theta}{dt}\right)$.

When the circuit is closed, we have, in addition, the induced current produced in the coil which varies as the rate of change of displacement $\dfrac{d\theta}{dt}$ and inversely as the total resistance of the circuit, hence the retarding couple is $\dfrac{m}{R}\dfrac{d\theta}{dt}$, where the constant m involves the magnetic flux due to the magnet and the area of the coil. Let I be the moment of inertia of the moving system and C, the couple per unit twist of the suspension fibre.

The equation of motion is then

$$I\frac{d^2\theta}{dt^2} + \left(\frac{m}{R} + p\right)\frac{d\theta}{dt} + C\theta = 0 \qquad ...(8)$$

or,

$$\frac{d^2\theta}{dt^2} + \frac{\left(\dfrac{m}{R} + P\right)}{I}\frac{d\theta}{dt} + \frac{C}{I}\theta = 0$$

or,

$$\frac{d^2\theta}{dt^2} + 2k\frac{d\theta}{dt} + \omega^2\theta = 0 \qquad ...(9)$$

where, $\quad 2k = \dfrac{\dfrac{m}{R} + p}{I} \text{ and } \omega^2 = \dfrac{C}{I} \qquad ...(10)$

Equation (9) is the equation of damped harmonic motion.

Solution:

Let a solution of this equation be

$$\theta = A\,e^{\alpha t} \qquad ...(11)$$

Then, $\qquad \dfrac{d\theta}{dt} = A\alpha\,e^{\alpha t}$

and $\qquad \dfrac{d^2\theta}{dt^2} = A\alpha^2 e^{\alpha t}$

Substituting these values in (9), we obtain

$$A\alpha^2\,e^{\alpha t} + 2k\,A\alpha\,e^{\alpha t} + \omega^2\,A\,e^{\alpha t} = 0$$

or, $\qquad \alpha^2 + 2k\alpha + \omega^2 = 0 \qquad ...(12)$

Hence, $\qquad \alpha = \dfrac{-2k \pm \sqrt{4k^2 - 4\omega^2}}{2} = -k \pm \sqrt{k^2 - \omega^2} \qquad ...(13)$

or, $\qquad \alpha_1 = -k + \sqrt{k^2 - \omega^2}$

and $\qquad \alpha_2 = -k - \sqrt{k^2 - \omega^2} \qquad ...(14)$

Case I: Non-oscillatory or dead-beat motion

When $k^2 > \omega^2$ or $k > \omega$, we have two real values of α. Hence, the solution of eqn. (9) is

$$\theta = A\,e^{\alpha_1 t} + B\,e^{\alpha_2 t} \qquad ...(15)$$

where, α_1 and α_2 are the two real roots of eqn. (12). The motion is non-oscillatory or *dead-beat*. The oscillation is said to be over-damped. This type of motion is also called *aperiodic* motion (Fig. 2.1a).

Case II: Critically damped motion

When $k^2 = \omega^2$ or $k = \omega$, the galvanometer is critically damped. Here, the system after displacement comes to rest in the minimum of time and the velocity never changes direction. The condition of oscillation is

$$\frac{I}{2I}\left(\frac{m}{R} + p\right) = \sqrt{\frac{C}{I}}$$

Case III: Oscillatory or ballistic motion

When $k^2 < \omega^2$ or $k < \omega$, the quantity $\sqrt{k^2 - \omega^2}$ becomes negative and hence the values of α are complex, i.e. partly real and partly imaginary. We can write

$$\alpha_1 = -k + i\sqrt{\omega^2 - k^2} = -k + ig$$

and
$$\alpha_2 = -k - ig, \text{ where } g = \sqrt{\omega^2 - k^2}$$

$$\therefore \quad \theta = A\,e^{(-k+ig)t} + B\,e^{(-k-ig)t}$$

$$= C_1\,e^{-kt}\,\sin(gt + \beta)$$

$$= C_1\,e^{-kt}\,\sin\left[\left(\sqrt{\omega^2 - k^2}\right)t + \beta\right] \qquad \text{...(16)}$$

The motion is oscillatory or ballistic (Fig. 2.1b). Here, C_1 is a new constant and β is the additional phase angle.

Time-period of Oscillation

The period of vibration is given by

$$T = \frac{2\pi}{g} = \frac{2\pi}{\sqrt{\omega^2 - k^2}} \qquad \text{...(17)}$$

In order that the whole charge may pass through the ballistic galvanometer before it has appreciably moved from its zero position, it is essential that its periodic time be large and hence I should be large and C should be small, we have

$$k < \omega$$

or,
$$\frac{1}{2I}\left(\frac{m}{R} + p\right) < \sqrt{\frac{C}{I}} < \frac{2\pi}{T}$$

Since
$$T = 2\pi\sqrt{\frac{I}{C}} \text{ as } \omega = \sqrt{\frac{C}{I}}$$

In order to satisfy the above condition, it is necessary to make k as small as possible. We have from eqn. (10),

$$k = \frac{I}{2I}\left(\frac{m}{R} + p\right)$$

and hence to make k small,

I   should be large,

R   should be large,

m   should be small and so the coil should be wound round a non-conducting frame, e.g. paper or bamboo,

p   should be small, and so the air resistance should be small and the fibre should be very fine.

Damping

The equation of motion of the coil of the ballistic galvanometer in damped harmonic motion, where the displacement θ is given by

$$\theta = C_1\, e^{-kt}\, \sin\left(\sqrt{\omega^2 - k^2}\cdot t + \beta\right)$$

Effort is, however, made in the ballistic galvanometer to make k as small as possible, so that the motion is SHM, but in actual practice k can never be zero and so the amplitude of vibration decreases with time. The motion is, thus, damped. The amplitude of motion is $C_1\, e^{-kt}$ and at $t = 0$.

The amplitude $\theta_0 = C_1$

at $t = \dfrac{T}{4} = \dfrac{\pi}{2\sqrt{\omega^2 - k^2}}$, the amplitude

$$\theta_1 = C_1\, e^{-\frac{k\pi}{2\sqrt{\omega^2 - k^2}}}$$

at $t = \dfrac{3T}{4} = \dfrac{3\pi}{2\sqrt{\omega^2 - k^2}}$, the amplitude

$$\theta_2 = C_1\, e^{-\frac{k\pi}{2\sqrt{\omega^2 - k^2}}}$$

at $t = \dfrac{ST}{4} = \dfrac{5\pi}{2\sqrt{\omega^2 - k^2}}$, the amplitude

$$\theta_3 = C_1\, e^{-\frac{5k\pi}{2\sqrt{\omega^2 - k^2}}}$$

$$\therefore \quad \frac{\theta_1}{\theta_2} = \frac{\theta_2}{\theta_3} = \frac{\theta_3}{\theta_4} = ... = e^{-\frac{k\pi}{\sqrt{\omega^2 - k^2}}}$$

$$d = e^\lambda$$

Hence, $$\lambda = \log_e d = \frac{k\pi}{\sqrt{\omega^2 - k^2}} \qquad \text{...(18)}$$

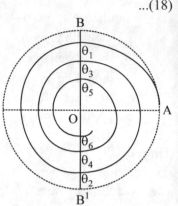

Fig. 2.6: Decreasing amplitudes (Damping)

Hence, it follows that the successive amplitudes θ_1, θ_2, θ_3, ..., etc. are continuously decreasing and are such that the ratio of one value to the next is constant. This constant ratio d is called the decrement $\log_e d = \lambda$ is called the logarithmic decrement. The successive amplitudes θ_1, θ_2, θ_3, ..., etc. can best be represented graphically with the help of an equiangular spiral. When the oscillations are simple harmonic, it may be represented by the projection of a rotation vector OA upon any fixed diameter BOB'. In this case, the successive amplitudes are OB, OB', ..., etc. but if the amplitudes get less in a constant ratio, the rotating vector OA shrinks at a constant rate, see Fig. 2.6.

Now the impulse is given to the coil at the mean position, if there were no damping, the observed deflection would have been $\theta_0 = C_1 = OA$. The first throw is observed a quarter period later and during all this time the shrinkage has taken place, so that

$$\theta_1 = C_1\, e^{-\frac{b\pi}{2\sqrt{\omega^2 - k^2}}} = C_1\, e^{-\frac{\lambda}{2}} \qquad \text{...(19)}$$

Hence, $$\frac{\theta_0}{\theta_1} = e^{\lambda/2} = 1 + \frac{\lambda}{2} + \frac{1}{2!}\left(\frac{\lambda}{2}\right)^2 + ... = 1 + \frac{\lambda}{2}$$

approximately, neglecting the higher powers of λ, which in the case of a ballistic galvanometer is always made as small as possible.

$$\therefore \qquad \theta_0 = \theta_1\left(1 + \frac{\lambda}{2}\right) \qquad \text{...(20)}$$

We can thus correct the effect of damping in a ballistic galvanometer, although we cannot avoid it. The corrected equation of the moving ballistic galvanometer becomes:

$$q = \frac{CT}{2\pi\, BA}\,\theta_1\left(1 + \frac{\lambda}{2}\right) \qquad \text{...(21)}$$

Experimental Determination of the Logarithmic Decrement λ

The coil of the ballistic galvanometer is made to oscillate by passing a discharge through it and the successive amplitudes on the right and left are recorded. Then,

$$\frac{\theta_1}{\theta_2} = \frac{\theta_2}{\theta_3} = \frac{\theta_3}{\theta_4} = ... = \frac{\theta_9}{\theta_{10}} = \frac{\theta_{10}}{\theta_{11}} = d = e^\lambda$$

Multiplying all these factors, we have

$$\frac{\theta_1}{\theta_{11}} = e^{10\lambda}$$

or, $10\lambda = \log_e\left(\frac{\theta_1}{\theta_{11}}\right)$,   or, $\lambda = \frac{1}{10}\log_e\left(\frac{\theta_1}{\theta_{11}}\right)$

or, $\lambda = \frac{e}{10}\log_{10}\left(\frac{\theta_1}{\theta_{11}}\right) = \frac{2.3026}{10}\log_{10}\left(\frac{\theta_1}{\theta_{11}}\right)$ ...(22)

Thus, knowing the value of λ, the first throw is corrected as in eqn. (22) above.

SOLVED EXAMPLES

Q.1. A damped vibrating system, starting from rest has an initial amplitude of 506 mm which reduces to 160 mm in that direction after 50 complete oscillations each of period 2.3 seconds. Find the damping coefficient and the correction for the initial displacement for damping.

Sol. The damping coefficient k is given by

$$k = 2.303\frac{f}{p}\log_{10}\left(\frac{a_0}{a_p}\right)$$

Here, $f = \frac{1}{2.3}; p - 50, a_0 = 506$

and $a_p = 160$

$\therefore$ $k = 2.303\frac{1}{2.3 \times 50}\log_{10}\frac{506}{160} = 0.01$

Initial amplitude is reached in one quarter, i.e. 0.25 of a vibration. Let its undamped value be 'a'. Then,

$$0.01 = \frac{2.303}{2.3 \times 0.25}\log_{10}\left(\frac{a}{506}\right)$$

$\therefore$ $\log_{10}\left(\frac{a}{506}\right) = 0.01 \times 0.25 = 0.0025$

whence $a = 509.0$ mm

$\therefore$ Correction required $= 509 - 506 = 3$ mm.

Q.2. An underdamped oscillator has its amplitude reduced to $\left(\dfrac{1}{10}\right)$th of its initial value after 100 oscillations. If time period is 2 seconds, calculate: (i) the damping constant, and (ii) the decay modulus.

Sol. (i) The amplitude 'a' of damped oscillator is given by

$$a = a_0 \, e^{-bt}$$

where a_0 is the amplitude of undamped oscillator. The first amplitude a_1 is observed after $\dfrac{T}{4}$ sec, where T is the periodic time. Hence,

$$a_1 = a_0 \, e^{-b(T/4)} \qquad \qquad ...(1)$$

Now, the successive amplitude occurs at internal of $\dfrac{T}{2}$. After 100 oscillations, the number of amplitudes will be 201 and time taken will be $\left(200\dfrac{T}{2} + \dfrac{T}{4}\right)$. Hence,

$$a_{201} = a_0 \, e^{-b\left(100 + \frac{T}{4}\right)} \qquad \qquad ...(2)$$

Dividing equation (2) by equation (1), we obtain

$$\frac{a_{201}}{a_1} = \frac{e^{-b\left(100T + \frac{T}{4}\right)}}{e^{-(T/4)}} = e^{-100\,bT}$$

According to the given problem

$$\frac{a_{201}}{a_1} = \frac{1}{10} \quad \text{and} \quad T = 2 \text{ sec}$$

$$\therefore \qquad 10^{-1} = e^{-100 \times b \times 2} \quad \text{or} \quad \log_e 10 = 200\,b$$

or, $\qquad 2.3 \log_{10} 10 = 200b$

or, $\qquad b = \dfrac{2.3}{200} = 0.0115 = 86.8 \text{ seconds.}$

Q.3. The amplitude of an oscillator of frequency 200 per second falls to $\left(\dfrac{1}{10}\right)$th of its initial value after 2000 cycles. Calculate: (i) its relaxation time, (ii) the quality factor, (iii) time in which its energy falls to $\dfrac{1}{10}$th of its initial value, and (iv) damping constant.

Sol. The instantaneous amplitude of damped oscillator $= a\, e^{-bt}$. Let at $t = 0$, initial amplitude $= a_0$. After 10 seconds, its amplitude $= \dfrac{a_0}{10}$.

$$\therefore \qquad \frac{a_0}{10} = a_0\, e^{-b \times 10} \quad \text{or} \quad 10 = e^{10b}$$

Taking log, we get $\log_e 10 = 10b$

$$\text{or,} \qquad 2.3 \log_{10} 10 = 10b$$

$$\text{or,} \qquad 2.3 = 10b \text{ or } b = 0.23$$

(i) Relaxation time $\qquad \tau = 2.174$ sec.

(ii) Quality factor $\qquad Q = \omega t = 2\pi n \times \tau = 2 \times 3.14 \times 200 \times 2.174 = 2730$

(iii) $\qquad\qquad\qquad E = E_0\, e^{-t/\tau}$

$$\text{or,} \qquad\qquad 10 = e^{-t/\tau}$$

$$\text{Now,} \qquad \log_e 10 = -t/\tau$$

$$\text{or,} \qquad\qquad t = \tau \log_e 10 = 2.174 \times 2.3 = 5 \text{ secs}$$

(iv) Damping constant $\quad b = 0.23$.

Q.4. Find the frequency of oscillation of a body executing damped harmonic motion if the frequency of SHM is 10 Hz and the damping coefficient is 0.02.

Sol. Here, $n = 10$ Hz, $k = 0.02$

So, $\qquad \omega = 2\pi n = 2 \times 3.14 \times 10 = 62.80$

The frequency of DHM is given by

$$\beta = \sqrt{\omega^2 - k^2}$$

$$= \sqrt{3943.84 - 0.0004}$$

$$= 62.79999 \text{ Hz}$$

$$f = \frac{\beta}{2\pi} = \frac{62.79999}{2 \times 3.14} = \frac{62.79999}{6.28} = 9.99 \text{ Hz} \cong 10 \text{ Hz}$$

QUESTIONS

1. Develop the theory of damped harmonic motion. Discuss the vibration of displacement with time when: (i) $k^2 > \omega^2$, (ii) $k^2 = \omega^2$, and (iii) $k^2 < \omega^2$, where k is the damping coefficient and ω is the angular frequency of oscillation.

2. Form the equation of damped oscillation. Solve the equation and discuss its general and particular solutions.

3. What do you mean by critical damping? Obtain the solution of DHM for this case and discuss on the variation of displacement with time.

4. Solve the equation of DHM when the frictional forces are small and the motion is oscillatory. Discuss on the vibration of amplitude with time.

5. Develop the general theory of DHM and discuss on the following cases:

    (i) Motion is dead-beat, (ii) Motion is critically damped, and (iii) Motion is oscillatory.

6. Investigate the effect of a damping force on the motion of a system where friction cannot be neglected.

7. Calculate the energy in damped vibration.

8. Discuss on the effect of damping on frequency of oscillation.

9. Find expression for the period and amplitude of DHM

10. Explain the terms:

    (i) Logarithmic decrement, (ii) Relaxation time, and (iii) Quality factor.

11. Develop the theory of moving coil galvanometer. When the galvanometer is dead-beat or ballistic?

Problems

1. A damped vibrating system starting from rest reaches a first amplitude of 500 mm which reduces to 50 mm in that direction after 100 oscillations each of period 2.3 seconds. Find the damping coeffcient and correction for the first displacement for damping.

    (**Ans.** k = 0.01, Correction = 2.9 mm)

2. The amplitude of an oscillator executing DHM is reduced to $\dfrac{1}{10}$ th of its initial value after 200 oscillations. If time period is 1 second, calculate: (i) the damping constant, and (ii) the decay modulus.

    (**Ans.** (i) 0.0115, and (ii) 86.8 sec)

3. The amplitude of an oscillator of frequency 100 cycles per second falls to $\dfrac{1}{10}$ th of its initial value after 1000 cycles. Calculate: (i) its relaxation time, (ii) its quality factor, (iii) time in which its energy falls to $\dfrac{1}{10}$ th of its initial value, and (iv) damping constant.

    (**Ans.** (i) 2.174 sec, (ii) 1365, (iii) 5 sec, and (iv) 0.23)

4. Find the logarithmic decrement of DHM when the damping coefficient is 0.23 and the time period is 2 seconds.

(**Ans.** 0.46)

5. Find the time period and frequency of DHM when the motion is oscillatory one, if $k = 0.01$ and $\omega = 628$ cycles/sec.

(**Ans.** 10^{-2} sec, 100 Hz)

6. What will be the value of the relaxation time if the damping coefficient (k) is 0.02?

(**Ans.** 25 secs)

7. Compute the quality factor of DHM, if the periodic time is 2 seconds and the damping coefficient is 0.005.

(**Ans.** 314)

8. Find the power dissipated in DHM if the energy stored in the system is 500 joules and relaxation time is 50 seconds.

(**Ans.** 10 J/sec)

9. Find the logarithmic decrement of DHM when the initial amplitude of vibration is 20 cm and the subsequent amplitude is 15 cm.

(**Ans.** 0.28727)

10. A condenser charged to 2 volts is discharged through a ballistic galvanometer. When the corrected deflection is 0.6 cm, current sensitivity is 2.2×10^{-8} amp/cm and the periodic time is 12 sec. Calculate the capacity of the condenser.

(**Ans.** 0.2015 µf)

11. A galvanometer has a rectangular coil which is 1.8×4.5 cm. It is suspended to move through a magnetic field that has a magnitude of 2.5×10^{-2} Wb/m^2. The coil has 90 turns. What maximum torque acts on the coil when it carries a current of 150 µA?

(**Ans.** 2.74 cm.dyne)

12. A galvanometer of moving coil type has a current sensitivity of 0.002 µa/mm. What current is necessary to produce a deflection of 20 cm on a scale 1 m distant?

(**Ans.** 0.4 µA)

13. If the moving coil of the galvanometer of the example above (Problem No. 12) has a resistance of 25 Ω, what is the potential difference across its terminals when the deflection is 20 cm?

(**Ans.** 10 µV)

14. A current of 2×10^{-4} A causes a deflection of 10 divisions on the scale of a portable type galvanometer. What is its current sensitivity?

(**Ans.** 20 µA/division)

15. What current will cause a full-scale deflection (100 divisions) of a portable-type galvanometer?

(**Ans.** 20×10^{-4}A)

Chapter 3

Forced Vibrations and Resonance

When a mass is subjected to an oscillatory external force such as a bridge which vibrates under the rhythm of marching soldiers, or the prongs of a turning fork which is exposed to regular impulses of a sound wave or when one sings near piano, the oscillations that ensure are called forced oscillations.

Use is very often made of an external periodic force to maintain the vibrations of a system. The impressed periodic force may be or may not have the same period as the free vibration. Note that in this case our vibrating system is subjected to damping and to an external periodic force. The friction tends to retard the motion and the impressed force tends to maintain it. Evidently the result should be a settle down to a certain definite amplitude and phase which will remain constant while the impressed periodic force continues, the frequency being that of the impressed force. In the special case when the period of the free vibrations coincides with the period of the force, we have the phenomenon of resonance, i.e. $n = f$.

Let the external periodic force be represented by F sin pt where F is the maximum value of the force of frequency $\dfrac{p}{2\pi}$, then we may write the equation of motion as follows:

$$m\frac{d^2y}{dt^2} + r\frac{dy}{dt} + \mu y = F \sin pt \qquad \qquad ...(1)$$

or, $\qquad \dfrac{d^2y}{dt^2} + \dfrac{r}{m}\dfrac{dy}{dt} + \dfrac{\mu}{m}y = \dfrac{F}{m}\sin pt$

Writing $\qquad \dfrac{r}{m} = 2k, \; \dfrac{\mu}{m} = \omega^2, \; \text{and} \; \dfrac{F}{m} = f, \text{we get}$

$$\frac{d^2y}{dt^2} + 2k\frac{dy}{dt} + \omega^2 y = f \sin pt \qquad \qquad ...(2)$$

The vibration maintained by a periodic force has ultimately the same period as that of the force, viz. $\dfrac{2\pi}{p}$ and the amplitude of the maintained vibration almost remains constant so long as the force continues to act. Therefore, we write for the maintained vibration

$$y = A \sin (pt - \theta) \qquad \qquad ...(3)$$

where A represents the amplitude of the maintained vibration θ the lag in phase of the same, i.e. θ represents the phase difference between the force and the resulting displacement of the system.

Substituting the value of y and its differential coefficients in equation is (2) and writing $\sin pt = \sin[(pt-\theta)+\theta]$ we obtain

$$- Ap^2 \sin (pt - \theta) + 2kpA \cos (pt - \theta) + \omega^2 A \sin (pt - \theta)$$

$$= f \sin[(pt - \theta) + \theta],$$

or, $\quad A (\omega^2 - p^2) \sin (pt - \theta) + 2kpA \cos (pt - \theta)$

$$= f \cos \theta \sin (pt - \theta) + f \sin \theta \cos (pt - \theta) \qquad ...(4)$$

Now equation (4) holds, good for all values of t and we may equate coefficients of sine and cosine of $(pt - \theta)$.

Hence, we get

$$A (\omega^2 - p^2) = f \cos \theta, \text{ and } 2kpA = f \sin \theta \qquad ...(5)$$

thus $\qquad A = \dfrac{f \sin \theta}{2kp}$ and $\tan \theta \, \dfrac{2kp}{\omega^2 - p^2}$ $\qquad ...(6)$

The solution of eqn. (2) is therefore

$$y = \frac{f \sin \theta}{2kp} \sin (pt - \theta) \qquad ...(7)$$

Also by squaring and adding the two components of eqn. (5) we get

$$A = \frac{f}{\sqrt{(\omega^2 - p^2)^2 + 4k^2 p^2}}$$

$$\therefore \qquad y = \frac{f}{\sqrt{(\omega^2 - p^2)^2 + 4k^2 p^2}} \sin (pt - \theta) \qquad ...(8)$$

Equation (8), however, does not represent the complete solution of eqn. (2) we may add to this another term for the value of y. The eqn. $y = a' e^{-kt} \sin (\beta t + \phi')$, which makes the left hand side of the eqn. (2) equal to zero without impairing its validity as a solution. Hence, we have the complete solution

$$y = \frac{f}{\sqrt{(\omega^2 - p^2)^2 + 4k^2 p^2}} \sin (pt - \theta) + a' e^{-kt} \sin (\beta t + \phi') \qquad ...(8a)$$

where $\tan \theta = \dfrac{2kp}{\omega^2 - p^2}$ and $\beta = \sqrt{\omega^2 - k^2}$, a' and ϕ' are arbitrary constants depending on the initial conditions.

Note that the vibration is made up of two terms, the 'first term' indicating an undamped vibration of the same frequency as the applied force and the 'second term' a damped vibration of almost natural frequency of the vibrating system which is only important in the early stages of the forced vibration. It opposes the forced oscillations but after some time the natural or free oscillations die away, only the forced oscillations remain, so that the 'first term' is the only important part of the solution.

Amplitude: The amplitude A of the forced vibration has been obtained by squaring and adding the two components of eqn. (5),

where $A^2\{(\omega^2 - p^2) + 4k^2p^2\} = f^2$

or, $\qquad A = \dfrac{f}{\sqrt{(\omega^2 - p^2)^2 + 4p^2k^2}}$ ...(9)

We notice from the expression given above that the amplitude with which the body is made to oscillate depends upon the relation of ω and p. The more they differ from each other, the greater $\omega^2 - p^2$ will be and the smaller the amplitude. It is maximum when $\omega = p$, i.e. when the free period of the body is the same as that of the impressed force. The maximum value of the amplitude is given by

$$A_m = \frac{f}{2k\omega}$$ ...(10)

Since $\qquad \omega = p$ (nearly)

This is the case of amplitude resonance. It is moreover, clear that if there is no damping, i.e. k is equal to zero, then for resonant vibration, i.e. when $p = \omega$, the amplitude will be infinite. Actually this never happen for the damping coefficient, k is never zero.

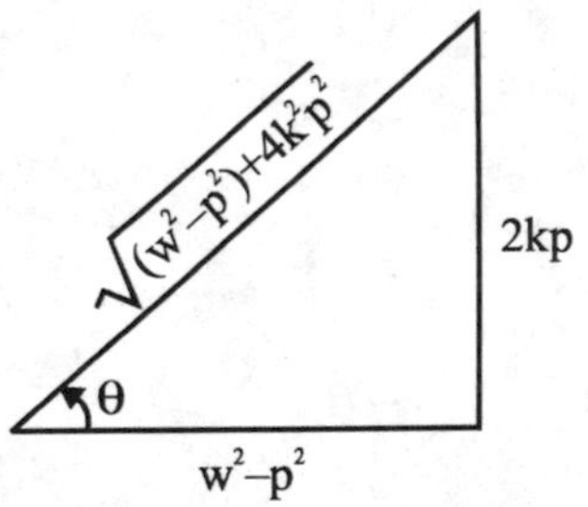

Phase: The meaning of the phase angle θ is best understood by reference to the right angled triangle of Fig. 3.1.

From it we see that

$$\sin \theta = \frac{2kp}{D}$$

$$\cos \theta = \frac{\omega^2 - p^2}{D}$$

and $\qquad \tan \theta = \dfrac{2kp}{\omega^2 - p^2}$ ...(11)

Provided $\qquad D = \sqrt{(\omega^2 - p^2)^2 + 4k^2p^2}$ ...(12)

Equation (8) may be written as

$$y = \frac{f}{D} \sin(pt - \theta)$$ ...(13)

Fig. 3.1

The phase angle θ depends upon the values of the damping coefficient k and the relative magnitudes of p and ω, that is, on the entire physical condition of the system. For all values of p smaller than n, tan θ is a small positive quantity and θ is very nearly zero, that is, the impressed force and the body are practically in phase. For all values of p greater than ω, however, tan θ is a small negative quantity, i.e. the forced vibration differs in phase by π nearly. For resonance, $p = \omega$, tan θ is infinite and θ is, therefore, $\dfrac{\pi}{2}$, i.e. the impressed force and the vibration differ in phase by quarter period.

Power Dissipation – Sharpness of Resonance Mathematically Treated

Let us see how the maintenance of vibrations under conditions of damping is influenced by the rate of supply of energy which is called power dissipation. The phase relationship between the impressed force and the forced vibration have a bearing on the problem. We know that the mechanical power or the rate of doing work is the product of force and velocity. In our case the force per unit mass is $f \sin pt$ and the velocity

$$\frac{dy}{dt} = \frac{f \sin \theta}{2k} \cos (pt - \theta),$$

which is obtained by differentiating eqn. (7), therefore the power at any instant supplied to maintain the oscillations

$$= f \sin pt \times \frac{dy}{dt}$$

$$= f \sin pt \frac{f \sin \theta}{2k} \cos (pt - \theta)$$

$$= \frac{f^2 \sin 0}{2k} \cos (pt - \theta) \sin pt$$

$$= \frac{f^2 \sin \theta}{4k} \{\sin \theta + \sin (2pt - \theta)\},$$

Since $2 \cos (pt - \theta) \sin pt = \sin \theta + \sin (2pt - \theta)$.

The average value of the second term is zero therefore the mean power supplied is given by

$$p = \frac{f^2}{4k} \sin^2 \theta \qquad \qquad ...(14)$$

the value of $\sin \theta$ is obtained from the phase triangle of Fig. 3.1 and is given by

$$\sin \theta = \frac{2kp}{\sqrt{(\omega^2 - p^2)^2 + 4k^2 p^2}} \qquad \qquad ...(15)$$

If $\theta = 0$ or π, the mean power supplied by the impressed force to maintain oscillations is zero.

If $\theta = \dfrac{\pi}{2}$, as in the case of resonance the displacement and force are quadrature apart, while the velocity is in phase with the force, the mean power supplied by the impressed force is maximum and is equal to

$$P_{max} = \frac{f^2}{4k} \qquad \qquad ...(16)$$

This quantity increases as the damping coefficient k decreases.

Closely connected with this problem is that of sharpness of resonance, we observed that on the case when $p = \omega$, it is the coefficient of damping which prevents the amplitudes of the vibration from becoming infinite. The damping has a relatively powerful influence when the resonance is approached. Let us see how the amplitude is approached. Let us see how the amplitude and the energy dissipation are related to each other in this case.

Expressing A as a fraction of the maximum, amplitude A_m, we have from eqns. (9) and (10),

$$\frac{A}{A_m} = \frac{2k\omega}{\sqrt{(\omega^2 - p^2)^2 - 4k^2 p^2}} \qquad \qquad ...(17)$$

Similarly, expressing the mean power supplied to maintain the oscillations in terms of the maximum power in the case of resonance position, we have

$$\frac{p}{p_{max}} = \sin^2 \theta \qquad \qquad ...(18)$$

Substituting the value of $\sin \theta$ from eqn. (15), we have

$$\frac{p}{p_{max}} = \frac{4k^2 p^2}{(\omega^2 - p^2)^2 + 4k^2 p^2} \qquad \qquad ...(19)$$

when $p = \omega$, $\dfrac{p}{p_{max}} = 1$, there is maximum energy in the forced vibration and at a frequency near resonance,

i.e., when $\qquad\qquad (\omega^2 - p^2)^2 = 4k^2 p^2,$

or, $\qquad\qquad \omega^2 - p^2 = \pm\, 2kp,$

or, $\qquad\qquad (\omega - p)(\omega + p) = \pm\, 2kp,$

or, $\qquad\qquad \omega - p = \pm\, k$ (Since $\omega + p = 2p$ approx)

or, $\qquad\qquad \dfrac{\omega - p}{\omega} = \pm\, \dfrac{k}{\omega}$

or
$$\frac{p}{\omega} = 1 \pm \frac{k}{\omega} \text{ approx} \qquad \qquad \ldots(20)$$

and
$$\frac{p}{p_{\max}} = \frac{4k^2 p^2}{8k^2 p^2} = \frac{1}{2} \qquad \qquad \ldots(21)$$

Thus it is clear that the power supplied by the force at resonance falls to half the value when the ratio of the frequency of the impressed force to that of the free vibration differs from one by a fraction $\dfrac{k}{\omega}$. This shows that smaller the damping coefficient and greater the natural frequency the more rapidly will the power supplied by the impressed force fall off on either side of the maximum. That is, the resonance will be sharper and the $\dfrac{A}{A_m}$ curve would be steeper. The ratio $\dfrac{k}{\omega}$ may therefore be taken as a measure of the sharpness of resonance and the reciprocal $\left(\dfrac{\omega}{k}\right)$ is sometimes referred to as persistence of vibrations.

Damping and its Effect on the Amplitude of Forced Vibrations

When an object is sounding, the energy of the sound waves is derived from the energy of vibration of the object. But a certain amount of energy is always dissipated in the form of heat due to viscosity of the particles of the object and friction with the result that the amplitude of the vibrations gradually decreases and the vibrations are said to be *damped*.

Fig. 3.2 exhibits the results of some simple cases for various degrees of damping indicated by curves 1, 2 and 3. The angular frequency (p) of the applied force in terms of the natural angular frequency (ω) of the body is plotted along the abscissae while the amplitude A of the forced vibration along the ordinates.

It is observed that as p increases, the amplitude increases more and more until $p = \omega$, and at $p = \omega$, the amplitude becomes infinitely large, shown by the dotted curve. This is the case of resonant vibration. When the natural frequency of vibration becomes equal to the frequency of the external force, resonance occurs and the amplitude of vibration becomes infinitely large. But, this is not always the case. Due to the presence of damping, an infinite amplitude is never attained. Actually, a curve like 1 is obtained for $p = \omega$.

As p is further increased, the amplitude of the forced vibration decreases again tending to zero for large frequencies of the applied force. Curves 1, 2 and 3 are obtained for $k =$ small, $k =$ medium and $k =$ large.

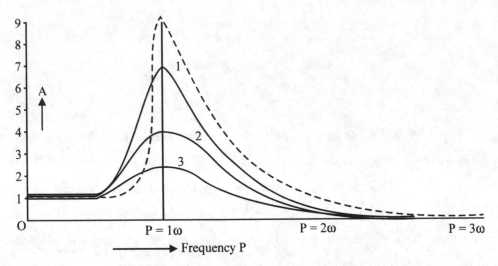

Fig. 3.2: Effect of damping on the amplitude of forced vibrations

Effect of Damping on the Phase of Forced Vibrations

If θ be the phase difference between the applied periodic force and the resulting displacement of the forced vibration, then

$$\tan\theta = \frac{2kp}{\omega^2 - p^2}$$

This relation shows that the phase between the impressed force and the resultant forced vibration depends upon the amount of damping and the relative magnitudes of p and ω. If damping be negligible, the phase of vibration coincides nearly with that of the impressed force for all values of p smaller than ω, since $\tan\theta$ is a small positive quantity. For all values of p greater than ω, however, $\tan\theta$ is a small negative quantity, i.e. the forced vibration differs in phase by π nearly, viz. half a period.

If $p = \omega$, $\tan\theta$ is infinite and θ is, therefore, $\frac{\pi}{2}$, i.e. the impressed force and the vibration differ in phase by quarter period while the force and velocity will be in the same phase. This is the case of exact resonance. For appreciable damping, it is observed that the phase difference gradually increases. For slow forced vibrations, where $p < \omega$, it is very small and increases with increasing frequency of the impressed force in a manner dependent upon the degree of damping and finally, always reaches the value $\frac{\pi}{2}$ at resonance. The forced vibration is in its position of greatest displacement when the impressed force is just passing through its equilibrium position. Fig. 3.3 shows phase differences as a function of impressed frequencies for various values of the coefficient of damping (k).

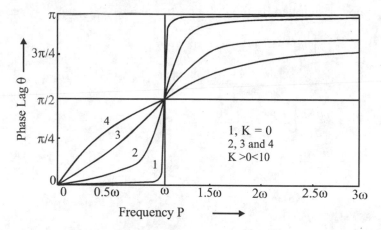

Fig. 3.3: Curves showing the effect of damping
on the phase of the forced vibrations

Power Absorption in Driven Oscillator

In presence of damping forces, there is a regular dissipation of energy of the harmonic oscillator. Therefore, for maintaining the energy or amplitude, the oscillator regularly requires some energy which is supplied by the driving force.

The driving force we have taken earlier is F sin *pt*. The solution of the equation of the forced harmonic motion is obtained as

$$y = A \sin (pt - \theta)$$

So, the velocity is

$$V = \frac{dy}{dt} = Ap \cos (pt - \theta)$$

The power absorbed by the forced harmonic oscillator or the power supplied by the driving force (externally applied force) is given by

$$P = \frac{dw}{dt} = \vec{F}.\vec{V} = FV \cos \delta$$

The maximum value of power is

$$P_m = FV \quad (\text{when } \delta = 0)$$

$$(P_m) = (F \sin pt) \, Ap \cos (pt - \theta)$$

$$= FAp \sin pt \cos (pt - \theta)$$

$$= mfAp \sin pt \cos (pt - \theta)$$

The average power supplied by the driving force is

$$(P_m)_{av} = \frac{1}{T} \int_0^T mfAp \sin pt \cos(pt - \theta)\, dt$$

$$= \frac{mfAp}{T} \int_0^T \frac{[\sin(2pt - \theta) + \sin\theta]\, dt}{2}$$

$$= \frac{mfAp}{2T} \left[-\frac{\cos(2pt - \theta)}{2p} + t\sin\theta \right]_0^T$$

$$= \frac{mfAp}{2T} \times T \sin\theta$$

$$= \frac{1}{2} mfAp \sin\theta$$

$$= \frac{1}{2} mAp \times f \sin\theta$$

$$= \frac{1}{2} mAp \times 2kAp \qquad [\because f\sin\theta = 2kAp]$$

$$= mkA^2p^2 \qquad\qquad\qquad ...(1)$$

Now, it can be shown that the average power supplied by the driving force is equal to the average power dissipated per cycle in the steady state and in the presence of damping.

The dissipative force

$$= r\frac{dy}{dt} \quad \left(\text{where } \frac{r}{m} = 2k \right)$$

The work done against this force per cycle or the power dissipated

$$P_m = \text{Force} \times \text{Velocity}$$

$$= \left(r\frac{dy}{dt} \right)\left(\frac{dy}{dt} \right)$$

$$= r\left(\frac{dy}{dt} \right)^2$$

$$= r\left[Ap\cos(pt - \theta) \right]^2$$

$$= r\,A^2p^2 \cos^2(pt - \theta)$$

$$= \left(\frac{r}{m} \right) m\,A^2p^2 \cos^2(pt - \theta)$$

$$= (2k)\, m\, A^2p^2 \cos^2(pt - \theta)$$

Average power dissipated is

$$(P_m)_{av} = \frac{1}{T} \int_0^T (2k\, m\, A^2p^2) \cos^2(pt - \theta)\, dt$$

$$= \frac{2km\, A^2p^2}{T} \cdot \frac{T}{2} = km\, A^2p^2 \qquad\qquad ...(2)$$

From eqns. (1) and (2), it is observed that the average power dissipated per cycle of vibration of the forced harmonic oscillator is equal to the average power supplied by the driving force.

Quality factor (Q)

$$Q = 2\pi \frac{\text{Average energy stored}}{\text{Energy dissipated per cycle}}$$

$$= 2\pi \frac{E_{av}}{T \times \text{Average power dissipated}}$$

Total energy $= KE + PE$

$$= \frac{1}{2} m \left(\frac{dy}{dt}\right)^2 + \frac{1}{2} \mu y^2$$

or,

$$E = \frac{1}{2} m A^2 p^2 \cos^2(pt - \theta) + \frac{1}{2} m \omega^2 A^2 \times \sin^2(pt - \theta)$$

or,

$$E_{av} = \frac{1}{2} m^2 A^2 p^2 \times \frac{1}{2} + \frac{1}{2} m \omega^2 A^2 \times \frac{1}{2}$$

$$= \frac{1}{4} m^2 A^2 (p^2 + \omega^2)$$

[Since the average value of $\sin^2(pt - \theta)$ and $\cos^2(pt - \theta)$ are each equal to $1/2$]

Hence,

$$Q = 2\pi \frac{\frac{1}{2} m A^2 (p^2 + \omega^2)}{\frac{2\pi}{p} \times km A^2 p^2}$$

$$= \frac{2\pi}{8\pi} \frac{p^2 + \omega^2}{kp}$$

$$= \frac{1}{4} \frac{p^2 + \omega^2}{kp}$$

$$= \frac{p}{4k} \left[1 + \left(\frac{\omega}{p}\right)^2\right] \qquad \qquad ...(1)$$

At resonance, $\omega = p$

Hence,

$$Q = \frac{p}{2k} \qquad \qquad ...(2)$$

Selectivity of Resonant Circuit

The power drops to half of its maximum value when the angular frequency changes, such that

$$\frac{1}{\left[\left(\omega^2 - p^2\right)^2 + 4k^2 p^2\right]} = \frac{1}{2}\left(\frac{1}{4k^2 p^2}\right)$$

or, $(\omega^2 - p^2) + 4k^2p^2 = 8k^2p^2$

or, $(\omega^2 - p^2) = 4k^2p^2$

or, $\sqrt{\omega^2 - p^2} = \pm 2kp$

or, $p^2 \pm 2kp - \omega^2 = 0$

or, $p = \pm \dfrac{(2k) \pm \sqrt{4k^2 + 4\omega^2}}{2}$

$= \pm \dfrac{2k}{2} \pm \dfrac{2}{2}\sqrt{k^2 + \omega^2}$

$= \pm k \pm \sqrt{k^2 + \omega^2}$

$\therefore \qquad p = \pm k \pm \omega \qquad$ (when $k^2 \ll \omega^2$)

Hence the change in angular frequency for which amplitude falls to half of its maximum value is given by

$$\Delta p = p \sim \omega = k \qquad \qquad ...(1)$$

The $2\Delta p$ is the full width of resonance curve at half maximum power.

Also another representation of Q value is given by

$$Q = \frac{\omega}{2k} = \frac{\omega}{2\Delta p} = \frac{\text{Angular frequency at resonance}}{\text{Full width at half maximum power}}$$

Thus, Quality factor (Q) measures the sharpness of tuning.

Effect of Mistuning on the Amplitude of Resonant Vibrations – Sharpness of Resonance

Experiment shows that a given amount of mistuning between the impressed and the natural frequencies has not always the same harmful effect on the intensity of resonant vibrations. When the frequencies agree, it is only the presence of damping which prevents the vibrations from attaining an infinite amplitude. The damping becomes more and more powerful in its influence as resonance is approached. Further, the greater the dependence of the intensity of forced vibration on the exact equality of the frequencies of the external force and the body, the sharper is the resonance said to be.

The mathematical investigation shows that sharpness of resonance varies inversely as the square of damping coefficient (k). In other words, the smaller the frictional damping, the more important does exact tuning become in order to obtain greater energy in the forced vibration.

For smaller values of damping coefficient, the peak value of the amplitude is greater and moreover, the amplitude falls rapidly on either side of resonant frequency as shown in curve 1 of Fig. 3.2. This corresponds to sharp resonance, i.e. here the tuning must be close to get powerful resonance. It is excellently illustrated by a tuning fork which may be excited through the medium of surrounding air by the vibration of a second tuning fork. When the two forks are closely in tune, there is a very marked response.

On the other hand, it is observed that in a sounding box or in an air column inside a pipe where the rate of damping is considerable, the resonant vibrations can be easily excited for a wide range of frequencies on either side of the resonant frequency as shown in Curves 2 and 3 of Fig. 3.2. It is very difficult to judge the exact moment of resonance. This type of resonance is said to be flat. In the case of a stretched string, the damping is as a rule greater than for a tuning-fork, but considerably less than for an air column; hence the sharpness of resonance is intermediate between that of the fork and of the air.

To illustrate this point further, let us suppose that the correct tuning length of a sonometer wire is l and the range on either side over which appreciable vibration of the wire is noticed is $\pm \delta l$, then the ratio $\delta l/l$ will give a measure of the sharpness of resonance. Similarly, if in the case of resonance of an air column the correct tuning length and the range are given by l_1 and $\pm \delta l_1$, the sharpness of resonance will be $\delta l_1/l_1$. On comparing these ratios, it will be seen that the former ratio is smaller than the latter. We therefore conclude that resonance in the case of sonometer wire is sharper whereas it is flat in a resonating air column.

Explanation

The explanation of these facts is very simple. Resonance is due to the cumulative effect of successive impulses. If the damping is very large as in the case of a vibrating air column or a sounding box, the natural vibrations will decay quickly, so that by the time, say, the 15th impulse is coming, the effect of the first impulse has already disappeared. This means there is practically no chance for the impulses to get appreciably out of place and thereby diminish the amplitude. For maximum amplitude, therefore, the body and impressed force have to be in phase for about 15 periods only, i.e. a slight amount of mistuning will not produce a harmful effect on the amplitude of resonant vibrations. If, on the other hand, the natural vibrations die away slowly as in a tuningfork, there is ample opportunity for the successive impulses to get out of phase soon and neutralize the effect of earlier impulses. In such a case, much closer tuning would therefore be necessary for maximum intensity of resonance, so that the impressed force and the natural vibration may practically always remain in phase.

Examples of Forced Vibration and Resonance

There are many cases of forced vibration and resonance. A few examples are given below:

(a) The prong of a tuning fork is set into vibration with its natural frequency and the stem of the tuning fork is pressed against a table top. The particles of the table are set into forced vibration and a sound is heard when resonance occurs. At resonance, the natural frequency of the particles of the table becomes equal to the frequency of tuning fork. This principle is utilised in sonometer experiments.

(b) Marching men while crossing a light bridge are asked to break steps to avoid dangerous oscillations causing the collapse of the bridge. This is because they exert a periodic force on the bridge which causes the bridge to vibrate. After some time, the natural frequency of vibration of the bridge becomes equal to the frequency of the periodic force exerted by the steps of the marching men. Thus, resonance occurs and the amplitude of vibration of the bridge increases, resulting the collapse of the bridge.

(c) The tuning circuit in a radio receiver is a L-C circuit whose frequency is given by

$$f = \frac{1}{2\pi\sqrt{LC}}$$

The antenna of the radio receiver receives modulated carrier of different frequencies from different radio transmitting stations. The tuning circuit selects a particular station to which it is tuned. At that time the frequency of the L-C circuit is equal to the frequency of the carrier received and then resonance occurs.

Thus, the phenomena of resonance play vital roles in acoustics, mechanics, electricity and other branches of physics.

Forced Oscillation of LCR AC Circuit

Let us consider one LCR circuit Fig. 3.4 connected by an alternating source

$$E = E_o \sin \omega t \qquad \qquad ...(1)$$

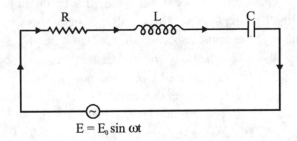

$$E = E_0 \sin \omega t$$

Fig. 3.4

Applying Kirchhoff's second law to this closed circuit, we obtain

$$V_R + V_L + V_C = E \qquad \qquad ...(2)$$

where, $\qquad \qquad V_R = Ri$

$$V_L = L\frac{di}{dt} \qquad \qquad ...(3)$$

$$V_c = \frac{q}{C}$$

and 'i' is the current flowing in the circuit.

From eqns. (1), (2) and (3), we have

$$Ri + L\frac{di}{dt} + \frac{q}{C} = E_o \sin \omega t \qquad \text{...(4)}$$

Let us consider a solution of this equation given by

$$i = A\,e^{i\omega t}, \qquad \text{...(5)}$$

We shall consider only the imaginary part of $e^{i\omega t}$, wherever necessary.

Hence, $$\frac{di}{dt} = i\omega A\,e^{i\omega t} \qquad \text{...(6)}$$

and $$q = \int i\,dt = \int Ae^{i\omega t}\,dt = \frac{Ae^{i\omega t}}{i\omega}$$

$$\therefore \qquad q = A\frac{e^{i\omega t}}{i\omega} \qquad \text{...(7)}$$

Substituting values from eqns. (5), (6) and (7) in the eqn. (4), we obtain,

$$R(A\,e^{i\omega t}) + L(i\omega A\,e^{i\omega t}) + \frac{1}{C}\left(\frac{Ae^{i\omega t}}{i\omega}\right) = E_o e^{i\omega t}$$

(as imaginary part of $E_o e^{i\omega t} = E_o \sin\omega t$)

Hence, equating the coefficients of $e^{i\omega t}$ from both the sides, we have

$$A\left\{R + i\omega L + \frac{1}{i\omega C}\right\} = E_o$$

or, $$A\left\{R + i\left(\omega L - \frac{1}{\omega C}\right)\right\} = E_o$$

or, $$A = \frac{E_o}{R + i\left(\omega L - \frac{1}{\omega C}\right)} \qquad \text{...(8)}$$

or, $$A = \frac{E_o}{z} \qquad \text{...(9)}$$

where $$z = R + i\left(\omega L - \frac{1}{\omega C}\right) \qquad \text{...(10)}$$

The complex number z is called electrical impedance of the circuit. It may be written in the polar form as

$$z = |z|\,e^{i\theta} \qquad \text{...(11)}$$

where, $$R = |z| \cos \theta \qquad \qquad ...(12)$$

$$\omega L - \frac{1}{\omega C} = |z| \sin \theta \qquad \qquad ...(13)$$

$$|z|^2 \left[\sin^2 \theta + \cos^2 \theta \right] = R^2 + \left(\omega L - \frac{1}{\omega C} \right)^2$$

or, $$|z| = \sqrt{R^2 + \left(\omega L - \frac{1}{\omega C} \right)^2} \qquad \qquad ...(14)$$

and $$\tan \theta = \frac{\omega L - \dfrac{1}{\omega C}}{R} \qquad \qquad ...(15)$$

$$A = \frac{E_o}{|z| e^{i\theta}} \qquad \qquad ...(16)$$

Hence, the eqn. (5) becomes

$$i = A e^{i\omega t} = \frac{E_o \, e^{i\omega t}}{|z| e^{i\theta}}$$

$$= \frac{E_o}{|z|} e^{i (\omega t - \theta)} \qquad \qquad ...(17)$$

Considering only imaginary parts of $e^{i(\omega t - \theta)}$, we obtain

$$i = \frac{E_o}{\sqrt{R^2 + \left(\omega L - \dfrac{1}{\omega C} \right)^2}} \sin (\omega t - \theta) \qquad \qquad ...(18)$$

or, $$i = i_o \sin (\omega t - \theta) \qquad \qquad ...(19)$$

where, the amplitude of the current in the circuit is i_o.

i.e., $$i_o = \frac{E_o}{\sqrt{R^2 + \left(\omega L - \dfrac{1}{\omega C} \right)^2}} = \frac{E_o}{|z|} \qquad \qquad ...(20)$$

Thus, the amplitude decreases as the impedance $|z|$ increases.

Amplitude in LCR AC Circuit

From eqn. (19), it is seen that the current lags behind the emf by an angle θ. We have

$$\tan \theta = \frac{\omega L - \dfrac{1}{\omega C}}{R} = \frac{X_L - X_C}{R} \qquad \qquad ...(21)$$

where X_L is the inductive reactance and X_C is the capacitive reactance of the circuit.

From equation (20), it is clear that the current amplitude i_o becomes maximum when $X_L = X_C$

or, $\qquad \omega L = \dfrac{1}{\omega C}$

or, $\qquad \omega^2 = \dfrac{1}{\omega L}$

or, $\qquad \omega = \dfrac{1}{\sqrt{LC}}$

or, $\qquad 2\pi f_r = \dfrac{1}{\sqrt{LC}}$

or, $\qquad f_r = \dfrac{1}{2\pi\sqrt{LC}}$ $\qquad\qquad$ ...(22)

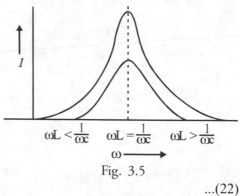

Fig. 3.5

where, f_r is the frequency at resonance. Thus, the value of i_o at resonance is

$$i_o = \frac{E_o}{R} \qquad\qquad ...(23)$$

This gives the height of the peak of a curve such as that of Fig. 3.5. When ω is either greater than or less than the resonance value, $|z|$ is greater than R and i_o is less than the value at resonance.

Phase in LCR AC Circuit

The phase 'θ' of the current in LCR circuit becomes zero at resonance.

Since, $\omega L = \dfrac{1}{\omega C}$ at resonance and the phase is given by

$$\tan\theta = \frac{\omega L - \dfrac{1}{\omega C}}{R} = 0$$

or, $\qquad \theta = 0°$

However, at lower frequencies, $\omega L < \dfrac{1}{\omega C}$ and $\tan\theta$ is negative, i.e. the phase θ approaches $-90°$. At higher frequencies, $\omega L > \dfrac{1}{\omega C}$, $\tan\theta$ is positive and the phase θ approaches $+90°$. Fig. 3.6 shows θ as a function of ω, the angular frequency. The phase difference between the emf and the current is θ. Fig. 3.6 illustrates the 'phase response' of the circuit. The smaller the value of R, the sharper is the change in θ, which varies from $-90°$ to $+90°$. The value of θ is $0°$ at resonance.

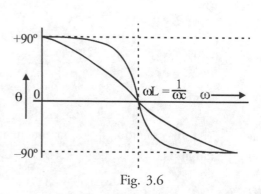

Fig. 3.6

Resonance in LCR AC Circuit

Distortion factor for electrical circuits is given by

$$D(\omega) = \frac{1}{|z|} = \frac{1}{\text{impedance}}$$

In case of LCR AC circuit, this factor is

$$D(\omega) = \frac{1}{|z|} = \frac{1}{\sqrt{R^2 + \left(\omega L - \dfrac{1}{\omega C}\right)^2}}$$

$$= \left[R^2 + \left(\omega L - \frac{1}{\omega C}\right)^2\right]^{-1/2}$$

For maximum value of D(ω), we must have

$$\frac{dD(\omega)}{d\omega} = 0$$

or, $$\frac{d}{d\omega}\left[R^2 + \left(\omega L - \frac{1}{\omega C}\right)^2\right]^{-1/2} = 0$$

or, $$-\frac{1}{2}\left[R^2 + \left(\omega L - \frac{1}{\omega C}\right)^2\right]^{-3/2} \times 2\left(\omega L - \frac{1}{\omega C}\right) \times \left(L + \frac{1}{\omega^2 C}\right) = 0$$

or, $$\left(\omega L - \frac{1}{\omega C}\right) = 0$$

or, $$\omega L = \frac{1}{\omega C}$$

or, $$\omega^2 = \frac{1}{LC}$$

or, $$\omega = \frac{1}{\sqrt{LC}}$$

or, $$2\pi f = \frac{1}{\sqrt{LC}}$$

or, $$f = \frac{1}{2\pi\sqrt{LC}} = f_r$$

$$\therefore \qquad \omega = \omega_r \qquad \qquad \qquad ...(1)$$

At this condition, $|z|$ is minimum and hence $D(\omega) = \dfrac{1}{|z|}$ is maximum. Then the current in the circuit is

$$i = \frac{E_o}{|z|} e^{i(\omega t - \theta)} = E_o D\, e^{i(\omega t - \theta)}$$

The amplitude of current is $E_o D$ having maximum value at resonance.

Sharpness of Resonance

As discussed above, the current amplitude is maximum at resonance and this is written as

$$(i_o)_{max} = \frac{E_o}{R} \quad \left(\because \omega L = \frac{1}{\omega C} \right)$$

Thus,

$$i = i_o \, e^{i(\omega t - \theta)} \qquad \qquad ...(2)$$

where

$$i_o = \frac{E_o}{\sqrt{R^2 + \left(\omega L - \dfrac{1}{\omega C} \right)}} \qquad \qquad ...(3)$$

and at resonance

$$i = (i_o)_{max} \, e^{i(\omega_r t - \theta)} \qquad \qquad ...(4)$$

A curve can be plotted between i and ω. The maximum height of curve (Fig. 3.7) gives the peak value of the current i, that is i_o at $\omega = \omega_r$. Now taking two values of ω, i.e. ω_1 and ω_2 on either side of ω, i.e., ω_1 and ω_2 on either side of ω_r, such that value of $i = i_o/\sqrt{2}$.

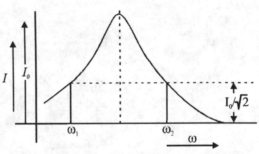

Fig. 3.7

The curve of Fig. 3.7 is sharper at resonant frequency f_r, indicating the maximum value of the current at $\omega = \omega_r$. The current falls from this peak value towards $\omega_1 < \omega_r$ and $\omega_2 > \omega_r$.

Quality Factor

The ratio of ω_r to $(\omega_2 - \omega_1)$ is termed as the quality factor of the circuit or simply 'Q' of the circuit. So,

$$Q = \frac{\omega_r}{\omega_2 - \omega_1} = \frac{f_r}{f_2 - f_1} \qquad \qquad ...(1)$$

where f_r is the resonant frequency, and f_1 and f_2 are the sideband frequencies. $f_2 - f_1$ is called the bandwidth. Thus,

$$Q = \frac{\text{Resonant frequency}}{\text{Bandwidth}} \qquad \qquad ...(2)$$

The power is half at sidebank frequencies. So, the points corresponding to ω_1 and ω_2 in the Fig. 3.7 are known as half-power points.

Since at half-power frequencies, the net reactance is equal to resistance, we have

$$|z| = R \qquad \qquad ...(3)$$

At ω_1, the capacitive reactance exceeds the inductive reactance. So,

$$\frac{1}{\omega_1 C} - \omega_1 L = R \qquad \qquad ...(4)$$

Similarly at ω_2, the inductive reactance exceeds the capacitive reactance, i.e.

$$\omega_2 L - \frac{1}{\omega_2 C} = R \qquad \qquad ...(5)$$

Now, multiplying eqn. (4) by $\dfrac{1}{\omega_2}$ and eqn. (5) by $\dfrac{1}{\omega_1}$ and adding the two, we obtain

$$L\left(\frac{\omega_2}{\omega_1} - \frac{\omega_1}{\omega_2}\right) + \frac{1}{C}\left(\frac{1}{\omega_1 \omega_2} - \frac{1}{\omega_1 \omega_2}\right) = R\left(\frac{1}{\omega_1} + \frac{1}{\omega_2}\right)$$

or,
$$L\left(\frac{\omega_2}{\omega_1} - \frac{\omega_1}{\omega_2}\right) = R\left(\frac{1}{\omega_1} + \frac{1}{\omega_2}\right)$$

or,
$$L\left(\frac{\omega_2^2 - \omega_1^2}{\omega_1 \omega_2}\right) = R\left(\frac{\omega_2 - \omega_1}{\omega_1 \omega_2}\right)$$

or,
$$\frac{(\omega_2 + \omega_1)(\omega_2 - \omega_1)}{\omega_1 \omega_2} = \frac{R}{L}\left(\frac{\omega_2 + \omega_1}{\omega_1 \omega_2}\right)$$

or,
$$\omega_2 - \omega_1 = \frac{R}{L}, \qquad \qquad ...(6)$$

Substituting this value of $(\omega_2 - \omega_1)$ in eqn. (1), we obtain

$$Q = \frac{\omega_r}{(R/L)} = \frac{\omega_r L}{R} \qquad \qquad ...(7)$$

But,
$$\omega_r = \frac{1}{\sqrt{LC}}$$

So,
$$Q = \frac{1}{\sqrt{LC}} \times \frac{L}{R} = \frac{1}{R}\sqrt{\frac{L}{C}} \qquad \qquad ...(8)$$

Again,
$$Q = \frac{\omega_r L}{R} \qquad \qquad ...(9)$$

or,
$$Q = \frac{X_L}{R} = \frac{X_L i_o}{R\, i_o} \qquad \qquad ...(10)$$

or,
$$Q = \frac{\text{Voltage drop across L}}{\text{Voltage drop across R}} \qquad \qquad ...(11)$$

Also squaring the relation

$$\omega_r = \frac{1}{\sqrt{LC}}$$

We obtain
$$L = \frac{1}{\omega_r^2 C} \qquad \text{...(12)}$$

From eqns. (9) and (12), we have

$$Q = \frac{\omega_r L}{R} = \frac{\omega_r}{R} \cdot \frac{1}{\omega_r^2 C}$$

$$= \frac{1}{\omega_r RC} \qquad \text{...(13)}$$

or,
$$Q = \frac{1}{\omega_r RC} \times \frac{i_o}{i_o}$$

$$= \frac{i_o / \omega_r C}{R i_o}$$

$$= \frac{\text{Voltage drop across C}}{\text{Applied voltage}} \qquad \text{...(14)}$$

Thus, as Q increases the $i - \omega$ curve becomes more and more sharp (Fig. 3.8) which means sharpness of resonance increases. Q, the quality factor of the circuit is a very important parameter and occurs constantly in any discussion of resonance. Equations (1) and (14) are only valid if Q is fairly large, which is usually the case. For example, in case of ordinary AC circuits, its value lies between 30 and 300; for a microwave cavity, its value becomes greater than 10,000; and its typical value for an optical spectrum line is 1 million.

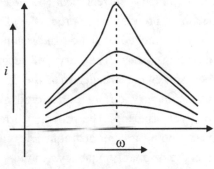

Fig. 3.8

Selectivity of the Resonant Circuit

The passage of frequency through a circuit depends upon the magnitude of impedance offered to it by the circuit. At $\omega = \omega_r$, the circuit has maximum value of $D(\omega)$ or the lowest value of $|z|$. Therefore, LCR circuit passes frequencies near resonant frequency more readily which shows its selective nature. The band of frequencies which is passed quite readily is called *passband* or *bandwidth*.

The passband is arbitrarily considered between ω_1 and ω_2 on both sides of ω_r. This passband is given as follows:

$$\Delta\omega = \omega_2 - \omega_1 = \frac{\omega_r}{Q} \qquad \text{...(1)}$$

Forced Oscillation of Coupled System

When two circuits are so arranged that the energy can be transferred from one to another, they are said to be coupled. If two circuits, resonant at the same frequency, are coupled together, the resulting behaviour largely depends upon the degree of coupling. We now consider following two cases:

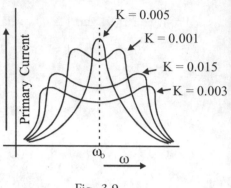

Fig. 3.9

(i) When coefficient of coupling is small: If the coefficient of coupling is very small ($k = 0.005$) then the effect of coupled impedance is negligible. Therefore, the curve of primary current as a function of frequency approximately resembles the series resonance curve (Fig. 3.9).

(ii) When coupling increases: As coupling is increased, the curve of primary current becomes somewhat broader and the peak value of primary current is reduced. At the same time, the secondary current peak becomes higher and the curve of secondary current somewhat becomes broader. This process will continue until a stage is reached at which the secondary current has maximum value. The curve of secondary current is then somewhat broader than the resonance curve of the secondary circuit considered alone, and has a relatively flat top. The primary current now has two peaks, being greater at frequencies just off resonance than the resonant frequency. All this will happen when the reflected resistance from secondary to the primary is equal to the resistance of the primary. This is known as *critical coupling* and causes the secondary current to have the maximum possible value (Fig. 3.10).

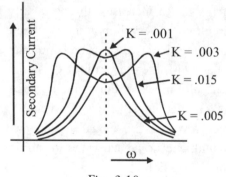

Fig. 3.10

SOLVED EXAMPLES

Q.1. **The forced harmonic oscillations have displacement amplitudes at frequencies $\omega_1 = 400$ sec^{-1} and $\omega_2 = 600$ sec^{-1}. Find the resonant frequency at which the displacement amplitude is maximum.**

Sol. The displacement in forced oscillation is given by

$$y = \frac{f}{\sqrt{(\omega^2 - p^2)^2 + 4k^2 p^2}}$$

The amplitudes at frequencies ω_1 and ω_2 are equal.

$\therefore \qquad \left(\omega^2 - \omega_1^2\right) + 4k^2\omega_1^2 = \left(\omega^2 - \omega_2^2\right)^2 + 4k^2\omega_2^2$

or, $\qquad \left(\omega^2 - \omega_1^2\right)^2 - \left(\omega^2 - \omega_2^2\right)^2 = 4k^2\left(\omega_2^2 - \omega_1^2\right)$

or, $\qquad \left(2\omega^2 - \omega_1^2 - \omega_2^2\right)\left(\omega_2^2 - \omega_1^2\right) = 4k^2\left(\omega_2^2 - \omega_1^2\right)$

$\therefore \qquad 4k^2 = 2\omega^2 - \omega_1^2 - \omega_2^2$

The resonant frequency is given by

$$f_r = \frac{P}{2\pi} \frac{\sqrt{(\omega^2 - 2k^2)}}{2\pi}$$

$$= \frac{\sqrt{\left[\omega^2 - \dfrac{\left(2\omega^2 - \omega_1^2 - \omega_2^2\right)}{2}\right]}}{2\pi}$$

$$= \frac{1}{2\pi}\sqrt{\frac{\omega_1^2 + \omega_2^2}{2}}$$

$$= \frac{1}{2\pi}\sqrt{\frac{(400)^2 + (600)^2}{2}}$$

$$= \frac{1}{2\pi}100\sqrt{26}$$

$$= \frac{50}{3.14} \times 5.099 = \frac{254.95}{3.14}$$

$$= 81.19 \simeq 81 \text{ cycles/sec.}$$

Q.2. **A body of mass 10 gm executes forced harmonic vibrations being impressed by an external periodic force 50 sin 20t N. Compute the amplitude of vibration of the body when $\omega = 200$ sec^{-1} and damping coefficient $k = 0.02$.**

Sol. Here,
$$m = 10 \text{ gm} = 10 \times 10^{-3} \text{ kg}$$
$$= 10^{-2} \text{ kg}$$
$$F' = 50 \sin 20t = F \sin pt$$

$\therefore$
$$f = \frac{F}{m} = \frac{50}{10^{-2}} = 50 \times 10^2 \text{ N/kg}$$

$$\omega = 200 \text{ sec}^{-1}, \ p = 20 \text{ sec}^{-1}, \ k = 0.02.$$

Hence, the amplitude of vibration is given by

$$A = \frac{f}{\sqrt{\left(\omega^2 - p^2\right)^2 + 4k^2 p^2}}$$

$$= \frac{50 \times 10^2}{\sqrt{\left[(200)^2 - (20)^2\right]^2 + 4.(0.02)^2 (20)^2}}$$

$$A = \frac{50 \times 100}{\sqrt{(40000 - 400)^2 + 4 \times 4 \times 10^{-4} \times 400}}$$

$$= \frac{5000}{\sqrt{(39600)^2 + 64 \times 10^{-2}}}$$

$$= \frac{5000}{\sqrt{\left(396 \times 10^2\right)^2 + 0.64}}$$

$$= \frac{5000}{\sqrt{(156816 \times 10^4) + 0.64}}$$

$$= \frac{5000}{\sqrt{1568160000.64}}$$

$$= \frac{5000}{39600}$$

$$= \frac{500}{3960} = 1.26 \text{ m}/10 = 0.126 \text{ m}.$$

Q.3. **Find the maximum amplitude at resonance of a particle of mass 0.02 kg executing forced harmonic vibration by an impressed force 10 sin 2t. Consider damping to be low, i.e. k = 0.01.**

Sol. Here,
$$m = 0.02 \text{ kg}$$
$$F' = 10 \sin 2t = F \sin pt$$
$$k = 0.01, \ p = 2 \text{ cycles/sec.}$$

$$f = \frac{F}{m} = \frac{10 \text{ N}}{0.02 \text{ kg}} = \frac{10}{2 \times 10^{-2}} \text{ N/kg} = 5 \times 10^2 \text{ N/kg}.$$

$$\therefore \quad A_{max} = \frac{f}{2k\sqrt{p^2 + k^2}}$$

$$= \frac{5 \times 10^2}{2 \times 0.01\sqrt{2^2 + (0.01)^2}}$$

$$= \frac{500}{0.02\sqrt{4 + 0.0001}}$$

$$= \frac{500}{0.02 \times 2} = \frac{500}{0.04}$$

$$= \frac{500}{4 \times 10^{-2}} = 125 \times 10^2 \text{ m.}$$

Q.4. **Taking the effect of low damping, find the maximum amplitude of FHM if $k = 0.02$, $p = 20$ Hz and $f = 200$ N/kg.**

Sol. For low damping,

$$A_{max} = \frac{f}{2kp} = \frac{200}{2 \times 0.02 \times 20} = \frac{5}{0.02}$$

$$= 2.5 \times 10^2 \text{ m} = 250 \text{ m.}$$

QUESTIONS

1. Develop the theory of forced harmonic motion. Find the amplitude of motion at resonant frequency.

2. Discuss on the sharpness of resonance.

3. Explain the various terms in the differential equation:

$$m\frac{d^2x}{dt^2} + r\frac{dx}{dt} + kx = F \sin pt$$

Solve the equation and discuss on its general and particular solutions.

4. What are forced vibrations? Give the theory of forced vibrations and discuss on the conditions of resonance. Distinguish between flat resonance and sharp resonance.

5. Develop the theory of forced vibration in a damped vibrating system when subject to periodic force. If damping is small, calculate the frequency of the driving force for which the displacement is maximum.

6. Explain the meaning of forced vibration. Sketch curves to show that the amplitude and phase of forced vibration varies with the damping and with the ratio of the impressed frequency to the natural frequency. Give a few examples of forced vibration.

7. Distinguish between free, forced and resonant vibrations. A body executing damped simple harmonic motion is subjected to an external periodic force. Investigate the forced vibration and obtain the condition of resonance.

8. Give the theory of forced vibration of a slightly damped oscillatory system with special reference to the case when the period of the exciting vibration approaches that of the natural vibration of the system.

9. Explain clearly the meaning of free, forced and sustained oscillations. Investigate mathematically the conditions under which forced oscillations become most vigorous.

10. Discuss the theory of the forced vibration of a damped system. What is sharpness of resonance? Give examples of: (i) a sharp, and (ii) a flat response curve and mention their utility.

11. Work out the theory of forced harmonic motion with resistance proportional to velocity. Explain how the amplitude of vibration and phase-lag alter with the forced frequency. What happens to the amplitude of vibration?

12. Develop the theory of oscillations in LCR AC circuit.

Problems

1. In an experiment on forced oscillations, the frequency of sinusoidal driving force is changed while its amplitude is kept constant. It is found that the amplitude of the vibrations is 0.01 mm at very low frequency of the driver and goes up to a maximum of 5 mm at driving frequency of 200 s^{-1}. Calculate: (i) Q of the system, and (ii) relaxation time.

    (**Ans.** (i) Q = 500, and (ii) τ = 0.4 sec)

2. If Q of an oscillator be 40, calculate the ratio of the amplitude of forced vibrations and the amplitude at resonance when the frequency of the applied force is: (i) 99%, (ii) 98%, and (iii) 97% of the natural frequency in the absence of damping.

    (**Ans.** (i) 0.78, (ii) 0.53, and (iii) 0.39)

3. A sinusoidal force is applied to an oscillator whose natural frequency is 100 Hz and relaxation time is 0.05 sec. Obtain the relative amplitudes of vibration as the frequency of the applied force is changed to: (i) 8 Hz, (ii) 95 Hz, (iii) 105 Hz, (iv) 120 Hz, and (v) 99 Hz.

    (**Ans.** (i) 0.032 A, (ii) 0.314 A, (iii) 0.30 A, (iv) 0.072 A, and (v) 0.85 A)

4. A particle of mass 1 gm vibrates under the action of the following forces: (a) elastic restoring force of 16 dynes/cm, (b) a frictional force of 2 dynes sec/cm, and (c) a periodically varying external force of maximum value 32 dynes. Find: (i) the frequency for amplitude resonance, (ii) the frequency for velocity resonance, and (iii) the maximum amplitude.

    (**Ans.** (i) 0.595 Hz, (ii) 0.636 Hz, and (iii) 4.13 cm)

5. The frequency of the applied periodic force is changed keeping its amplitude constant. The amplitude of vibration of the body executing forced oscillation is found to vary from 0.01 cm to 10 cm when the frequency increases to a maximum value of 500 Hz. Calculate: (i) the Q of the system, and (ii) the relaxation time.

    (**Ans.** (i) 100, and (ii) 0.318 sec)

6. A sinusoidal force is applied to a body whose natural frequency is 100 sec^{-1} and relaxation time is 0.5 sec. Compute the relative amplitude of vibration at resonance as the frequency of the impressed force is changed to 120 sec^{-1}.

$$\left[\text{Ans. } \left(\frac{5}{6}\right)\text{A}\right]$$

7. Find the maximum value of the mean power supplied by the impressed force of 20 sin 4t to a DHM having damping coefficient k = 0.01 and m = 0.05 kg.

(**Ans.** P$_{max}$ = 10^4 watts)

8. A body executing damped harmonic motion is impressed upon by a force of 40 sin 5t. If the mass of the body is 0.02 kg and the coefficient of damping is 0.02, find the maximum value of the mean power supplied by the impressed force.

(**Ans.** P$_{max}$ = 25 × 10^3 watts)

9. Across a 110 volts 25 Hz line, there are connected in series a resistance of 200 ohms, an inductance of reactance 100 ohms and a resistance of 40 ohms, and a capacitance of 25 μf. Compute:

   (i)   the impedance of the circuit,
   (ii) the current, and
   (iii) the angle by which the current leads or lags in phase the applied emf.

(**Ans.** (i) 285.5 ohms, (ii) 0.385 amp, and (iii) 32°48′)

10. One LCR AC circuit contains an inductance of 10 mH, a capacitance of 1 μf and a resistance of 10 ohms in series. If the rms value of applied emf is 100 volts, then calculate:

    (i)   resonant angular frequency,
    (ii) resonant current,
    (iii) Q-value of the circuit, and
    (iv) angular frequency width of the resonant curve.

(**Ans.** (i) 10^4 rad/sec, (ii) 10 amp, (iii) 10, and (iv) 10^3 rad/sec)

●●●

Chapter 4

Coupled Oscillators

In this chapter, we shall take into consideration the coupled vibrations of mechanical systems (with a finite number of degrees of freedom) in the neighbourhood of an equilibrium position or a state of uniform motion. In mechanical systems, the frequency of vibrations is more important than the solutions for the amplitudes of vibrations, subject to the initial condition of the system.

Coupled Vibration

We have observed in the case of forced harmonic oscillation that the driving force is no way affected by the driven system. The amplitude and frequency of the driving force remains unaltered.

But the role of the driver and the driven will be changed when the two systems are comparable. Motion of each system will react on that of the other, when two systems have equal masses or inertia. Such type of vibration is known as *Coupled Vibration*.

This phenomenon can be demonstrated as follows:

Two similar simple pendulums are suspended from an India rubber cord EF, attached between two rigid uprights. Now the pendulum B is set in motion by initial displacement. The oscillation of the pendulum B provides a periodic force to the elastic support EF. The vibration of the cord supplies periodic forces to other pendulum D at its point of suspension. The pendulum D begins to oscillate. The amplitude of D will gradually increase while the amplitude of B will gradually diminish and almost B comes to rest. This goes on to a certain point when reverse process sets in, i.e. amplitude of D slows down momentarily, D comes to rest and the amplitude of B increases with time. This process will be repeated. Here, B acts as a driver and D is driven. Next, D acts as a driver and B as a driven. Fig. 4.1 represents the phenomenon of coupled oscillation.

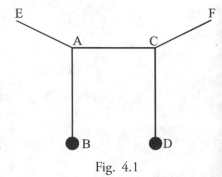

Fig. 4.1

There is feedback of energy from driven to the driver. The feedback energy is lesser if the difference between the frequency of oscillations of the two is large and the coupling between the two is said to be 'loose'. If the feedback is large, the coupling is said to be 'tight'.

The time-displacement curves of these two pendulums are shown in Fig. 4.2. The time-displacement curve of B is represented by the (Fig. 4.2a) and that of pendulum D is shown in (Fig. 4.2b).

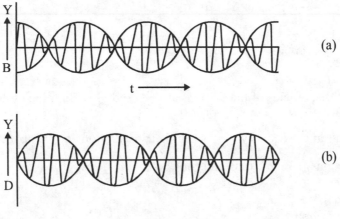

Fig. 4.2

Analytical Treatment of Coupled Oscillation

Let us consider two systems coupled together, oscillating in a direction y. Let their masses be m_1 and m_2; and their forces of restitution be μ_1 and μ_2 for unit displacement; and μ_{12} is the force of restitution on the first or the second for the unit displacement of the second or the first, which may be called coupling coefficient.

The equation of motions of two systems are given by

$$m_1 \frac{d^2 y_1}{dt^2} + \mu_1 y_1 + \mu_{12} y_2 = 0 \qquad \qquad ...(1a)$$

$$m_2 \frac{d^2 y_2}{dt^2} + \mu_2 y_2 + \mu_{12} y_1 = 0 \qquad \qquad ...(1b)$$

Dividing the eqn. (1a) by $\sqrt{m_1}$ and (1b) by $\sqrt{m_2}$, we have

$$\sqrt{m_1} \frac{d^2 y_1}{dt^2} + \frac{\mu_1}{\sqrt{m_1}} y_1 + \frac{\mu_{12}}{\sqrt{m_1}} y_2 = 0 \qquad \qquad ...(2a)$$

$$\sqrt{m_2} \frac{d^2 y_2}{dt^2} + \frac{\mu_2}{\sqrt{m_2}} y_2 + \frac{\mu_{12}}{\sqrt{m_2}} y_1 = 0 \qquad \qquad ...(2b)$$

Let us take

and
$$\left. \begin{array}{c} \sqrt{m_1}\, y_1 = Y_1 \\ \sqrt{m_2}\, y_2 = Y_2 \end{array} \right\} \qquad \qquad ...(3)$$

Hence, from eqns. (2) and (3), we obtain,

$$\frac{d^2Y_1}{dt^2} + \omega_1^2 Y_1 + \omega^2 Y_2 = 0 \qquad \qquad ...(4a)$$

$$\frac{d^2Y_2}{dt^2} + \omega_2^2 Y_2 + \omega^2 Y_1 = 0 \qquad \qquad ...(4b)$$

where, $\qquad \omega_1^2 = \dfrac{\mu_1}{m_1}, \omega_2^2 = \dfrac{\mu_2}{m_2}$ and $\omega^2 = \dfrac{\mu_{12}}{\sqrt{m_1}\sqrt{m_{12}}}$

The natural frequencies of the systems are $\dfrac{\omega_1}{2\pi}$ and $\dfrac{\omega_2}{2\pi}$. Both the systems are assumed

to oscillate with the same frequency $\dfrac{p}{2\pi}$, when their oscillations are simple harmonic.

Let the solutions of eqns. (4a) and (4b) be

$$\left.\begin{array}{l} Y_1 = A\, e^{ipt} \\ Y_2 = B\, e^{ipt} \end{array}\right\} \qquad \qquad ...(5)$$

Substituting these values in eqn. (4) we have

$$\left(\omega_1^2 - p^2\right) A = -\,\omega^2 B \qquad \qquad ...(6a)$$

$$\left(\omega_2^2 - p^2\right) B = -\,\omega^2 A \qquad \qquad ...(6b)$$

Multiplying eqns. (6a) and (6b), we obtain

$$\left(\omega_1^2 - p^2\right)\left(\omega_2^2 - p^2\right) = \omega^4$$

or, $\qquad p^4 - p^2\left(\omega_1^2 + \omega_2^2\right) + \left(\omega_1^2\omega_2^2 - \omega^4\right) = 0$

Solving, we have

$$p^2 = \frac{1}{2}\left(\omega_1^2 + \omega_2^2\right) \pm \sqrt{\left(\omega_1^2 + \omega_2^2\right)^2 - 4\left(\omega_1^2\omega_2^2 - \omega^4\right)}$$

or, $\qquad p^2 = \dfrac{1}{2}\left(\omega_1^2 + \omega_2^2\right) \pm \dfrac{1}{2}\sqrt{\left(\omega_1 - \omega_2\right)^2 + 4\omega^4} \qquad \qquad ...(7)$

When the natural frequencies of two systems are equal

$$\omega_1 = \omega_2$$

and $\qquad p^2 = \omega_1^2 \pm \omega^2$

or, $\qquad p = \sqrt{\omega_1^2 \pm \omega^2}$

or, $\qquad p_1 = \sqrt{\omega_1^2 + \omega^2}, p_2 = \sqrt{\omega_1^2 - \omega^2}$

Now, we can express p_1 as

$$p_1 = \omega_1\left(1 + \frac{\omega^2}{\omega_1^2}\right)^{1/2}$$

$$= \omega_1 + \frac{1}{2}\frac{\omega^2}{\omega_1} \text{ approx.} \qquad \qquad ...(8a)$$

and
$$p_2 = \omega_1\left(1 - \frac{\omega^2}{\omega_1^2}\right)^{1/2}$$

$$= \omega_1 - \frac{1}{2}\frac{\omega^2}{\omega_1} \text{ approx.} \qquad \qquad ...(8b)$$

From (8a) and (8b), we have

$$\left.\begin{array}{l} p_1 + p_2 = 2\omega_1 \\ p_1 - p_2 = \omega^2/\omega_1 \end{array}\right\} \qquad \qquad ...(8c)$$

where $\dfrac{p_1}{2\pi}$ and $\dfrac{p_2}{2\pi}$ are two modified frequencies of the coupled systems when coupled together.

So, the general solution can be written as

$$\left.\begin{array}{l} Y_1 = A_1\,e^{ip_1t} + A_2\,e^{ip_2t} \\ Y_2 = B_1\,e^{ip_1t} + B_2\,e^{ip_2t} \end{array}\right\} \qquad \qquad ...(9)$$

From eqns. (3) and (9), we have

$$y_1 = \frac{Y_1}{\sqrt{m_1}} = \frac{1}{\sqrt{m_1}}\left[A_1(\cos p_1t + i\sin p_1t) + A_2(\cos p_2t + i\sin p_2t)\right]$$

or, $\quad y_1 = a_1\,\cos p_1t + a_2\,\sin p_1t + a_3\,\cos p_2t + a_4\,\sin p_2t \qquad ...(10a)$

Similarly,

$$y_2 = b_1\,\cos p_1t + b_2\,\sin p_1t + b_3\,\cos p_2t + b_4\,\sin p_2t \qquad ...(10b)$$

The constants a_1, a_2, a_3, a_4 and b_1, b_2, b_3, b_4 are evaluated from the initial conditions. If the particle m_1 starts motion from rest by initial displacement,

i.e., at $\quad t = 0,$

$$y_1 = (y_1)_0, y_2 = 0, \frac{dy_1}{dt} = 0, \frac{dy_2}{dt} = 0$$

we have, $\quad a_1 + a_3 = (y_1)_0$

At the beginning, there is no coupling and $p_1 = p_2$ as a special case.

$$a_1 = a_3 = \frac{(y_1)_0}{2}$$

and $\quad a_2 = a_4 = 0$

Likewise, $\quad b_1 + b_3 = 0$

or, $\quad b_1 = -b_3 = b \text{ (say)}$

and $\quad b_2 = b_4 = 0$

From eqn. (10a) we obtain,

$$y_1 = \frac{(y_1)_0}{2}(\cos p_1t + \cos p_2t)$$

$$y_2 = b\,(\cos p_1t - \cos p_2t)$$

or,
$$y_1 = (y_1)_0 \cos \frac{p_1 + p_2}{2} t . \cos \frac{p_1 - p_2}{2} t \qquad \text{...(11a)}$$

and
$$y_2 = 2b \sin \frac{p_1 + p_2}{2} t . \sin \frac{p_1 - p_2}{2} t \qquad \text{...(11b)}$$

Now, from (8c), (11a) and (11b), we obtain

$$\left. \begin{array}{l} y_1 = (y_1)_0 \cos \omega_1 t . \cos \dfrac{\omega^2}{2\omega_1} t \\[3mm] y_2 = 2b \sin \omega_1 t . \sin \dfrac{\omega^2}{2\omega_1} t \end{array} \right\} \qquad \text{...(12)}$$

Substituting the values of p_1^2 and p_2^2 from eqns. (8a) and (8b) in eqns. (6a) and (6b), we obtain

$$\frac{A}{B} = +1 \text{ or } A = B$$

In other words, the amplitude of Y_1 is equal to the amplitude of Y_2, i.e.,

$$Y_1 = \sqrt{m_1} \, y_1, \, Y_2 = \sqrt{m_2} \, y_2$$

Hence, the amplitude of $y_1 = \sqrt{\dfrac{m_2}{m_1}}$ (amplitude of y_2)

But the amplitude of $y_1 = \dfrac{(y_1)_0}{2}$ and $y_2 = b$

So,
$$\frac{(y_1)_0}{2} = \sqrt{\frac{m_2}{m_1}} . b$$

or,
$$b = \frac{(y_1)_0}{2} \sqrt{\frac{m_2}{m_1}}$$

Therefore,

$$\left. \begin{array}{l} y_1 = (y_1)_0 \cos \dfrac{\omega^2}{2\omega_1} t . \cos \omega_1 t \\[3mm] y_2 = \sqrt{\dfrac{m_1}{m_2}} (y_1)_0 \sin \dfrac{\omega^2}{2\omega_1} t . \sin \omega_1 t \end{array} \right\} \qquad \text{...(13)}$$

These two equations represent the oscillation of two systems where they are coupled together. The time-displacement curves of the two systems are shown in Fig. 4.2.

Oscillating Systems with only One Degree of Freedom

Let us consider oscillating systems as shown in Fig. 4.3.

Here system (a) represents a mass M, that is constrained to move in a linear path. It is attached to a spring of force constant K and is acted upon by a dash point mechanism that introduces a frictional constraint proportional to velocity of the mass. An external force $F_o \sin \omega t$ acts on the mass.

By Newton's law of motion, we have

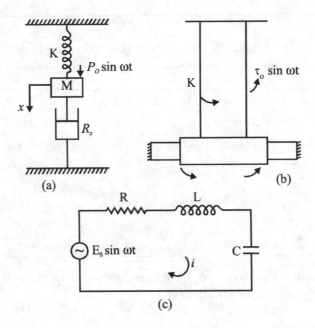

Fig. 4.3: Oscillating systems with one degree of freedom

$$M\frac{d^2x}{dt^2} = -Kx - R\dot{x} + F_o \sin \omega t$$

where R is the frictional coefficient at the dash point.

$$M\ddot{x} + R\dot{x} + Kx = F_o \sin \omega t, \qquad \qquad ...(1)$$

Fig. 4.3b represents a system undergoing torsional oscillations. It consists of a disc of moment of inertia I attached to a shaft of torsional constant K. The disc undergoes torsional damping, proportional to its angular velocity $\dfrac{d\theta}{dt}$. The disc is exerted upon by an oscillatory torque $\tau_0 \sin \omega t$. Hence, by Newton's law, we have

$$I\ddot{\theta} + R\dot{\theta} + K\theta = \tau_o \sin \omega t \qquad \qquad ...(2)$$

Fig. 4.3c is a series electrical circuit having inductance L, resistance R and elastance $S = 1/c$, attached to a source $E_o \sin \omega t$. By applying Kirchhoff's law to the circuit, we obtain

$$L\frac{di}{dt} + Ri + \frac{q}{c} = E_o \sin \omega t$$

or, $\qquad L\dfrac{d^2 q}{dt^2} + R\dfrac{dq}{dt} + \dfrac{1}{c}q = E_o \sin \omega t$

or, $\qquad L\ddot{q} + R\dot{q} + Sq = E_o \sin \omega t$ ...(3)

Analogy between linear, torsional and electrical systems

By comparing equations (1), (2) and (3), we can have the following table:

Linear		Torsional		Electrical	
Mass	M	Moment of inertia	I	Inductance	L
Stiffness	K	Torsional stiffness	K	Elastance	S = 1/c
Damping	R	Torsional damping	R	Resistance	R
Impressed force	$F_o \sin \omega t$	Impressed torque	$\tau_o \sin \omega t$	Impressed potential	$E_o \sin \omega t$
Displacement	x	Angular displacement	θ	Capacitor charge	q
Velocity	$\dot{x} = v$	Angular velocity	$\dot{\theta} = \omega$	Current	$i = \dot{q}$

From the above table, it is evident that one can analyse only one system and then can obtain the corresponding solution of other systems.

Free Vibrations

Let us consider the system of (Fig. 4.3a) where the force $F_o \sin \omega t = 0$, the equation gives

$$M\ddot{x} + R\dot{x} + Kx = 0 \qquad ...(4)$$

This is the equation of motion for damped harmonic vibrations. Solution of this equation has already been discussed in details, in Chapter 2. In the oscillatory case, when $(R^2 - 4MK) < 0$, we obtain

$$\omega_s = \sqrt{\dfrac{K}{M} - \dfrac{R^2}{4M^2}} \qquad ...(5)$$

where ω_s is the angular frequency of the motion.

If there is no damping present, then R = 0 and we have harmonic oscillations with natural frequency

$$\omega_o = \sqrt{\dfrac{K}{M}} \qquad ...(6)$$

System with Two Degrees of Freedom

Let us consider the motion of two systems as shown in Fig. 4.4. The state of both these systems is determined by two quantities. These are the linear displacements of the two masses of the mechanical system (x_1, x_2) or the mesh charge of the electrical system (q_1, q_2).

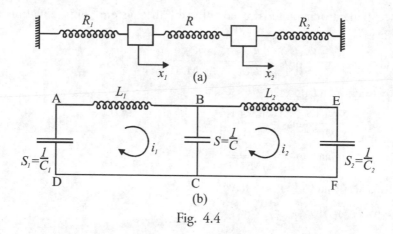

Fig. 4.4

A system whose motion and position are characterised by two independent quantities, is known as a system having two degrees of freedom. Now, applying Kirchhoff's law to the mesh ABCD and BCFE separately, we obtain

$$\left.\begin{array}{l} L_1\ddot{q}_1 + S_1 q_1 + S(q_1 - q_2) = 0 \\ L_2\ddot{q}_2 + S_2 q_2 + S(q_2 - q_1) = 0 \end{array}\right\} \qquad ...(7)$$

and

For the system of (Fig. 4.4a), these equations are

$$\left.\begin{array}{l} M_1\ddot{x}_1 + K_1 x_1 + K(x_1 - x_2) = 0 \\ M_2\ddot{x}_2 + K_2 x_2 + K(x_2 - x_1) = 0 \end{array}\right\} \qquad ...(8)$$

Let the solutions of the equations of these two systems be

$$x_1 = A_1 e^{\alpha t} \text{ and } x_2 = A_2 e^{\alpha t} \qquad ...(9)$$

where A_1, A_2 and α may be real or complex. Then

$$\ddot{x}_1 = A_1 \alpha^2 e^{\alpha t} \text{ and } \ddot{x}_2 = A_2 \alpha^2 e^{\alpha t} \qquad ...(10)$$

Substituting the values of x_1, x_2, $\ddot{x}_1$ and $\ddot{x}_2$ in the eqn. (8), we have

$$M_1\alpha^2 A_1 e^{\alpha t} + K_1 A_1 e^{\alpha t} + K(A_1 - A_2) e^{\alpha t} = 0$$

or,

$$\left.\begin{array}{l} A_1(M_1\alpha^2 + K_1 + K) - A_2 K = 0 \\ -A_1 K + A_2(M_2\alpha^2 + K_2 + K) = 0 \end{array}\right\} \qquad ...(11)$$

and

Equations represented by (11) have solutions other than the trivial one $A_1 = A_2 = 0$, if

$$\Delta(\alpha) = \begin{vmatrix} M_1\alpha^2 + K_1 + K & -K \\ -K & M_2\alpha^2 + K_2 + K \end{vmatrix} = 0 \qquad ...(12)$$

Expanding the determinant, we have

$$\Delta(\alpha) = (M_1\alpha^2 + K_1 + K)(M_2\alpha^2 + K_2 + K) - K^2 = 0 \qquad ...(13)$$

This is called the characteristic equation of the system. This equation may also be written in the form

$$\left(\alpha^2 + \frac{K_1 + K}{M_1}\right)\left(\alpha^2 + \frac{K_2 + K}{M_2}\right) - \frac{K^2}{M_1 M_2} = 0 \qquad ...(14)$$

For convenience, let us take

$$\left.\begin{aligned}\omega_{11}^2 &= \frac{K_1 + K}{M_1}, \omega_{22}^2 = \frac{K_2 + K}{M_2} \\ \text{and } \omega_{12}^2 &= \frac{K}{\sqrt{M_1 M_2}}\end{aligned}\right\} \qquad ...(15)$$

From eqns. (14) and (15), we have

$$\left(\alpha^2 + \omega_{11}^2\right)\left(\alpha^2 + \omega_{22}^2\right) - \omega_{12}^4 = 0 \qquad ...(16)$$

or, $\qquad \alpha^4 + \alpha^2\left(\omega_{11}^2 + \omega_{22}^2\right) + \omega_{11}^2 \omega_{22}^2 - \omega_{12}^4 = 0 \qquad ...(17)$

Therefore, we obtain

$$\alpha^2 = -\frac{1}{2}\left(\omega_{11}^2 + \omega_{22}^2\right) \pm \frac{1}{2}\sqrt{\left(\omega_{11}^2 - \omega_{22}^2\right) + 4\omega_{12}^2} \qquad ...(18)$$

The expression under the radical is positive and absolute value of the second term is less than the first term. Hence, α^2 is real and negative. Accordingly, let us write

$$\alpha^2 = -\omega^2 \text{ or } \alpha = \pm j\omega \qquad ...(19)$$

Hence,

$$\omega^2 = \frac{\omega_{11}^2 + \omega_{22}^2}{2} \pm \frac{1}{2}\sqrt{\left(\omega_{11}^2 - \omega_{22}^2\right) + 4\omega_{12}^2} \qquad ...(20)$$

Equation (20) gives two values of ω^2, let us call them ω_1 and ω_2. From eqn. (19), we obtain four values of α, given by

$$j\omega_1, \ -j\omega_1, \ j\omega_2 \text{ and } -j\omega_2$$

Substituting these values of α in equation (9), we obtain

$$\left.\begin{aligned}x_1 &= A_{11}e^{j\omega_1 t} + \overline{A}_{11}e^{-j\omega_1 t} + A_{12}e^{j\omega_2 t} + \overline{A}_{12}e^{-j\omega_2 t} \\ x_2 &= A_{21}e^{j\omega_1 t} + \overline{A}_{21}e^{-j\omega_1 t} + A_{22}e^{j\omega_2 t} + \overline{A}_{22}e^{-j\omega_2 t}\end{aligned}\right\} \qquad ...(21)$$

Since x_1 and x_2 are real, $\overline{A}_{11}$ must be conjugate of A_{11}, and $\overline{A}_{12}$ the conjugate of A_{12}. If we write

$$\left.\begin{aligned}A_{11} &= \frac{C_{11} \, e^{j\theta_1}}{2}, \overline{A}_{11} = \frac{C_{11} \, e^{-j\theta_1}}{2} \\ A_{12} &= \frac{C_{12} \, e^{j\theta_2}}{2}, \overline{A}_{12} = \frac{C_{12} \, e^{-j\theta_2}}{2} \\ A_{21} &= \frac{C_{21} \, e^{j\theta_1}}{2}, \overline{A}_{21} = \frac{C_{21} \, e^{-j\theta_1}}{2} \\ A_{22} &= \frac{C_{22} \, e^{j\theta_2}}{2}, \overline{A}_{22} = \frac{C_{22} \, e^{-j\theta_2}}{2}\end{aligned}\right\} \qquad ...(22)$$

where C_{11}, C_{12}, C_{21}, C_{22}, θ_1 and θ_2 are new arbitrary constants. We then write the solution (21) in the form of

$$x_1 = C_{11} \cos (\omega_1 t + \theta_1) + C_{12} \cos (\omega_2 t + \theta_2) \Big\}$$
$$x_2 = C_{21} \cos (\omega_1 t + \theta_1) + C_{22} \cos (\omega_2 t + \theta_2) \Big\} \qquad ...(23)$$

From these equations, it is observed that the most general solution of the coupled system is made up of the superposition of two pure harmonic oscillations. In each of these oscillations, the two masses oscillate with the same frequency and in the same phase. The amplitudes of oscillation are in a definite ratio given by the equation (12).

The pure harmonic oscillations are called principal oscillations or the principal modes of oscillations of the system.

Normal Modes

Now, we consider the case when

$$M_1 = M_2 = M \qquad ...(24)$$
$$K_1 = K_2 = K_0 \qquad ...(25)$$

This is a very symmetric case. Now, equation (8) reduces to

$$M\ddot{x}_1 + (K_0 + K)x_1 - Kx_2 = 0 \Big\}$$
$$M\ddot{x}_2 + (K_0 + K)x_2 - Kx_1 = 0 \Big\} \qquad ...(26)$$

Adding these two equations, we obtain

$$M(\ddot{x}_1 + \ddot{x}_2) + K_0(x_1 + x_2) = 0 \qquad ...(27)$$

Let us take $(x_1 + x_2) = Q_1$, a new co-ordinate. Then, we have

$$M\ddot{Q}_1 + K_0 Q_1 = 0 \qquad ...(28)$$

This is the equation of simple harmonic motion where Q_1 is the displacement. Hence, the solution of this equation is

$$Q_1 = A_1 \sin \omega_1 t + B_1 \cos \omega_1 t \qquad ...(29)$$

where

$$\omega_1 = \sqrt{\frac{K_0}{M}} \qquad ...(30)$$

This is the angular frequency of a motion where the two masses swing to left and right with equal amplitudes in such a manner that the coupling spring is not stretched. If now we subtract the given equations (26), we obtain

$$M(\ddot{x}_1 - \ddot{x}_2) + (K_0 + 2K)(x_1 - x_2) = 0 \qquad ...(31)$$

We can write

$$x_1 - x_2 = Q_2 \qquad ...(32)$$

where Q_2 is another co-ordinate.

From eqns. (31) and (32), we have

$$M\ddot{Q}_2 + (K_0 + 2K)Q_2 = 0 \qquad \qquad ...(33)$$

This equation has a solution

$$Q_2 = A_2 \sin \omega_2 t + B_2 \cos \omega_2 t \qquad \qquad ...(34)$$

where

$$\omega_2 = \sqrt{\frac{K_0 + 2K}{M}} \qquad \qquad ...(35)$$

This case represents the one in which the two masses move in opposite directions with the same amplitude.

The transformation form of x-co-ordinate to Q-co-ordinates can be written in the matrix form as

$$\begin{bmatrix} Q_1 \\ Q_2 \end{bmatrix} = \begin{bmatrix} 1 & 1 \\ 1 & -1 \end{bmatrix} \begin{bmatrix} X_1 \\ X_2 \end{bmatrix} \qquad \qquad ...(36)$$

Also

$$\begin{bmatrix} X_1 \\ X_2 \end{bmatrix} = \begin{bmatrix} 1 & 1 \\ 1 & -1 \end{bmatrix}^{-1} \begin{bmatrix} Q_1 \\ Q_2 \end{bmatrix}$$

or,

$$\begin{bmatrix} X_1 \\ X_2 \end{bmatrix} = \frac{1}{2}\begin{bmatrix} 1 & 1 \\ 1 & -1 \end{bmatrix} \begin{bmatrix} Q_1 \\ Q_2 \end{bmatrix} \qquad \qquad ...(37)$$

Thus, we see that by a linear transformation of the co-ordinates (x_1, x_2) to the co-ordinates (Q_1, Q_2), we have effected a separation of the variables, so that the motion of the Q_1 and Q_2 co-ordinates are uncoupled. These new co-ordinates Q_1 and Q_2 are called normal co-ordinates and the modes with which the systems vibrate are called *normal modes*.

Lagrange's Equation

Let us consider a mass M attached to a spring of force constant K. X is a linear co-ordinate measured from the equilibrium position of the mass. If the mass is moving with a velocity $v = \dot{x}$, its kinetic energy is given by

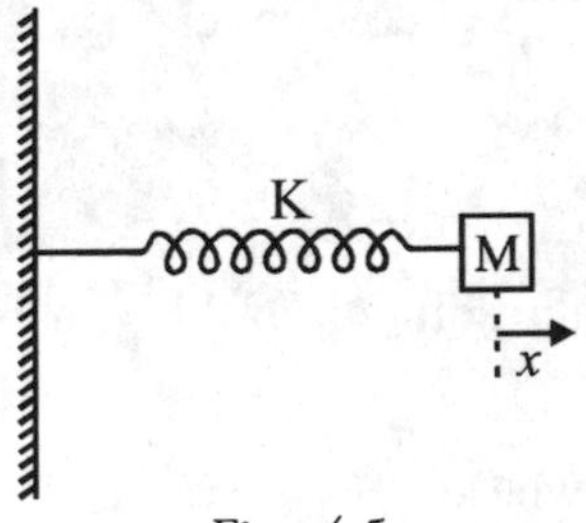

Fig. 4.5

$$T = \frac{1}{2}Mv^2 = \frac{1}{2}M(\dot{x})^2 \qquad \qquad ...(38)$$

When the spring is stretched by a distance x from its equilibrium position, then the elastic or potential energy stored in the spring is

$$V = \int_0^x F\,dx = \int_0^x Kx\,dx = \frac{1}{2}Kx^2 \qquad \qquad ...(39)$$

Now, by Newton's second law of motion, we have

$$Ma = -F, \qquad \qquad ...(40)$$

Acceleration $\quad a = \dfrac{dv}{dt} = \dfrac{d}{dt}\left(\dfrac{dx}{dt}\right)$

$$= \dfrac{d^2 x}{dt^2} = \ddot{x}$$

Hence, $\qquad M\ddot{x} = -F$ \hfill ...(41)

The kinetic energy is

$$T = \frac{1}{2}mv^2 = \frac{1}{2}m\dot{x}^2 \qquad\qquad ...(42)$$

$$\dfrac{\partial T}{\partial \dot{x}} = m\dot{x} = mv \qquad\qquad ...(43)$$

Similarly, differentiating eqn. (39) with respect to x, we have

$$\dfrac{\partial V}{\partial x} = Kx = F \qquad\qquad ...(44)$$

Equation (41) can be written as

$$M\ddot{x} = -F$$

or, $\qquad \dfrac{d}{dt}(M\dot{x}) = -F$ \hfill ...(45)

From eqns. (43), (44) and (45), we obtain

$$\dfrac{d}{dt}\left(\dfrac{\partial T}{\partial \dot{x}}\right) = -\dfrac{\partial V}{\partial x}$$

or, $\qquad \dfrac{d}{dt}\left(\dfrac{\partial T}{\partial \dot{x}}\right) + \dfrac{\partial V}{\partial x} = 0$ \hfill ...(46)

This form of equation of motion is called Lagrange's equation.

N-Coupled Oscillators

Let us consider vibration of a molecule having N-atoms. There are 3N degrees of freedom for this molecule. Let X, Y, Z be the co-ordinates of the nth atom in terms of a co-ordinate system and a, b, c be the values of the co-ordinates of the equilibrium position of the nth atom, i.e. the values assumed by x, y, z. When the molecule is at rest in its equilibrium position, displacements from the equilibrium position will be given by

$$\left.\begin{array}{l} \Delta x = x - a \\ \Delta y = y - b \\ \Delta z = z - c \end{array}\right\} \qquad\qquad ...(47)$$

Let us calculate the kinetic and potential energy of the molecule having N-atoms.

(a) **Kinetic energy:** The kinetic energy is given by

$$T = \frac{1}{2}\sum_{n=1}^{N} m\left[\left(\frac{d\Delta x}{dt}\right)^2 + \left(\frac{d\Delta y}{dt}\right)^2 + \left(\frac{d\Delta z}{dt}\right)^2\right] \qquad ...(48)$$

Let us replace the co-ordinates Δx_1, ... Δx_N by a new set of co-ordinates q_1, ...q_{3N} defined as follows:

$$q_1 = \sqrt{m_1\Delta x_1}, \quad q_2 = \sqrt{m_1 y_1},$$
$$q_3 = \sqrt{m_1 \Delta z_1}, \quad q_4 = \sqrt{m_2 \Delta x_2}, \text{ etc.} \qquad ...(49)$$

The velocities corresponding to these co-ordinates are given by $\dot{q}_1, \dot{q}_2, \dot{q}_3 .... \dot{q}_{3N}$. Then we can express the kinetic energy of the molecule having N-atoms as

$$T = \frac{1}{2}\sum_{i=1}^{3N} (\dot{q}_i)^2 \qquad ...(50)$$

(b) The potential energy for a displacement is some function of the displacement. Here, the potential energy is the function of q's. For all values of displacements, the potential energy V may be expressed in a power series in the displacement q_i as

$$V = V_0 + \sum_{i=1}^{3N}\left(\frac{\partial V}{\partial q_i}\right)_0 q_i + \frac{1}{2}\sum_{i,j=1}^{3N}\left(\frac{\partial^2 V}{\partial q_i \partial q_j}\right)_0 + \text{higher terms} \qquad ...(51)$$

or,

$$V = V_0 + \sum_{i=1}^{3N} f_i q_i + \frac{1}{2}\sum_{i,j=1}^{3N} f_{ij} q_i q_j + \text{higher terms} \qquad ...(52)$$

Taking the initial energy V_0 as zero, we have

$$\left(\frac{\partial V}{\partial q_i}\right)_0 = f_i = 0, \; i = 1, 2, ..., 3N$$

For sufficiently small amplitudes of vibration, the higher terms can be neglected. So, equation (52) becomes

$$V = \frac{1}{2}\sum_{i,j=1}^{3N} f_{ij} q_i q_j \qquad ...(53)$$

in which the f_{ij}'s are constants, given by

$$f_{ij} = \left(\frac{\partial^2 V}{\partial q_i \partial q_j}\right)_0 \qquad ...(54)$$

with $\qquad f_{ij} = f_{ji}$

(c) **Lagrange equation:** Lagrange equation can be written as

$$\frac{d}{dt}\left(\frac{\partial T}{\partial \dot{q}_j}\right) + \frac{\partial V}{\partial q_j} = 0 \qquad ...(55)$$

where $\qquad j = 1, 2, ..., 3N.$

Since T is a function of the velocity only and V is a function of co-ordinates only, substituting from eqns. (50) and (53), in equation (55), we obtain

$$\frac{d}{dt}\frac{\partial}{\partial \dot{q}_j}\left[\frac{1}{2}\sum_{j=1}^{3N}\dot{q}_j^2\right]+\frac{\partial}{\partial q_j}\left[\frac{1}{2}\sum_{ij}^{3N}f_{ij}q_j\right]=0$$

or,

$$\sum_{j=1}^{3N}\frac{d}{dt}(\dot{q}_j)+\frac{1}{2}\sum_{i=1}^{3N}f_{ij}q_i=0$$

or,

$$\ddot{q}_j+\frac{1}{2}\sum_{i=1}^{3N}f_{ij}q_i=0 \qquad \qquad ...(56)$$

This is a set of 3N simultaneous second-order linear differential equations. One possible solution of this equation is

$$q_i = A_i \cos\left(\sqrt{\lambda}\,t+\theta\right) \qquad \qquad ...(57)$$

where A_i, λ and θ are properly chosen constants. If this expression is substituted in the differential equation (56), we obtain

$$\sum_{i=1}^{3N}(f_{ij}-\delta_{ij}\lambda)\,A_i = 0 \qquad \qquad ...(58)$$

where δ_{ij} is the Kronecker delta.

$$\left.\begin{array}{l}\delta_{ij}=1, \text{ for } i=j\\ \qquad =0, \text{ for } i\neq j\end{array}\right\} \qquad \qquad ...(59)$$

Equation (58) is a set of simultaneous homogeneous linear algebraic equation in the 3N unknown amplitudes A_i.

Only for special values of λ, equation (58) has non-vanishing solutions and for all other values of λ, the solution is the trivial one.

$$A_i = 0, \ i = 1, 2, \ ...., \ 3N.$$

corresponds to no Variation.

The special values of λ are those which satisfy the determinant

$$\begin{vmatrix} f_{11}-\lambda & f_{12} & f_{13} & f_{1.3N} \\ f_{21} & f_{22}-\lambda & f_{23} & f_{2.3N} \\ .... & .... & .... & .... \\ f_{3N,1} & f_{3N,2} & f_{3N,3} & f_{3N,3N-\lambda} \end{vmatrix} = 0 \qquad \qquad ...(60)$$

The elements of this determinant are the coefficients of the unknown amplitudes A_i in the set of equation (58). When a fixed value of λ, say λ_k, is chosen so as to cause the determinant to vanish, the coefficients of unknown A_i in eqn. (58) become fixed, and it is then possible to obtain a solution A_{ik}, for which additional subscript k will be used to indicate the correspondence with the particular value of λ_k. Such a system of equations does not determine the A_{ik} uniquely, but gives only their ratios: an arbitrary set A_{ik} may be obtained by putting $A_{ik} = 1$.

A convenient and unique mathematical solution may be designed by the quantities like those which are defined in terms of an arbitrary solution, A'_{ik}, by the formula

$$l_{ik} = \frac{A'_{ik}}{\left[\sum_i (A_{ik})^2\right]^{1/2}} \qquad \qquad ...(61)$$

These amplitudes are normalised in the sense that

$$\sum l_{ik}^2 = 1 \qquad \qquad ...(62)$$

The solution of the actual physical problem, then can be obtained by putting

$$A_{ik} = K_k \, l_{ik} \qquad \qquad ...(63)$$

where K_k is the constant determined by the initial values of the co-ordinates q_i and velocities $\dot{q}_i$.

Normal Modes of Vibration of a Molecule

Equation (58) shows that each atom of the molecule is oscillating about its equilibrium position with a simple harmonic motion of amplitude $A_{ik} = K_k \, l_{ik}$, frequency $\dfrac{\lambda^{1/2}}{2\pi}$ and phase θ_k. Furthermore, corresponding to a given solution λ_k of the secular equation, the frequency and phase of motion of each co-ordinate is the same, but the amplitudes may be, and usually are different for each co-ordinate. On account of the equality of phase and frequency, each atom reaches its position of maximum displacement at the same time, and each atom passes through its equilibrium position at the same time. A mode of vibration having these characteristics is called normal mode of vibration and its frequency is known as normal or fundamental frequency of the molecule.

Applications

(i) Water molecule: Fig. 4.6 represents the three modes of vibrations of a water molecule. The arrows indicate the relative displacements of the atoms, in the co-ordinate system mentioned above, and when the molecule is vibrating in the particular mode of vibration.

Fig. 4.6: Normal modes of vibration of a water molecule

From the nature of normal mode, the displacements of the different atoms remain in the same ratio to one another throughout the motion. The atoms are also constrained to move back and forth along straight lines.

(ii) Diatomic molecule or two masses attached by a spring: Let us consider two masses m_1 and m_2 connected by a spring of force constant k. When two masses are displaced from their equilibrium positions by contracting or extending the spring, both the masses vibrate harmonically about their mean positions (Fig. 4.7).

Fig. 4.7: A diatomic molecule (Two masses attached by a spring)

Let us suppose that the unstretched length of the spring is l_0 and at any instant, its length is l. Hence, the force acting on either mass is given by

$$F = - k(l - l_0) \qquad \text{...(64)}$$

If l_1 and l_2 are the position vectors of the two masses with respect to centre of mass, their equations of motions are

$$\frac{d^2 \vec{l}_1}{dt^2} = \frac{-k}{m_1} (l - l_0) \hat{l} \qquad \text{...(65)}$$

and

$$\frac{d^2 \vec{l}_2}{dt^2} = \frac{k}{m_2} (l - l_0) \hat{l} \qquad \text{...(66)}$$

where $\hat{l}$ is the unit vector along $\vec{l}$.

Opposite sign shows that the forces are acting on the two particles in opposite directions. Subtracting equation (66) from (65), we have

$$\frac{d^2}{dt^2} (\vec{l}_1 - \vec{l}_2) = - k \left(\frac{1}{m_1} + \frac{1}{m_2} \right) (l - l_0) \hat{l} \qquad \text{...(67)}$$

or,

$$\frac{d^2 \vec{l}}{dt^2} = -\frac{k}{\mu} (l - l_0) \hat{l} \qquad \text{...(68)}$$

where,

$$\left(\frac{1}{\mu} \right) = \frac{1}{m_1} + \frac{1}{m_2} = \frac{m_1 + m_2}{m_1 m_2}$$

or

$$\mu = \frac{m_1 m_2}{m_1 + m_2} \qquad \text{...(69)}$$

and

$$\vec{l}_1 - \vec{l}_2 = \vec{l} \qquad \text{...(70)}$$

If the system is not rotating, the direction of $\vec{l}$ is fixed. Equation (68) represents the motion of a single harmonic oscillator of mass μ and force constant k.

The time-period of oscillation is given by

$$T = 2\pi \sqrt{\frac{\mu}{k}} \qquad \text{...(71)}$$

The frequency is

$$f = \frac{1}{T} = \frac{1}{2\pi}\sqrt{\frac{\mu}{k}} \qquad \ldots(72)$$

Coupled Vibration of two Loads Attached to a String Under Tension

Let two loads of mass m_1 and m_2 are attached at B and C of the string ABCD fixed at A and D. The string is taken weightless compared to the masses attached. The action due to gravity on the loads is neglected. The length of the string is l stretched under tension T which assumed to remain constant during the motion of the load. Let AB = a, BC = b and CD = c, such that

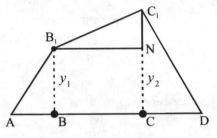

Fig. 4.8

$$a + b + c = l$$

Let y_1 and y_2 are the displacements of the loads at any time t which is small. B_1 and C_1 represent the displaced position of the load. Algebraic sum of the total vertical component of tension at B_1 is

$$- T \sin B_1AB + T \sin C_1B_1N$$

$$= - T\frac{y_1}{a} + T\frac{(y_2 - y_1)}{b} \quad \text{as angles are small.}$$

$$= \frac{T}{b}y_2 - Ty_1\left(\frac{1}{a} + \frac{1}{b}\right) \qquad \ldots(1)$$

Likewise, the algebraic sum of the vertical components of tensions at C_1 is

$$\frac{- Ty_2}{c} - T\frac{(y_2 - y_1)}{b}$$

$$= \frac{Ty_1}{b} - Ty_2\left(\frac{1}{c} + \frac{1}{b}\right) \qquad \ldots(2)$$

These are the restoring forces on the masses m_1 and m_2. Hence, the equations of motion of the two masses are

$$m_1\frac{d^2y_1}{dt^2} + T\left(\frac{1}{a} + \frac{1}{b}\right)y_1 - \frac{T}{b}y_2 = 0 \qquad \ldots(3)$$

$$m_2\frac{d^2y_2}{dt^2} + T\left(\frac{1}{c} + \frac{1}{b}\right)y_2 - \frac{T}{b}y_1 = 0 \qquad \ldots(4)$$

Let us put

$$T\left(\frac{1}{a} + \frac{1}{b}\right) = \mu_1$$

$$T\left(\frac{1}{c} + \frac{1}{b}\right) = \mu_2$$

and $-\dfrac{T}{b} = \mu_{12}$, in the equations (3) and (4). Then, we have

$$m_1 \frac{d^2 y_1}{dt^2} + \mu_1 y_1 + \mu_{12} y_2 = 0 \qquad \text{...(5)}$$

$$m_2 \frac{d^2 y_2}{dt^2} + \mu_2 y_2 + \mu_{12} y_1 = 0 \qquad \text{...(6)}$$

These two equations can be solved as per the procedure laid down earlier in the beginning of this chapter (Analytical treatment of coupled oscillation).

Special Case: If the masses of the two loads are equal, then we can write

$$m_1 = m_2 = m$$

and if $\ a = b = c = \dfrac{l}{3}$

then eqns. (3) and (4) yield

$$m \frac{d^2 y_1}{dt^2} + \frac{6T}{l} y_1 - \frac{3T}{l} y_2 = 0 \qquad \text{...(7)}$$

$$m \frac{d^2 y_2}{dt^2} + \frac{6T}{l} y_2 - \frac{3T}{l} y_1 = 0 \qquad \text{...(8)}$$

Adding eqns. (7) and (8), we have

$$m \frac{d^2}{dt^2} (y_1 + y_2) + \frac{3T}{l} (y_1 + y_2) = 0 \qquad \text{...(9)}$$

Subtracting the two, we obtain

$$m \frac{d^2}{dt^2} (y_1 - y_2) + \frac{9T}{l} (y_1 - y_2) = 0 \qquad \text{...(10)}$$

The solution of eqns. (9) and (10) are

$$y_1 + y_2 = a_1 \cos \sqrt{\frac{3T}{ml}} \, t + b_1 \sin \sqrt{\frac{3T}{ml}} \, t \qquad \text{...(11)}$$

$$y_1 - y_2 = a_2 \cos \sqrt{\frac{9T}{ml}} \, t + b_2 \sin \sqrt{\frac{9T}{ml}} \, t \qquad \text{...(12)}$$

From eqns. (11) and (12), we obtain

$$y_1 = a'_1 \cos \sqrt{\frac{3T}{ml}} \, t + b'_1 \sin \sqrt{\frac{3T}{ml}} \, t + a'_2 \cos \sqrt{\frac{9T}{ml}} \, t + b'_2 \sin \sqrt{\frac{9T}{ml}} \, t \qquad \text{...(13)}$$

$$y_2 = a'_1 \cos \sqrt{\frac{3T}{ml}} \, t + b'_1 \sin \sqrt{\frac{3T}{ml}} \, t - a'_2 \cos \sqrt{\frac{9T}{ml}} \, t - b'_2 \sin \sqrt{\frac{9T}{ml}} \, t \qquad \text{...(14)}$$

where, $\quad \dfrac{a_1}{2} = a'_1, \qquad \dfrac{b_1}{2} = b'_1,$

$$\frac{a_2}{2} = a'_2, \qquad \frac{b_2}{2} = b'_2$$

These constants a'_1, a'_2, b'_1 and b'_2 can be determined from boundary conditions. Thus, the masses will vibrate with frequencies

$$f_1 = \frac{1}{2\pi} \sqrt{\frac{3T}{ml}}$$

and

$$f_2 = \frac{1}{2\pi} \sqrt{\frac{9T}{ml}}$$

which depend upon the signs of y_1 and y_2, the displacements of the loads (Fig. 4.8).

SOLVED EXAMPLES

Q.1. **Two masses 2 gm and 0.5 gm are connected with a massless spring of force constant 10^3 dynes/cm. Calculate:**

(i) the frequency of oscillations,

(ii) energy of the oscillations of the small mass which is vibrating with amplitude 1 cm, and

(iii) maximum velocity of the smaller mass.

Sol. (i) $\qquad f = \frac{1}{2\pi} \sqrt{\frac{k}{\mu}}$

Here, $\qquad m_1 = 2$ gm, $m_2 = 0.5$ gm

and $\qquad k = 10^3$ dynes/cm

So, $\qquad \mu = \frac{m_1 \times m_2}{m_1 + m_2}$

$$= \frac{2 \times 0.5}{2 + 0.5} = \frac{1}{2.5} = \frac{2}{5}$$

$\therefore \qquad f = \frac{1}{2\pi} \sqrt{\frac{1000 \times 5}{2}} = 8$ cycles/sec.

(ii) Amplitude of the larger mass

$$= \frac{5}{2} \times \frac{1}{10} = \frac{1}{4} \text{ cm}$$

Total displacement $= 1 + \frac{1}{4} = \frac{5}{4}$ cm

For maximum displacement, total energy is potential. Hence

$$E = \frac{1}{2}kX_{max}^2 = \frac{1}{2} \times 10^3 \times \left(\frac{5}{4}\right)^2$$

$$= \frac{1}{2} \times 1000 \times \frac{25}{16} = \frac{125 \times 25}{4}$$

$$= 781 \text{ ergs.}$$

(iii) $$(V_2)_{max} = \frac{(V_1)_{max}}{4}$$

because $(V)_{max} = \omega A$

$$(E_k)_{max} = \frac{1}{2}m_1(V_1)_{max}^2 + \frac{1}{2}m_2(V_2)_{max}^2$$

$$= \frac{1}{2} \times 0.5 \times (V_1)_{max}^2 + \frac{1}{2} \times 2 \times \frac{(V_1)_{max}^2}{16}$$

or, $$781 = (V_1)_{max}^2 \left[\frac{1}{4} + \frac{1}{16}\right]$$

$$= (V_1)_{max}^2 \left(\frac{5}{16}\right)$$

or, $$(V_1)_{max}^2 = \frac{781 \times 16}{5}$$

or, $$(V_1)_{max} = \sqrt{\frac{781 \times 16}{5}}$$

$$= 50 \text{ cm/sec}$$

Q.2. **Two discs of masses 200 gm and 100 gm are connected by a weightless spring of force constant 10^5 dynes/cm as shown in the Fig. 4.9. Compute:**

(i) **the frequency of oscillation of the spring when the system is resting on a table,**

(ii) **when the table is removed and the system is falling freely, and**

Sol. (i) When the system is resting on the table, the equation of motion of the spring is given by

$$m_1 \frac{d^2 y}{dt^2} = -ky$$

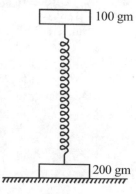

Fig. 4.9

where y is the displacement of mass m_1 at any instant t. Here, $m_1 = 100$ gm and $k = 10^5$ dynes/cm. Hence, frequency of oscillation is given by

$$f = \frac{1}{2\pi}\sqrt{\frac{k}{m_1}} = \frac{1}{2\pi}\sqrt{\frac{10^5}{100}} = 5\,\text{Hz}.$$

(ii) When the table is removed and the system falls freely under the gravity, the equation of motion of the spring about the centre of mass of the system is given by

$$\mu\frac{d^2y}{dt^2} = -ky$$

where μ is the reduced mass of the system and is equal to

$$\mu = \frac{m_1 m_2}{m_1 + m_2} = \frac{200 \times 100}{200 + 100} = \frac{200}{3}$$

So, the frequency of oscillation is

$$f = \frac{1}{2\pi}\sqrt{\frac{k}{\mu}} = \frac{1}{2\pi}\sqrt{\frac{10^5 \times 3}{200}}$$

$$= \frac{1}{2 \times 3.14}\sqrt{1500}$$

$$= \frac{38.7298}{6.28} = 6\,\text{Hz}.$$

Q.3. **The fundamental vibrational angular frequency of the HF molecule is found to be 7.55×10^{14} rad/sec. Calculate:**

(i) force constant of HF molecule, and

(ii) work done in stretching the atoms by 0.5 Å.

(Given: atomic mass of hydrogen $m_H = 1.66 \times 10^{-24}$ gm, atomic mass of fluorine $m_F = 19 \times 1.66 \times 10^{-24}$ gm and normal internuclear distance is 1 Å)

Sol. (i) Reduced mass

$$\mu = \frac{1 \times 19}{1 + 19} \times 1.66 \times 10^{-24}$$

Angular frequency

$$\omega = \sqrt{\frac{k}{\mu}}$$

or, $$\omega^2 = \frac{k}{\mu}$$

or, $$k = \mu\omega^2$$

$$= \frac{19 \times 1.66 \times 10^{-24} \times (7.55 \times 10^{14})^2}{20}$$

$$= \frac{19 \times 1.66 \times (7.55)^2 \times 10^4}{20}$$

$$\simeq 9 \times 10^5 \text{ dynes/cm.}$$

(ii) Work done in stretching the atoms is given by

$$W = \frac{1}{2} k (r - r_0)^2$$

$$= \frac{1}{2} \times 9 \times 10^5 \times (0.5 \times 10^{-8})^2$$

$$\simeq 9 \times 10^{-11} \text{ ergs.}$$

$$\simeq 6 \text{ eV.}$$

Q.4. **In HCl molecule, the force needed to alter the distance between the atoms from equilibrium is 5.4 × 10⁻⁵ dynes/cm. Calculate the fundamental frequency of oscillation of the HCl molecule.**

(Given: m_H = 1.66 × 10⁻²⁴ gm, m_{Cl} = 35.5 × 1.66 × 10⁻²⁴ gm)

Sol. The reduced mass of the molecule is

$$\mu = \frac{m_H \times m_{Cl}}{m_H + m_{Cl}} = \frac{35.5 \times (1.66 \times 10^{-24})^2}{(1 + 35.5)(1.66 \times 10^{-24})}$$

$$= 1.6 \times 10^{-24} \text{ gm}$$

The frequency of oscillation of HCl molecule is

$$f = \frac{1}{2\pi} \sqrt{\frac{k}{\mu}}$$

$$= \frac{1}{2 \times 3.14} \sqrt{\frac{5.4 \times 10^{-5}}{1.6 \times 10^{-24}}}$$

$$= \frac{1}{6.28} \sqrt{\frac{54}{16} \times 10^{19}}$$

$$= \frac{10^9}{6.28} \sqrt{\frac{540}{16}}$$

$$= \frac{10^9}{6.28 \times 4} \sqrt{540}$$

$$= \frac{10^9 \times 23.2379}{25.12} = 0.925 \times 10^9$$

$$= 925 \times 10^6 \text{ Hz.}$$

Q.5. For HCl molecule, one vibrational frequency lie at 2000 cm^{-1}. Find the force constant. (Avogadro's number N = 6.0 × 10^{23} per gm mole)

Sol. Here, $\dfrac{1}{\lambda} = 2000$ cm^{-1}

So, $f = c \times \dfrac{1}{\lambda} = 3 \times 10^{10} \times 2000$

$= 6 \times 10^{13}$ Hz

$\mu = \dfrac{1 \times 35}{1 + 35} \times \dfrac{1}{6 \times 10^{23}} = 1.6 \times 10^{-24}$ gm

$\therefore \quad k = \mu\omega^2 = \mu(2\pi f)^2 = 4\pi^2 \mu f^2$

$= 4 \times 9.87 \times 1.6 \times 10^{-24} \times 36 \times 10^{26}$

$= 20.8 \times 10^4$ dyne/cm.

QUESTIONS

1. What is a coupled harmonic oscillator? Give three examples of coupled vibrating system.

2. Develop the general theory of vibration of two coupled oscillators.

3. Write down the equations of motion of linear, torsional and electrical coupled oscillators. Mention the analogy among them.

4. Develop the theory of vibration of a system with two degrees of freedom.

5. Develop the theory of normal modes of vibration of the two coupled oscillators.

6. Deduce Lagrange's equation for the vibration of a string attached to a mass.

7. Find the kinetic energy, potential energy and Lagrange's equation for vibration of a molecule having N-atoms. What are normal modes of vibration of the molecule?

8. Deduce the expression for the time-period of oscillation of a diatomic molecule.

9. Develop the theory of coupled vibration of two loads attached to a string under tension.

10. Two loads of equal masses are attached to a string under tension. Find the frequencies of oscillation of this coupled system.

Problems

1. Two loads of masses 9 gm and 10 gm are connected with a massless rod of length 50 cm. If they are suspended so as to keep this rod horizontal and the couple per unit twist of this suspension is 1000 dyne-cm^2 per radian. Find:

   (i) the frequency of vibration, and

   (ii) the amplitude when the total energy is 10^5 ergs.

$$\left[\text{Hint: (i)} f = \frac{1}{2\pi}\sqrt{\frac{C}{I}}, \text{ and (ii) E} = \frac{1}{2}C\theta \right]$$

2. For oxygen molecule, let one vibrational frequency lie at 1000 cm^{-1}. Deduce the force constant. (**Ans.** 5×10^5 dynes/cm)

3. Two springs of force constants k_1 and k_2 are joined to a mass m as shown in the figure below. Find an expression for the time-period of oscillation of the system.

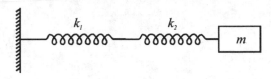

$$\left[\textbf{Ans. } T = 2\pi \sqrt{\frac{(k_1 + k_2)m}{k_1 k_2}} \right]$$

4. Find the electrical analogue of the above problem. What is the frequency of oscillation in such a case? (**Hint:** Series combination of two capacitors)

5. The vibrational frequency of H^{1}Cl^{35} molecule is 8990×10^{10} cycles/sec. Find the vibrational frequency of H^{2}Cl^{35} molecule, assuming the force constant to be the same in the two cases.

$$\left[\textbf{Hint : } \frac{f_1}{f_2} = \sqrt{\frac{\mu_1}{\mu_2}} \right]$$

● ● ●

Chapter 5

Wave Motion

Propagation of Longitudinal Waves in Elastic Medium

Sound is a longitudinal wave motion in material media-solids, liquids and gases and the velocity with which it is propagated, depends upon the two characteristics of the medium: (i) its elasticity – a property which indicates that when the particles of a medium suffer relative displacements, a force is set up which tends to bring back to their original configuration, (ii) its inertia, therefore on the kind of critical impulses given from outside, the elasticity and the mass contents of the medium.

Newton first showed that in a homogeneous medium the velocity of a longitudinal wave is given by $v = \sqrt{\dfrac{E}{\rho}}$, where E is the modulus of elasticity for the particular type of strain set-up and ρ, the density of the medium.

Velocity of a Plane Longitudinal Wave

We have to find the velocity of compressional waves in terms of the characteristics of the medium namely its 'elasticity' and 'inertia'.

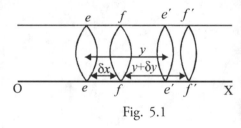

Fig. 5.1

Let us consider plane longitudinal waves travelling in air (or any elastic fluid) in a tube of cross-section α (Fig. 5.1). Suppose the waves are travelling from left to right along OX. Let us consider two planes perpendicular to OX at e and f, and at a small distance δx apart. Let e' and f' be the displaced positions of the planes originally at e and f. If the displacement of the plane at e be y and that of the plane at f be $y + \delta y$, then the rate of change of displacement is $\dfrac{dy}{dx}$. The displacement at f therefore may be written as $y + \left(\dfrac{dy}{dx}\right)\delta x$ and the distance apart of the planes has been changed by $y + \left(\dfrac{dy}{dx}\right)\delta x - y = \dfrac{dy}{dx}\delta x$.

Thus, the waves have altered the volume of the fluid between the planes and caused a variation of pressure from point to point in the tube.

The change in volume of the layer $ef = \alpha \dfrac{dy}{dx} \delta x$

Now volume of the layer $ef = \alpha \delta x$

$\therefore$ Volumetric strain $= \dfrac{\alpha \dfrac{dy}{dx} \delta x}{\alpha \, \delta x} = \dfrac{dy}{dx}$ ...(1)

If now p be the excess pressure on the plane e, the total excess force on the plane $e = \alpha p$,

and that on the plane $f = \alpha p + \alpha \left(\dfrac{dp}{dx} \right) \delta x$.

The represent force on the element δx in the direction f' to e'

$$= \alpha p + \alpha \left(\dfrac{dp}{dx} \right) \delta x - \alpha p$$

$$= \alpha \left(\dfrac{dp}{dx} \right) \delta x \qquad \qquad ...(2)$$

Mass of air between e and f

$$= \alpha \delta x \rho, \text{ where } \rho \text{ is the density of air.}$$

$$\text{Acceleration of air particles} = \dfrac{d^2 y}{dt^2}$$

$\therefore$ Mass $\times$ Acceleration $= \alpha \, \delta x \, \rho \, \dfrac{d^2 y}{dt^2}$ ...(3)

By Newton's second law of motion, we have

$$\alpha \delta x \rho \dfrac{d^2 y}{dt^2} = \alpha \dfrac{dp}{dx} \delta x$$

or, $$\rho \dfrac{d^2 y}{dt^2} = \dfrac{dp}{dx} \qquad \qquad ...(4)$$

Now volume elasticity

$$E = \dfrac{\text{stress}}{\text{strain}} = \dfrac{p}{dy/dx}$$

$\therefore$ $$p = E \dfrac{dy}{dx} \text{ and } \dfrac{dp}{dx} = \dfrac{d}{dx} \left(E \dfrac{dy}{dx} \right) = E \dfrac{d^2 y}{dx^2}$$

Setting the value of $\dfrac{dp}{dx}$ in eqn. (4), we have

$$\rho \dfrac{d^2 y}{dt^2} = E \dfrac{d^2 y}{dx^2}$$

or, $$\dfrac{d^2 y}{dt^2} = \dfrac{E}{\rho} \dfrac{d^2 y}{dx^2} \qquad \qquad ...(5)$$

Comparing it with the eqn. $\dfrac{d^2y}{dt^2} = v^2 \dfrac{d^2y}{dx^2}$, we have

$$v^2 = \frac{E}{\rho}$$

or,
$$v = \sqrt{\frac{E}{\rho}}$$

Calculation for Velocity in a Homogeneous Medium: Let us Apply Newton's Formula to the Case of Air

Now for a fluid, $E = V\dfrac{dP}{dV}$

where dP is the change in pressure P and dV is the change in volume V. Newton assumed that the temperature during compression and rarefactions in a sound wave remains practically constant and therefore Boyle's law (PV = a constant) holds good.

It can be easily shown that for a gas obeying Boyle's law E = P, where P is its pressure. Suppose that pressure is increased from P to P + dP and the volume in consequence is changed from V to V − dV.

$\therefore$
$$PV = (P + dP)\,(V - dV)$$
$$= PV - PdV + VdP - dVdP$$

Neglecting the product $dV \times dP$ being small, we have

$$PdV = VdP \text{ or } P = V\frac{dP}{dV} = E_\theta$$

where E_θ is the isothermal elasticity.

or,
$$V = \sqrt{\frac{P}{\rho}} \qquad\qquad\qquad ...(6)$$

Now in the case of air at NTP,
$P = h\rho_m g = 76 \times 13.6 \times 981$ dynes/cm^2, and $\rho = 0.001293$ gm/cm^3

$\therefore$ Velocity of sound at NTP $= \sqrt{\dfrac{76 \times 13.6 \times 981}{0.001293}}$ cm/sec.

$$= 280 \text{ m/sec.}$$

This result falls short of the experimental value according to which V = 332 m/sec, by about 10%. Newton, however, by using imperfect data for P and ρ found the discrepancy to be about 12%. He accounted for it by assuming that the molecules of air occupied about $\dfrac{1}{8}$th of the space traversed by sound and that sound passed instantly through these, only taking time to go through the interspaces. He also supposed that the presence of water vapour had no share in conductivity of the sound. Both the assumptions, however, are unjustified and the true explanation was given by Laplace.

Laplace's Correction

The discrepancy between the calculated and the observed value for the velocity of sound was first explained by Laplace in 1816. He said that compressions and rarefactions accompanying the passage of a sound wave in air occur so rapidly that the rise of temperature during compression and the fall of temperature during rarefaction do not get enough time to be equalised by radiation or conduction to or from the surrounding air, as air is a medium of very low thermal conductivity. Hence the assumption that the changes in volume and pressure are isothermal and obey Boyle's law, is not valid. The temperature changes in this case take place under adiabatic conditions and we, have therefore to utilize the adiabatic elasticity of the gaseous medium in evaluating the velocity of sound with the help of eqn. $v = \sqrt{\dfrac{E}{\rho}}$. For an adiabatic change the relation between pressure and volume of a gas, is given by

$$PV^{\gamma} = \text{const.}$$

where $\gamma = \dfrac{c_P}{c_V}$ is the ratio between the specific heat at constant pressure and the specific heat at constant volume.

Let us find the value of Adiabatic Elasticity from this relation. Differentiating it, we get

$$V\gamma dP + \gamma PV^{\gamma-1} dV = 0$$

or $\qquad \gamma PV^{\gamma-1} dV = -V^{\gamma} dP$

or $\qquad \gamma P = -V\dfrac{dP}{dV} = E_\phi$

where E_ϕ is the Adiabatic Elasticity, substituting this value for E in the eqn.

$$v = \sqrt{\dfrac{E}{\rho}}, \text{ we get}$$

$$v = \sqrt{\dfrac{\gamma P}{\rho}} \qquad\qquad ...(7)$$

for air, $\gamma = \dfrac{E_\phi}{E} = 1.414$. Using the data as before, velocity of sound in air at or NTP is given by

$$v = \sqrt{\dfrac{1.414 \times 70 \times 13.6 \times 981}{0.001293}} \text{ cm/sec}$$

$$= 332.5 \text{ m/sec.}$$

This is in close agreement with the value obtained experimentally and therefore establishes the truth of Laplace's correction.

Relation Between Velocity and Temperature

The relation (7)

$v = \sqrt{\dfrac{\gamma P}{\rho}}$ is sometimes expressed in an alternative form. If R be the gas constant for

a gram molecule of the gas, then for a perfect gas $PV = RT$, where V is its gm molecular

volume at Kelvin temperature T. Now, density $\rho = \dfrac{M}{V} = \dfrac{MP}{RT}$, where M is the molecular weight.

$$v = \sqrt{\frac{\gamma P}{\rho}} = \sqrt{\frac{\gamma P}{MP/RT}} = \sqrt{\frac{\gamma RT}{M}} \qquad \text{...(8)}$$

or $$\sum v = \sqrt{\gamma RT} \qquad \text{...(8a)}$$

where R is the gas constant for one gram of the gas.

From eqn. (8a) we find that velocity of sound is proportional to the square root of the Kelvin temperature.

Discussion of the Velocity Formula

We shall now use the formula, $v = \sqrt{\gamma P/\rho}$, to find how the velocity of sound in a gas depends upon the nature of the sound waves and on the pressure, density, temperature and the molecular structure of the gas which affects the value of γ.

(i) Effect of Amplitude and Wavelength: The expression for velocity is independent of the amplitude and the wavelength of the note, and therefore we can safely conclude that velocity does not depend upon the amplitude and the wavelength, but this holds so long as the amplitude is small as is generally the case. For waves of large amplitude the relationship is not true and it has been found that loud sounds such as those of a cannon travel faster than ordinary sounds in the immediate neighbourhood of the source. In such a wave, pressure is not proportional to density and the volume elasticity increases as the density is increased by compression and diminishes as the density is reduced by refraction. Consequently, a compression in a change of waveform. Moreover, due to the rapid compression in the early stages, there will probably be a rise in temperature which will cause an increase of velocify. At a distance from the source the sound velocity settles down to normal value corresponding to that for small amplitudes.

(ii) Effect of Pressure: The change in the pressure of a given mass of gas would produce a corresponding change in volume. From Boyle's law, $PV = $ constant.

Since $\quad V = \dfrac{m}{\rho}$, we have

$\therefore \quad \dfrac{Pm}{\rho} = $ a constant

or, $\quad \dfrac{P}{\rho}$ remains unaltered.

The velocity is thus independent of any change in atmospheric pressure so long as that change is sufficiently slow and Boyle's law holds good.

This has been verified experimentally and it has been found that the velocity of sound at higher altitudes is the same as at sea-level, although the atmospheric pressure at the two places is different.

(iii) Effect of Density: The pressure remaining the same; the velocities of sound in two gases are inversely proportional to the square root of their densities, provided 'γ' has the same value for both of them, e.g. oxygen and hydrogen. Oxygen is 16 times denser than hydrogen, therefore the velocity of sound in hydrogen is 4 times that in oxygen.

$$\therefore \quad \frac{\text{Velocity in hydrogen}}{\text{Velocity in oxygen}} = \sqrt{\frac{\gamma P}{\rho_h}} \div \sqrt{\frac{\gamma P}{\rho_o}}$$

$$= \sqrt{\frac{\rho_o}{\rho_h}} = \sqrt{\frac{16}{1}} = 4$$

(iv) Effect of Temperature: When the temperature increases, the density of the gas decreases without affecting the pressure. As the numerator in the expression $\sqrt{\gamma P/\rho}$ would remain unchanged, a rise of temperature would increase the velocity and vice-versa.

If v_0 is the velocity of sound in air at 0°C and v_t is the velocity at t°C, and ρ_0 and ρ_t are the corresponding densities at these temperatures, then if the pressure remains unchanged,

$$v_0 = \sqrt{\frac{\gamma P}{\rho_0}} \text{ and } v_t = \sqrt{\frac{\gamma P}{\rho_t}}$$

$$\therefore \qquad \frac{v_t}{v_0} = \sqrt{\frac{\rho_t}{\rho_0}} = \sqrt{1 + \alpha t} \qquad \text{[By Charles's law]}$$

where 'α' is the coefficient of volume expansion.

or, $$\frac{v_t}{v_0} = \sqrt{1 + \frac{t}{273}}, \text{ since } \alpha = \frac{1}{273}$$

whence $$\frac{v_t}{v_0} = \sqrt{\frac{273 + t}{273}} = \sqrt{\frac{T}{T_0}} \qquad \qquad ...(9)$$

where T and T_0 are kelvin temperatures, i.e. velocities are directly proportional to the square root of kelvin temperatures. If, however the variation of t is not too large, we have

$$\frac{v_t}{v_0} = (1 + \alpha t)^{1/2} = \left(1 + \frac{1}{2}\alpha t\right) \text{ approx.}$$

expanding by binomial theorem and neglecting the higher powers.

or $$\frac{v_t}{v_0} = \left(1 + \frac{1}{2} \cdot \frac{t}{273}\right) = \left(1 + \frac{t}{546}\right) \text{ approx.}$$

Taking the value of $v_0 = 332$ m/sec, we have

$$v_t = 332\left(1 + \frac{t}{546}\right)$$

or $$v_t = (332 + 0.616) \text{ m/sec} \qquad \qquad ...(10)$$

Thus the velocity of sound at $0°$ C increases by 0.61 m/sec for $1°$ C rise of temperature. This is called the temperature coefficient of velocity of sound and agrees very well with the experimental value. Note that this approximation holds good for air and at ordinary temperatures.

Velocity of Plane Progressive Sound Wave Through a Thin Solid Rod

When a thin rod of large length is stretched in the direction of its length, there will be increase in length with contraction in cross section and the rod is set in longitudinal vibration.

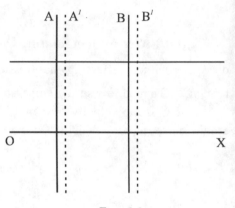

Fig. 5.2

We consider such a rod whose length is large compared with the diameter. Let ρ be the density, Y be the Young's modulus of the material of the rod. Let α be the cross section of the rod.

Let us consider an elementary since AB in the rod bounded by two planes A and B at x and $x + dx$ respectively at right angles to the axis of the rod. Let these planes be displaced by the amount ξ and $\xi + d\xi$. The distance between the planes A, B, which was initially dx becomes after displacement

$$(x + dx + \xi + d\xi) - (x + \xi) = dx + d\xi$$

So increase in length is $d\xi$, and increase in length per unit length is $\frac{d\xi}{dx}$.

If f is the total force acting on the area α.

The longitudinal stress $= \frac{f}{\alpha}$.

So, $$Y = \frac{f/\alpha}{d\xi/dx} \text{ or } f = \alpha Y \frac{d\xi}{dx} \qquad \qquad ...(1)$$

The force acting on the second face is

$$f + \frac{df}{dx} dx \qquad \qquad ...(2)$$

Eqn. (2) with the help of eqn. (1)

$$\alpha Y \frac{d\xi}{dx} + \frac{d}{dx}\left(\alpha Y \frac{d^2\xi}{dx}\right)dx \qquad \text{...(3)}$$

The resultant force is given by the difference of expressions (2) and (1),

$$\alpha Y \frac{d\xi}{dx} + \frac{d}{dx}\left(\alpha Y \frac{d\xi}{dx}\right)dx - \alpha Y \frac{d\xi}{dx} = \alpha Y \frac{d^2\xi}{dx^2} \qquad \text{...(4)}$$

But the mass of the element dx is $\rho\alpha dx$ and its acceleration is

$$\frac{d^2\xi}{dt^2} \qquad \text{...(5)}$$

Hence, $\qquad \rho\alpha \dfrac{d^2\xi}{dt^2} dx = \rho Y \dfrac{d^2\xi}{dx^2} dx$

or, $\qquad \dfrac{d^2\xi}{dt^2} = \dfrac{Y}{\rho} \dfrac{d^2\xi}{dx^2} \qquad \text{...(6)}$

This is a wave equation, so solution should be of the form

$$\xi = f(ct - x) + F(ct + x)$$

Proceeding in the similar way, we get velocity of longitudinal wave in a solid rod

$$c = \sqrt{\frac{Y}{\rho}}$$

Vibration of String

Velocity of transverse wave propagation along a string stretched under tension

The string is taken to be uniform in cross-section having constant mass ρ per unit length. It is taken to be perfectly flexible, its stiffness is supposed negligible and it is not subjected to appreciable change in length while in vibration.

A long thin light string stretched under tension T, between two massive and well clamped support.

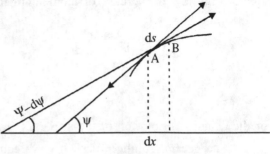

Fig. 5.3

We consider that the undisturbed position of the string as x-axis and the motion is confined in y-axis in the *xy* plane.

An infinitesimally small element of length AB = ds is considered. Its transverse rate of change of momentum is

$$\rho \, dx \, \frac{d^2y}{dt^2} \qquad \qquad ...(1)$$

and the resultant stretching force of A and B acts along the tangent at these points.

The transverse component of the stretching force at A is

$$- \, T \, \sin \, \psi = - \, T \frac{dy}{ds} \qquad \qquad ...(2)$$

ψ is the angle which tangent at A makes with the axis of A, Fig. 5.3.

The transverse component of the stretching force at B is

$$T \sin \psi_2 = \left\{ T \frac{dy}{ds} - \frac{d}{ds} \left(T \frac{dy}{ds} \, ds \right) \right\}, \text{ where } \psi_2 = \psi - d\psi \qquad ...(3)$$

Hence the resultant force is from eqns. (2) and (3)

$$T \sin \psi - T \sin (\psi - d\psi) = T \frac{d^2y}{ds^2} \, ds \qquad \qquad ...(4)$$

Since the displacement is small and ψ and $(\psi - d\psi)$ are small,

$$\sin \psi = \psi = \tan \psi = \frac{d\psi}{dx}$$

$$\sin (\psi - d\psi) = \tan (\psi - d\psi) = \frac{dy}{dx} - \frac{d^2y}{dx^2} \, dx$$

and we may write $\dfrac{dy}{dx}$ instead of $\dfrac{dy}{ds}$ in the expression (3). Hence, the resultant force on the element AB is written as

$$T \frac{d^2y}{dx^2} \, dx \qquad \qquad ...(5)$$

Equating the rate of increase of momentum from eqns. (1) and (5), we have

$$\rho \frac{d^2y}{dt^2} = T \frac{d^2y}{dx^2}$$

$$\frac{d^2y}{dx^2} = \frac{T}{\rho} \frac{d^2y}{dx^2} \qquad \qquad ...(6)$$

for the equation of motion of the string.

The general solution of such wave equation

$$y = f(ct - x) + F(ct + x) \qquad \qquad ...(7)$$

where c stands for the velocity of propagation of transverse wave along the string.

Substituting the value of $\dfrac{d^2y}{dt^2}$ and $\dfrac{d^2y}{dx^2}$ from eqn. (7) in eqn. (6)

$$c^2 \left(f'' + F''\right) = \frac{T}{\rho} \left(f'' + F''\right)$$

or, $$c = \sqrt{\frac{T}{\rho}} \qquad \qquad ...(8)$$

Hence eqn. of motion is

$$\frac{d^2 y}{dt^2} = c^2 \frac{d^2 y}{dx^2} \qquad \qquad ...(9)$$

Transverse vibration in stretched strings

A string is a chord, of any material, whose length is very large as compared to its diameter. Let a string be stretched between two points X and Y. The equilibrium position of the string is shown in Fig. 5.4 (i). Let the string be pulled to a position (ii). Tension acting in the string, in this position, has a resultant in the upward direction.

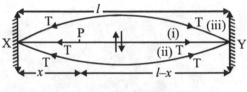

Fig. 5.4

As soon as the string is released it gets pulled upward. On reaching the mean position it acquires a velocity in the upward direction and overshoots the mark. Now tension acting in the string has a downward resultant due to which its velocity decreases and comes to rest in position (iii). Downward resultant pulls the string downward. As it reaches the mean position, it acquires a velocity and overshoots in downward direction. This way it keeps on vibrating continuously. The vibrations are called transverse vibrations in stretched strings. Fig. 5.4 shows a string vibrating in one segment. In fact, it is a case of setting up of stationary waves, obtained due to superimposition of two similar waves travelling, through the string with same speeds in opposite directions. Let 'P' be a point situated at a distance 'x' from X and '*l* – *x*' from Y where '*l*' is the length of the string. At any instant '*p*' is acted upon by two waves, one incident and the other reflected.

Considering '*x*' to be the origin of disturbance, equation of incident wave can be written as,

$$y_1 = a \sin \frac{2\pi}{\lambda} (ct - x)$$

Equation of reflected wave is,

$$y_2 = -a \sin \frac{2\pi}{\lambda} (ct - x')$$

where x' is the distance travelled by the reflected wave in reaching P. Amplitude is changed from '*a*' to '*–a*', assuming that no loss of energy takes place on reflection.

$$x' = l + (l - x) = 2l - x$$

or,
$$y_2 = -a \sin \frac{2\pi}{\lambda}\left[ct - (2l - x)\right]$$

As the wave interfere, net displacement 'y' is

$$y = y_1 + y_2$$

∴
$$y = a\sin\frac{2\pi}{\lambda}(ct - x) - a\sin\frac{2\pi}{\lambda}\left[ct - (2l - x)\right]$$

Since, $\sin A - \sin B = 2\sin\left(\dfrac{A - B}{2}\right)\cos\left(\dfrac{A + B}{2}\right)$

∴
$$y = 2a\sin\left[\frac{2\pi}{\lambda}\frac{(ct - x - ct + 2l - x)}{2}\right] \times \cos\left[\frac{2\pi}{\lambda}\left(\frac{ct - x + ct - 2l + x}{2}\right)\right]$$

or,
$$y = 2a\sin\frac{2\pi}{\lambda}(l - x)\cos\frac{2\pi}{\lambda}(ct - l) \qquad \qquad ...(1)$$

Since the string is rigidly clamped at 'X' and 'Y', the resultant amplitude of vibrations at X and Y should always be zero, i.e. ends X and Y should be nodes. At Y,
$$x = L$$

∴
$$y = 0 \qquad \qquad \left[\because \sin\frac{2\pi}{\lambda}(l - x) = \sin 0 = 0\right]$$

at X, $\qquad \qquad x = 0$

'y' as given by eqn. (1) can only vanish if,

$$\sin\frac{2\pi l}{\lambda} = 0 = \sin(k\pi)$$

where $k = 0, 1, 2, 3, ...$

∴
$$\frac{2\pi l}{\lambda} = k\pi$$

or,
$$l = \frac{k\lambda}{2}$$

Case I: When $k = 1$, $\lambda = \lambda_0$

$$l = \frac{\lambda_0}{2} \text{ or } \lambda_0 = 2l$$

where 'y_0' is the wavelength of wave in the string.

But $\dfrac{\lambda_0}{2}$ is the distance between the consecutive nodes. Therefore, two consecutive nodes must be apart a distance equal to the length of string, i.e. the string should vibrate in one segment as shown in Fig. 5.4. This mode of vibration of the string is said to be the fundamental mode of vibrations. The frequency f_0 of the string vibrating in this mode is called fundamental frequency or first harmonic. It is given by

$$f_0 = \frac{c}{\lambda_0}$$

Since velocity 'c' of the transverse wave is,

$$c = \sqrt{\frac{T}{m}}$$

where T is the tension in the string and m is its mass per unit length, substituting for 'c' and λ

$$f_0 = \frac{1}{2l}\sqrt{\frac{T}{m}} \qquad \qquad ...(2)$$

Case II: When $\qquad k = 2, \lambda = \lambda_1$

$$\lambda = 2\frac{\lambda_1}{2}$$

where λ_1 is the wavelength of wave.

$$\therefore \qquad \frac{\lambda_1}{2} = \frac{l}{2}$$

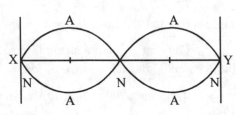

Fig. 5.5

i.e., the distance between two consecutive nodes is $\frac{l}{2}$. So the string must vibrate in two segments as shown in Fig. 5.5. In this mode the string is said to be vibrating with first overtone or second harmonic. If its frequency of vibration is f_1.

$$f_1 = \frac{c}{\lambda_1}$$

i.e., $\qquad\qquad f_1 = \frac{1}{l}\sqrt{\frac{T}{m}}$

or, $\qquad\qquad f_1 = \frac{2}{2l}\sqrt{\frac{T}{m}}$

or, $\qquad\qquad f_1 = 2f_0$

i.e., first overtone is two times the fundamental frequency.

Case III:

When $\qquad k = 3, \lambda = \lambda_2$

$$l = 3\frac{\lambda_2}{2}$$

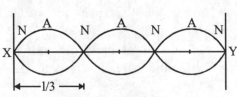

Fig. 5.6

or, $\qquad \frac{\lambda_2}{2} = \frac{l}{3}$ or $\lambda_2 = \frac{2l}{3}$

when λ_2 is the wavelength of sound in the string in this case. The string vibrates in three loops as shown on Fig. 5.6. The string is said to be vibrating with second overtone or third harmonic. If 'f_2' is its frequency

$$f_2 = \frac{c}{\lambda_2}$$

$$= \frac{1}{2l/3} \sqrt{\frac{T}{m}}$$

or,

$$f_2 = \frac{3}{2l} \sqrt{\frac{T}{m}}$$

or,

$$f_2 = 3f_0$$

i.e., the second overtone is three times the fundamental frequency.

Thus, in general, a string is capable of vibrating in a number of loops, depending upon the tension T and mass per unit length 'm'.

If p = number of loops into which a string vibrates.

$$f_{p-1} = \frac{p}{2l} \sqrt{\frac{T}{m}}$$

The f_{p-1} is the frequency of $(p-1)$th overtone or pth harmonic.

Laws of Transverse Vibrations in Stretched Strings

Statements telling 'how' and 'which' are factors affecting frequency of vibrations of a stretched string are termed as laws of transverse vibrations in stretched strings.

From eqn. (2) fundamental frequency of a vibrating string is

$$f = \frac{1}{2l} \sqrt{\frac{T}{m}} \qquad \qquad ...(3)$$

(i) Law of length: It states that the frequency of vibrations of a stretched string varies inversely as its length 'l', i.e.

$$f \propto \frac{1}{l}$$

(ii) Law of tension: It states that the frequency of vibrations of a stretched string varies directly as the square root of the tension 'T' in the string, i.e.

$$f \propto \sqrt{T}$$

(iii) Law of mass per unit length: It states that the frequency of vibrations of a stretched string varies inversely as the square root of its mass per unit length m, i.e.

$$f \propto \frac{1}{\sqrt{m}}$$

Combining all the factors together,

$$f \propto \frac{1}{l}\sqrt{\frac{T}{m}}$$

or,

$$f = \frac{k}{l}\sqrt{\frac{T}{m}}$$

Experimentally, it has been confirmed that value of k, when the string vibrates with fundamental frequency is $\frac{1}{2}$.

$$f = \frac{1}{2l}\sqrt{\frac{T}{m}}$$

Let us consider a string of unit length ($l = 1$) having density 'ρ' and diameter 'D' (Fig. 5.7).

Area of cross section

$$a = \frac{\pi D^2}{4}$$

Volume of string, $V = a \times l = \dfrac{\pi D^2}{4} \times 1 = \dfrac{\pi D^2}{4}$

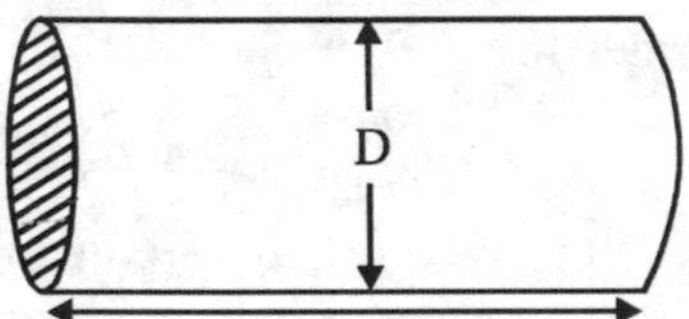

$\therefore$ Mass per unit length, $m = V \times \rho$

or,

$$m = \frac{\pi D^2}{4}\rho$$

Fig. 5.7

Substituting for 'm' in eqn. (5)

$$f = \frac{1}{2l}\sqrt{\frac{T}{\dfrac{\pi D^2}{4}\rho}} = \frac{1}{2l}\sqrt{\frac{4T}{\pi D^2 \rho}}$$

or,

$$f = \frac{1}{2l}\frac{2}{D}\sqrt{\frac{T}{\pi\rho}} \quad \text{or} \quad \boxed{f = \frac{1}{lD}\sqrt{\frac{T}{\pi\rho}}}$$

Thus, law of mass per unit length can broken into two sub-laws:

(i) The law of diameter: It states that the frequency of vibrations of a stretched string varies inversely as the diameter 'D' of the string, i.e.

$$f \propto \frac{1}{D}$$

(ii) The law of density: It states that the frequency of vibrations of a stretched string varies inversely as the square root of the density of the material of the string, i.e.

$$f \propto \frac{1}{\sqrt{\rho}}$$

where, 'ρ' is the density of the string.

Experimental Verification of the Laws

Laws of vibrations of stretched strings can be verified by means of 'sonometer' shown in Fig. 5.8. It consists of a hollow wooden box 'B'. A string has one of its ends connected to a rigid support s, passes over a pulley and has a hanger suspended from its other end. Two wedges ω_1 and ω_2 are placed below the wire and a small paper rider 'R' is placed over the wire.

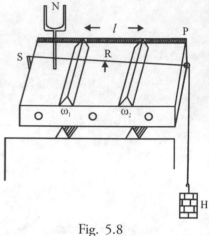

Fig. 5.8

A vibrating tuning fork of frequency 'n' has its own stem pressed against the board. Vibrations from the tuning fork impressed upon the wire. The experimental length of the wire can be changed by adjusting the wedges. A particular length of wire vibrates in resonance with the vibrating tuning fork. This is indicated when the paper rider flies off.

(i) Law of length ('T' and 'm' kept constant):

Take tuning fork of frequency say 258. Find the length 'l' of the string which vibrates in resonance with the tuning fork. Replace the tuning fork with another one of frequency 512. It will be observed that the length of same wire which vibrates in resonance with tuning fork

of frequency 512 is '$\dfrac{l}{2}$', indicating that

$$f \propto \frac{1}{l}$$

(ii) Law of tension (keeping 'l' and 'm' constant):

Let 'l' be the length of wire which vibrates in resonance with a tuning fork of frequency 256 when tension in the string is T. Increase the tension of '4T'. It will be observed that the same length of wire now vibrates in resonance with tuning fork of frequency 512, indicating that

$$f \propto \sqrt{T}$$

(iii) Law of mass per unit length (keeping 'l' and 'T' constant):

Let 'l' be the length, 'm' be the mass and tension T, which vibrates in resonance with a tuning fork of frequency 256. Replace the wire with another one of mass per unit length $\dfrac{'m'}{4}$ and put same tension 'T' upon it. It will be observed that length 'l' of this wire vibrates in resonance with a tuning fork of frequency 512, indicating that

$$f \propto \frac{1}{\sqrt{m}}.$$

Experimental Determination of Frequency of a Tuning Fork

Frequency of a tuning fork can also be determined by means of a sonometer. Strike the tuning fork gently against a rubber pad and press its stem against board of the sonometer. Adjust the wedges till the paper rider flies off. Note the length '*l*' known length (say 50 cm) of wire and find its mass using a balance. If 'M' is the mass put on the hanger, the tension, T is

$$T = Mg$$

substituting for T, *l* and *m*, we get frequency '*f*' of vibration of string

$$f = \frac{1}{2l}\sqrt{\frac{Mg}{m}}$$

Since the string vibrates in resonance with tuning fork of unknown frequency *f*,

$$\therefore \quad f' = f$$

$$\therefore \quad f' = \frac{1}{2l}\sqrt{\frac{Mg}{m}}$$

Superposition of Waves

When two or more waves propagate simultaneously in the same medium, each wave travels through the medium as though the other waves are not present. In the sense of propagation through the medium neither wave affects the progress of the other. However, when two set of waves are made to cross each other, as for example, the waves created by dropping two stones simultaneously in a quiet pool, very interesting and complicated effects are observed. In the region of crossing there are places where the disturbance is less or more than that due to either wave alone. A very simple law is used to explain these effects which is known as the principle of superposition.

Principle of superposition: The resultant displacement of a particle of the medium at any instant of time is the vector sum of the displacements due to each wave separately.

Different phenomena of superposition of waves

1. Two waves of the same frequency moving in the same direction – Interference of waves.

2. Two waves of slightly different frequencies moving in the same direction – Beats.

3. Two waves of the same frequency moving in the opposite direction – Stationary waves.

Interference of waves: When two waves of same frequency moving in the same direction overlap with each other, interference occurs. Consequently a resultant wave is obtained.

At some points, the crests or the troughs of two waves arrive simultaneously and combine to produce greater amplitude. The energy becomes maximum at these points. At some other points the crests of one wave fall upon the troughs of the other wave and vice-versa and combine to produce zero amplitude. The energy at these points becomes minimum. The points where the waves arrive in phase, the energy becomes maximum and the points where the waves arrive out of phase, the energy becomes minimum. Thus, there is a modification in the distribution of energy due to superposition of waves.

This modification in the distribution of energy due to superposition of two or more waves of same phase or of constant phase difference travelling simultaneously in the same medium is called interference.

There are 2 types of interference, viz.

(a) Constructive interference, and

(b) Destructive interference.

(a) Constructive interference: The interference in which the amplitude are added or the waves arrive in phase, is called constructive interference. Here the energy becomes maximum.

(b) Destructive interference: The interference in which the waves arrive out of phase to produce zero resultant amplitude is called destructive interference. The energy becomes minimum in this case.

Demonstration of interference: A simple method of demonstrating interference effects in wave motion is by dropping two stones into a quiet pond simultaneously at two neighbouring points as shown in Fig. 5.9. The stones are dropped at points S_1 and S_2 simultaneously so that S_1 and S_2 behave as two sources of disturbances sending out waves in phase. The waves propagated from the two sources are represented as areas. The solid lines represent the crests and the dotted lines represent the troughs of the water waves. The concentric areas are half a wavelength apart (i.e., the phase difference between two consecutive wave fronts =

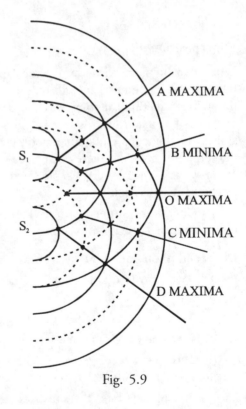

Fig. 5.9

$\dfrac{2\pi}{\lambda} \times \dfrac{\lambda}{2} = \pi$ radian). The crests can be compared to areas of maximum rarefaction in case of sound waves that are longitudinal. The region where the two crests or two troughs coincide, the interference is constructive (maximum) and the region where crest of one coincides with the trough of the other and vice-versa, the interference is destructive (minima). Along the lines indicated by A, O, D, the two waves meet in phase and we get maxima whereas along the lines B, C, the two waves meet 180° out of phase and we get minima.

Analytical treatment of interference: Let us consider 2 simple harmonic waves having same frequency, same wavelength and travelling in same direction with same velocity. These waves are represented by eqns:

$$y_1 = a_1 \sin \omega t \qquad \qquad \text{...(1)}$$

$$y_2 = a_2 \sin \omega t \qquad \qquad \text{...(2)}$$

i.e., to start with, the waves are of equal wavelength 'λ', equal phases 'ωt' and amplitudes a_1 and a_2. But on reaching the point P, a phase difference of

$$\phi = \frac{2\pi}{\lambda} \times (S_2 P - S_1 P) \text{ is introduced.}$$

$(S_2 P - S_1 P)$ = path difference = p.

So, one of the displacements, say y_2 on reaching P is given by

$$y_2 = a_2 \sin (\omega t + \phi) \qquad \qquad \text{...(3)}$$

Hence, the resultant displacement at P is given by

$$
\begin{aligned}
y = y_1 + y_2 &= a_1 \sin \omega t + a_2 (\omega t + \phi) \\
&= a_1 \sin \omega t + a_2 \sin \omega t \cos \phi + a_2 \cos \omega t \sin \phi \\
&= \sin \omega t \, (a_1 + a_2 \cos \varphi) + \cos \omega t \, (a_2 \sin \phi) \\
&= \sin \omega t \, (A \cos \theta) + \cos \omega t \, (A \sin \theta) \\
&= A \sin (\omega t + \theta) \qquad \qquad \text{...(4)}
\end{aligned}
$$

where we have substituted

$$A \cos \theta = a_1 + a_2 \cos \phi$$

and $\qquad A \sin \theta = a_2 \sin \phi.$

$$A^2 = a_1^2 + a_2^2 \cos^2 \phi + 2a_1 a_2 \cos \phi + a_2^2 \sin^2 \phi$$

or, $\qquad A = \sqrt{a_1^2 + a_2^2 \, 2a_1 a_2 \cos \phi} \qquad \qquad \text{...(5)}$

and $\qquad \tan \theta = \dfrac{a_2 \sin \phi}{a_1 + a_2 \cos \phi} \qquad \qquad \text{...(6)}$

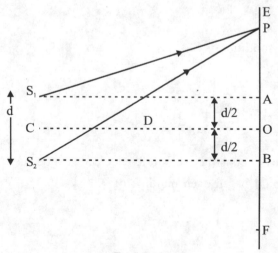

Fig. 5.10

So, we find that the component displacements have amplitudes α_1 and α_2 whereas the resultant displacement has the amplitude A. The components are of equal phase 'ωt' but the phase of resultant differs from the components by 'θ'.

Special Cases

Case I: If $\phi = 0, 2\pi, 4\pi, ...$

i.e., in general when $\phi = n \times 2\pi$

where $\qquad n = 0, 1, 2, 3, ...$

$$\cos \phi = +1$$

$$A = \sqrt{a_1^2 + a_2^2 + 2a_1 a_2 \cos 2n\pi}$$

$$= \sqrt{a_1^2 + a_2^2 + 2a_1 a_2}$$

$$= \sqrt{(a_1 + a_2)^2}$$

or, $\qquad A = a_1 + a_2$

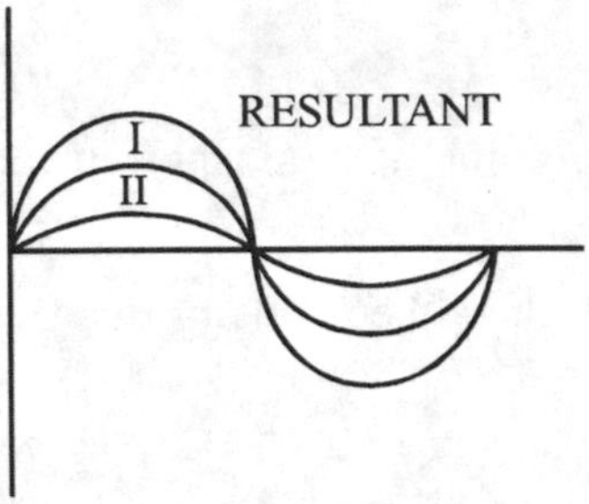

Fig. 5.11

i.e., net amplitude is equal to the sum of individual amplitudes. This type of interference is called constructive interference. In this case crest of one wave coincides with the crest of the other. The intensity of a point where this type of interference takes place is maximum and the point is known as maxima.

Thus, condition for constructive interference is that

$$\phi = 2n\pi \text{ where } n = 0, 1, 2, 3, ...$$

i.e., the phase difference between two waves should be an even multiple of π radian.

If a phase difference of 'ϕ' is equivalent to a path difference 'p' then

$$\phi = \frac{2\pi}{\lambda} p$$

$\because$ The required condition is

$$\frac{2\pi}{\lambda} p = 2 n\pi$$

or, $\qquad \boxed{p = 2n.\frac{\lambda}{2}}$ $\qquad$ ...(7)

So, the condition for constructive interference can also be stated as 'the path difference between two waves should be an even multiple of $\lambda/2$'.

Case II: If $\phi = \pi, 3\pi, 5\pi, ...$

i.e., in general of $\phi = (2n + 1)\pi$

where $\qquad n = 0, 1, 2, 3, ...$

$$\cos \phi = -1$$

$\therefore$
$$\Lambda = \sqrt{a_1^2 + a_2^2 + 2a_1a_2(-1)}$$
$$= \sqrt{(a_1 - a_2)^2}$$
$$= a_1 - a_2$$

i.e., the net amplitude is equal to the difference between the individual amplitudes. This type of interference is called destructive interference. In this case, crest of one wave coincides with the trough of the other. The intensity of a point where this type of interference takes place minimum and the point is known as minima.

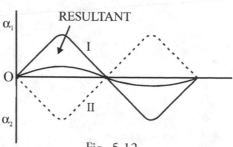

Fig. 5.12

Thus, the condition for 'destructive interference' is
$$\phi = (2n + 1)\pi$$
where $\qquad n = 0, 1, 2, 3, ...$

i.e., the phase difference between two waves should be an odd multiple of π radian.

If p is the equivalent path difference

$$\frac{2\pi}{\lambda} p = (2n+1)\pi$$

or, $\qquad \boxed{p = (2n+1)\frac{\lambda}{2}}$ $\qquad$ ...(8)

So, for destructive interference, the path difference between two waves should be an odd multiple of '$\lambda/2$'.

Expression for intensity: The intensity is directly proportional to the square of the amplitude and hence intensity of maxima is
$$I_{max} \propto (a_1 + a_2)^2 \qquad ...(9)$$
and that of minima is
$$I_{min} \propto (a_1 - a_2)^2 \qquad ...(10)$$

Again $\qquad \dfrac{I_{max}}{I_{min}} = \dfrac{(a_1 + a_2)^2}{(a_1 - a_2)^2} = \dfrac{(r+1)^2}{(r-1)^2} \qquad ...(11)$

where $\qquad r = \dfrac{a_1}{a_2}$

Intensity distribution: If the two waves have equal amplitudes.
$$a_1 = a_2 = a \text{ (say)}$$

Now, we get
$$A = \sqrt{a^2 + a^2 + 2a^2 \cos\phi}$$

$$= \sqrt{2a^2 + 2a^2 \cos \phi}$$

$$= \sqrt{2a^2 (1 + \cos \phi)}$$

$$= \sqrt{2a^2 \times 2 \cos^2 \frac{\phi}{2}}$$

$$= \sqrt{4a^2 \cos^2 \frac{\phi}{2}}$$

or,
$$A^2 = 4a^2 \times \cos^2 \frac{\phi}{2}$$

For maxima,
$$\phi = 2n\pi$$

$\therefore \qquad \cos \frac{\phi}{2} = \cos \frac{2n\pi}{2} = \cos n\pi = \pm 1$

or,
$$\cos^2 \frac{\phi}{2} = 1$$

$\therefore \qquad (A_{max})^2 = 4a^2$

For minima,
$$\phi = (2n + 1) \pi$$

$\therefore \qquad \cos \frac{\phi}{2} = \cos (2n - 1) \frac{\pi}{2} = 0$

$\because \qquad (A_{min})^2 = 0$

Since intensity is proportional to A^2, $I_{max} \times K \times 4a^2$

$\because \qquad I_{min} = 0.$

A graphical representation of variation of amplitude (or intensity) with 'ϕ' is shown in Fig. 5.13.

Fringe width: Now, to find out the fringe width, let us put $S_1S_2 = d$, $S_1A = S_2B = CO = D =$ perpendicular distance between S_1 or S_2 and the screen EF. 'O' is the central point on the screen.

$$S_1O = S_2O \text{ and } AO = BO = \frac{d}{2}$$

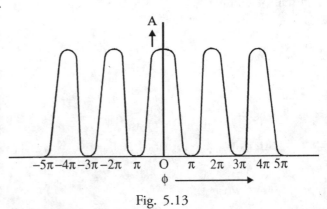

$$-5\pi -4\pi -3\pi -2\pi \quad \pi \qquad O \qquad \pi \quad 2\pi \ 3\pi \ 4\pi \ 5\pi$$

$$\phi \longrightarrow$$

Fig. 5.13

So sound waves starting with equal phase from S_1 and S_2 reach the point O, also in phase. Let OP $= x$, where P is the point of observation on the screen EF (Fig. 5.10).

Now

$$(S_2P)^2 = D^2 + \left(x + \frac{d}{2}\right)^2 \text{ in } \Delta S_2PB$$

$$(S_1P)^2 = D^2 + \left(x - \frac{d}{2}\right)^2 \text{ in } \Delta S_1PA$$

$$(S_2P)^2 - (S_1P)^2 = 4x \times \frac{d}{2} = 2xd$$

or, $$S_2P - S_1P = \frac{2xd}{(S_2P + S_1P)} = \frac{2xd}{2D} = \frac{xd}{D}$$

(Using, $S_2P \cong S_1P = CO = D$, which is true when x is much smaller than D.)
So, path difference,

$$p = S_2P - S_1P = \frac{xd}{D} \qquad \qquad ...(12)$$

For maxima $$p = 2n \times \frac{\lambda}{2} = n\lambda$$

So, $$\frac{xd}{D} = n\lambda, \text{ or, } [x]_{\text{maxima}} = n\left(\frac{\lambda D}{d}\right)$$

For central maxima, $n = 0$, $x_0 = 0$

For 1st order maxima, $n = 1$, $x_1 = \frac{\lambda D}{d}$

For 2nd order maxima, $n = 2$, $x_1 = \frac{2\lambda D}{d}$ and so on.

Hence, fringe width = Fringe separation between successive maxima

$$= \left(\frac{\lambda D}{d} - 0\right) = \left(\frac{2\lambda D}{d} - \frac{\lambda D}{d}\right) = ... = \frac{\lambda D}{d}$$

It can be shown that the fringe separation between successive minima is also equal to $\lambda D/d$. Hence, we find that the separation between successive maxima is same as that between two successive minima, which has been defined as fringe width and is given as:

$$\omega = \frac{\lambda D}{d} \qquad \qquad ...(13)$$

Conditions for interference: Waves must fulfil the following conditions for the interference to take place:

1. The two sources must emit continuous waves having same frequency and wavelength.

2. The amplitudes of waves must either be equal or nearly equal.

3. The two sources should be situated close to each other.

4. The two sources should be coherent one.

Coherent sources are those sources which have either no phase difference between them or a constant difference of phase between them.

Phenomenon of beats: When two sets of waves of slightly different frequencies are superposed, the amplitude of the resultant wave varies from maximum to minimum with time.

The periodic variations of the intensity of wave resulting from the superposition of two waves of slightly different frequencies is known as phenomenon of beats.

One maxima and a following minima combined together is called one beat. If f is number of beats per second, it can be seen that

$$f = f_1 - f_2$$

where 'f_1' and 'f_2' are the frequencies of two waves producing beats.

Analytical treatment: Let us consider two waves of frequencies f_1 and f_2 having same amplitude and phase travelling through a medium in same direction. These waves can be represented by equation

$$y_1 = a \sin 2\pi f_1 t \qquad \qquad ...(1)$$
$$y_2 = a \sin 2\pi f_2 t \qquad \qquad ...(2)$$

If 'y' is the net displacement of particle vibrating under the simultaneous effect of both of them, according to principle of superposition.

$$y = y_1 + y_2$$

or,
$$y = a \sin 2\pi f_1 t + a \sin 2\pi f_2 t$$

$$= a\left[\sin 2\pi f_1 t + \sin 2\pi f_2 t\right]$$

$$= 2a \sin 2\pi\left(\frac{f_1 + f_2}{2}\right)t \cos 2\pi\left(\frac{f_1 - f_2}{2}\right)t$$

or,
$$y = 2a \cos 2\pi\left(\frac{f_1 - f_2}{2}\right)t \sin 2\pi\left(\frac{f_1 + f_2}{2}\right)t \qquad \qquad ...(3)$$

Eqn. (3) is the eqn. of a simple-harmonic wave having amplitude

$$A = 2a \cos 2\pi\left(\frac{f_1 - f_2}{2}\right)t \qquad \qquad ...(4)$$

and frequency

$$f' = \left(\frac{f_1 + f_2}{2}\right) \qquad \qquad ...(5)$$

Presence of 't' in eqn. (4) indicates that the resultant amplitude A does not remain constant but oscillates between maximum and zero (due to 'cos' factor) with time.

(i) Maxima: Intensity of resultant wave is maximum

when, $\cos 2\pi \left(\dfrac{f_1 - f_2}{2} \right) t = \pm 1 = \cos(n\pi)$

where, $\qquad n = 0, 1, 2, 3, ...$

$\therefore \qquad 2\pi \left(\dfrac{f_1 - f_2}{2} \right) t = n\pi$

or, $\qquad t = \dfrac{n}{f_1 - f_2}$

Thus maximas will be obtained at times,

$$0, \frac{1}{f_1 - f_2}, \frac{2}{f_1 - f_2}, \frac{3}{f_1 - f_2}, ...$$

(ii) Minima: Intensity of resultant wave is minimum when,

$\cos 2\pi \left(\dfrac{f_1 - f_2}{2} \right) t = 0 = \cos(2n+1)\dfrac{\pi}{2}$

where $\qquad n = 0, 1, 2, 3, ...$

$\therefore \qquad 2\pi \left(\dfrac{f_1 - f_2}{2} \right) t = (2n+1)\dfrac{\pi}{2}$

or, $\qquad t = \dfrac{2n+1}{2(f_1 - f_2)}$

Thus, minimas are obtained at times,

$$\frac{1}{2(f_1 - f_2)}, \frac{3}{2(f_1 - f_2)}, \frac{5}{2(f_1 - f_2)}, ...$$

If t_b = beat period.

Beat period is defined as time interval between two consecutive beats or it is the time between two consecutive maximas or minimas. It can be easily seen that

$$t_b = \frac{1}{f_1 - f_2}$$

If f = number of beats per second
(i.e., frequency of beats)

$$f = \frac{1}{\text{beat period}} = \frac{1}{\dfrac{1}{f_1 - f_2}}$$

or, $\qquad f = f_1 - f_2$

i.e., the number of beats per second is equal to the difference in frequencies of two waves.

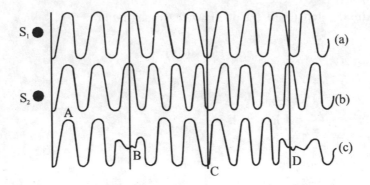

Fig. 5.14: Formation of beats 'Graphical method'

Stationary waves: If two sinusoidal waves of the same amplitude and frequency travel in opposite directions through a medium, the two waves will be superposed in such a manner that stationary waves will be formed. These waves are also called standing waves.

Formation of stationary waves: When one incident wave and a reflected wave of same amplitude and frequency travelling through a medium overlap with each other, stationary waves are formed.

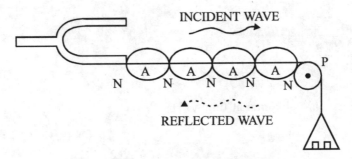

Fig. 5.15: Formation of stationary waves

Reflection of waves occurs either from a fixed boundary or from a free boundary.

(a) Conditions for a fixed boundary:

  (i) There will be a phase change of 180°.

  (ii) Nodes are formed at the fixed boundary. Nodes are points of zero amplitude.

  (iii) The reflected wave is inverted at the fixed boundary.

(b) Conditions for a free boundary:

  (i) There will be no change of phase.

  (ii) Antinodes are formed at the free boundary. Antinodes are points of maximum amplitude.

  (iii) The reflected wave is erect.

Analytical Treatment

Case I. Reflection from a free boundary: Let us consider a progressive wave travelling in the positive direction of X-axis. Displacement of any particle at a distance a (from origin) and at an instant t is given by

$$y_1 = a \sin \frac{2\pi}{\lambda} (vt - x) \qquad \qquad ...(1)$$

A reflected wave travels along negative direction of X-axis and has all other characteristics similar to incident one. Displacement 'y_2', due to this wave, at same point and at same instant, can be

$$y_2 = a \sin \frac{2\pi}{\lambda} (vt + x) \qquad \qquad ...(2)$$

As the two waves superimpose each other, net displacement 'y', according to the principle of superposition is

$$y = y_1 + y_2$$

$$y = a \sin \frac{2\pi}{\lambda} (vt - x) + a \sin \frac{2\pi}{\lambda} (vt + x)$$

$$= a \left[\sin \frac{2\pi}{\lambda} (vt - x) + \sin \frac{2\pi}{\lambda} (vt + x) \right]$$

$$y = 2a \sin \frac{2\pi}{\lambda} vt \cos \frac{2\pi}{\lambda} x$$

or, $\qquad \qquad y = 2a \cos \frac{2\pi}{\lambda} x \sin \frac{2\pi}{\lambda} vt \qquad \qquad ...(3)$

A close scrutiny of equation (3) reveals that it represents SHM of amplitude 'A' given by

$$A = 2a \cos \frac{2\pi}{\lambda} x$$

Evidently, 'A' depends upon 'x', i.e. the spatial positions of the points. Since '$\cos \frac{2\pi}{\lambda} x$' is a periodic function varying between $+ 2a$, 0 and $- 2a$. Amplitude, at any point, is independent of time.

For antinodes: All the points where

$$\cos \frac{2\pi}{\lambda} x = \pm 1,$$

the amplitude is maximum. These points of 'maximum amplitude' are known as antinodes.

Hence, $\cos\dfrac{2\pi}{\lambda} = \pm\,1 = \cos\,(n\pi)$

where $n = 0,\ 1,\ 2,\ 3,\ ....$

$\therefore$ $\dfrac{2\pi}{\lambda}x = n\pi$

or, $x = n\dfrac{\lambda}{2}$

Thus the amplitudes are formed at distances,

$$0,\dfrac{\lambda}{2},2\dfrac{\lambda}{2},3\dfrac{\lambda}{2},....\ \text{from the origin.}$$

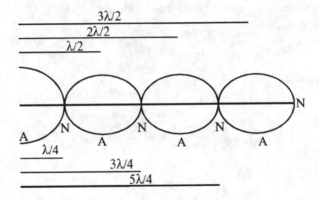

Fig. 5.16: Formation of nodes and antinodes

For nodes: All the points where

$$\cos\dfrac{2\pi}{\lambda}x = 0,$$

the amplitude is minimum. These points of 'minimum amplitude' are known as nodes.

Hence, $\cos\dfrac{2\pi}{\lambda}x = 0 = \cos\,(2n+1)\dfrac{\pi}{2}$

where, $n = 0,\ 1,\ 2,\ 3,\ ....$

$\dfrac{2\pi}{\lambda}x = (2n+1)\dfrac{\pi}{2}$

or, $x = (2n+1)\dfrac{\lambda}{4}$

Nodes are formed at distances

$$\dfrac{\lambda}{4},\dfrac{3\lambda}{4},\dfrac{5\lambda}{4},....\ \text{from the origin.}$$

Case II. Reflection from a fixed boundary: Let us consider a progressive wave travelling along the +ve direction of X. The displacement of any particle at a distance X from the origin and at an instant 't' is given by

$$y_1 = a \sin \frac{2\pi}{\lambda} (vt - x) \qquad \qquad ...(4)$$

The reflected wave from the rigid boundary is given by

$$y_2 = a \sin \left\{ \frac{2\pi}{\lambda} (vt + x) + \pi \right\} \qquad \qquad ...(5)$$

because by Stokes' theorem reflected wave reflected from a denser medium or backed by a denser medium carries an extra phase π.

From eqn. (5), we have

$$y_2 = - a \sin \frac{2\pi}{\lambda} (vt - x) \qquad \qquad ...(6)$$

Applying principle of superposition, the resultant displacement is given by

$$y = y_1 + y_2$$

$$= a \sin \frac{2\pi}{\lambda} (vt - x) - a \sin \frac{2\pi}{\lambda} (vt + x)$$

$$= a \left[\sin \frac{2\pi}{\lambda} (vt - x) - \sin \frac{2\pi}{\lambda} (vt + x) \right]$$

$$= a \left[\sin \frac{2\pi vt}{\lambda} \cos \frac{2\pi x}{\lambda} - \cos \frac{2\pi vt}{\lambda} \sin \frac{2\pi x}{\lambda} \right.$$
$$\left. - \sin \frac{2\pi vt}{\lambda} \cos \frac{2\pi x}{\lambda} - \cos \frac{2\pi vt}{\lambda} \sin \frac{2\pi x}{\lambda} \right]$$

$$= - 2a \cos \frac{2\pi vt}{\lambda} \sin \frac{2\pi x}{\lambda}$$

$$= - 2a \sin \frac{2\pi x}{\lambda} \cos \frac{2\pi vt}{\lambda}$$

$$= A \cos 2\pi ft$$
$$= A \cos \omega t \qquad \qquad ...(7)$$

where the amplitude of the resultant wave is 'A'.

$$A = - 2a \sin \frac{2\pi x}{\lambda} \qquad \qquad ...(8)$$

Evidently, 'A' depends upon x. The value of amplitude varies between $+ 2a$, 0 and $- 2a$. But amplitude at any point is independent of time.

(i) For antinodes: When, $\sin \frac{2\pi x}{\lambda} = \pm 1$, the amplitude is maximum being equal to $\mp 2a$.

These points of maximum amplitude are called antinodes. Hence,

$$\sin \frac{2\pi x}{\lambda} = \pm\, 1 = \sin (2n+1)\frac{\pi}{2}$$

where, $n = 0, 1, 2, 3, ...$

or, $\dfrac{2\pi x}{2} = (2n+1)\dfrac{\pi}{2}$

or, $x = (2n+1)\dfrac{\lambda}{4}$ ...(9)

Thus, the distance between two consecutive antinodes is

$$\frac{3\lambda}{4} - \frac{\lambda}{4} = \frac{2\lambda}{4} = \frac{\lambda}{2}.$$

(ii) For nodes: When $\sin \dfrac{2\pi x}{\lambda} = 0$, the amplitude is minimum, i.e. zero. These points of minimum amplitude are called nodes. Hence,

$$\sin \frac{2\pi x}{\lambda} = 0 = \sin n\pi$$

where, $n = 0, 1, 2, 3, ...$

or, $\dfrac{2\pi x}{\lambda} = n\pi$

or, $x = \dfrac{n\lambda}{2}$ ...(10)

Thus, the distance between two consecutive nodes is

$$\frac{\lambda}{2} - 0 = \frac{\lambda}{2}$$

Analytical Treatment of Stationary Waves Formed in an Open Organ Pipe and in Closed Organ Pipe

Stationary waves are formed when an incident wave and a reflected wave of same amplitude and frequency overlap with each other. The reflection of a wave occurs either from a free boundary or from a fixed boundary. An open organ pipe has a free boundary whereas a closed organ pipe (closed at one end, open at the other) serves the case of fixed boundary.

Analytical Treatment

Case I. Open organ pipe: Let us consider a progressive wave travelling in the positive direction of X-axis in an open organ pipe. Displacement of any particle at a distance x (from origin) and at an instant t is given by

$$y_1 = a \sin \frac{2\pi}{\lambda}(vt - x)$$ ...(1)

A reflected wave travels along –ve direction of X-axis and has all other characteristics similar to incident one. Displacement 'y_2', due to this wave, at same point and at same instant, can be

$$y_2 = a \sin \frac{2\pi}{\lambda} (vt + x) \qquad \qquad \text{...(2)}$$

As the two waves superimpose each other, net displacement 'y', according to the principle of superposition is

$$y = y_1 + y_2$$

$$y = a \sin \frac{2\pi}{\lambda} (vt - x) + a \sin \frac{2\pi}{\lambda} (vt + x)$$

$$= a \left[\sin \frac{2\pi}{\lambda} (vt - x) + \sin \frac{2\pi}{\lambda} (vt + x) \right]$$

$$= 2a \sin \frac{2\pi}{\lambda} vt \cos \frac{2\pi}{\lambda} x$$

or, $$y = 2a \cos \frac{2\pi}{\lambda} x \sin \frac{2\pi}{\lambda} vt \qquad \qquad \text{...(3)}$$

A close scrutiny of equation (3) reveals that it represents SHM of amplitude 'A' given by

$$A = 2a \cos \frac{2\pi}{\lambda} x$$

Evidently, 'A' depends upon 'x', i.e. the spatial positions of the points. Since '$\cos \frac{2\pi}{\lambda} x$' is a periodic function varying between $+ 2a$, 0 and $- 2a$. Amplitude, at any point, is independent of time.

For antinodes: All the points where

$$\cos \frac{2\pi}{\lambda} x = \pm 1,$$

the amplitude is maximum. These points of 'maximum amplitude' are known as antinodes.

Hence, $$\cos \frac{2\pi}{\lambda} = \pm 1 = \cos (n\pi)$$

where, $$n = 0, 1, 2, 3, ...$$

$$\therefore \qquad \frac{2\pi}{\lambda} x = n\pi$$

or, $$x = n \frac{\lambda}{2}$$

Thus the amplitudes are formed at distances,

$0, \dfrac{\lambda}{2}, 2\dfrac{\lambda}{2}, 3\dfrac{\lambda}{2}, ...$ from the origin.

For nodes: All the points where

$$\cos \dfrac{2\pi}{\lambda} x = 0,$$

the amplitude is minimum. These points of 'minimum amplitude' are known as nodes.

Hence, $\cos \dfrac{2\pi}{\lambda} x = 0 = \cos (2n + 1) \dfrac{\pi}{2}$

where, $n = 0, 1, 2, 3, ...$

$\therefore$ $\dfrac{2\pi}{\lambda} x = (2n + 1) \dfrac{\pi}{2}$

or, $x = (2n + 1) \dfrac{\lambda}{4}$

Nodes are formed at distances,

$\dfrac{\lambda}{4}, \dfrac{3\lambda}{4}, \dfrac{5\lambda}{4}, ...$ from the origin.

Harmonics: In an open pipe when a compression reaches the far end, a rarefied wave reflects back to the mouth (initial end). This rarefied wave is again reflected back as a compression wave to the other end. This is because both the ends are open and the air has utmost freedom of vibration. Within the tube the reflected pulses superpose as a result of which longitudinal waves are formed having antinodes and nodes at definite intervals. At the open ends of the tube, antinodes are formed because air at the open ends vibrate with maximum amplitudes and nodes are formed in between where the amplitude of vibration is minimum (may be zero).

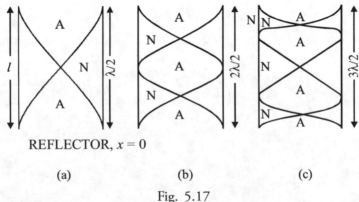

Fig. 5.17

Let us consider (Fig. 5.17a) in which two antinodes (A) are formed at $x = 0$ and $x = \dfrac{\lambda}{2}$ and a node (N) is formed at $x = \dfrac{\lambda}{4}$. For this

$$v = f_0\lambda = f_0 \times 2L \qquad\qquad \left(\because L = \dfrac{\lambda}{2}\right)$$

$$\Rightarrow \quad f_0 = \dfrac{v}{2L}$$

which is called fundamental or first harmonic.

In (Fig. 5.17b), three antinodes are formed at $x = 0, \dfrac{\lambda}{2}, \dfrac{2\lambda}{2}$ and two nodes are formed at $x = \dfrac{\lambda}{4}, \dfrac{3\lambda}{4}$.

For this, $V = f_1\lambda = f_1 \times L \qquad\qquad \left(\because L = 2\dfrac{\lambda}{2}\right)$

$$\Rightarrow \quad f_1 = \dfrac{V}{L} = \dfrac{2V}{2L} = 2\left(\dfrac{V}{2L}\right) = 2f_0$$

This is called harmonic.

In (Fig. 5.17c), four antinodes are formed at $x = 0, \dfrac{\lambda}{2}, \dfrac{2\lambda}{2}, \dfrac{3\lambda}{2}$ and three nodes are formed at $x = \dfrac{\lambda}{4}, \dfrac{3\lambda}{4}$ and $\dfrac{5\lambda}{4}$.

For this, $\qquad V = f_2\lambda = f_2 \times \dfrac{2L}{3} \qquad\qquad \left(\because L = \dfrac{3\lambda}{2}\right)$

so that, $\qquad f_2 = \dfrac{3V}{2L} = 3\left(\dfrac{V}{2L}\right) = 3f_0$

The frequency f_0 is called as fundamental and the other frequencies f_1, f_2, etc. are called overtones (harmonics). The ratio of these frequencies is given by

$$f_0 : f_1 : f_2 = 1 : 2 : 3.$$

That means in case of open pipe both odd and even harmonics are present. (When the frequencies of the overtones are exact multiples of the frequency of the fundamental they are called as harmonics.)

Table 1 below gives a clear idea about the harmonics produced in an open pipe.

Table-1
Harmonics of an open pipe of length 'L'

No.	Name	Wavelength in air	Frequency of the tone produced	Relation with the fundamental frequency
1.	1st harmonic	$2L$	$f_0 = \dfrac{V}{2L}$	Equal to fundamental
2.	2nd harmonic	$\dfrac{2L}{2}$	$f_1 = \dfrac{2V}{2L}$	$f_1 = 2f_0$
3.	3rd harmonic	$\dfrac{3L}{3}$	$f_2 = \dfrac{3V}{2L}$	$f_2 = 3f_0$
etc.	etc.	etc.	etc.	etc.

Case II. Closed organ pipe: Let us consider a progressive wave travelling along the +ve direction of X in a fixed boundary. The displacement of any particle at a distance X from the origin and at an instant 't' is given by

$$y_1 = a \sin \frac{2\pi}{\lambda} (vt - x) \qquad \qquad ...(4)$$

The reflected wave from the rigid boundary is given by

$$y_2 = a \sin \left\{ \frac{2\pi}{\lambda} (vt + x) + 2\pi \right\} \qquad \qquad ...(5)$$

because, by Stokes' theorem, reflected wave reflected from a denser medium or backed by a denser medium carries an extra phase 'π'.

From eqn. (5), we have

$$y_2 = - a \sin \frac{2\pi}{\lambda} (vt + x) \qquad \qquad ...(6)$$

Applying principle of superposition, the resultant displacement is given by

$$y = y_1 + y_2$$

$$= a \sin \frac{2\pi}{\lambda} (vt - x) - a \sin \frac{2\pi}{\lambda} (vt + x)$$

$$= a \left[\sin \frac{2\pi}{\lambda} (vt - x) - \sin \frac{2\pi}{\lambda} (vt + x) \right]$$

$$= a \left[\sin \frac{2\pi vt}{\lambda} \cos \frac{2\pi x}{\lambda} - \cos \frac{2\pi vt}{\lambda} \sin \frac{2\pi x}{\lambda} \right.$$

$$\left. - \sin\frac{2\pi vt}{\lambda}\cos\frac{2\pi x}{\lambda} - \cos\frac{2\pi vt}{\lambda}\sin\frac{2\pi x}{\lambda}\right]$$

$$= -2a\cos\frac{2\pi vt}{\lambda}\sin\frac{2\pi x}{\lambda}$$

$$= -2a\sin\frac{2\pi x}{\lambda}\cos\frac{2\pi vt}{\lambda}$$

$$= A\cos 2\pi ft$$

$$= A\cos\omega t \qquad\qquad ...(7)$$

where the amplitude of the resultant wave is 'A'

$$A = -2a\sin\frac{2\pi x}{\lambda} \qquad\qquad ...(8)$$

Evidently, 'A' depends upon x. The value of amplitude varies between $+2a$, 0 and $-2a$. But amplitude at any point is independent of time.

(i) For antinodes: When

$$\sin\frac{2\pi x}{\lambda} = \pm 1,$$

the amplitude is maximum being equal to $\mp 2a$. These points of maximum amplitude are called antinodes. Hence,

$$\sin\frac{2\pi x}{\lambda} = \pm 1 = \sin(2n+1)\frac{\pi}{2}$$

where $n = 0, 1, 2, 3, ...$

$$\therefore \qquad \frac{2\pi x}{2} = (2n+1)\frac{\pi}{2}$$

or, $$x = (2n+1)\frac{\lambda}{4} \qquad\qquad ...(9)$$

Thus, the distance between two consecutive antinodes is

$$\frac{3\lambda}{4} - \frac{\lambda}{4} = \frac{2\lambda}{4} = \frac{\lambda}{2}.$$

(ii) For nodes: When $\sin\frac{2\pi x}{\lambda} = 0$, the amplitude is minimum, i.e. zero. These points of minimum amplitude are called nodes. Hence,

$$\sin\frac{2\pi x}{\lambda} = 0 = \sin n\pi$$

where $n = 0, 1, 2, 3, ...$

or, $\dfrac{2\pi x}{\lambda} = n\pi$

or, $x = \dfrac{n\lambda}{2}$ ...(10)

Thus, the distance between two consecutive nodes is

$$\dfrac{\lambda}{2} - 0 = \dfrac{\lambda}{2}$$

Harmonics: Closed organ pipe is usually a hollow tube open at one end and closed at the other, see Fig. 5.18. When air is blown at the mouth a compressed wave travels to the closed end which is a rigid reflector and hence the air near the end is compressed to a pressure greater than the atmospheric pressure. This compressed air forces back the air behind and hence a compressed wave returns along the pipe. Thus a compressed wave is reflected from the closed end as a compressed wave and returns to the mouth. At the mouth the air is free to vibrate and the compressed reflected wave changes to a rarefied wave which travels inside the tube and gets reflected as a rarefied wave from the closed end. In this way vibrations of various frequencies are set up and due to superposition of the reflected pulse with the direct incident pulse, stationary longitudinal waves are set up with fixed nodes and antinodes at regular intervals. The closed end being a rigid reflector is always a seat of node and open end always a seat of antinode.

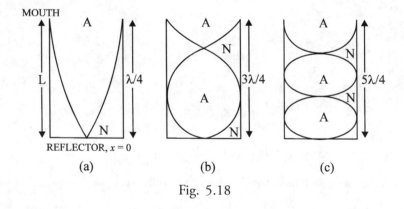

(a)　　　　　(b)　　　　　(c)

Fig. 5.18

Let us consider (Fig. 5.18a) in which a node is formed at $x = 0$ and an antinode is formed at $x = \dfrac{\lambda}{4}$. This is the simplest mode of vibration. For this, $V = f'_0 \lambda = f'_0 \times 4L$ so that

$$f'_0 = \dfrac{V}{4L} \qquad \left(\because L = \dfrac{\lambda}{4} \right)$$

In (Fig. 5.18b), two nodes are formed at $x = 0$ and $\lambda/2$ and two antinodes are formed at $x = \lambda/4$ and $3\lambda/4$. For this, $V = f'_1 \lambda = f'_1 \times (4L/3)$ $\left(\because L = \dfrac{3\lambda}{4} \right)$

Then $\qquad f_1' = \dfrac{3V}{4L}$

In (Fig. 5.18c), three nodes are formed at $x = 0$, $\lambda/2$, $2\lambda/2$ and three antinodes are formed at $x = \lambda/4$, $3\lambda/4$ and $5\lambda/4$.

For this, $\qquad V = f_2' \lambda = f_2'\left(\dfrac{4L}{5}\right)$, so that

$$f_2' = \dfrac{5V}{4L}$$

Hence f_0' is the fundamental frequency and f_1', f_2', etc. are the overtones. The ratio of these frequencies are given by

$$f_0' : f_1' : f_2' = 1 : 3 : 5$$

That means in case of closed pipe only odd harmonics are present.

Table 2 below gives a clear idea about the harmonics produced in a closed pipe.

Table-2
Harmonics of a closed pipe of length 'L'

No.	Name	Wavelength in air	Frequency of the tone produced	Relation with the fundamental frequency
1.	1st harmonic	$4L$	$f_0' = \dfrac{V}{4L}$	Equal to fundamental
2.	2nd harmonic	$\dfrac{4L}{3}$	$f_1' = \dfrac{3V}{4L}$	$f_1' = 3f_0'$
3.	3rd harmonic	$\dfrac{4L}{5}$	$f_2' = \dfrac{5V}{4L}$	$f_2' = 5f_0'$
etc.	etc.	etc.	etc.	etc.

Definitions, Laws and Formulae

Definitions and Laws

Wave: A wave is a disturbance that moves through a medium in such a manner that at any point the displacement is a function of time, while at any instant the displacement is a function of the position of the point.

A wave can travel only in elastic medium.

Longitudinal wave: The wave in which the particles of the medium vibrate in a direction parallel to the direction of propagation of wave is known as longitudinal wave.

Transverse wave: The wave in which the particles of the medium vibrate in a direction at right angles to the direction of propagation of the wave is known as transverse wave.

Characteristics of a Wave: Various characteristics of a wave are:

(i)	Amplitude	(ii)	Wavelength
(iii)	Phase	(iv)	Frequency
(v)	Time period	(vi)	Wave velocity
(vii)	Wave number	(viii)	Angular wave number
(ix)	Wave front		

Amplitude: The amplitude of a wave is the maximum displacement of the particle of the medium from its mean position. It is denoted by A and measured in the units of length.

Wavelength: The wavelength is the distance between two adjacent particles (of the medium) that are in the same phase. It is denoted by X and measured in the units of length.

Phase: The phase of a particle of the medium represents its state of vibration. The relative positions of the particles of the medium represent the phase of the motion.

Frequency: It is the number of complete vibrations made by a particle of the medium in one second. It is denoted by 'f' and measured in cycles/sec or Hertz.

Time period: It is the time taken by the particle of the medium to complete one vibration. It is denoted by T and measured in secs.

Wave velocity: It is the distance travelled by a wave through a medium in 1 second. It is denoted by V and measured in m/sec, cm/sec and ft/sec.

Wave number: It is the reciprocal of wavelength and is defined as the number of waves in unit distance. It is denoted by $1/\lambda$.

Angular wave number: It is the change of phase per unit distance. So, it is the phase difference between two particles of the medium when path difference between them is unity. It is denoted by K and measured in rad/m or rad/cm.

Wave front: It is the locus of the particles of the medium that are in the same phase, due to passage of a wave through it.

Progress wave: The wave which advances or progresses through the medium is called a progressive wave.

Intensity of a wave: The transfer of energy per unit time per unit area perpendicular to the direction of motion of wave is called the intensity of the wave. It is denoted by I.

Principle of Superposition: The resultant displacement of a particle of the medium at any instant of time is the vector sum of the displacement due to each wave separately.

Different phenomena of superposition of waves

(i) Two waves of the same frequency moving in the same direction – **Interference of waves**.

(ii) Two waves of slightly different frequencies moving in the same direction – **Beats**.

(iii) Two waves of the same frequency moving in the opposite direction – **Stationary waves**.

Constructive interference: The interference in which the amplitudes are added or the waves arrive in phase is called constructive interference. Here the energy becomes maximum.

Destructive interference: The interference in which the waves arrive out of phase to produce zero resultant amplitude is called destructive interference. The energy becomes minimum in this case.

Nodes: The points of zero amplitude in a stationary wave are called the nodes.

Antinodes: The points of maximum amplitude in a stationary wave are called the antinodes.

Fundamental frequency: The lowest frequency out of possible modes of vibration is called the fundamental frequency.

Harmonics: The frequencies which are integral multiples of the fundamental are called harmonics or overtones.

Intensity: The average power per unit area transmitted in a wave is called the intensity.

Laws of transverse vibration of string

(i) Law of length: Tension remaining constant and the frequency of transverse vibration of a particular string varies inversely as its length.

(ii) Law of tension: Length remaining constant and the frequency of transverse vibration of a particular string varies directly as the square root of tension.

(iii) Law of mass: Tension and length remaining constant and the frequency varies inversely as the square root of the mass per unit length.

(iv) Law of diameter: Tension and length remaining constant and the frequency varies inversely as the diameter.

(v) Law of density: Tension, length and diameter remaining constant and the frequency is inversely proportional to the square root of density.

Doppler effect: The apparent change in frequency due to the relative motion between the source and the observer is called Doppler effect.

Formulae

Phase: $\phi = \left(\dfrac{2\pi}{\lambda}\right) x$

where, x is the path difference and λ is the wavelength.

Velocity of a wave:

$$v = f\lambda = \frac{\omega}{k}$$

where, f is the frequency, ω is the angular frequency and k is the propagation vector.

Velocity of longitudinal wave:

$$v = \sqrt{\frac{E}{\rho}}$$

where, E is the elastic modulus and ρ is the density of the medium.

Velocity of transverse wave in a stretched string:

$$v = \sqrt{\frac{T}{\mu}}$$

where, μ is the mass per unit length of the medium and T is the tension applied.

Equations of progressive wave:

$$Y = A \sin \frac{2\pi}{\lambda} (vt \pm x)$$

$$= A \sin (\omega t \pm Kx)$$

Energy of progressive wave:

$$\text{Energy} = \frac{1}{2} m\omega^2 A^2$$

where, m is the mass of the vibrating particle and A is the amplitude of vibration.

Energy density:

$$u = \frac{U}{v_0 l} = \frac{1}{2} \rho \omega^2 A^2$$

where, ρ is the density of the medium.

Intensity of wave:

$$\dot{I} = \frac{1}{2} \rho \omega^2 A^2 \times \text{volume} = 2\pi^2 \rho v f^2 A^2$$

where, v is the volume of the medium.

Equation of standing wave (Reflection from free boundary):

$$Y = 2A \cos \left(\frac{2\pi x}{\lambda} \right) \sin \omega t$$

Equation of standing wave (Reflection from fixed boundary):

$$Y = 2A \sin \left(\frac{2\pi x}{\lambda} \right) \cos \omega t$$

Velocity of sound in solids:

$$v = \sqrt{\frac{Y}{\rho}}$$

where, Y is the Young's modulus of the solid considered.

Velocity of sound in liquids:

$$v = \sqrt{\frac{K}{\rho}}$$

where, K is the bulk modulus of the liquid considered.

Newton's formula for velocity of sound in air or gases:

$$V = \sqrt{\frac{P}{\rho}}$$

where, P is the pressure of the gas or air.

Laplace formula for velocity of sound in air or gases:

$$v = \sqrt{\frac{\gamma P}{\rho}}$$

where, γP is called the adiabatic bulk modulus of elasticity.

$$\gamma = \frac{c_p}{c_v}$$

and $\gamma = 1.41$ for air or diatomic gases.

Velocity of sound in a gas at any temperature $t°C$:

$$v_t = \sqrt{\frac{\gamma RT}{M}}$$

where, $\qquad T = t°C + 273$

and $\qquad v_t = v_0 \left[1 + \frac{t}{546} \right]$

Taking $\qquad v_0 = 332$ m/sec at NTP,

$$\frac{v_0}{546} = \frac{332}{546} = 0.61 \text{ m/sec}$$

Hence, the velocity of sound in air increases or decreases by 0.01 m/sec for every 1°C rise or fall of temperature.

Effect of humidity on the velocity of sound

$$v_m = v_d(1 + 0.189 \, f/p)$$

where, v_m is the velocity of sound in moist air, v_d is the velocity of sound in dry air, f is the saturation vapour pressure in cm of mercury.

Also, $\dfrac{v_m}{v_d} = \sqrt{\dfrac{\rho_d}{\rho_m}}$

where, ρ_m is the density of moist air and ρ_d is the density of dry air.

Beat frequency:

$$f = f_1 - f_2$$

where, f_1 and f_2 are the frequencies of individual waves.

Harmonics of open pipe:

$$f = n\frac{v}{2L}$$

where, $n = 1, 2, 3, 4, ...$

Harmonics of closed pipe:

$$f = n\frac{v}{4L}$$

where, $n = 1, 3, 5, ...$

Frequency of transverse vibrations in a stretched string (Sonometer):

$$f = \frac{1}{2L}\sqrt{\frac{T}{m}}$$

where, T is the tension, m is the mass per unit length of the string and L is resonating length of the string.

Interference:

$$A^2 = A_1^2 + A_2^2 + 2A_1A_2\cos\phi$$

$$\tan\theta = \frac{A_2\sin\phi}{A_1 + A_2\cos\phi}$$

where, A is the resultant amplitude and θ is the phase of the resultant wave.

Intensity level and loudness:

$$\beta = 10\log\left(\frac{I}{I_0}\right)$$

where, β is the intensity level of sound wave, I_0 is the arbitrary reference intensity which corresponds to the faintest sound that can be heard. Intensity level is labelled as decibels, abbreviated as **db**. If the intensity is 10^{-12} w/m^2, the intensity level is 0 db. The maximum intensity the ear can tolerate is 1 w/m^2, which corresponds to 120 db.

Doppler effect:

    (i) Source moving, observer and the medium stationary:

$$f_o = f_s \left(\frac{v}{v \mp v_s} \right)$$

    (ii) Observer moving, source and the medium stationary:

$$f_o = f_s \left(\frac{v \mp v_0}{v} \right)$$

    (iii) Both the source and the observer are moving with the medium stationary:

$$f_o = \left(\frac{v - v_0}{v - v_s} \right) f_s$$

if both are moving in the same direction.

$$f_o = \left(\frac{v + v_0}{v - v_s} \right) f_s$$

if both are approaching each other.

    (iv) Source, observer and the medium are in motion:

$$f_o = f_s \left(\frac{v \pm v_m \pm v_o}{v \pm v_m \pm v_s} \right)$$

where, v is the velocity of sound, v_m is the velocity of the medium, v_o is the velocity of the observer, v_s is the velocity of the source, f_o is the frequency heard by the observer and f_s is the frequency of the source.

SOLVED EXAMPLES

Q.1. A broadcasting station radiates at a frequency of 710 KHz. What is the wavelength in metre? Given the velocity of waves = 3×10^8 m/sec.

Sol. $f = 710$ KHz $= 710 \times 10^3$ Hz

$$v = 3 \times 10^8 \text{ m}, \quad v = f\lambda$$

$\Rightarrow$
$$\lambda = \frac{v}{f} = \frac{3 \times 10^8 \text{ m/sec}}{710 \times 10^3 \text{ Hz}}$$

$$= 422.5 \text{ m}.$$

Q.2. Compressional wave impulses are sent to the bottom of sea from a ship and the echo is heard after 1.5 sec. Calculate the depth of the sea assuming the mean temperature of water 4°C and its bulk modulus 2.2×10^{10} dyne/cm².

Sol. Density of water at 4°C (ρ) = 1 gm/cm^3

Bulk modulus of elasticity,

$$E = 2.2 \times 10^{10} \text{ dyne/cm}^2$$

∴ Velocity of sound in water

$$v = \sqrt{\frac{E}{\rho}} = \sqrt{\frac{2.2 \times 10^{10}}{1}}$$
$$= 1.48 \times 10^5 \text{ cm/sec}$$

If $x =$ depth of sea.

Distance travelled by the wave = 2x.

∴   Time taken = 1.5 sec

$$1.5 = \frac{2x}{1.48 \times 10^5}$$

or, $$x = \frac{1.5 \times 1.48 \times 10^5}{2} \text{ cm} = 1110 \text{ m}.$$

Q.3. **A train approaching a tunnel surrounded by a cliff and the driver sends a short whistle when 1.6 km away. The echo reaches him 9 sec later. Find the speed of the train, velocity of sound = 340 m/sec.**

Sol. The train sends a whistle at A and receives the echo while at B.

If v is the velocity of the train.

$$AB = 9v$$

Distance travelled by the sound = 2 × 1600 – 9v

$$= 3200 - 9v$$

∴   3200 – 9v = 340 × 9

⇒   3200 – 9v = 3060

⇒   9v = 140

⇒ $$v = \frac{140}{9} \text{ m/sec}$$

$$= \frac{140}{9} \times \frac{18}{5} \text{ km/h} = 56 \text{ km/h}.$$

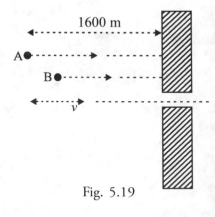

Fig. 5.19

Q.4. **A simple harmonic wave is represented by:**

$$Y = 5 \sin 2\pi \left(\frac{s}{0.05} - 0.05 \, w \right)$$

Find the wavelength, amplitude, frequency and velocity of the wave.

Sol. The general equation of wave

$$y = A \sin \frac{2\pi}{\lambda} (vt - x)$$

$$= A \sin 2\pi \left(\frac{vt}{\lambda} - \frac{x}{\lambda} \right) \qquad \ldots(1)$$

The equation of progressive wave given in question is

$$Y = 5 \sin 2\pi \left(\frac{t}{0.05} - 0.05x \right) \qquad \ldots(2)$$

Comparing eqns. (1) and (2), we get

$$A = 5 \text{ cm,}$$

$$\therefore \qquad \frac{vt}{\lambda} = \frac{t}{0.05}$$

$$\frac{x}{\lambda} = 0.05x$$

$$\Rightarrow \qquad \lambda = \frac{1}{0.05} = \frac{10^2}{5} = 20 \text{ cm}$$

$$\Rightarrow \qquad \frac{v}{\lambda} = \frac{1}{0.05}$$

$$\Rightarrow \qquad \frac{v}{20} = \frac{1}{0.05}$$

$$\Rightarrow \qquad v = \frac{20 \times 100}{5} = 400 \text{ cm/sec}$$

$$f = \frac{v}{\lambda} = \frac{400}{20} = 20 \text{ cm/sec.}$$

Q.5. The equation of a transverse wave travelling in a rope is given by:

$$y = 10 \sin \pi (0.01x - 2.00t)$$

in which 'y' and 'x' are expressed in cm at 't' in sec. Find the amplitude, frequency, velocity and wavelength of the wave.

Sol. The general equation of wave is represented by

$$Y = A \sin \frac{2\pi}{\lambda} (vt - x)$$

$$= A \sin \left(\frac{2vt}{\lambda} - \frac{2x}{\lambda} \right) \qquad \ldots(1)$$

The equation of progressive wave given in the question is

$$y = 10 \sin \pi (0.01x - 2t)$$

$$= -10 \sin \pi (2t - 0.01x) \qquad \ldots(2)$$

Comparing eqns. (1) and (2), we get

$$A = 10 \text{ cm}$$

$$\frac{2vt}{\lambda} = 2t$$

$$\frac{2x}{\lambda} = 0.01x$$

$$\Rightarrow \qquad \lambda = \frac{2}{0.01} = 200 \text{ cm}$$

$$\Rightarrow \qquad \frac{v}{\lambda} = 1$$

$$\Rightarrow \qquad v = \lambda$$

$$\Rightarrow \qquad v = 200 \text{ cm/sec}$$

$$f = \frac{v}{\lambda} = \frac{200}{200} = 1 \text{ cm/sec.}$$

Q.6. **A simple harmonic wave of amplitude 8 units transverses a line of particles in the direction of the z-axis. At any given instant of time, for a particle at a distance of 10 cm from the origin, the displacement is 6 units, and for a particle at a distance of 25 cm from the origin, the displacement is +4 units. Calculate the wavelength.**

Sol. A = 8 units, $x_1 = 10$ cm, $y_1 = 6$ units,
$x_2 = 25$ cm, $y_2 = 4$ units

$$y_1 = A \sin 2\pi \left(\frac{t}{T} - \frac{x_1}{\lambda} \right) \qquad \qquad \text{...(1)}$$

$$y_2 = A \sin 2\pi \left(\frac{t}{T} - \frac{x_2}{\lambda} \right) \qquad \qquad \text{...(2)}$$

From eqn. (1),

$$6 = 8 \sin 2\pi \left(\frac{t}{T} - \frac{x_1}{\lambda} \right)$$

$$\Rightarrow \sin 2\pi \left(\frac{t}{T} - \frac{x_1}{\lambda} \right)$$

$$= \frac{6}{8} = \frac{3}{4} = 0.75 = \sin 48.6°$$

$$\Rightarrow \qquad 2\pi \left(\frac{t}{T} - \frac{x_1}{\lambda} \right) = 48.6°$$

$\Rightarrow \qquad \dfrac{2\pi t}{T} - \dfrac{2\pi x_1}{\lambda} = 48.6° \qquad\qquad …(3)$

From eqn. (2),

$$4 = 8 \sin 2\pi \left(\dfrac{1}{T} - \dfrac{x_2}{\lambda} \right)$$

$$\Rightarrow \ \sin 2\pi \left(\dfrac{t}{T} - \dfrac{x_2}{\lambda} \right)$$

$$= \dfrac{4}{8} = \dfrac{1}{2} = \sin 30°$$

$$\Rightarrow \qquad 2\pi \left(\dfrac{1}{T} - \dfrac{x_2}{\lambda} \right) = 30°$$

$\Rightarrow \qquad \dfrac{2\pi t}{T} - \dfrac{2\pi x_2}{\lambda} = 30° \qquad\qquad …(4)$

Subtracting eqn. (4) from (3),

$$\dfrac{2\pi t}{T} - \dfrac{2\pi x_2}{\lambda} = 48.6°$$

$$\dfrac{2\pi t}{T} - \dfrac{2\pi x_2}{\lambda} = 30°$$

$$\underline{\quad - \qquad + \qquad - \quad}$$

$$\underline{\dfrac{2\pi}{\lambda}(x_2 - x_1) = 48.6° - 30°}$$

$$\Rightarrow \qquad \dfrac{2\pi}{\lambda}(x_2 - x_1) = 18.6°$$

$$\Rightarrow \qquad \dfrac{2\pi}{\lambda}(25 - 10) = 18.6°$$

$$\Rightarrow \qquad \dfrac{\pi}{\lambda} \times 2 \times 15 = 18.6°$$

$$\Rightarrow \qquad \dfrac{\pi}{\lambda} = \dfrac{18.6°}{30}$$

$$\Rightarrow \qquad \lambda = \dfrac{30}{18.6} \times 3.142$$

$$= \dfrac{30 \times 3.142}{18.6°}$$

$$= \frac{30 \times 3.142}{2 \dfrac{\pi}{180} \times 18.6° \text{ rad}}$$

$$= \frac{180 \times 30 \times 1.342}{2 \times 3.142 \times 18.6}$$

$$= \frac{180 \times 15}{18.6} = \frac{180 \times 15 \times 10}{186}$$

$$= \frac{4500}{31} = 145.16 \text{ cm.}$$

Q.7. **A simple harmonic motion is represented by:**

$$x = 10 \sin\left(\frac{2\pi s}{T} + \alpha\right)$$

and the time period is 30 sec. At the time $t = 0$, the displacement is 5 cm. Calculate: (i) phase angle at $t = 7.5$ sec and (ii) phase distance between two positions at a time interval of 60 sec.

Sol. $t = 30$ sec
When $t = 0$, $y = 5$ cm

$$y = 10 \sin\left(\frac{2\pi t}{T} + \alpha\right)$$

$\Rightarrow \qquad 5 = 10 \sin \alpha$

$\Rightarrow \qquad \sin \alpha = \dfrac{1}{2} = \sin 30°$

$\Rightarrow \qquad \alpha = 30$

$\Rightarrow \qquad \alpha = 30 \times \dfrac{2\pi}{180} \text{ rad}$

$\qquad \alpha = \dfrac{\pi}{3} \text{ rad}$

(i) $t = 7.5$ sec

$\qquad T = 30$ sec, $t = 0$, $y_1 = 5$ cm

$$y = 10 \sin\left(\frac{2\pi t}{T} + \alpha\right)$$

$$5 = 10 \sin (\alpha)$$

$$\sin \alpha = \frac{5}{10} = \frac{1}{2} = \sin 30°$$

$$\alpha = 30° = \frac{\pi}{6} \text{ rad}$$

$$\phi = \frac{2\pi t}{T} = \frac{2\pi \times 75}{30} = \frac{\pi}{2}$$

$$\phi + \alpha = \left(\frac{\pi}{6} + \frac{\pi}{2}\right) = \frac{\pi + 3\pi}{6} = \frac{4\pi}{3}$$

$$= \frac{2\pi}{3} \text{ rad}$$

(ii) $$\Delta\phi = \frac{2\pi}{T} \times t_1 = \frac{2\pi}{30} \times 60 \text{ where } t_1 = 60 \text{ secs.}$$

$$= \frac{2\pi}{5} \text{ rad.}$$

Q.8. **A string 1 m long with a mass of 0.1 gm/cm is under the tension of 400 N. Find the fundamental frequency.**

Sol. $l = 1$ m,

$$m = 0.1 \text{ gm/cm}$$
$$= 0.1 \times 10^{-3} \times 10^2 \text{ kg/m}$$
$$= 0.01 \text{ kg/m}$$
$$T = 400 \text{ N}$$

Fundamental frequency,

$$f = \frac{1}{2l}\sqrt{\frac{T}{m}} = \frac{1}{2 \times 1}\sqrt{\frac{400}{(0.01)}} = 100 \text{ Hz.}$$

Q.9. **A stone hangs in air from a wire which is stretched over a sonometer. The bridges of sonometer are 0.45 m apart and the wire is immersed in water, the length between the bridges must be altered to 0.36 m to bring it again in unison with the fork. Calculate the density of stone.**

Sol. M = Mass of the stone

In air, $T_1 = Mg$, $l_1 = 0.45$ m, $l_2 = 0.36$ m

In water, $T_2 = Mg - v$

$$= Mg - \frac{M}{\rho} \times 1 \times g \quad \text{[Density of water = 1 g/cm}^3\text{]}$$

$$= Mg\left(1 - \frac{1}{\rho}\right)$$

Since, $\dfrac{\sqrt{T}}{l} = \text{constant}$

$$\frac{\sqrt{T_1}}{l_1} = \frac{\sqrt{T_2}}{l_2}$$

$$\sqrt{\frac{T_1}{T_2}} = \frac{l_1}{l_2} \text{ or } \sqrt{\frac{T_2}{T_1}} = \frac{l_2}{l_1}$$

$$\sqrt{\frac{Mg\left(1-\dfrac{1}{\rho}\right)}{Mg}} = \frac{0.36}{0.45}$$

$$\therefore \qquad \left(1 - \frac{1}{\rho}\right) = \left(\frac{36}{45}\right)^2 = \frac{16}{25}$$

$$\frac{1}{\rho} = 1 - \frac{16}{25} = \frac{9}{25}$$

$$\rho = \frac{25}{9} = 2.78 \text{ g cm}^{-3}.$$

Q.10. **A wire under tension vibrates with a frequency of 450 Hz. What would be fundamental frequency if the wire were half long twice as thick and under one fourth of tension?**

Sol. $f_1 = 450 = \dfrac{1}{l_1 D_1} \sqrt{\dfrac{T_1}{\pi\rho}}$

Let the length, tension and diameter of the string be l_1, T_1 and D_1 respectively. 'ρ' is the density of wire.
If the parameters of the wire are changed to l_2, T_2 and D_2, its frequency is

$$f_2 = \frac{1}{l_2 D_2} \sqrt{\frac{T_2}{\pi\rho}}$$

$$\frac{f_2}{f_1} = \frac{l_1}{l_2} \cdot \frac{D_1}{D_2} \sqrt{\frac{T_2}{T_1}}$$

Here, $\qquad \dfrac{l_2}{l_1} = \dfrac{1}{2}, \ \dfrac{D_2}{D_1} = 2 \text{ and } \dfrac{T_2}{T_1} = \dfrac{1}{4}$

$$\frac{f_2}{450} = 2 \times \frac{1}{2} \times \sqrt{\frac{1}{4}}$$

$$f_2 = 400 \times \frac{1}{2} = 225 \text{ Hz.}$$

Q.11. **A wire of length 40 cm vibrates 250 times/sec. If the length of the wire is increased to 50 cm and the stretching force is reduced to 1/4th of its original value, find the new frequency.**

Sol. $l_1 = 40$ cm, $f_1 = 250$, $T_1 = T$

$$f_1 = \frac{1}{2l_1}\sqrt{\frac{T_1}{m}}$$

$l_2 = 50$ cm, $f_2 = ?$, $T_2 = T_{1/4}$

$$f_2 = \frac{1}{2l_2}\sqrt{\frac{T_2}{m}}$$

$$\frac{f_2}{f_1} = \frac{l_1}{l_2}\sqrt{\frac{T_2}{T_1}}$$

$$= \frac{40}{50}\sqrt{\frac{T_2}{l_1} \times \frac{l_1}{T_1}}$$

$$f_2 = 250 \times \frac{4}{5} \times \frac{1}{2} = 100 \text{ Hz.}$$

Q.12. **A particle with a mass of 0.2 kg moves according to the law:**

$$y = 0.08 \cos (20\pi t + \pi/4)$$

Find the velocity of the particle, its acceleration and the acting force.

Sol. $y = 0.08 \cos\left(2\pi t + \dfrac{\pi}{4}\right)$

$$\frac{dy}{dt} = 0.08 - \sin\left(20\pi t + \frac{\pi}{4}\right) \times 20\pi$$

$$= -\sin\left(20\pi t + \frac{\pi}{4}\right) \times 1.6\pi$$

$$= \cos\left(20\pi t + \frac{\pi}{4} + \frac{\pi}{2}\right) \times 1.6\pi$$

$$\Rightarrow \quad v = \cos\left(20\pi t + \frac{3\pi}{4}\right) \times 1.6\pi$$

$$\frac{dv}{dt} = -\sin\left(20\pi t + \frac{3\pi}{4}\right) \times 1.6\pi \times 20\pi$$

$$= \cos\left(20\pi t + \frac{3\pi}{4} + \frac{\pi}{2}\right) \times 32\pi^2$$

$$\Rightarrow \quad a = \cos\left(20\pi t + \frac{5\pi}{4}\right) \times 32\pi^2$$

$$F = ma = 0.2 \text{ kg} \times \cos\left(20\pi t + \frac{5\pi}{4}\right) \times 32\pi^2$$

$$= 6.4\pi^2 \cos\left(20\pi t + \frac{5\pi}{4}\right)$$

Q.13. A particle vibrates harmonically at a frequency of 0.5 Hz. At the initial moment it is in equilibrium position moving at a speed of 20 cm/sec. Write down the equation of vibrations.

Sol. $f = 0.5$ Hz

$$\frac{1}{f} = T = \frac{1}{0.5} = 2 \text{ sec}$$

$$v_{max} = 20 \text{ cm/sec}$$

$$\Rightarrow \qquad \frac{2\pi A}{\lambda} v = 20 \text{ cm/sec}$$

$$\Rightarrow \qquad 2\pi A f = 20 \text{ cm/sec}$$

$$\Rightarrow \qquad 2\pi \times 0.5 \, A = 20 \text{ cm/sec}$$

$$\Rightarrow \qquad A = \frac{20}{\pi} \text{ cm} = \frac{20}{3.142} = 6.36 \text{ cm}$$

$$\omega = 2\pi f t = 2\pi \times 0.5 t = \pi t$$

$$\phi = \frac{2\pi}{\lambda} x = 0, \pi, 2\pi$$

$$y' = A \sin \omega t$$
$$= 6.36 \sin \left[(\omega t + \pi) - \pi\right]$$
$$= 6.36 \sin \left\{- \left[(\pi - (\omega t + \pi))\right]\right\}$$
$$= - 6.36 \sin (\omega t + \pi)$$

$$= 6.36 \cos\left(\omega t + \frac{3\pi}{2}\right)$$

Q.14. The waves of sound of wavelength 1 m and 1.01 m in a gas produce 30 beats in 10 sec. Calculate the velocity of sound in the gas.

Sol. Let 'v' be the velocity of sound

$$f_1 = \frac{v}{\lambda_1} = \frac{v}{1}$$

$$f_2 = \frac{v}{\lambda_2} = \frac{v}{1.01}$$

Number of beats per second

$$m = \frac{30}{10} = 3$$

$$f_1 - f_2 = 3$$

$$\frac{v}{1} - \frac{v}{1.01} = 3$$

$$\Rightarrow \quad v\left[1 - \frac{1}{1.01}\right] = 3$$

$$\Rightarrow \quad v\left[\frac{1.01 - 1}{1.01}\right] = 3$$

$$\Rightarrow \quad v = \frac{3 \times 1.01}{0.01} = 303 \text{ m/sec.}$$

Q.15. Two tuning forks make 4 beats per second. The frequency of one is 288. When the prong of the other is weighed with a little of wax, the number of beats per second is decreased. Find the original frequency.

Sol. Number of beats per second, $m = 4$

Frequency, $f_1 = 288$

Unknown frequency,

$$f = m \pm f = 288 \pm 4$$
$$= 292 \text{ or } 284$$

$\therefore$ On loading the tuning fork the frequency f number of beats with 288 would increase, if f is loaded. This is contrary to the expression of observed. Hence, this is wrong.

If $f = 292$ on loading it decreases and comes closer to 288, thereby decreasing the number of beats per second. This agrees with experimental result.

Q.16. Two tuning forks A and B produce 4 beats per second. On loading B with a little of wax, 6 beats per second is produced. If the quantity of wax is reduced the number of beats drop to four. If frequency of A be 256. Find the natural frequency of B.

Sol. Let f' = Frequency of B

Frequency of A = 256

Number of beats per second,

$$m = 4$$
$$f = 256 \pm 4 = 260 \text{ or } 252$$
$$f = 260.$$

Q.17. At what temperature, the velocity of sound in air is 3/2 times its value at 0°C?

Sol. Let t be the required temperature

$$\frac{v_1}{v_0} = \frac{3}{2}$$

$$\frac{v_1}{v_0} = \sqrt{\frac{273+t}{273}}$$

$$\sqrt{\frac{273+t}{273}} = \frac{3}{2}$$

$$\frac{273+t}{273} = \frac{9}{4}$$

$$\Rightarrow \qquad 1092 + 4t = 2457$$

$$4t = 1365$$

$$t = 341.25°C.$$

Q.18. **Find the temperature at which sound travels in hydrogen with same velocity as in oxygen of 800°C. Density of oxygen is 16 times that of hydrogen.**

Sol. Let the velocity of hydrogen $= v_H$
The velocity of oxygen at 800°C $= v_{Ox}$

$$v_H = v_{Ox}$$

$$v_H \sqrt{\frac{273+t}{273}} = v_{Ox} \sqrt{\frac{273+800}{273}}$$

$$\frac{v_{Ox}}{v_H} = \sqrt{\frac{273+t}{1073}}$$

$$\frac{v_{Ox}}{v_H} = \sqrt{\frac{\rho_H}{\rho_{Ox}}} = \sqrt{\frac{d}{16d}} = \frac{1}{4}$$

$$\therefore \qquad \sqrt{\frac{273+t}{1073}} = \frac{1}{16}$$

$$4368 + 16t = 1073$$

$$16t = -3295$$

$$t = -205.9°C.$$

Q.19. **What will be pitch of the note emitted by a closed pipe 0.324 m long at 4°C if velocity of sound at 0°C is 332 ms⁻¹?**

Sol. $v_0 = 332$ m/sec, $l = 0.324$ m

$$v_H = v_0 \sqrt{\frac{273+t}{273}}$$

$$= 332 \sqrt{\frac{273+4}{273}} = 332 \times \sqrt{\frac{277}{273}}$$

In case of a closed pipe,

$$f = \frac{v}{4l} = \frac{332}{4 \times 0.324} \times \sqrt{\frac{277}{273}} = 258 \text{ vib/sec.}$$

Q.20. **What must be ratio of length of closed and open pipe in order that the second overtone of the closed pipe shall be in unison with the fourth overtone of the open pipe?**

Sol. Let l_c and l_o be the length of closed and open pipe respectively.
Fundamental frequency of closed pipe,

$$f = \frac{v}{4l_c}$$

The overtones emitted by closed pipe are f, $3f$, $5f$, ...
Frequency of second overtone of closed pipe,

$$= \frac{5v}{4l_c}$$

Fundamental frequency of open pipe,

$$f = \frac{v}{2l_o}$$

The overtones emitted by an open pipe are f', $2f'$, $4f'$, ...
Frequency of the fourth overtone by open pipe,

$$= 5\frac{v}{2l_o}$$

$$5\frac{v}{2l_o} = 5\frac{v}{4l_c}$$

$$\frac{l_c}{l_o} = \frac{1}{2}.$$

Q.21. **Two closed organ pipes whose sound together produce 12 beats in 4 sec at a temperature of 27°C. Find the temperature at which the number of beats produced are 16 during the same period.**

Sol. Let l_1 and l_2 be the lengths of two pipes. Their fundamental frequencies are given by

$$f_1 = \frac{v_{27}}{4l_1}, f_2 = \frac{v_{27}}{4l_2}$$

where v_{27} is the velocity of sound at 27°C.

$$\frac{v_{27}}{4l_1} - \frac{v_{27}}{4l_2} = \frac{12}{4}$$

$$v_{27}\left(\frac{1}{4l_1} - \frac{1}{4l_2}\right) = 3 \qquad \qquad \text{...(1)}$$

At $t°$C,

$$v_t\left(\frac{1}{4l_1} - \frac{1}{4l_2}\right) = \frac{16}{4} = 4 \qquad \qquad \text{...(2)}$$

Dividing eqn. (2) by (1)

$$\frac{v_t}{v_{27}} = \frac{4}{3}$$

$$\sqrt{\frac{273+t}{273+27}} = \frac{4}{3}$$

$$\frac{273+t}{300} = \frac{16}{9}$$

$$2457 + 9t = 4800$$

$$t = 260.3°C.$$

Q.22. **A policeman on duty detects a drop of 15% in the pitch of the horn of a motor car as it crosses him. If the velocity of sound is 330 m/s. Calculate the speed of car.**

Sol. Let n be the pitch of the note of the horn as the car approaches with velocity a,

$$n' = n\frac{v}{v-a}$$

If n'' be the pitch of the note when car recede away with velocity 'a' after crossing,

$$n'' = n\frac{v}{v+a}$$

$$\frac{n''}{n'} = \frac{nv}{v+a} \cdot \frac{v-a}{nv}$$

$$\frac{n''}{n'} = \frac{v-a}{v+a}$$

Since, there is a drop of 15%

$$\frac{n''}{n'} = \frac{85}{100}$$

$$\frac{v-a}{v+a} = \frac{85}{100}$$

$$\frac{330 - a}{330 + a} = \frac{17}{20}$$

$$5610 + 17a = 6600 - 20a$$

$$37a = 990$$

$$a = \frac{990}{37} = 26.2 \text{ m/sec.}$$

Q.23. **Two railway locomotives are moving on parallel tracks in opposite directions each with a speed of 72 km/h. One engine blows producing a note of frequency 576. Calculate the frequency of sound heard by the driver of the second engine before and after crossing of the engines. Given: velocity of sound = 350 m/s.**

Sol. Let 'a' and 'b' be the velocities of two engines. If 'n' is the apparent frequency of the note before they cross each other.

$$n' = n\frac{v + b}{v - a}$$

$$a = b = 72 \text{ km/h} = 20 \text{ m/sec}$$

$$= 576\frac{350 + 20}{350 - 20}$$

$$= 576 \times \frac{36}{33} = 645.9 \text{ c/s}$$

Let 'n'''' be the apparent frequency of the note after they have crossed each other.

$$n' = n\frac{v - b}{v + b} = n.\frac{350 - 20}{350 + 20}$$

$$= 576 \times \frac{33}{37} = 51.37 \text{ c/s.}$$

Q.24. **A wave is represented by the equation: $y = 0.20 \sin (0.40) \pi (x - 60t)$, where all distances are measured in m and time in second. Find: (i) the amplitude, (ii) the wavelength, (iii) the frequency, and (iv) the displacement at $x = 5.5$ cm and $t = 0.02$ sec.**

Sol. The equation of a progressive wave is given by

$$y = A \sin\frac{2\pi}{\lambda}(x - vt) \qquad \ldots(1)$$

The equation of a progressive wave given in the question is

$$y = 0.20 \sin 0.40\pi (x - 60t) \qquad \ldots(2)$$

Comparing eqns. (1) and (2),

(i) A = 0.2 cm, v = 60 cm/sec

(ii)
$$\frac{2\pi}{\lambda} = 0.40\pi$$

$$\Rightarrow \qquad \lambda = \frac{2}{0.4} = 5 \text{ cm}$$

(iii)
$$f = \frac{v}{\lambda} = \frac{60}{5} = 12 \text{ sec}$$

(iv)
$$x = 5.5 \text{ cm}, \ t = 0.02 \text{ sec}$$
$$y = 0.2 \sin 0.40\pi \ (5.5 - 60 \times 0.2)$$
$$= 0.2 \sin 1.32\pi$$
$$= 0.2 \times (- 0.86) = -0.17 \text{ cm}.$$

Q.25. **A wave is represented by:**

$$y = 0.25 \times 10^{-3} \cos (500t - 0.025x)$$

where y and x are in cm and t in seconds. Find: (i) the amplitude, (ii) period, (iii) angular frequency, (iv) wavelength, and (v) amplitude of particle velocity.

Sol. The general wave equation is

$$y = A \cos 2\pi \left(\frac{t}{T} - \frac{x}{\lambda} \right) \qquad \qquad ...(1)$$

The given equation is

$$y = 0.25 \times 10^{-3} \cos (500t - 0.025x) \qquad \qquad ...(2)$$

From eqns. (1) and (2)

(i) $A = 0.25 \times 10^{-3}$ cm

(ii)
$$\frac{2\pi t}{T} = 500t$$

$$\Rightarrow \qquad T = \frac{2\pi}{500} = \frac{\pi}{250} \text{ sec}$$

(iii)
$$f = \frac{1}{T} = \frac{1}{2\pi/500}$$

$$= \frac{500}{2\pi}$$

or, $\quad \omega = 2\pi f = 500$ rad/sec

(iv)
$$\lambda = \frac{2\pi}{0.025} \text{ cm} = \frac{20\pi}{25} \text{ m} = \frac{4\pi}{5} \text{ m}$$

(v)
$$\frac{dy}{dt} = 0.25 \times 10^{-3} \times 500 \sin (500t - 0.025x)$$

$\Rightarrow$ Amplitude of particle velocity is
$$0.25 \times 10^{-3} \times 500 = 0.125 \text{ cm/sec}.$$

Q.26. The equation of progressive wave in a string is given by $y = 5 \sin \pi (4t - 0.002x)$, where x and y are in cm and t in secs. Find the amplitude, frequency, speed and wavelength of the wave.

Sol. Equation of a progressive wave is given by

$$y = A \sin 2\pi \left(\frac{t}{T} - \frac{x}{\lambda} \right) \qquad \qquad ...(1)$$

The given equation is

$$y = 5 \sin \pi (4t - 0.002x)$$
$$= 5 \sin 2\pi (2t - 0.001x) \qquad \qquad ...(2)$$

Comparing eqns. (1) and (2)

$$A = 5 \text{ cm}, \quad \frac{1}{T} = 2/\text{sec}$$

$$\text{Frequency} = \frac{1}{T} = 2 \text{ Hz}$$

$$\frac{1}{\lambda} = 0.001$$

$$\Rightarrow \qquad \lambda = \frac{1}{0.001} = 1000 \text{ cm}$$

$$v = f\lambda = 2 \times 1000 = 2000 \text{ cm/sec.}$$

Q.27. Find the phase difference between two particles of a medium 10 cm apart when a progressive wave of wavelength 80 cm pass through it.

Sol. Phase difference $= \dfrac{2\pi}{\lambda} \times$ Path difference

$$= \frac{2\pi}{80} \times 10 = \frac{\pi}{4} \text{ rad.}$$

Q.28. Find the wavelength of the longitudinal waves passing through air at 0°C, if the wave has a frequency of 1000 Hz and the wave velocity is 330 m/sec.

Sol. $f = 1000$ Hz, $v = 330$ m/sec

$$v = f\lambda$$

$$\Rightarrow \qquad \lambda = \frac{v}{f} = \frac{330}{1000} = 0.33 \text{ m.}$$

Q.29. A vibrating source sends out waves of wavelength 100 cm through medium A and 150 cm through medium B. What is the velocity of the waves in medium B, if that in A is 150 cm/sec?

Sol. $\lambda_A = 100$ cm, $\lambda_B = 150$ cm

$$v_A = 150 \text{ cm/sec}, \ v_B = ?$$

$$f = \frac{v_A}{\lambda_A} = \frac{150 \text{ cm/sec}}{100 \text{ cm}} = \frac{3}{2} \cdot \frac{1}{\text{sec}} = \frac{v_B}{\lambda_B}$$

$$v_B = \lambda_B \times \frac{3}{2} = \frac{150 \times 3}{2}$$

$$= 75 \times 3 = 225 \text{ cm/sec.}$$

Q.30. **The speed of sound waves in air is 330 m/sec at NTP. A point source of sound has a frequency of 500 per second. It radiates energy uniformly in all directions at the rate of 11 watt. What is the amplitude and intensity of the waves at a distance of 5 m from the source? (density of air at NTP = 1.29 kg/m³)**

Sol. The energy from the point source spread over a sphere of surface area $4\pi r^2$.
Here, $r = 5$ m

$$I = \frac{E}{St} = \frac{P}{S}$$

$$= \frac{11 \text{ watt}}{\dfrac{4 \times 22}{7} \times (5 \text{ m})^2}$$

$$= 3.5 \times 10^{-2} \text{ watt/m}$$

Again $I = 2\pi^2 f^2 A^2 v\rho$

So, $$A = \sqrt{\frac{I}{2\pi^2 f^2 v\rho}}$$

$$= \sqrt{\frac{3.5 \times 10^{-2}}{2 \times (3.14)^2 \times (500)^2 \times 330 \times 1.29}}$$

$$= 1.27 \times 10^{-4} \text{ m.}$$

Q.31. **A drop of water 2 mm in diameter, falling from a height of 50 cm in a bucket, generates sound which can be heard from a distance of 5 m. Take all the gravitational energy difference as going into the sound from the transformations being speed over 0.2 sec. Find the average intensity at the listener's end.**

Sol. Given, diameter of drop = 2 mm = 2×10^{-3} m
So, radius of the drop = R = 10^{-3} m
Height of the fall of the drop = h = 50 cm = 0.5 m

So, loss of PE = $mgh = \frac{4}{3}\pi R^3 \rho g h$

$$= \frac{4}{3} \times 3.14 \times (10^{-3})^3 \times 10^3 \times 9.8 \times 0.5$$

$$(\because \rho = 10^3 \text{ kg/m}^3)$$

$$= 20.6 \times 10^{-6} \text{ J}$$

This is also the sound energy generated. Time of spread of the energy up to a distance of 5 m = 0.2 sec. The surface area of the sphere of spread

$$= 4\pi r^2 = 4 \times 3.14 \times 5^2$$
$$= 3.14 \times 10^2 \text{ m}^2$$

Intensity at the listener's end

$$I = \frac{\text{Energy}}{\text{Area} \times \text{Time}}$$

$$= \frac{20.6 \times 10^{-6}}{3.14 \times 10^2 \times 0.2}$$

$$= 3.28 \times 10^{-2} \text{ w/m}^2.$$

Q.32. The equation of a stationary wave is

$$y = 4 \cos \frac{\pi x}{5} \sin 60 \, \pi t$$

where, x in cm and t in seconds. Find: (a) the amplitude, (b) the velocity of component waves, and (c) the distance between nodes.

Sol. Equation of a standing wave is given by,

$$y = 2A \cos \frac{2\pi x}{\lambda} . \sin \frac{2\pi t}{T} \qquad \qquad ...(1)$$

The given eqn. is

$$y = 4 \cos \frac{\pi x}{5} . \sin 60 \, \pi t$$

$$= 4 \cos \frac{2\pi x}{10} . \sin 2\pi.30t \qquad \qquad ...(2)$$

Comparing equations (1) and (2)

$$2A = 4$$
$$\Rightarrow \qquad A = 2 \text{ cm}$$

$$\frac{2\pi t}{T} = 2\pi \, 30t$$

$$\Rightarrow \qquad \frac{1}{T} = 30 \text{ Hz}$$

$$\Rightarrow \qquad f = 30 \text{ Hz}$$
$$\lambda = 10 \text{ cm}$$

Velocity of component waves = 30 × 10 = 300 cm/sec

Distance between consecutive nodes

$$= \frac{\lambda}{2} = \frac{10}{2} = 5 \text{ cm.}$$

Q.33. **A resonance air column resonates with a tuning fork of frequency 512 Hz at the length 17.4 cm. Neglecting the end correction, find the speed of sound in air.**

Sol. $f = 512$ Hz,

$l = 17.4$ cm

This is a case of closed pipe.

We know, $f = \dfrac{v}{4l}$

$\Rightarrow$ $v = f \times 4l$

$= 512 \times 4 \times 17.4$

$= 356.35$ m/sec.

Q.34. **Find the length of an open air pipe if a tuning fork of frequency 480 is in resonance with it. The speed of sound is 340 m/sec.**

Sol. $f = 480$ Hz,

$v = 340$ m/sec

This is a case of open pipe.

We know, $f = \dfrac{v}{2l}$

$\Rightarrow$ $l = \dfrac{v}{2f}$

$$= \frac{340}{2 \times 480} = 35.4 \text{ cm.}$$

Q.35. **Find the frequencies of the first, second and third harmonics in an open air column and closed air column of length 34 cm. The speed of sound at room temperature = 340 m/sec.**

Sol. $v = 340$ m/sec $= 340,00$ cm/sec,

$\lambda = 34$ cm

We know that for open air column,

$$f_0 = \frac{v}{2l}, f_1 = \frac{2v}{2l}, f_2 = \frac{3v}{2l}$$

So, 1st harmonics $= \dfrac{34000}{2 \times 34} = 500$ Hz

2nd harmonics $= 2 \times 500 = 1000$ Hz

3rd harmonics $= 3 \times 500 = 1500$ Hz

For closed pipe,

$$f_0 = \frac{v}{4l}, f_1 = \frac{3v}{4l}, f_2 = \frac{5v}{4l}$$

So, 1st harmonics $= \dfrac{34000}{4 \times 34} = 250$ Hz

2nd harmonics $= 3 \times 250 = 750$ Hz

3rd harmonics $= 5 \times 250 = 1250$ Hz.

Q.36. **A resonance column when vibrated with a tuning fork of frequency 480 Hz produces maximum sound with air columns of length 16 cm and 51 cm respectively. Find: (i) the speed of sound, and (ii) end correction.**

Sol. $f = 480$ Hz

l_1, for 1st resonance $= 16$ cm

l_2, for 2nd resonance $= 51$ cm

(i) We know that

$$v = 2f \, (l_2 - l_1)$$
$$= 2 \times 480 \, (51 - 61)$$
$$= 2 \times 480 \times 35 \text{ cm/sec}$$
$$= 336 \text{ m/sec}$$

(ii) If Δl the end correction

$$3(4 + \Delta l) = l_2 + \Delta l$$

$$\Rightarrow \qquad \Delta l = \frac{(l_2 - 3l_1)}{2} = \frac{51 - 48}{2} = 1.5 \text{ cm}.$$

Q.37. **A flexible wire 80 cm long has a mass of 0.4 gm. It is stretched by a force 500 N. Find the velocity of the transverse wave in the string.**

Sol. Mass of wire $= 0.4$ gm $= 0.4 \times 10^{-3}$ kg

Length $= 80$ cm $= 0.8$ m

Tension $= 500$ N

So, $\qquad m = \dfrac{0.4 \times 10^{-3}}{0.8}$ kg/m

$$v = \sqrt{\frac{T}{m}} = \sqrt{\frac{500 \times 0.8}{0.4 \times 10^{-3}}}$$

$$= 10^3 \text{ m/sec.}$$

Q.38. **The fundamental frequency of vibration of a wire is 200 Hz. Find the frequency of the first overtone if the tension increases four times.**

Sol. $f_0 \alpha \sqrt{T}, \quad f'_0 \alpha \sqrt{T'}$

$$\frac{f'_0}{f_0} = \sqrt{\frac{T'}{T}}$$

$$= \sqrt{\frac{4T}{T}} = 2 \, (\because T' = 4T)$$

$\Rightarrow$
$$f'_0 = 2f_0 = 2 \times 200$$
$$= 400 \text{ Hz}.$$

Q.39. **A wire of density 9×10^3 kg/m^3 is stretched between two clamps one metre apart and elongates by 0.05 cm. What is the lowest frequency of transverse vibrations in the wire? Young's modulus of the wire is $Y = 9 \times 10^{10}$ N/m^2.**

Sol. $l = 1$ m,

$\Delta l = 0.05$ cm $= 0.05 \times 10^{-2}$ m,

$\rho = 9 \times 10^3$ kg/m^3

$Y = 9 \times 10^{10}$ N/m^2

$$Y = \frac{Mg/\pi r^2}{\Delta l/l}$$

$$\frac{mg}{\pi r^2} = Y\left(\frac{\Delta l}{l}\right)$$

where, M = mass of the wire of length 1 m

$\dfrac{M}{L}$ = mass per unit length = m

Now, f_0 = frequency of the lowest mode, i.e. fundamental frequency

$\Rightarrow$
$$f_0 = \frac{1}{2l}\sqrt{\frac{T}{m}} = \frac{1}{2l}\sqrt{\frac{Mg}{\pi r^2 \rho}}$$

$$= \frac{1}{2l}\sqrt{\frac{Y(\Delta l/l)}{\rho}}$$

$$= \frac{1}{2 \times 1}\sqrt{\frac{9 \times 10^{10} \times \left[\dfrac{0.05 \times 10^{-2}}{1}\right]}{9 \times 10^3}}$$

$$= \frac{1}{2}\sqrt{5 \times 10^3} = \frac{70.7}{2} = 35.35 \text{ Hz}$$

Q.40. **What tension would be required to create a standing wave with four segments in a string 100 cm long weighing 0.5 gm if frequency of transverse waves along the string is 100 Hz?**

Sol. We know,

$$f_{P-1} = \frac{P}{2l}\sqrt{\frac{T}{m}}$$

Here, $f_{P-1} = 100$ Hz

$P = 4,$

$l = 100$ cm

$$m = \frac{0.5 \text{ gm}}{100 \text{ cm}} = 5 \times 10^{-3} \text{ gm/cm}$$

$$T = \left(\frac{f_{P-1} \times 2l}{P}\right)^2 m$$

$$= \left[\frac{100 \text{ cycles/sec} \times 2 \times 100 \text{ cm}}{4}\right]^2 \times 5 \times 10^{-3} \text{ gm/cm}$$

$$= \left(\frac{100 \times 2 \times 100}{4}\right)^2 \times 5 \times 10^{-3} \text{ dyne}$$

$$= 1.25 \times 10^5 \text{ dyne.}$$

Q.41. **What is the tension in a 3 m cord, whose mass is 0.15 kg, if a transverse wave is observed to travel in the cord with a speed of 20 m/sec?**

Sol. m = mass per unit length

$$= \frac{0.015 \text{ kg}}{3 \text{ m}} = 5 \times 10^{-2} \text{ kg/m}$$

Now, $v = \sqrt{\frac{T}{m}}$

$\Rightarrow$ $T = v^2 m$

$$= \left(20 \frac{\text{m}}{\text{sec}}\right)^2 \times 5 \times 10^{-2} \text{ kg/m}$$

$$= 400 \times 5 \times 10^{-2}$$

$$= 2000 \times 10^{-2} = 20 \text{ N.}$$

Q.42. On increasing the weight in stretching a given string by 2.5 kg, the frequency is altered in the ratio 3:2. Find the original stretching weight.

Sol. Let the original stretching weight be x kg wt

So, $$T_1 = x \text{ kg wt}$$
$$T_2 = (x + 2.5) \text{ kg wt}$$

The ratio of frequencies is 3:2

or, $$\frac{f_2}{f_1} = \frac{3}{2}$$

On applying the law of tension, we get

$$\frac{f_2}{f_1} = \sqrt{\frac{T_2}{T_1}}$$

$$\Rightarrow \qquad \frac{3}{2} = \sqrt{\frac{x + 2.5}{x}}$$

Squaring,

$$\Rightarrow \qquad \frac{9}{4} = \frac{(x + 2.5)}{x}$$

$$\Rightarrow \qquad 9x = 4x + 10.0$$
$$\Rightarrow \qquad 5x = 10$$
$$\Rightarrow \qquad x = 2 \text{ kg.}$$

Q.43. A tuning fork A of frequency 256/sec produces 4 beats when sounded together with another tuning fork B of unknown frequency. Find out the frequency of fork B.

Sol. $$f_A = 256/\text{sec, beats} = 4$$
$$f_B = f_A \pm 4$$
$$= 256 \pm 4$$
$$= 260 \text{ or } 252/\text{sec.}$$

Fork B is loaded with wax. So, f_B decreases. If the beat frequency now decreases, then f_B shall be equal to 260 (higher value) and if the beat frequency increases, f_B shall be equal to 252 (lower value).

Q.44. Two tuning forks when sounded together produce 5 beats/sec. The frequency of one fork is 512 Hz. One arm of the other fork is filed and the number of beats is found to be 7/sec, find the frequency of the other.

Sol. Let f_A and f_B represent the frequencies of the forks A and B respectively, and f be the beat frequency.

$$\therefore \qquad f = f_A \sim f_B$$
If $$f_A = 512 \text{ Hz}$$

then $\qquad f_B = 512 \pm 5 = 517$ or 507 $(\because f = 5)$

Fork B is filed. So its frequency is now f_B' such that $f_B' > f_B$ (on filing the frequency increased)

Now the new beat is

$$f = 7,$$

$$f' = f + 2$$

So, $f_B' = f_B + 2$ because f_A is fixed.

Now, $f_B' = f_A \pm f' = 512 \pm 7 = 519$ or 505

So, $f_B = 519 - 2 = 517$ or $505 - 2 = 503$

But f_B can be either 517 or 503. Hence, the frequency of unknown fork is 517 Hz.

Q.45. **Two tuning forks A and B are sounded together to have a beat frequency of 4 Hz. Fork A is loaded and the new beat frequency is found to be 3 Hz. If the fork A has a frequency of 288 Hz, find that of fork B.**

Sol. $f_A = 288$ Hz, $f = 4$ Hz

$f_B = f_A \pm 4 = 288 \pm 4 = 292$ Hz or 284 Hz

Fork A is loaded and hence its new frequency f_A' is less than old frequency f_A.

New beat $f' = 3 = f - 1$

So, $f_A' = f_A - 1 = 288 - 1 = 287$ Hz

Now, frequency of the unknown fork B is $f_B = 284$ Hz.

Q.46. **Two open pipes give 5 beats/sec when sounded together in their fundamental notes, the lengths of the pipes are 100 cm and 103 cm. Find the velocity of sound.**

Sol. $l_1 = 100$ cm, $l_2 = 103$ cm

Number of beats per second = 5

Since the frequency of the fundamental mode in the open air column is $f = \dfrac{v}{2l}$

$$f_1 = \frac{v}{2 \times l_1} = \frac{v}{2 \times 100} = \frac{v}{200}$$

$$f_2 = \frac{v}{2 \times l_2} = \frac{v}{2 \times 103} = \frac{v}{206}$$

But $\qquad f_1 - f_2 = 5$

$\Rightarrow \qquad \dfrac{v}{200} - \dfrac{v}{206} = 5$

$$\Rightarrow \qquad v\left(\frac{6}{200 \times 206}\right) = 5$$

$$\Rightarrow \qquad v = \frac{5 \times 200 \times 206}{6} \text{ cm/sec}$$

$$= 343 \text{ m/sec.}$$

Q.47. **Two tuning forks A and B are sounded together and 8 beats/sec are heard. A is in resonance with a column of air 32 cm long in a pipe closed at one end and B is in resonance when the length of the column is increased by 1 cm. Find the frequency of each fork.**

Sol. Let f_1 and f_2 be the frequencies of tuning forks A and B. Given that,

$f_1 - f_2 = 8, f_1 = f_2 + 8$

Also $l_1 = 32$ cm and $l_2 = 33$ cm

We know that

$$f_1 l_1 = f_2 l_2$$

$$\Rightarrow \qquad (f_2 + 8)\, 32 = f_2 \times 33$$

$$\Rightarrow \qquad f_2 = 256 \text{ Hz}$$

and $f_1 = f_2 + 8 = 256 + 8 = 264$ Hz.

Q.48. **Find the speed of longitudinal sound wave through silver, given that Young's modulus of silver = 7.75 × 10¹⁰ N/m² and density of silver = 1.05 × 10³ kg/m³.**

Sol. Y = 7.75 × 10¹⁰ N/m²,

$$\rho = 1.05 \times 10^3 \text{ kg/m}^3$$

Speed of longitudinal wave

$$= \sqrt{\frac{Y}{\rho}} = \sqrt{\frac{7.75 \times 10^{10}}{1.05 \times 10^3}} \text{ m/sec}$$

$$= 8.59 \times 10^3 \text{ m/sec.}$$

Q.49. **What must be the stress in a stretched steel wire for the speed of longitudinal waves to be equal to 100 times the speed of transverse waves?**

(Y for steel = 20 × 10¹⁰ N/m²)

Sol. As per question,

$$v_1 = 100 v_t$$

$$\sqrt{\frac{vP}{\rho}} = 100\sqrt{\frac{T}{m}}$$

m = mass per unit length = $\pi r^2 \rho$

or,
$$\frac{Y}{\rho} = 10^4 \frac{T}{\pi r^2 \rho}$$

or,
$$\frac{T}{\pi r^2} = \text{stress} = \frac{Y}{10^4}$$

$$= \frac{20 \times 10^{10} \text{ N/m}^2}{10^4}$$

∴ Stress needed $= 20 \times 10^6$ N/m².

Q.50. The speed of sound in water at 0°C is 1346 m/sec. Compute the adiabatic bulk modulus of water.

Sol. $v = 1346$ m/sec,
$$\rho = 10^3 \text{ kg/m}^3$$

$$v = \sqrt{\frac{K}{\rho}}$$

$\Rightarrow$ $\qquad K = v^2\rho = (1346)^2 \times 10^3$ N/m²

($\because$ density of water $= 10^3$ kg/m³)
$$K = 18.1 \times 10^8 \text{ N/m}^2.$$

Q.51. The density of oxygen is 16 times that of hydrogen. For both, $\gamma = 1.4$. The speed of sound is 317.5 m/sec in oxygen at 0°C. What is the speed in hydrogen?

Sol. $\qquad v - \sqrt{\frac{vP}{\rho}} \quad v_{\text{H}} - \sqrt{\frac{vP}{\rho_{\text{H}}}}$

$$v_0 = \sqrt{\frac{vP}{\rho_0}}, \frac{v_{\text{H}}}{v_0} = \sqrt{\frac{\rho_0}{\rho_{\text{H}}}} = \sqrt{16} = 4$$

$v_{\text{H}} = \text{velocity in hydrogen} = 4 \times v_0$
$$= 4 \times 317.5 = 1270 \text{ m/sec.}$$

Q.52. A train passes railway station with a speed of 36 km/hr and blowing continually a whistle of frequency 340 Hz. What shall be the frequency noted by a man waiting on the platform when the train is: (i) approaching, and (ii) departing the station? (Speed of sound = 330 m/sec).

Sol. $v = 330$ m/sec $= 33 \times 10^3$ cm/sec
$$v_s = 36 \text{ km/hr} = 1 \times 10^3 \text{ cm/sec}$$
$$f = 340 \text{ Hz}$$

(i)
$$f_0 = f\left(\frac{v}{v - v_s}\right)$$

$$= 340 \text{ Hz} \left[\frac{33 \times 10^3 \text{ cm/sec}}{33 \times 10^3 \text{ cm/sec} - 1 \times 10^3 \text{ cm/sec}} \right]$$

$$= 340 \times \frac{33}{32} = 350.6 \text{ Hz}$$

(ii)
$$f_0 = f \left(\frac{v}{v + v_s} \right)$$

$$= 340 \text{ Hz} \left[\frac{33 \times 10^3 \text{ cm/sec}}{(33 \times 10^3 + 1 \times 10^3) \text{ cm/sec}} \right]$$

$$= 330 \text{ Hz}.$$

Q.53. Find out the velocity at which a source of frequency 10,000 Hz should approach an observer at rest in order to produce a Doppler shift of 200 Hz. (Velocity of sound = 340 m/sec).

Sol. $f_0 = f \left(\dfrac{v}{v - v_s} \right)$

Here $f_0 > f$,

$$f = 10,000 \text{ Hz}$$

Given, Doppler shift $= f_0 - f = 200$ Hz

$$f_0 = 10,200 \text{ Hz}$$

$$v = 340 \text{ m/sec}$$

$$= 34 \times 10^3 \text{ cm/sec}$$

$$f_0 v - f_0 v_s = f v$$

$$\Rightarrow \qquad v_s = \frac{f_0 v - f v}{f_0} = \frac{(f_0 - f)v}{f_0}$$

$$= \frac{200 \times 34 \times 10^3}{10,200}$$

$$= \frac{68}{102} \times 10^3 \text{ cm/sec}$$

$$= 24 \text{ km/hr}.$$

Q.54. A car moving at a speed of 48 km/hr passes by a siren which is sounding at a frequency of 300 Hz. Find out the apparent frequency of the siren noted by the driver of the car as the car passes the siren. ($v = 340$ m/sec).

Sol. When the car approaches the siren

$$v_0 = 48 \text{ km/hr}$$
$$= 1.33 \times 10^3 \text{ cm/sec}$$

$$f_0 = f\left(\frac{v + v_0}{v}\right)$$

$$f_0 = 300 \,\text{Hz}\left(\frac{34 \times 10^3 + 1.33 \times 10^3}{34 \times 10^3}\right)$$

$$= 311.7 \text{ Hz}$$

When the car recede away from the observer

$$f_0 = f\left(\frac{v - v_0}{v}\right)$$

$$= 300 \,\text{Hz}\left(\frac{34 \times 10^3 - 1.33 \times 10^3}{34 \times 10^3}\right)$$

$$= 288.3 \text{ Hz.}$$

Q.55. **Two trains are approaching a station from opposite sides with speeds of 90 km/hr and 72 km/hr respectively. The train moving with a speed of 90 km/hr emits a whistle of frequency 500 Hz continually. Find the frequency of the sound noted by a passenger in the other train (i) before the trains meet, and (ii) after the trains meet. (Velocity of sound = 340 m/sec).**

Sol. $v_s = 90$ km/hr $= 25$ m/sec,
$$v_0 = 72 \text{ km/hr} = 20 \text{ m/sec}$$

(i) Here the observer and the source both move towards each other

$$f_0 = \frac{f(v + v_0)}{v - v_s}$$

$$= \frac{500\,(340 + 20)}{340 - 25} = 571.4 \text{ Hz}$$

(ii) Here observer and source move away from each other

$$f_0 = \frac{f(v - v_0)}{v + v_s}$$

$$= \frac{500\,(340 - 20)}{340 + 25} = 438.3 \text{ Hz.}$$

Q.56. **An ultrasonic wave of frequency 45000 Hz is sent out by sonar shows a frequency rise of 300 Hz on reflector from an approaching submarine. If velocity of sound in water is 1500 m/sec, calculate the speed of the submarine.**

Sol. $f = 45000$ Hz

Increase in frequency, $\Delta f = 300$ Hz

Speed of sound in water $= 1500$ m/sec

Here observer is at rest, so $v_0 = 0$

We know that

$$f_0 = \frac{v - v_0}{v - v_s} f = \frac{v}{v - v_s} f$$

Here v_s is taken to be +ve because the source is the image of the ship formed by the reflected surface of the submarine. It is approaching the observer with twice the speed of the submarine (when mirror is moving with speed v, the image moves with speed $= 2v$)

So,

$$f_0 - f = f\left(\frac{v}{v - v_s} - 1\right) = f\frac{V_S}{(v - v_s)}$$

or,

$$300 \times 1500 - 300v_s = 45000v_s$$

or,

$$v_s = \frac{300 \times 1500}{45300} = 10 \text{ m/s}$$

But v_s is twice the speed of the submarine.

∴ Speed of the submarine = 5 m/s.

Q.57. **A source and an observer are approaching one another with the relative velocity of 40 m/sec. If the true source frequency is 1200 Hz, find the observed frequency under the following conditions:**

  (i) All velocity is in the source alone.

  (ii) All velocity is in the observer alone.

  (iii) The source moves in air at 100 m/sec towards the observer, but the observer also moves with a velocity v_o in the direction.

Sol. For (i) $v_s = 40$ m/s

So,

$$f_o = \frac{1200 \times 340}{(340 - 40)}$$

$$= 1360 \text{ Hz.}$$

(ii) $v_o = 40$ m/sec

$$f_o = \frac{1200 \times (340 + 40)}{340} = 1340 \text{ Hz}$$

For (iii) $v_s = 100$ m/sec

Again $v_s - v_o = 40$ m/sec

So, $v_o = 60$ m/sec

$$f_o = \frac{1200\,(340 - 60)}{340 - 100}$$

$$= 1400 \text{ Hz.}$$

QUESTIONS

1. Distinguish between longitudinal and transverse wave.

2. Derive an expression for the velocity of plane longitudinal wave in an elastic medium.

3. Find an expression for the velocity of sound in a homogeneous medium.

4. Derive Newton's formula for the velocity of sound in a gas. How did Laplace modify this formula?

5. Explain the effect of various factors which influence the velocity of sound in air.

6. Derive an expression for the velocity of plane progressive sound wave through a thin solid rod.

7. Derive an expression for the velocity of transverse wave along a string stretched under tension.

8. Develop the theory of transverse vibration in a stretched string.

9. State and explain the laws of transverse vibration in a stretched string.

10. How can you experimentally verify the laws of transverse vibrations in a stretched string?

11. How can the frequency of a tuning-fork be determined in the laboratory with the help of a sonometer?

12. What are various characteristics of a wave?

13. Derive an expression for the velocity of a longitudinal wave in a solid medium.

14. What are stationary waves and how are they produced? Analytically explain their characteristic properties.

15. Give the theory of formation of stationary waves. Discuss their formation in a pipe which is closed at one end and in one which is open at both ends.

16. What are nodes and antinodes? How will you demonstrate their existence?

17. Explain the manner in which reflection of a transverse wave occurs when the wave reaches a fixed point in the string. Show with the help of diagrams, how stationary waves are formed in a string?

18. Show how stationary vibrations are produced in air? What are their properties?

    Explain in a general way, how reflection of sound takes place at the open and closed ends of a tube.

Problems

1. Calculate the velocity of sound in a gas at 0°C for which specific heat at constant pressure is 0.24 and specific heat at constant volume is 0.173. Given that J = 4.2 Joules per calorie.

   (**Ans.** 326.4 m/sec)

2. Find the temperature at which sound travels in hydrogen with the same velocity as in oxygen at 1000°C.

   (**Ans.** −193°C)

3. Compare the velocity of sound in saturated air at 30°C with that in dry air at the same temperature, the pressure being equal to 752 mm in both cases.

(**Ans.** $v_{dry\ air}$ = 349.7 m/sec, $v_{saturated\ air}$ = 352.6 m/sec)

4. Calculate the velocity of sound in air at 20°C and 1.1×10^6 dynes/cm² pressure, given that density of air at 0°C and 10^6 dynes/cm² pressure is 0.00122 gm/cm³. (c_p = 0.242 and c_v = 0.171).

(**Ans.** v_{20} = 352.2 m/sec)

5. If the velocity of sound in hydrogen at 0°C is 4200 ft/sec, what will be the velocity of sound (at the same temperature) in the mixture of two parts by volume of hydrogen to one of oxygen? (The density of oxygen is 16 times that of hydrogen).

(**Ans.** v_m = 1715 ft/sec)

6. Find the velocity of sound in air at 50°C and 70 cm of Hg pressure, given that velocity at NTP is 331 m/sec.

(**Ans.** 359.9 m/sec)

7. Calculate the velocity of sound in air at 0°C and pressure of 76 cm of Hg; given that the corresponding density of air is 0.001293 gm/cm³; g = 981 cm/sec² and specific gravity of Hg is 13.6.

(**Ans.** 331.2 m/sec)

8. At what temperature is the velocity of sound in oxygen equal to its velocity in nitrogen at 30°C? Given that the atomic weights of the gases are 16 and 14 respectively.

(**Ans.** –8°C)

9. Calculate the velocity of sound in air at 27°C; given that the mean molecular weight of air is 28.8, γ = 1.40 and R = 8.31×10^7 ergs/mole.deg.

(**Ans.** 348 m/sec)

10. The wavelength of the note emitted by a tuning fork, frequency 512 vibrations per second in air at 29°C is 68.3 cm. If the density of air at NTP is 1.293 gm per litre, calculate c_p/c_v of air. Density of Hg = 13.6 gm/cm³.

(**Ans.** r = 1.41)

11. At what temperature the speed of sound will be double the speed in air at 0°C?

12. A string is under tension of 32 N. If 10 metre length of the string has a mass of 2 gram, then what is the velocity of transverse waves along the string?

(**Ans.** 400 m/sec)

13. An increment of 20 N in the tension of a string increases the pitch by 25%. What was the original tension?

(**Ans.** 35.5 N)

14. A wire under tension vibrates with a fundamental frequency of 512 Hz. What would be the frequency of second harmonic if the wire were half as long, twice as thick and acted upon by one-fourth tension?

(**Ans.** 768 Hz)

15. Stationary waves of frequency 200 Hz are formed in air. If the velocity of the waves is 1120 ft/sec, what is the shortest distance between:

    (i) two nodes, (ii) two antinodes, and (iii) a node and an antinode?

(**Ans.** (i) 8, (ii) 2.8 ft, and (iii) 1.4 ft)

16. A string vibrates according to the equation

$$y = -5 \sin \frac{\pi x}{3} \cos 40\pi t$$

    where x and y are in cm and t in seconds.

    (i) What is the distance between the nodes?

    (ii) What is the speed of the particle of the string at a position $x = 1.5$ cm, when $t = \dfrac{9}{8}$ sec?

(**Ans.** (i) 3 cm, (ii) zero)

17. A sonometer string and a tuning fork of frequency 256, when vibrating together, produce 4 beats per second. When the fork is slightly loaded, 6 beats are heard per second. What is the frequency of the string?

(**Ans.** 260 Hz)

18. Two tuning forks A and B produce 6 beats per second when sounded together. If B is slightly loaded with wax, the beats reduce to 4 per second. The frequency of A is 512/sec, what is the frequency of B?

(**Ans.** 518 cycles/sec)

19. Two similar sonometer wires of the same material under the same tension produce 2 beats per second. The length of one is 0.5 m and the length of the other is 0.501 m. Compute the frequency of vibrations produced in the two wires.

(**Ans.** 1002 Hz, 1000 Hz)

20. An open pipe has a length of 0.3 m. What are the possible frequencies of vibration of the air column in it? (Velocity of sound in air = 350 m/s)

(**Ans.** $1133n$ Hz where, $n = 1, 2, 3, ...$)

Chapter 6

Reflection and Transmission of Waves

A wave is the disturbance that travels in a medium. When a progressive wave meets the boundary of two media, reflection and refraction occur. A part is reflected back to the first medium and another part is refracted to the second medium (Fig. 6.1).

That means when a wave travelling through a given medium meets another medium, a part is transmitted to the second medium and a part is reflected back to the first medium. Some of the characteristics of the reflected and transmitted waves are different from those of the incident wave. The changes in the characteristics of a wave at the boundary are described as boundary behaviour of the wave. The phenomena of reflection of a wave from a free boundary and fixed boundary differ from each other.

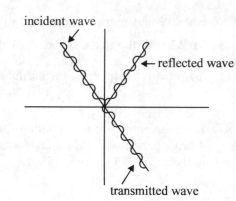

Fig. 6.1

Let us examine the changes suffered by a wave pulse travelling along a string when it meets another medium. In Fig. 6.2, a thin string AB is joined with a thick string BC. The strings form two media (one dimensional) with boundary at B.

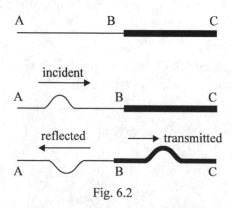

Fig. 6.2

(a) Wave pulse originating in the thin string

Imagine a transverse pulse in the thin string AB moving towards the boundary at B.

As the pulse reaches the boundary, a pulse travels along the thick string BC (called the transmitted pulse) and a pulse moves back along the thin string (called the reflected pulse). The following features are observed in the reflected and transmitted waves.

(i) The amplitudes of both the reflected and transmitted waves are less than that of the incident wave.

(ii) The speed of the reflected wave is the same as that of the incident wave. But the speed of the transmitted wave in the thicker string is less than that of the incident wave. This difference in the speed of propagation can be explained by the formula.

$$v = \sqrt{\frac{T}{m}}$$

Since the strings AB and BC are joined at B, the tension T is the same in both of them. The value of m is greater in the thicker string BC and hence the speed is less. Here BC is a denser medium and AB is a rarer medium.

(iii) The frequency v (and hence the angular frequency ω) of the reflected and transmitted waves are the same as that of incident wave.

(iv) The wavelength of the reflected wave is the same as that of the incident wave. But the wavelength of the transmitted wave is reduced.

(v) The reflected wave pulse is inverted with respect to the incident pulse but transmitted pulse suffers no inversion. Thus, a phase change of π occurs when reflection takes place at a denser medium.

(b) Wave pulse originating in the thick string

When a pulse originating in thicker string BC moves towards the junction at B, similar reflection and transmission occur (Fig. 6.3). But one important difference in this case is that, the reflected pulse is not inverted, i.e. there is no phase change when the reflection is from the rarer medium.

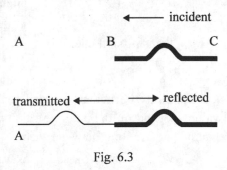

Fig. 6.3

These boundary behaviours of waves can be summarized as follows:

- The amplitudes (and hence the intensities) of the reflected and transmitted waves are less than that of the incident wave.

- The frequency (and hence angular frequency and time period) of the wave remains unchanged by reflection or transmission.

- The speed of propagation and wavelength decrease when the wave enters a denser medium from a rarer medium. Similarly, they increase when the wave enters a rarer medium from a denser medium.

- A phase change of π takes place due to reflection from a denser medium, but no such phase change occurs when reflection is from a rarer medium.

This boundary behaviour of one dimensional waves is also shared by two dimensional and three dimensional waves. When two dimensional water waves on the surface of shallow water meets deep water, reflection and transmission occurs at the boundary (Fig. 6.4). When disturbances occurring in the deep sea (such as Tsunami) comes to the sea shore (shallow water) the speed and wavelength decrease. This cause devastation as waves coming from the deep sea rides over those near the shore and waves of large height rushes into the lands. Similarly, light waves (three dimensional) moving in air is partly reflected and partly transmitted when it meets other media such as water or glass.

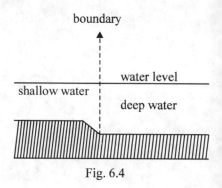

Fig. 6.4

Reflection and transmission coefficients

The intensity of the incident wave is distributed between the reflected and transmitted waves at the boundary (if there is no absorption). The fractions of the incident intensity carried by the reflected and transmitted waves are called reflection coefficient and transmission coefficient, respectively.

$$\text{Reflection coefficient} = \frac{\text{intensity of reflected wave}}{\text{intensity of incident wave}}$$

$$\text{Transmission coefficient} = \frac{\text{intensity of transmitted wave}}{\text{intensity of incident wave}}$$

The sum of the reflection coefficient and transmission coefficient is equal to 1. Their relative values depend on the nature of the two media and the manner of coupling between them.

When a three dimensional wave such as light meets another medium, such as water or glass it is divided into a reflected wave (in the first medium) and transmitted wave (in the second medium) at the boundary. The relative intensities of the two components depend on the speeds of the wave in the two media and the angle of incidence. In Fig. 6.5, a light beam AB

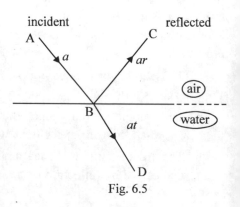

Fig. 6.5

travelling in air meets water surface. The directions of the reflected beam BC and transmitted beam BD can be obtained by using the laws of reflection and the laws of refraction. Let r and t be the fractions of the incident amplitude that are reflected and transmitted, respectively. If the incident amplitude is represented by a, the reflected and transmitted amplitudes will be ar and at, respectively.

Now let us apply the principle of reversibility of light to this situation. The beam DB and CB will combine to form the beam BA. Let r' and t' be the fraction of amplitude reflected and transmitted when the light wave from water meets air surface. The incident ray CB with amplitude ar will be partly reflected along BA with amplitude arr and partly transmitted along BE with amplitude art. Similarly, the incident beam DB in water with amplitude at will be transmitted to air along BA with amplitude atr' and reflected back to water along BE with amplitude atr' (Fig. 6.6). Since the beams CB and DB produce only the beam BA and the beam BE does not exist, we have

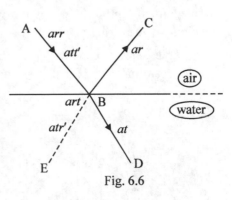

Fig. 6.6

$$arr + att' = a \qquad ...(1)$$

and $$art + atr' = 0 \qquad ...(2)$$

These lead to the relations

$$r^2 + tt' = 1 \qquad ...(3)$$

and $$r' = -r \qquad ...(4)$$

The fractions t and t' are, in general, different. Their values depend on the velocities of the wave in the two media. The negative sign in eqn. (4) signifies that there is a phase difference of π between the waves reflected from the boundary on the two sides. In fact, the wave in the rarer medium reflected from the denser medium undergoes a phase change of π, whereas a wave in the denser medium reflected from the rarer medium does not undergo any phase change. The fraction of change in amplitude of the incident wave due to reflection is the same on both sides.

Characteristic impedance of a string

Characteristic impedance of a string corresponds to the ratio of the transverse force applied to the string to generate wave motion and the transverse velocity of the particles of the string. So mathematically

$$\text{Characteristic impedance} = \frac{\text{Transverse force}}{\text{Transverse velocity}}$$

or, $$\vec{Z}_m = \frac{\vec{F}}{\vec{v}}$$

Expression for characteristic impedance

Consider a string with one end fixed rigidly at origin ($x = 0$) and let the other end be at a long distance (say ∞). Let the transverse wave be generated in the string with a harmonic force $F = F_0 e^{i\omega t}$ applied at the origin ($x = 0$). Let the external force be constrained in xy plane ⊥ to the length of the string. The string vibrates in the xy plane and will act as a forced oscillator and medium for propagation of the wave moving with a velocity C. Fig 6.7 shows wave behaviour of the string at $x = 0$.

Transverse Component $T \sin \theta$ acts ⊥ to the equilibrium position of the string and provides restoring force for the deformed string due to the propagation of waves.

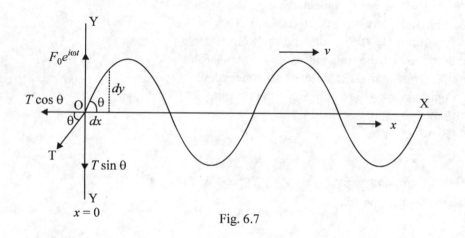

Fig. 6.7

Let us analyse the behaviour of the string at the origin which originates the wave motion. At the origin external force acts as deforming force and for sustained oscillations, this must be equal and opposite to restoring force developed due to elasticity. So we have

$$\vec{F} = F_0 e^{i\omega t} = -T \sin \theta$$

For small values of θ, $\sin \theta = \tan \theta\ \partial y / \partial x$

So we may get $\vec{F} = F_0 e^{i\omega t} = -T \left(\dfrac{\overline{\partial y}}{\partial x} \right)$ ...(5)

where, we have substituted the displacement of the string as a complex quantity. The real part will correspond to the external force $F_0 \cos \omega t$ and the imaginary part will correspond to $F_0 \sin \omega t$.

Let us write the wave equation $y = \vec{A}\, e^{i(\omega t - kx)}$ ...(6)

The displacement amplitude $\vec{A}$ may also be a complex quantity. Putting in the equations (5) and (6) we get

$$\vec{F} = F_0 e^{i\omega t} = -T[-ik\,\vec{A}\,e^{i(\omega t - kx)}] \qquad \ldots(7)$$

At $x = 0$, the above equation will become

$$F_0 e^{i\omega t} = -T[ik\,\vec{A}\,e^{i\omega t}]$$

From which we get

$$\vec{A} = \frac{F_0}{ikT} \qquad \ldots(8)$$

Also, from equation (6) we have the particle velocity as

$$\vec{v} = \frac{\overline{\partial y}}{\partial t} = i\omega\,\vec{A}\,e^{i(\omega t - kx)} \qquad \ldots(9)$$

which at $x = 0$, becomes

$$\vec{v} = i\omega\,\vec{A}\,e^{i\omega t} \qquad \ldots(10)$$

Putting the value of $\vec{A}$ from (8), we get

$$\vec{v} = i\omega\left[\frac{F_0}{ikT}\right]e^{i\omega t}$$

or, $\qquad \vec{v} = \frac{\omega}{kt}F_0 e^{i\omega t} = \frac{\omega}{kT}\vec{F}$

So $\qquad \dfrac{\vec{F}}{\vec{v}} = \dfrac{k}{\omega}T \qquad \ldots(11)$

But $\dfrac{\vec{F}}{\vec{v}}$ gives the complex impedance. For the string we find that $\dfrac{\vec{F}}{\vec{v}}$ is not a complex quantity. So, we denote it by Z. Here Z is the characteristic impedance of the string.

$$Z = \frac{k}{\omega}T$$

$$= \frac{2\pi/\lambda}{2\pi v}T = \frac{T}{v\lambda}$$

Hence $\qquad Z = \dfrac{T}{c} \qquad \ldots(12)$

where c is the wave velocity. Also $c = \sqrt{\dfrac{T}{\rho}}$

$$\therefore \qquad T = \rho c^2$$

or, $\qquad Z = \dfrac{\rho c^2}{c}$

So, $\qquad Z = \rho c$ $\hfill ...(13)$

Significance of Characteristic Impedance

As already stated, the string behaves both as medium for **propagation of waves** as well as system of **forced oscillations**; so opposition to propagation of wave or **impedance** is offered by the string.

As per equation (13), $Z = \rho c$ and c depends on inertia as well as on elasticity (From the equation $c = \sqrt{\dfrac{T}{\rho}}$). It means that characteristic impedance is the property of the string due to its inertia and elasticity).

Also we have from eqns. (7) and (9) as

$$\frac{\vec{F}}{\vec{v}} = \frac{-T[-ik\,\vec{A}\,e^{i(\omega t - kx)}]}{i\omega\,\vec{A}\,e^{i(\omega t - kx)}} = \frac{k}{\omega}\,T$$

which means that the expression is true for all values of x.

Also from eqn. (13), we find that characteristic impedance is not a complex quantity but is a real quantity. In case of a mechanical oscillator, **complex impedance** is given by

$$\overline{Zm} = r + i(m\omega - S/\omega)$$

For a string neither s nor m is zero. Also $m\omega \neq S/\omega$ for all values of ω.

So in order that impedance be not a complex quantity; $r = 0$ which means that there is no dissipative damping in the string when wave propagation takes place. It shows that there is no loss of energy in the wave propagation. However, in case some dissipative force is present, then the impedance will acquire **complex character**.

Reflection and Refraction of Waves in a String

The phenomenon of reflection and refraction is familiar in light as well as in sound waves as every medium offers a characteristic impedance to the propagation of waves. As a wave reaches the surface of separation of two media, the characteristic impedance also changes suddenly across the medium. It means that reflection or refraction may attribute to the sudden change of characteristic impedance. We can analyse the reflection or refraction of waves in a string and it will be seen that the results are applicable to all types of waves.

Fig. 6.8 shows the two strings *ab* and *bc* joined at say $x = 0$.

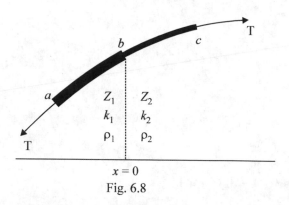

Fig. 6.8

Let Z_1, k_1 and ρ_1 be the characteristic impedance, wave number and mass/unit length of the string *ab* and Z_2, k_2 and ρ_2 are corresponding values for the string *bc*. The strings are stretched under a tension say T and let the harmonic wave be incident through the string *ab*.

Equation for the incident wave may be written as

$$y_i = \vec{A_i}\, e^{i(\omega t - k_1 x)} \qquad\qquad ...(14)$$

As the wave reaches the junction *b* at $x = 0$, a part of the wave may be reflected or returned along the string *ba* and the equation for the reflected wave may be written as

$$\vec{y_r} = \vec{A_r}\, e^{i(\omega t + k_1 x)} \qquad\qquad ...(15)$$

where positive sign with $k_1 x$ shows that the wave travels along *ba* (i.e., negative *x*-direction). Also a part of the wave may be transmitted (refracted) at the junction *b* and will move along the string *bc*. So we may write the equation for refracted or transmitted wave as

$$y_t = A_t\, e^{i(\omega t - k_2 x)} \qquad\qquad ...(16)$$

It may be noted that the following conditions have to be satisfied at the junction (*b*).

1. The displacement must be continuous across the junction which means

$$\vec{y_i} + \vec{y_r} = \vec{y_t}$$

or, $$\vec{A_i}\, e^{i(\omega t - k_1 x)} + \vec{A_r}\, e^{i(\omega t + k_1 x)} = \vec{A_t}\, e^{i(\omega t - k_2 x)}$$

As at the junction, $x = 0$, which means

$$\vec{A_i} + \vec{A_r} = \vec{A_t} \qquad\qquad ...(17)$$

2. At the junction, the transverse force $T(\partial \vec{y}/\partial x)$ should be continuous which may be written as

$$T\frac{\partial}{\partial x}\cdot(\vec{y_i}+\vec{y_r})=T\frac{\partial y_t}{\partial x}$$

Simplifying we get

$$-k_1\,\vec{A_i}\,e^{(\omega t-k_1 x)}+ik_1\,\vec{A_r}\,e^{t(\omega t+k_1 x)}=-ik_2\,\vec{A_t}\,e^{i(\omega t-k_2 x)}$$

Again, because at the junction; $x = 0$, so we have

$$k_1(\vec{A_i}-\vec{A_r})=k_2\,\vec{A_t}$$

or, $$\vec{A_i}-\vec{A_r}=\frac{k_2}{k_1}\vec{A_t} \qquad \qquad ...(18)$$

Also $$k_1=\frac{2\pi}{\lambda_1}=\frac{2\pi v}{v\lambda_1}=\frac{\omega}{c_1}$$

where c_1 is the wave velocity in the string *ab*.
But $c_1 = T/Z_1$, where Z_1 is the impedance of the string *ab*.

Hence $$k_1=\frac{\omega}{T}Z_1$$

Similarly, $$k_2=\frac{\omega}{T}Z_2$$

$$\therefore \qquad \frac{k_2}{k_1}=\frac{Z_2}{Z_1}$$

Putting in equation (18), we have

$$\vec{A_i}-\vec{A_r}=\frac{Z_2}{Z_1}\vec{A_t} \qquad \qquad ...(19)$$

Adding the equations (17) and (19), we get

$$\vec{A_i}=\frac{1}{2}\left[1+\frac{Z_2}{Z_1}\right]\vec{A_t} \qquad \qquad ...(20)$$

Subtracting the equation (19) from equation (17), we get

$$\vec{A_r}=\frac{1}{2}\left[1-\frac{Z_2}{Z_1}\right]\vec{A_t} \qquad \qquad ...(21)$$

Reflection coefficient for amplitude (r_a)

It is defined as the ratio of the amplitude of the reflected wave to that of the incident wave. So we have

$$r_a = \frac{\vec{A_r}}{A_i} = \frac{\frac{1}{2}\left[1 - \frac{Z_2}{Z_1}\right]\vec{A_t}}{\frac{1}{2}\left[1 + \frac{Z_2}{Z_1}\right]\vec{A_t}}$$

or, $\qquad r_a = \dfrac{Z_1 - Z_2}{Z_1 + Z_2}$ ...(22)

Transmission coefficient for amplitude (t_a)

It is defined as the ratio of the amplitude of transmitted wave to the amplitude of incident wave, which may be written as

$$t_a = \frac{\vec{A_t}}{\vec{A_i}}$$ ...(23)

From equation (20), we have

$$t_a = \frac{\vec{A_t}}{\vec{A_i}} = \frac{2Z_1}{Z_1 + Z_2}$$ ...(24)

From the above relations (22) and (24) we can make the following conclusions:

(a) Both r_a and t_a are real quantities as Z_1 and Z_2 are real quantities.

(b) Both r_a and t_a depend only on Z_1 and Z_2.

(c) The above relations are valid for all the frequencies.

(d) The relations are valid for all types of waves.

QUESTIONS

1. What do you mean by reflection and transmission of waves?
2. How does the wave pulse originate in a thin string and in a thick string?
3. Define reflection coefficient and transmission coefficient.
4. Define the characteristic impedance of a string.
5. Derive the expression for the characteristic impedance of a string.
6. Discuss on the significance of the characteristic impedance of a string.
7. Develop the theory of reflection and refraction of waves in a string.

●●●

Chapter 7

Interference

Interference of Light

The phenomenon of interference of light has proved the validity of the wave theory of light. It occurs when two waves of same frequency, same amplitude or nearly equal amplitude travel in the same direction. Due to their superposition, the resultant intensity at different points of the medium undergoes change from point to point. The change in the intensity of light in a medium from point to point is called interference of light.

Thus, the modification or change in the uniform distribution of light intensity in a medium due to superposition of two light waves is called interference of light.

or

The redistribution of energy when two light waves superimpose upon each other is known as interference.

Types of Interference:

There are two types of interference:

(i) Constructive Interference, and

(ii) Destructive Interference.

(i) Constructive Interference

In constructive interference, the crest of one wave falls on the crest of other and the trough of one wave coincides with the trough of other, as shown in (Fig. 7.1a). As a result maximum displacement occurs. Hence a maxima is seen or found. In other words, in the region of superposition there are certain points at which waves superimpose in such a way that resultant intensity is greater than the sum of intensities due to separate waves. Interference at such points is called constructive interference.

(ii) Destructive Interference

In destructive interference, the crest of one wave falls on the trough of other and vice-versa as shown in (Fig. 7.1b). As a result the resultant displacement is minimum producing minima. Therefore, interference consists of a series of maxima and minima known as interference fringes. In other words, in the region of superposition there are certain points at which waves superimpose in such a way that the resultant intensity is less than the sum of intensities due to separate wave and interference at these points is called destructive interference.

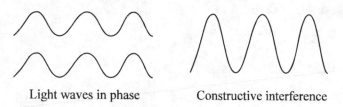

Light waves in phase   Constructive interference

Fig. 7.1(a)

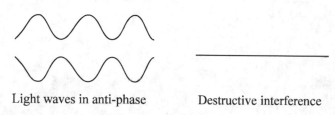

Light waves in anti-phase   Destructive interference

Fig. 7.1(b)

Mathematical Analysis of Interference

Let two waves of same frequency and same wavelength moving in same direction be represented by

$$y_1 = a_1 \sin \omega t \qquad \qquad ...(1)$$

$$y_2 = a_2 \sin(\omega t - \phi) \qquad \qquad ...(2)$$

where ϕ = phase difference between the two waves, and a_1 and a_2 are the amplitude of waves.

The resultant displacement is given by

$$y = y_1 + y_2$$

$$= a_1 \sin \omega t + a_2 \sin(\omega t + \phi)$$

$$= a_1 \sin \omega t + a_2 \sin \omega t \cos \phi + a_2 \cos \omega t \sin \phi$$

$$\Rightarrow \quad y = \sin \omega t (a_1 + a_2 \cos \phi) + a_2 \sin \phi \cos \omega t \qquad \qquad ...(3)$$

Let $\qquad y = A \cos \theta = a_1 + a_2 \cos \phi \qquad \qquad ...(4)$

$$A \sin \theta = a_2 \sin \phi \qquad \qquad ...(5)$$

So, $\qquad y = A \sin \omega t \cos \theta = A \cos \omega t \sin \theta$

$$y = A \sin(\omega t + \theta) \qquad \qquad ...(6)$$

So the resultant wave is also a simple harmonic motion with an amplitude A and phase difference θ.

Expression for Resultant Amplitude

Squaring and adding eqns. (4) and (5) we get

$$A^2\left(\sin^2\theta + \cos^2\theta\right) = \left(a_1 + a_2\cos\phi\right)^2 + a_2^2\sin^2\phi$$

$$A^2 = a_1^2 + a_2^2\cos^2\phi + 2a_1a_2\cos\phi + a_2^2\sin^2\phi$$

$$= a_1^2 + a_2^2\left(\cos^2\phi + \sin^2\phi\right) + 2a_1a_2\cos\phi$$

$$A^2 = a_1^1 + a_2^2 + 2a_1a_2\cos\phi \qquad \qquad ...(7)$$

or, $\qquad A = \sqrt{a_1^2 + a_2^2 + 2a_1a_2\cos\phi} \qquad \qquad ...(8)$

Condition for Maxima

Amplitude 'A' will be maximum if $\cos\phi = +1 = \cos 2n\pi$

$$\Rightarrow \qquad \phi = 2n\pi \qquad \qquad ...(9)$$

where $\qquad n = 0, 1, 2, ...$

Again $\qquad \phi = \dfrac{2\pi}{\lambda}$ path difference

$$2n\pi = \frac{2\pi}{\lambda} \text{ path difference}$$

$$n = \frac{1}{\lambda} \text{ path difference}$$

$$\Rightarrow \qquad \text{Path difference} = n\lambda = \frac{2n\lambda}{2} \qquad \qquad ...(10)$$

So, for a maxima or constructive interference the path difference between two waves should be even multiple of $\dfrac{\lambda}{2}$.

So, maxima amplitude becomes,

$$A_{\max} = \sqrt{a_1^2 + a_2^2 + 2a_1a_2} \qquad \qquad \left[\because \cos\phi = 1\right]$$

$$A_{\max} = \sqrt{\left(a_1 + a_2\right)^2}$$

$$A_{\max} = a_1 + a_2 \qquad \qquad ...(11)$$

So, intensity of maxima

$$I_{\max} \propto A_{\max}^2$$

or, $\qquad I_{\max} \propto \left(a_1 + a_2\right)^2 \qquad \qquad ...(12)$

Condition for Minima

Amplitude 'A' will be minimum, if

$$\cos\phi = -1 = \cos(2n+1)\pi$$

$$\phi = (2n+1)\pi \qquad \qquad ...(13)$$

where $n = 0, 1, 2, 3, ...$

Again $\phi = \dfrac{2\pi}{\lambda}$ path difference

$$(2n+1)\pi = \dfrac{2\pi}{\lambda} \text{ path difference}$$

$$\text{Path difference} = (2n+1)\dfrac{\lambda}{2} \qquad \qquad ...(14)$$

So, for minima or destructive interference path difference should be odd multiple of $\dfrac{\lambda}{2}$.

So minima amplitude become

$$A_{min} = \sqrt{a_1^2 + a_2^2 - 2a_1a_2} \qquad (\because \cos\phi = -1)$$

$$= \sqrt{(a_1 - a_2)^2}$$

$$A_{min} = a_1 - a_2 \qquad \qquad ...(15)$$

So, intensity of minima is

$$I_{min} \propto (a_1 - a_2)^2$$

So, $\dfrac{I_{max}}{I_{min}} = \dfrac{(a_1 + a_2)^2}{(a_1 - a_2)^2}$

Special Cases

If amplitudes are equal, i.e. $a_1 = a_2 = a$

Then, $A_{max} = a + a = 2a$

$$A_{max} = 2a$$

$\therefore \qquad I_{max} = K(4a^2)$

$$A_{min} = 0$$

$\therefore \qquad I_{min} = 0$

Now before interference

$$I_1 = ka^2$$

$$I_2 = ka^2$$

$\therefore$ Total intensity $= (2a^2)\,K$

where, K is a constant of proportionality.

Now after interference,

$$I_{max} = K(4a^2)$$

$$I_{min} = 0$$

$\therefore$ Average intensity $= \dfrac{I_{max} + I_{min}}{K^2(2a^2)}$

Therefore, interference obeys law of conservation of energy and is therefore called redistribution of energy, i.e. same amount of energy $(2a^2)$ has been transferred from minima to maxima.

Intensity Distribution Curve

A graph showing variation of intensity with phase difference is known as intensity distribution graph or curve.

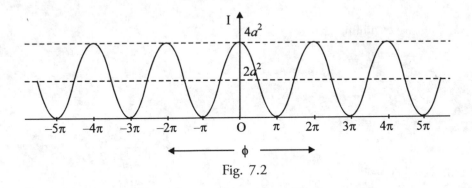

Fig. 7.2

Methods of Producing Interference Pattern of Light

The methods used for obtaining interference are divided into the following two classes:

Division of wavefront

In this method, the wavefront produced by a source is divided into two parts by reflection or refraction or diffraction and later the two parts are brought near so that they can combine to interfere. In this class, the Young's double slit method, the Fresnel's biprism method, the Fresnel's double mirror method and the Lloyd's mirror method are included.

Division of amplitude

In this method, the amplitude of the wave front is divided into two amplitudes by reflection and refraction and later the two are combined to interfere. In this class, the Newton's rings method and the Michelson's method are included.

Conditions for sustained interference

The various conditions for sustained interference are:

(i) The two sources must be monochromatic, i.e. they must emit light of same wavelength or frequency.

(ii) The two sources must have either no phase difference or constant phase difference with respect to time.

(iii) The two interfering sources should be coherent.

(iv) The amplitudes of the interfering waves should be equal or nearly equal.

Coherent Sources

Two sources are said to be coherent if they emit light waves of the same frequency, nearly the same amplitude and are always in phase with each other. It means that the two sources must emit radiations of the same colour (wavelength). For experimental purposes, two virtual sources formed from a single source can act as coherent sources. Some of the coherent sources are laser sources, monochromatic lights, etc.

The resultant intensity at a point due to the superposition of two waves depends on the amplitudes of the individual waves and the phase difference between them. In case of two incoherent waves, the phase difference is not constant but varies rapidly with time. The variation of phase results in rapid variation of intensity at the point of superposition. Thus the intensity at the point does not remain steady. On the other hand if the sources are coherent, they have a constant phase difference which results in a constant intensity at the point of superposition, thus giving a steady interference pattern.

Two independent sources of light cannot be coherent because light is produced due to transition of electrons from higher to lower energy levels in atoms of the source material. Electronic transitions in atoms of different sources or even in different atoms of the same source take place at random. Thus light emitted by two different sources have no definite phase relationship between them. Thus two independent sources cannot be coherent, because coherence implies a constant phase difference.

Incoherent Sources

The two sources of light whose frequencies are not same and the phase difference between the wave emitted by them does not remains constant with respect to time are defined as incoherent sources. In other words, these are light waves which are out of phase with each other. Some of the incoherent sources are sources of light such as sun, star, electric bulb, candle, glowing solid, etc.

Huygens' Wave Theory or Wave Theory of Light (1678)

(i) Light motion is a wave motion.

(ii) Each point in a light source out waves in all directions. These waves fall upon the retina of the eye and cause the sensation of light.

(iii) These waves move in all imaginary, isotropic and hypothetical medium called 'ether'. Ether was assumed to have large elasticity and low density.

(iv) Light waves in the ethereal medium are transverse like water waves.

(v) Light waves lengths are very small.

Huygens' Principle

It explains how a light wave is propagated in any medium. According to it:

(i) When waves spread out from a light source, other particles present around it start to vibrate. The continuous focus of all ether particles vibrating in same phase is called the wavefront.

(ii) Every ether particle on a given wavefront becomes the centre of new waves, called secondary wavelets.

(iii) The secondary wavelets from each point spread out in all direction with the velocity of light.

(iv) The envelope of these wavelets drawn in the forward direction at any instant gives the new wavefront at that instant.

Explanation:

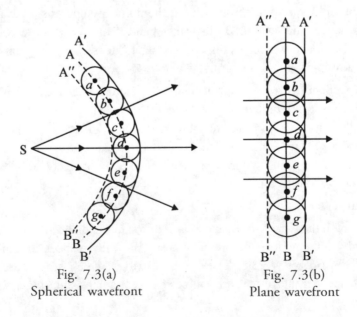

Fig. 7.3(a)
Spherical wavefront

Fig. 7.3(b)
Plane wavefront

According to Huygens, a source S of light sends out waves in all directions, through a hypothetical medium called ether. S is a source of light energy in the form of waves in all directions. After any instant of time '*t*', all the particles of the medium on the surface AB will be vibrating in phase. If υ be the velocity of light then the distance traversed by each secondary wavelet is equal to υ*t*. So circles of radii υ*t* are drawn with centre a, b, c, d, e, f, g, etc. Here AB is the primary wavefront. A wavefront can be defined as the locus of all the points of the medium which are vibrating in phase and are also displaced at the same time. The surfaces A″B′ and A″B″ refer to the secondary wavefront. A′B′ is the forward wavefront and A″B″ is the backward wavefront.

If the distance of the source is small (Fig. 7.3a), the wavefront is spherical. When the source is at a large distance, then any small portion of the wavefront can be considered plane (Fig. 7.3b). Thus rays of light diverging from or converging to a point give rise to a spherical wavefront and a parallel beam of light give rise to a plane wavefront.

Young's Double Slit Experiment

Thomas Young in 1801 reported his experiment on the interference of light. He demonstrated the wave nature of light by performing the famous double slit experiment. He showed that when two wave trains of light superimpose on each other, the resultant intensity in the region of superposition is in general different from the sum of intensities due to each wave separately. This modification in the distribution of intensity of light in the region of superposition is called interference.

Experimental Setup

The interference pattern consists of alternate dark and bright fringes. The formation of bright and dark fringes or bands can be explained on the basis of wave theory. S is a single slit illuminated by a monochromatic source of light, which send out waves in all directions. S_1 and S_2 are two narrow slits placed at a distance 'd' from each other. The two slits are placed near the sources. The wavefront starting from S falls on S_1 and S_2 become the source of secondary wavelets. As secondary wavelets proceed from S_1

and S_2, they superimpose on each other to produce the interference pattern. In other words, the two waves interfere constructively or destructively depending on the nature of path difference between them. The interference fringes are observed on a screen placed at a distance 'D' from the two slits. 'P' is the point of observation where the fringe may be bright dark depending on phase difference.

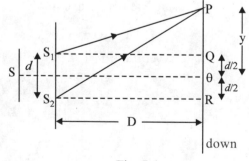

Fig. 7.4

Mathematical Analysis

In Fig. 7.4 $\Delta S_1 PQ$,

$$(S_1P)^2 = (S_1Q)^2 + (PQ)^2$$

$$= D^2 + \left(y - \frac{d}{2}\right)^2$$

$$= D^2 \left[1 + \frac{(y-d/2)^2}{D^2}\right]^{1/2} \qquad \qquad ...(1)$$

In $\Delta S_2 PR$

$$(S_2P)^2 = (S_2R)^2 + (PR)^2$$

$$= D^2 + \left(y + \frac{d}{2}\right)^2$$

$$= D^2 \left[1 + \frac{(y+d/2)^2}{D^2}\right]$$

$$S_2P = D\left[1 + \frac{(y+d/2)^2}{D^2}\right]^{1/2} \qquad \qquad ...(2)$$

Path difference $= (S_2P - S_1P)$

So, $\quad (S_2P - S_1P) = D\left[1 + \frac{(y+d/2)^2}{D^2}\right]^{1/2} - D\left[1 + \frac{(y-d/2)^2}{D^2}\right]^{1/2}$

By using Binomial expansion

$$(1 \pm x)^n \approx 1 \pm nx$$

if n is very less than 1

$$(S_2P - S_1P) = D\left(1 + \frac{(y+d/2)^2}{2D^2}\right) - D\left[1 + \frac{(y-d/2)^2}{2D^2}\right]$$

$$= \cancel{D} + \frac{(y+d/2)^2}{2D} - \cancel{D} - \frac{(y-d/2)^2}{2D} = \frac{(y+d/2)^2}{2D} - \frac{(y-d/2)^2}{2D}$$

$$= \frac{y^2 + d^2/4 + yd - y^2 - d^2/4 + yd}{2D} = \frac{2yd}{2D}$$

$$(S_2P - S_1P) = \frac{yd}{D} \qquad\qquad ...(3)$$

(i) Condition for Bright Fringes

The point of observation 'P' will be bright if path difference is equal to even multiple of $\lambda/2$.

i.e., $\qquad S_2P - S_1P = 2n\dfrac{\lambda}{2}$

$$\frac{yd}{D} = 2n\frac{\lambda}{2}$$

$$y = \frac{n\lambda D}{d} \qquad\qquad ...(4)$$

where n = 0, 1, 2, 3, ...

Eqn. (4) gives the distances of the bright fringes from the point O. At O, the path difference is zero and a bright fringe is formed.

When $\quad n = 1, \qquad\qquad y_1 = \dfrac{\lambda D}{d}$

$\qquad\qquad n = 2, \qquad\qquad y_2 = \dfrac{2\lambda D}{d}$

$\qquad\qquad n = 3, \qquad\qquad y_3 = \dfrac{3\lambda D}{d}$

Similarly, $\qquad\qquad\qquad y_n = \dfrac{n\lambda D}{d}$

(ii) Condition for Dark Fringes

P will be dark if path difference is equal to odd multiple of $\lambda/2$.

$$y = (2n+1)\frac{AD}{2d} \qquad\qquad ...(5)$$

where n = 0, 1, 2, 3, ...

This equation gives the distances of dark fringes from the point O.

When $\quad n = 0, \qquad\qquad y_0 = \dfrac{\lambda D}{2d}$

$$n = 1, \qquad y_1 = \frac{3\lambda D}{2d}$$

$$n = 2, \qquad y_2 = \frac{5\lambda D}{2d}$$

and similarly $\qquad y_n = (2n+1)\dfrac{\lambda D}{2d}$

Fringe Width

The distance between any two consecutive bright fringes or that of dark fringes is called fringe width.

Distance between two consecutive bright fringes

Let y_n and y_{n+1} be the distance of nth and $(n + 1)$th bright fringes.

So, $\qquad \beta = y_{n+1} - y_n$

$$= (n+1)\frac{\lambda D}{d} - \frac{n\lambda D}{d}$$

$$= \frac{n\lambda D}{d} + \frac{\lambda D}{d} - \frac{n\lambda D}{d}$$

$$\Rightarrow \qquad \beta_b = \frac{\lambda D}{d} \qquad\qquad\qquad ...(6)$$

which is independent of n.

Distance between two consecutive dark fringes

$$\beta = y_{n+1} - y_n$$

$$= \frac{[2(n+1)]\lambda D}{2d} - \frac{(2n+1)\lambda D}{2d}$$

$$= \frac{2n\lambda D + 2\lambda D + \lambda D - 2n\lambda D - \lambda D}{2d}$$

$$= \frac{2\lambda D}{2d} = \frac{\lambda D}{d} \qquad\qquad\qquad ...(7)$$

which is independent of n.

So, from eqns. (6) and (7) we have

$$\beta_d = \beta_b \qquad\qquad\qquad ...(8)$$

So the width of the bright fringe is equal to the width of the dark fringe. All the fringes are equispaced, i.e. equal in width and are independent of the order of the fringe 'n'. The breadth of a bright or a dark fringe is however equal to half the fringe width and is equal to $\dfrac{\lambda D}{2d}$. The fringe width is $\dfrac{\lambda D}{d}$. The fringe width is directly proportional to the wavelength of light and distance between screen and the two slits or sources. It is inversely proportional to the distance between the two slits or sources.

i.e., $\qquad \beta \propto \lambda$

$$\beta \propto D$$

Thus, the width of the fringe increases:

(a) with increase in wavelength,

(b) with increase in the distance 'D', and

(c) by bringing the two slits or sources S_1 and S_2 close to each other, i.e. if 'd' is decreased.

Newton's Ring

Newton's ring is an excellent example of interference in an air film of variable thickness. When a plano-convex lens of large focal length is placed in contact with a plane glass plate, we get a thin air film between the lower surface of the lens and the upper surface of the glass plate. The thickness of this film is very small at the point of contact and gradually increases as we proceed outwards from the centre (i.e., from the point of contact). When monochromatic light falls normally on such a film, we get an inner dark spot surrounded by alternatively bright and dark circular rings when seen by reflected light and vice-versa by transmitted light. If the incident light is white, a series of concentric coloured rings are seen round about the point of contact in both the reflected and transmitted light. The air film is wedge-shaped and the interference fringes or rings of equal thickness so formed were first observed by Newton and hence are called Newton's rings.

Experimental Setup

S is a source of monochromatic light that sends light waves through a convex lens L_1 such that 'S' is at its focus. As a result parallel beam of light emerges from the lens and falls on a plane glass plate G_2 inclined at an angle $45°$ to the incident beam. It acts like a partially silvered mirror that reflects the incident beam downwards. A plano-convex lens L_2 of large radius of curvature is placed on another plane glass plate G_1 such that an air film of varying thickness is produced between top surface of glass plate G_1 and the bottom of the plano-convex lens L_2. So the reflected beam proceeding downward gets reflected again at the top and bottom surface of air film. These reflected rays again move upward and passes through the glass plate G_2 to enter into a microscope (M), where interference fringes in the form of alternate bright and dark circular rings are observed. These rings are known as Newton's rings.

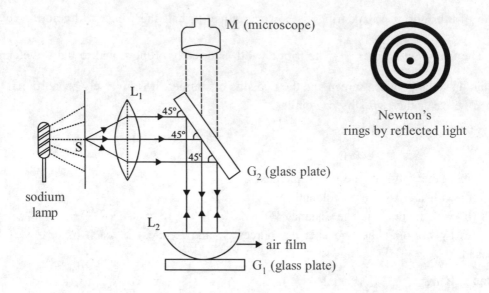

Newton's
rings by reflected light

Fig. 7.5

Theory

Newton's rings are formed due to interference between the waves reflected from the top and bottom surfaces of the air film. It is based on the principle of interference by division of amplitude.

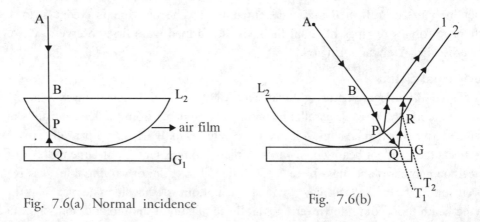

Fig. 7.6(a) Normal incidence

Fig. 7.6(b)

The incident ray AB falls on the air film normally and suffers reflections first at point P which is on the upper surface of the air film and then again at point Q, on the lower portion of the air film. But when a ray of light suffers reflection or refraction at a point backed by a denser medium, then the ray undergoes a phase change of π or a path difference of $\dfrac{\lambda}{2}$. So in this case the ray being reflected at Q backed by glass medium has an additional path difference of $\dfrac{\lambda}{2}$ than the ray reflected at point P. These two reflected rays then produce interference which are in the form of concentric circles.

So from the theory of interference in the thin film, the path difference is given by,

$$\text{Path difference} = 2\mu t \cos r + \frac{\lambda}{2} \qquad \qquad ...(1)$$

where μ = refractive index of the medium of the thin film

        t = thickness of the film

        r = angle of refraction

        λ = wavelength of light.

For normal incidence, $r = 0$

and for air film $\mu = 1$

So, path difference $= \left(2t + \dfrac{\lambda}{2}\right)$

For constructive interference or bright ring

Condition for brightness is that path difference must be even multiple of $\dfrac{\lambda}{2}$.

i.e., $\Delta = 2n\dfrac{\lambda}{2}$

$$2t + \frac{\lambda}{2} = 2n\frac{\lambda}{2}$$

$$2t = n\lambda - \frac{\lambda}{2} \quad \Rightarrow \quad 2t = \frac{2n\lambda - \lambda}{2}$$

$$\Rightarrow \qquad 2t = (2n - 1)\frac{\lambda}{2} \qquad \qquad ...(2)$$

where $n = 1, 2, 3, ...$

So, eqn. (2) is the condition for maxima.

For destructive interference or dark ring

For dark ring, $\Delta = (2n + 1)\dfrac{\lambda}{2}$

$$\Rightarrow \qquad 2t + \frac{\lambda}{2} = (2n + 1)\frac{\lambda}{2}$$

$$\Rightarrow \qquad 2t = (2n + 1 - 1)\frac{\lambda}{2}$$

$$\Rightarrow \qquad 2t = 2n\frac{\lambda}{2}$$

$$2t = n\lambda \qquad \qquad ...(3)$$

Eqn. (3) is the condition for minima.

Hence it is clear from eqns. (2) and (3) that for a bright or dark ring of any particular order, t should be constant. But here in the air film the locus of points having the same thickness is a circle with its centre at the point of contact. Hence, the fringes are circular rings having the common centre at the point of contact.

Relation Between Radius of Curvature of Plano-Convex Lens and Thickness of the Film

Here   OP = OB = Radius of curvature of plano-convex lens.

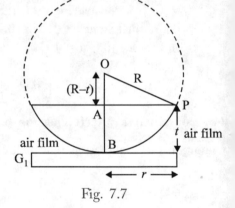

r = radius of Newton's

t = thickness of air film

Now, In Δ OAP

$$(OP)^2 = (OA)^2 + (AP)^2$$

$$R^2 = (R - t)^2 + r^2$$

$$R^2 = R^2 + t^2 - 2Rt + r^2$$

$$\Rightarrow \qquad t^2 - 2Rt + r^2 = 0$$

$$\Rightarrow \qquad r^2 = 2Rt - 4^2 \qquad \qquad ...(4)$$

Fig. 7.7

But the thickness of the film is very small and r is very large, so t^2 is neglected in comparison to 2 Rt.

So,   eqn. (4) becomes

$$r^2 = 2\,Rt \quad \Rightarrow \quad 2t = \frac{r^2}{R} \qquad \qquad ...(5)$$

(i) For bright ring

Using eqns. (5) and (2), we get

$$\frac{r^2}{R} = (2n - 1)\frac{\lambda}{2} \qquad \Rightarrow \qquad r^2 = (2n - 1)\frac{\lambda R}{2} \qquad \qquad ...(6)$$

where $r = \dfrac{D}{2}$, where D is the diameter of the Newton's ring.

$$\Rightarrow \quad \frac{(D/2)^2}{R} = (2n-1)\frac{\lambda}{2}$$

$$\Rightarrow \quad \frac{D^2}{4R} = (2n-1)\frac{\lambda}{2}$$

$$\Rightarrow \quad D^2 = (2n-1)\frac{\lambda}{2} \times 4R = (2n-1)2\lambda R$$

$$\Rightarrow \quad D_n = \sqrt{2(2n-1)\lambda R} \qquad \qquad ...(7)$$

$$\Rightarrow \quad D_n \propto \sqrt{(2n-1)}, \qquad \text{where} \quad n = 1, 2, 3, ...$$

Here n stands for nth ring of bright light. Eqn. (7) gives diameter of nth bright fringe. So diameter of nth bright ring is directly proportional to $\sqrt{2n-1}$, i.e. to the square root of odd natural numbers.

(ii) For dark ring

The value of t in eqn. (3) gives

$$\frac{r^2}{R} = 2n\frac{\lambda}{2} \qquad \Rightarrow \qquad r^2 = n\lambda R \qquad \qquad ...(8)$$

$$\Rightarrow \quad \frac{(D/2)^2}{R} = 2n\frac{\lambda}{2} \qquad \Rightarrow \qquad \frac{D^2}{4R} = n\lambda$$

$$\Rightarrow \quad D^2 = 4n\lambda R \qquad \Rightarrow \qquad D_n = \sqrt{4n\lambda R}$$

$$\Rightarrow \quad D_n = 2\sqrt{n\lambda R} \qquad \qquad ...(9)$$

Since λ and R are constant so,

$$D_n \propto \sqrt{n}$$

Eqn. (9) gives the diameter of nth dark fringe. So diameter of nth dark ring is directly proportional to square root of natural numbers. Also the rings get narrower as diameter increases.

Central Spot in Newton's Ring is Dark for Reflected Light

At the centre of the Newton's ring, thickness of air film is zero as the plano-convex lens and glass plate touch each other at this point, i.e. $t = 0$. So, the path difference between the two reflected beams is given as,

$$2\mu t \cos r + \frac{\lambda}{2} = n\lambda \qquad \qquad [\because \ 2t = n\lambda]$$

$$0 + \frac{\lambda}{2} = n\lambda$$

$$n\lambda = \frac{\lambda}{2} \qquad \qquad ...(1)$$

where $n\lambda$ is the path difference.

So, the path difference is the odd multiple of $\frac{\lambda}{2}$. This is the condition for destructive interference which gives rise to minima. So the central spot is Newton's ring appears dark when seen by reflected light.

Applications of Newton's Ring

(1) Determination of wavelength of light using Newton's ring

The wavelength of a monochromatic source of light can be easily determined in the laboratory by Newton's ring apparatus.

Theory:

For nth dark ring, $\qquad D_n^2 = 4n\lambda R$ $\qquad \qquad ...(1)$

where $\qquad \lambda$ = wavelength of source of light

$\qquad \qquad$ R = radius of curvature of the plano-convex lens

$\qquad \qquad$ D_n = diameter of nth dark ring.

Similarly for $(n + p)$th dark ring, relation (1) becomes,

$$D_{n+p}^2 = 4(n+p)\lambda R \qquad \qquad ...(2)$$

where $\qquad P = 1, 2, 3, ...$

Subtracting eqn. (1) from (2)

$$D_{n+p}^2 - D_n^2 = 4(n+P)R\lambda - 4n\lambda R$$

$$= 4nR\lambda + 4P\lambda R - 4n\lambda R$$

$$= 4P\lambda R$$

$$\Rightarrow \qquad \lambda = \frac{D_{n+p}^2 - D_n^2}{4PR} \qquad \qquad ...(3)$$

From eqn. (3) the value of λ can be easily found out.

R can be found out by using a spherometer.

$$R = \frac{d^2}{6h} + \frac{h}{2} \qquad \qquad ...(4)$$

where d = distance between the legs of the spherometer.

$$d = \frac{d_1 + d_2 + d_3}{3}$$

h = height of curve surface.

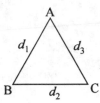

Fig. 7.8

Plotting of graph:

A graph is plotted between square of the diameters of various dark rings and the order of dark rings. The graph obtained is shown in the Fig. 7.9.

$$\text{slope} \quad = \tan\theta = \frac{BC}{AC} = \frac{D_{n+p}^2 - D_n^2}{n + p - n}$$

$$\text{slope} \quad = \frac{D_{n+p}^2 - D_n^2}{P} \qquad \qquad ...(5)$$

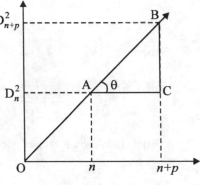

Now using eqns. (4) and (5) in eqn. (3), the value of λ is found out experimentally.

Fig. 7.9

i.e., $\qquad \lambda = \frac{\text{slope}}{4R} \qquad \qquad ...(6)$

(2) Determination of refractive index of liquid by using Newton's ring

When air film is used, then a diameter of nth dark ring is given by

$$D_n^2 = 4n\lambda R \qquad \qquad ...(1)$$

So, $(n + p)$th dark ring diameter is given by

$$D_{n+p}^2 = 4(n + p)\lambda R \qquad \qquad ...(2)$$

Subtracting eqn. (1) from (2), we get

$$D_{n+p}^2 - D_n^2 = 4(n + p)\lambda R - 4n\lambda R$$

$$\Rightarrow \qquad D_{n+p}^2 - D_n^2 = 4P\lambda R \qquad \qquad ...(3)$$

Let μ be the refractive index of liquid. Now if the air film is replaced by liquid film then the diameter of nth and $(n + p)$th dark rings are given by

$$D'^2_n = \frac{4nR\lambda}{\mu} \qquad \qquad ...(4)$$

$$D'^2_{n+p} = \frac{4(n+p)R\lambda}{\mu} \qquad \qquad ...(5)$$

$$\therefore \qquad D'^2_{n+p} - D'^2_n = \frac{4PR\lambda}{\mu} \qquad \qquad ...(6)$$

Dividing eqn. (3) and eqn. (6) we have

$$\frac{D'^2_{n+p} - D'^2_n}{D^2_{n+p} - D^2_n} = \frac{4PR\lambda}{\mu} \times \frac{1}{4P\lambda R}$$

$$\Rightarrow \qquad \frac{D'^2_{n+p} - D'^2_n}{D^2_{n+p} - D^2_n} = \frac{1}{\mu}$$

$$\Rightarrow \qquad \mu = \frac{D^2_{n+p} - D^2_n}{D'^2_{n+p} - D'^2_n} \qquad \qquad ...(7)$$

Eqn. (7) shows the relation to determine the refractive index of liquid by using Newton's ring.

Interference in thin film

Newton and Hooke observed and developed the interference phenomenon due to multiple reflections from the surface of thin transparent materials. The interference in case of thin films takes place due to:

(1) reflected light, and (2) transmitted light

(1) Interference in a thin film by reflected light

Consider a thin transparent film of thickness 't' and refractive index μ as shown in (Fig. 7.10a). When a light ray SA is incident on the film, it is partly reflected along AR_1 and partly refracted along AB. At point B of lower surface of thin film, the light ray AB is reflected along BC. Finally, it emerges along CR_2.

Since the reflected rays AR_1 and CR_2 are derived from the same incident ray SA, they act as coherent light rays and produce an interference pattern. Let i and r be the angles of incidence and refraction. Draw the normals BM and CN on AC and AR_1 from B and C respectively. The path difference between the reflected rays AR_1 and CR_2 is then given by

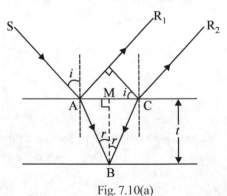

Fig. 7.10(a)
Interference in a thin film by reflected light

Δ = path ABC in thin film–path AN in air

i.e., $\quad \Delta = \mu(AB + BC) - AN$ ...(1)

In right-angled triangles AMB and CMB, we have

$$\cos r = \frac{BM}{AB} = \frac{BM}{BC}$$

$\Rightarrow \qquad AB = BC = \dfrac{t}{\cos r} \qquad \left[\because \; BM = t\right]$ ...(2)

Also in right-angled triangle ANC, we have

$$\sin i = \frac{AN}{AC}$$

$\Rightarrow \qquad AN = AC \sin i = (AM + MC) \sin i$ ...(3)

Now again in right-angled Δ AMB and Δ CMB, we have

$$\tan r = \frac{AM}{MB} = \frac{CM}{MB}$$

$\Rightarrow \qquad AM = CM = MB \tan r = t \tan r$ ...(4)

Using eqn. (4) in eqn. (3), we have

$$AN = (AM + MC)\sin i$$

$$= (t \tan r + t \tan r) \sin r$$

$$= (2t \tan r)\sin i$$

$$= 2t\left(\frac{\sin r}{\cos r}\right)\sin i$$

$$= 2t\left(\frac{\sin r}{\cos r}\right)\mu \sin r \qquad \left[\because \; \mu = \frac{\sin i}{\sin r}\right]$$

$$AN = 2\mu t \frac{\sin^2 r}{\cos e} \qquad \text{...(5)}$$

Using eqns. (2) and (5) in eqn. (1), we have

$$\Delta = \mu(AB + BC) - AN$$

$$= \mu\left(t/\cos r + t/\cos r\right) - 2\mu t\left(\sin^2 r/\cos r\right)$$

$$= \mu\left(2t/\cos r\right) - 2\mu t\left(\sin^2 r/\cos r\right)$$

$$= 2\mu t/\cos r - 2\mu t \sin^2 r/\cos r$$

$$= \frac{2\mu t}{\cos r}\left[1 - \sin^2 r\right]$$

$$= \frac{2\mu t}{\cos r}\left(\cos^2 r\right)$$

$$\Delta = 2\mu r \cos r \qquad\qquad ...(6)$$

Since the ray SA is reflected from a denser medium, an additional path difference of $\lambda/2$ is produced between the reflected rays AR_1 and CR_2. Hence eqn. (6) becomes

$$\Delta = \left(2\mu t \cos r\right) - \frac{\lambda}{2}$$

This is the actual path difference produced between the interfering reflected rays AR_1 and CR_2.

Conditions for constructive interference and destructive interference

For constructive interference, the path difference will be an even multiple of $\lambda/2$, i.e.

$$\Delta = 2n\left(\lambda/2\right) \qquad \text{where,} \quad n = 0, 1, 2, 3, ...$$

$$\Rightarrow \qquad 2\mu t \cos r - \frac{\lambda}{2} = 2n\left(\frac{\lambda}{2}\right)$$

$$\Rightarrow \qquad 2\mu t \cos r = 2n\frac{\lambda}{2} + \frac{\lambda}{2}$$

$$\Rightarrow \qquad 2\mu t \cos r = \left(2n+1\right)\frac{\lambda}{2} \qquad\qquad ...(7)$$

This is the required condition for constructive interference, i.e. the film will appear bright.

For destructive interference, the path difference will be an odd multiple of $\lambda/2$, i.e.

$$\Delta = \left(2n-1\right)\left(\frac{\lambda}{2}\right) \qquad\qquad \text{where} \quad n = 1, 2, 3, 4, ...$$

$$\left(2\mu t \cos r\right) = -\frac{\lambda}{2} = \left(2n-1\right)\frac{\lambda}{2}$$

$$2\mu t \cos r = \left(2n-1\right)\frac{\lambda}{2} + \frac{\lambda}{2} = \left(2n-1+1\right)\frac{\lambda}{2}$$

$$2\mu t \cos r = 2n\frac{\lambda}{2} \quad 2\mu t \cos r = n\lambda \qquad\qquad ...(8)$$

This is the required condition for destructive interference, i.e. the film will appear dark.

(2) Interference in a thin film by transmitted light

Consider a thin transparent film of uniform thickness 't' and refractive index μ as shown in (Fig. 7.10b). When a light ray SA is incident on the film, it is partly transmitted along AB. At point B of lower surface of thin film, the light ray is partly reflected along BC and partly transmitted along BT_1. At point C of upper surface of thin film, the light ray BC is partly reflected along CD. Finally, it emerges along DT_2. Since the transmitted rays BT_1 and DT_2 are derived from the same incident ray SA, they act as coherent light rays and produce interference pattern. Let i and r be the angles of incidence and refraction, respectively.

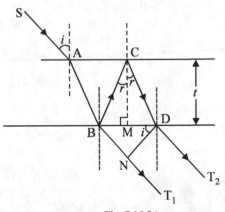

Fig. 7.10(b)
Interference in a thin film by transmitted light

Draw the normals CM and DN on BD and BT_1 from C and D respectively. The path difference between the transmitted rays BT_1 and DT_2 is given by

$$\Delta = \text{Path BCD in thin film} - \text{Path BN in air}$$

$$= \mu(BC + CD) - BN \qquad \qquad ...(1)$$

In the right-angled triangles CMB and CMD, we have

$$\cos r = CM/BC = CM/CD$$

$$BC = CD = t/\cos r \qquad [\because \quad CM = t] \qquad \qquad ...(2)$$

Also, in the right-angled triangle BND, we have

$$\sin i = BN/BD$$

$$BN = BD \sin i$$

$$\Rightarrow \qquad BN = (BM + MD)\sin i \qquad \qquad ...(3)$$

Now, again in right-angled triangles CMB and CMD, we have

$$\tan r = BM/CM = MD/CM$$

$$\Rightarrow \qquad BM = MD = CM \tan r = t \tan r$$

Using eqns. (3) in (2) we have

$$BN = (t \tan r + t \tan r)\sin i$$

$$= (2t \tan r)\sin i$$

$$= 2t(\sin r/\cos r)\sin i$$

$$= 2t(\sin r/\cos r)\mu \sin r \qquad \left[\because \ \mu = \frac{\sin i}{\sin r}\right]$$

$$BN = 2\mu t \frac{\sin^2 r}{\cos r} \qquad \qquad ...(4)$$

Using eqns. (4) and (2) in eqn. (1) we have

$$\Delta = \mu(BC + CD) - BN$$

$$= \mu(t/\cos r + t/\cos r) - 2\mu t \sin^2 r/\cos r$$

$$= \mu(2t/\cos r) - 2\mu t \sin^2 r/\cos r$$

$$= \frac{2\mu t}{\cos r} - \frac{2\mu t \sin^2 r}{\cos r}$$

$$= \frac{2\mu t}{\cos r}\left[1 - \sin^2 r\right]$$

$$= \frac{2\mu t}{\cos r}\cos^2 r$$

$$\Delta = 2\mu t \cos r \qquad \qquad ...(5)$$

This is the actual path difference produced between the interfering transmitted rays BT_1 and DT_2.

Conditions for constructive interference and destructive interference

For constructive interference, the path difference will be an even multiple of $\lambda/2$, i.e.

$$\Delta = 2n\left(\frac{\lambda}{2}\right) \qquad \qquad \text{where,} \quad n = 0, 1, 2, 3, ...$$

$$2\mu t \cos r = 2n\left(\frac{\lambda}{2}\right)$$

$$2\mu t \cos r = n\lambda \qquad \qquad ...(6)$$

This is the required condition for constructive interference, i.e. the film will appear bright. For destructive interference, the path difference will be an odd multiple of $\lambda/2$, i.e.

$$\Delta = (2n-1)(\lambda/2) \qquad \qquad \text{where,} \quad n = 1, 2, 3, ...$$

$$\Rightarrow \qquad 2\mu t \cos r = (2n-1)\frac{\lambda}{2} \qquad \qquad ...(7)$$

This is the required condition for destructive interference. In this case, the film will appear dark.

(3) Interference in a wedge-shaped thin film by reflected light

Consider a thin transparent wedge-shaped film of varying thickness and refractive index μ as shown in (Fig. 7.10c). When a light ray SA is incident on the film, it is partly reflected along AR_1 and partly refracted along AB. At point B of lower surface of thin film, the light ray AB is reflected along BC. Finally, it emerges along CR_2. Since the reflected rays AR_1 and CR_2 are derived from the same incident ray SA, they act as coherent light rays and produce an interference pattern.

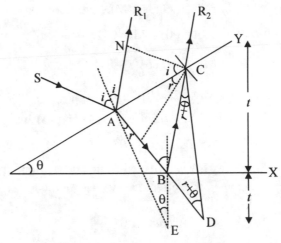

Fig. 7.10(c)
Interference in a wedge-shaped thin film by reflected light

Draw the normals CM and CN on AB and AR_1 respectively from C.
The path difference between the reflected rays AR_1 and CR_2 is given by

$$\Delta = \text{Path ABC in thin film} - \text{Path AN in air}$$

$$= \mu(AB + BC) - AN$$

$$= \mu(AM + MB + BC) - AN \qquad \text{...(1)}$$

Let i and r be the angles of incidence and refraction respectively. From Snell's law, we have

$$\mu = \frac{\sin i}{\sin r} = \frac{AN/AC}{AM/AC} = \frac{AN}{AM}$$

$$\Rightarrow \qquad AN = \mu \, AM \qquad \text{...(2)}$$

Using eqn. (2) in eqn. (1), we have

$$\Delta = \mu(AM + MB + BC) - \mu \, AM$$

$$\Delta = \mu(MB + BC) \qquad \text{...(3)}$$

Now draw a normal CP on lower surface OX. Extending CP and AB, they meet at a point D.

In right-angled triangles CPB and DPB, we have

$$CP = PD = t, \quad BC = BD$$

and $\quad \angle BPD = \angle BPC = 90°$

Hence from geometry, $\angle BDP = \angle BCP = r + \theta$.

Then, Eqn. (3) becomes

$$\Delta = \mu(MB + BD)$$

$\Rightarrow$ $\Delta = \mu\ MD$ ...(4)

Also, in right triangle CMD, we have

$\cos(r+\theta) = MD/CD$

$MD = CD\cos(r+\theta)$

$MD = 2t\cos(r+\theta)$ ...(5)

Eliminating MD from eqns. (4) and (5) we get

$\Delta = 2\mu t\cos(r+\theta)$ ...(6)

Since, ray SA is reflected from a denser medium, an additional path difference of $\lambda/2$ is produced between the reflected rays AR_1 and CR_2. Hence, eqn. (6) becomes

$$\Delta = 2\mu t\cos(r+\theta) - \frac{\lambda}{2}$$ ...(7)

This is the actual path difference produced between the interfering reflected rays AR_1 and CR_2.

Conditions for constructive interference and destructive interference

For constructive interference, the path difference will be an even multiple of $\lambda/2$, i.e.

$$\Delta = 2n\left(\frac{\lambda}{2}\right)$$ where, $n = 0, 1, 2, 3, ...$

$$2\mu t\cos(r+\theta) - \frac{\lambda}{2} = 2n\left(\frac{\lambda}{2}\right)$$

$\Rightarrow$ $2\mu t\cos(r+\theta) = 2n(\lambda/2) + \lambda/2$

$\Rightarrow$ $2\mu t\cos(r+\theta) = (2n+1)\lambda/2$ ...(8)

This is the required condition for constructive interference. In this case, the film will appear bright.

For destructive interference, the path difference will be an odd multiple of $\lambda/2$, i.e.

$$\Delta = (2n=1)\frac{\lambda}{2}$$ where, $n = 1, 2, 3, 4, ...$

$$2\mu t\cos(r+\theta) - \frac{\lambda}{2} = (2n-1)\frac{\lambda}{2}$$

$$2\mu t\cos(r+\theta) = (2n-1)\frac{\lambda}{2} + \frac{\lambda}{2}$$

$$2\mu t\cos(r+\theta) = 2n\frac{\lambda}{2}$$

$$2\mu t \cos(r + \theta) = n\lambda \qquad ...(9)$$

This is the required condition for destructive interference. In this case, the film will appear dark.

Fringe width in a wedge-shaped film

Consider a thin transparent wedge-shaped film of varying thickness and refractive index μ as shown in (Fig. 7.10d). Suppose that AR_1 and CR_2 are the light rays formed by the reflection of an incident ray SA from the upper and the lower surfaces of the film. Draw a normal AM on OX. Let AM = t.

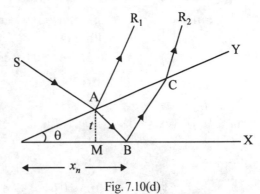

Fig. 7.10(d)
Fringe width in a wedge-shaped film

The condition for dark fringe (destructive interference) at A is given by

$$2\mu t \cos(r + \theta) = n\lambda \qquad ...(1)$$

Let the nth dark ring be located at M at a distance x_n for the wedge O of the film. Then in $\triangle$AMO, we have

$$\tan\theta = \frac{t}{x_n}$$

$$\Rightarrow \quad t = x_n \tan\theta \qquad ...(2)$$

Using eqn. (2) in eqn. (1), we have

$$2\mu t\, x_n \tan\theta \cos(r + \theta) = n\lambda \qquad ...(3)$$

Similarly, if $(n + 1)$ the dark ring is located at a distance x_{n+1} from the wedge O of the film, then from eqn. (3), we have

$$2\mu x_{n+1} \tan\theta \cos(r + \theta) = (n+1)\lambda \qquad ...(4)$$

Subtracting eqn. (3) from eqn. (4), we get

$$2\mu(x_{n+1} - x_n)\tan\theta \cos(r + \theta) = (n+1)\lambda - n\lambda$$

$$\Rightarrow \quad 2\mu(x_{n+1} - x_n)\tan\theta \cos(r + \theta) = \lambda \qquad ...(5)$$

Hence the separation between the two successive dark fringes, i.e. fringe width of bright fringe is given by

$$W = x_{n+1} - x_n = \frac{\lambda}{2\mu \tan\theta \cos(r + \theta)} \qquad ...(6)$$

This eqn. is independent of n. It shows that all the bright fringes have the same fringe width. Similarly, we can show that all the dark fringes have the same fringe width given by the above expression.

SOLVED EXAMPLES

Q.1. Light from a narrow slit passes through two parallel slits 0.4 mm apart and the fringes when measured at a distance of 40 cm from the slit are 0.5 mm apart. Find the wavelength of light.

Sol. d = 0.4 mm = 0.04 cm

D = 40 cm

β = 0.5 mm = 0.05 cm

$$\beta = \frac{\lambda D}{d}$$

$$\Rightarrow \quad \lambda = \beta \frac{d}{D}$$

$$= \frac{0.05 \times 0.04}{40} = \frac{5 \times 4 \times 10^{-4}}{40}$$

$$= 0.5 \times 10^{-4} \text{cm} = 5 \times 10^{-5} \text{cm}.$$

Q.2. Two straight and narrow parallel slits 0.3 mm apart are illuminated by a monochromatic source of wavelength 5.9×10^{-5}cm. Fringes are obtained at a distance of 30 cm from the slit. Find the width of the fringes.

Sol. d = 0.3 mm = 0.03 cm

λ = 5.9 × 10⁻⁵cm

D = 30 cm

$$\beta = \frac{\lambda D}{d} = \frac{5.9 \times 10^{-5} \times 30}{0.03}$$

$$= \frac{5.9 \times 3 \times 10^{-5}}{3 \times 10^{-2}}$$

$$= 59 \times 10^{-3} = 0.059 \text{ cm}.$$

Q.3. Light from a narrow slit passes through two parallel slits 0.2 cm apart. Find the wavelength of the light if the interference fringes on a screen one metre distant from the two slits are 0.295 mm apart.

Sol. d = 0.2 cm,

λ = ?

D = 1 m = 100 cm

β = 0.295 mm = 0.295 × 10⁻¹cm

$$\beta = \frac{\lambda D}{d}$$

$$\Rightarrow \quad \lambda = \beta \frac{d}{D}$$

$$= \frac{0.295 \times 10^{-1} \times 0.2}{100}$$

$$= \frac{295 \times 2 \times 10^{-1} \times 10^{-3} \times 10^{-1}}{100}$$

$$= 590 \times 10^{-5} \times 10^{-2} = 59 \times 10^{-6} \text{ cm}.$$

Q.4. **In a Young's double slit experiment the slits are 0.01 cm apart. The interference pattern appears at a screen 1 m away. If the wavelength of incident light is 4000 Å, compute:**

 (i) angular displacement of the first minimum from the line of symmetry, and

(ii) width of the central bright fringes.

Sol. $d = 0.01 \text{ cm} = 1 \times 10^{-2} \text{cm}$

   $D = 1 \text{ m} = 100 \text{ cm}$

   $\lambda = 4000 \text{ Å} = 4000 \times 10^{-8} \text{cm}$

       $= 4 \times 10^{-5} \text{cm}$

(i) $y = (2n+1) \dfrac{\lambda D}{2d}$

Here, $n = 0$

       $y = \dfrac{\lambda D}{2d}$

    $\tan \theta = \dfrac{(y - d/2)}{D}$

As θ is small, $\tan \theta = \theta$

    $\theta = \dfrac{y - d/2}{D}$

       $= \dfrac{\lambda D}{2d D} - \dfrac{d}{2D}$

$$= \frac{\lambda}{2d} - \frac{d}{2D}$$

$$= \frac{1}{2}\left[\frac{\lambda}{d} - \frac{d}{D}\right]$$

$$= \frac{1}{2}\left[\frac{4 \times 10^{-5}}{10^{-2}} - \frac{10^{-2}}{10^{2}}\right]$$

$$= \frac{1}{2}\left[4 \times 10^{-3} - 10^{-4}\right]$$

$$= \frac{1}{2}\left[40 \times 10^{-4} - 10^{-4}\right]$$

$$= \frac{1}{2} \times 10^{-4}\left[40 - 1\right] = \frac{1}{2} \times 10^{-4} \times 39$$

$$= 19 \times 10^{-4} = 0.0019 = 0.002 \text{ rad.}$$

(ii) $\qquad \beta = \dfrac{\lambda D}{d}$

$$= \frac{4 \times 10^{-5} \times 10^{2}}{10^{-2}}$$

$$= 4 \times 10^{-1} = 0.4 \text{ cm.}$$

Q.5. **In Young's experiment while using a source of light of 5000 × 10⁻¹⁰m wavelength the fringe width obtained is 0.60 cm. If the distance between screen and the slit is reduced to half, what should be the wavelength of light source to get fringes 0.40 cm wide?**

Sol. $\qquad \lambda_1 = 5000 \times 10^{-10} \text{m}$

$$= 5000 \times 10^{-8} \text{cm} = 5 \times 10^{-5} \text{cm}$$

$$\beta_1 = 0.6 \text{ cm} = 6 \times 10^{-1} \text{cm}$$

The distance between screen and the slit = D_1

$$\beta_1 = \frac{\lambda_1 D_1}{d} \qquad\qquad\qquad ...(1)$$

For the 2nd case, $\lambda_2 = ?$

$$\beta_2 = 0.4 \text{ cm} = 4 \times 10^{-1} \text{cm}$$

$$D_2 = \frac{D_1}{2}$$

$$\beta_2 = \frac{\lambda_2 D_2}{d} \qquad \qquad ...(2)$$

Dividing eqn. (1) by eqn. (2) we get,

$$\frac{\beta_1}{\beta_2} = \frac{\lambda_1 D_1}{d} \times \frac{d}{\lambda_2 D_2}$$

$$= \frac{\lambda_1 D_1}{\lambda_2 D_2}$$

$$= \frac{\lambda_1 D_1 2}{\lambda_2 D_1} = \frac{2\lambda_1}{\lambda_2}$$

$$\Rightarrow \qquad \lambda_2 = \frac{2\lambda_1 \beta_2}{\beta_1}$$

$$= \frac{2 \times 5 \times 10^{-5} \times 4 \times 10^{-1}}{6 \times 10^{-1}}$$

$$= \frac{20}{3} \times 10^{-5}$$

$$= 6.66 \times 10^{-5} \text{cm}$$

$$= 6.66 \times 10^{-2} \text{m}.$$

Q.6. **A source of red light ($\lambda = 7000$ Å) produces interference through two narrow slits spaced at a distance of 0.01 cm. At what distance from the slits should a screen be placed so that first few interference bands are spaced 1 cm apart?**

Sol. $\lambda = 7000 \text{ Å}$

$$= 700 \times 10^{-8} \text{cm} = 7 \times 10^{-5} \text{cm}$$

$d = 0.01 \text{ cm} = 10^{-2} \text{cm}$

$D = ?$

$\beta = 1 \text{ cm}$

$$\beta = \frac{\lambda D}{d}$$

$$\Rightarrow \quad D = \frac{\beta\,d}{\lambda} = \frac{1\times 10^{-2}}{7\times 10^{-5}}$$

$$= \frac{1}{7}\times 10^{-2}\times 10^{5}$$

$$= \frac{1}{7}\times 10^{3} = \frac{1000}{7}$$

$$= 142.85\,\text{cm} = 142.9\,\text{cm}.$$

Q.7. **A source of light of wavelength 500 Å is placed at one end of a table 200 cm long and 5 mm above its flat well-polished top. Find the fringe width of the interference bands located on a screen at the end of the table.**

Sol.   Distance of source s from the table = 5 mm = 0.5 cm

Distance of s' from table = 0.5 cm.

If 'd' is the distance between s and s'.

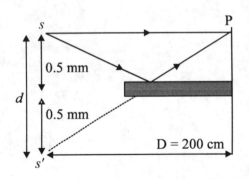

$$d = 0.5 + 0.5 = 1\,\text{cm}$$

$$D = 200\,\text{cm}$$

$$\lambda = 500\,\text{Å} = 500\times 10^{-8}\,\text{cm}$$

$$= 5\times 10^{-6}\,\text{cm}$$

$$\beta = \frac{\lambda D}{d}$$

$$\beta = \frac{5\times 10^{-6}\times 200}{1}$$

$$= \frac{1000\times 10^{-6}}{1} = 10^{-3}\,\text{cm}.$$

Q.8. **Interference bands are produced by a Fresnel's biprism in the focal plane of a reading microscope. The focal plane is 100 cm distant from the slit. A lens is inserted between the biprism and microscope and gives two images of the slit for two postilions of lens. In one separation between them is 4.05 mm and in other 2.90 mm. If sodium light is used, find the distance between interference bands. 'λ' for sodium light is 5886×10^{-8} cm.**

Sol. $\lambda = 5886\times 10^{-8}\,\text{cm}$

$$D = 100\,\text{cm}$$

$$x = 4.05\,\text{mm} = 0.405\,\text{cm}$$

$$y = 2.90 \text{ mm} = 0.290 \text{ cm}$$

$$d = \sqrt{xy} = \sqrt{0.405 \times 0.290}$$

$$\beta = \frac{\lambda D}{d} = \frac{5886 \times 10^{-8} \times 100}{\sqrt{0.405 \times 0.290}}$$

$$= 0.017 \text{ cm}.$$

Q.9. **If two sources of intensity ratio 100 : 1 interfere, deduce the ratio of intensities of the maxima and minima of interference.**

Sol. Given $\dfrac{I_1}{I_2} = \dfrac{100}{1} = 100$

So, $\quad r = \sqrt{I_1/I_2} = \sqrt{100} = 10$

$\therefore \quad \dfrac{I_{max}}{I_{min}} = \dfrac{(r+1)^2}{(r-1)^2}$

$$= \frac{121}{81} = \frac{3}{2}.$$

Q.10. **In the interference with two sources of intensities I_0 and $4I_0$, find the intensities at points where phase differences are: (i) 0, and (ii) $3\pi/2$.**

Sol. $\qquad I_1 = I_0$ and $I_2 = 4I_0$

So, $\qquad a_1 = a_0$ and $a_2 = 2a_0 \qquad$ ($\because$ Intensity square of the amplitude)

Intensity at any point of phase difference φ

$$I_1 = a_1^2 + a_2^2 + 2a_1 \, a_2 \cos\varphi$$

(i) So, when $\varphi = 0$

$$I = (a_1 + a_2)^2 = (a_0 + 2a_0)^2$$

$$= (3a_0)^2 = 9a_0^2 = 9I_0$$

(ii) When, $\varphi = 3\pi/2$

$$I = a_1^2 + a_2^2 + 2a_1 a_2 \cos(3\pi/2)$$

$$= a_1^2 + a_2^2 = I_0 + 4I_0 = 5I_0.$$

Q.11. **Light from a narrow slit passes through two parallel slits 0.2 cm apart. The interference fringes on a screen 100 cm away are 0.295 mm apart. What is the wavelength of light?**

Sol. $\qquad d = 0.2 \text{ cm}, \quad D = 100 \text{ cm}.$

The fringe width,

$$W = 0.295 \text{ mm} = 0.0295 \text{ cm}$$

$$\lambda = \frac{Wd}{D}$$

$$= \frac{0.0295 \, \text{cm} \times 0.2 \, \text{cm}}{100 \, \text{cm}}$$

$$= 5900 \times 10^{-8} \, \text{cm}$$

$$= 5900 \, \text{Å}.$$

Q.12. **In Young's experiment with green mercury line (λ = 5460 Å), the fringes are measured with a micrometer eyepiece 0.8 m behind the double slit. It is found that 20 of them occupy a distance of 0.01092 m. What is the distance between the two slits?**

Sol. $\lambda = 5460 \, \text{Å}$

20 fringes occupy a distance of 0.01092 m.

1 fringe occupy a distance of

$$= \frac{0.01092}{20} \, \text{m}$$

$$W = \frac{2 \times 0.01092 \, \text{m}}{20} = 0.001092 \, \text{m}$$

Now, $\quad D = 0.8 \, \text{m}$

and $\quad \lambda = 5460 \, \text{Å} = 5460 \times 10^{-8} \, \text{cm}$

$$= 5460 \times 10^{-10} \, \text{m}$$

Hence distance between two slits,

$$d = \frac{\lambda \, D}{W} = \frac{5460 \times 10^{-10} \, \text{m} \times 0.8 \, \text{m}}{0.001092 \, \text{m}}$$

$$= 4 \times 10^{-8} \, \text{m}.$$

Q.13. **The distance between two coherent sources is 3 × 10^{-2} cm and the fringes are observed on a screen 1 m away. It is found that with a certain monochromatic source of light, the 12th bright fringe is situated at a distance of 20 mm from the central fringe. Find the wavelength of light.**

Sol. $\quad D = 1 \, \text{m} = 100 \, \text{cm},$

$$d = 3 \times 10^{-2} \, \text{cm}$$

$$n = 12, \, x = 20 \, \text{mm} = 2 \, \text{cm}.$$

$$\lambda = \frac{xd}{nD} = \frac{2 \, \text{cm} \times 3 \times 10^{-2} \, \text{cm}}{12 \times 100 \, \text{cm}}$$

$$= 0.5 \times 10^{-4} \, \text{cm}$$

$$= \frac{0.5 \times 10^{-4}}{10^4} \times 10^4$$

$$= 0.5 \times 10^4 \times 10^{-8}$$

$$= 5 \times 10^5 \, \text{Å}$$

$$= 5000 \, \text{Å}.$$

Q.14. In a Lloyd's mirror interference experiment the slit image is formed 0.6 cm the slit. The slit is illuminate by wavelength of 6000 Å. At what distance from slit a screen is to be placed so that the fringe separation is 0.03 cm ?

Sol.
$$W = \frac{\lambda D}{d}$$

$$\Rightarrow \quad D = \frac{Wd}{\lambda}$$

Here,
$$W = 0.03 \, \text{cm} = 3 \times 10^{-2} \, \text{cm},$$

$$d = 0.6 \, \text{cm} = 6 \times 10^{-1} \, \text{cm}$$

$$\lambda = 6000 \, \text{Å} = 6000 \times 10^{-8} \, \text{cm}$$

Putting these values,

$$D = \frac{3 \times 10^{-3} \, \text{cm} \times 6 \times 10^{-1} \, \text{cm}}{6000 \times 10^{-8} \, \text{cm}} = 300 \, \text{cm}.$$

Q.15. A biprism is placed at a distance of 5 cm in front of a narrow slit illuminated by a light of wavelength 5000 Å. The distance between the virtual images produced by the biprism Is 0.05 cm. Find the fringe width of the interference fringe pattern observed on a screen placed 95 cm in front of the biprism.

Sol.
$$\lambda = 5000 \, \text{Å} = 5000 \times 10^{-8} \, \text{cm},$$

$$d = 0.05 \, \text{cm} = 5 \times 10^{-2} \, \text{cm}$$

$$D = 5 \, \text{cm} + 95 \, \text{cm} = 100 \, \text{cm}$$

So fringe width,

$$W = \frac{\lambda D}{d}$$

$$= \frac{5000 \times 10^{-8} \times 10^2}{5 \times 10^{-2}} \, \text{cm} = 0.1 \, \text{cm}.$$

QUESTIONS

1. What do you mean by interference of light waves?

2. Distinguish between constructive and destructive interference.

3. Mention the conditions for production of a sustained interference pattern.

4. Two harmonic waves of the same amplitude phase superpose. Obtain expression for amplitude and phase of the resultant wave.

5. Obtain an expression for the fringe width in the interference pattern due to Young's double slit.

6. Draw a schematic diagram of the experimental arrangement to form Newton's rings.

7. With the help of a suitable ray diagram, describe the production of Newton's rings.

8. Explain how the wavelength of monochromatic light is determined using Newton's ring arrangement.

9. In a Newton's ring system, the centre is bright. Is the ring observed in reflected or transmitted light? Justify your answer.

10. Why very narrow slits are taken in Young's double slit interference experiment?

11. Prove analytically that the shapes of the interference fringes obtained in Young's double slit experiment are hyperbolic.

12. Explain with necessary theory how you will determine refractive index of water by using Newton's ring apparatus?

13. Explain how the law of conservation of energy holds good in light interference pattern?

14. What is the condition for the destructive interference in terms of phase difference between the two interfering waves?

15. What are the conditions to obtain interference pattern having good contrast in Young's double slit experiment?

16. (i) What is the coherent source? Explain the different methods to obtain coherent sources.

    (ii) What is the colour of the central fringe in Young's double slit experiment when white light source is used?

Problems

1. The radii of the 10th and 50th rings in Newton's ring experiment are 0.2 cm and 0.8 cm respectively. Find the wavelength of light if the radius of curvature of the convex surface of the plano-convex lens is 100 cm. **(Ans. $\lambda = 1.925 \times 10^{-4}$ cm)**

2. In Newton's ring experiment, the diameter of nth and $(n + 8)$th bright rings are 4.6 mm and 7.4 mm respectively. The radius of curvature of the lower surface of the lens is 2 cm. Determine the wavelength of light. **(Ans. $\lambda = 5.25 \times 10^{-5}$ cm)**

3. In Young's double slit experiment, the screen is 1 m away from the slits. The separation between the slits is 1 mm and the distance between the 1st and 4th fringe is 1.2 mm. Find the wavelength of light used. **(Ans. $\lambda = 4000$ Å)**

4. In Young's double slit experiment, the intensities on the screen due to the individual slits are 1 and 2 units respectively. What is the difference between the maximum intensity in fringe pattern? Derive the formula you will use. **(Ans.** $\Delta I = 4\sqrt{2}$ **)**

5. In a Newton's ring system, the diameters of the 5th and 10th dark ring are 0.122 cm and 0.150 cm respectively. What is the diameter of the 15th dark ring?

   (Ans. $D_{15} = 0.173$ cm**)**

6. In a Newton's ring arrangement, the diameter of a bright ring is 0.5 cm. What would be the diameter of the ring if the lens placed on the plane glass plate is replaced by another one having double the radius of curvature? **(Ans.** $D_n' = 0.707$ cm**)**

7. In Newton's ring experiment if illuminating source of light having two wavelengths 6000 Å and 4500 Å is used, it is found that nth dark ring due to 6000 Å coincides with (n + 1)th dark ring due to 4500 Å. Calculate the radii of nth dark rings due to 6000 Å and 4500 Å if radii of curvature of the plano-convex lens is 100 cm.

   (Ans. $R = 0.134$ cm**)**

8. In a Young's double slit experiment, interference pattern is obtained having the fringe width 0.1 mm. What would be the fringe width if the wavelength of light is increased by 10% and the distance between the source and the screen is reduced by 10% keeping other factors constant? **(Ans.** $\beta' = 0.099$ mm**)**

● ● ●

APPENDIX - I

Example 1:

In a two-slit interference pattern, at a point we observe the 10th order maximum for $\lambda = 7000$ Å. What order will be visible here, if the source of light is replaced by light of wavelength 5000 Å?

Solution:

Let β_1 and β_2 be the widths of the fringes for the wavelength $\lambda_1 = 7000$ Å and $\lambda_2 = 5000$ Å. If y is the distance of the point of observation from the central bright fringe then,

$$y = n_1 \beta_1 = n_2 \beta_2$$

where n_1 and n_2 are order of fringes for λ_1 and λ_1.

Again, $\beta = \dfrac{\lambda D}{d}$

So, $n_1 \beta_1 = n_2 \beta_2$

$\Rightarrow \quad n_1 \dfrac{\lambda_1 D}{d} = n_2 \dfrac{\lambda_2 D}{d} \qquad \Rightarrow \qquad n_1 \lambda_1 = n_2 \lambda_2$

$\Rightarrow \quad n_2 = \dfrac{n_1 \lambda_1}{\lambda_2} = \dfrac{10 \times 7000}{5000} = 14$.

Example 2:

The coherent sources of monochromatic light of wavelength 6000 Å produce an interference pattern on a screen kept at a distance of 1 metre from them. The distance between the two consecutive bright fringes on the screen is 0.5 mm. Find the distance between the two coherent sources.

Solution:

The fringe width is given by the expression

$$\beta = \dfrac{\lambda D}{d}$$

$$\lambda = 6000 \text{ Å} = 6000 \times 10^{-10} \text{ m} = 6.0 \times 10^{-7} \text{ m}$$

$$\beta = 0.5 \text{ mm} = 0.5 \times 10^{-3} \text{ m} = 5 \times 10^{-4} \text{ m}$$

Hence, $d = \dfrac{\lambda D}{\beta} = \dfrac{6.0 \times 10^{-7} \text{ m} \times 1 \text{ m}}{5 \times 10^{-4} \text{ m}} = 1.2 \times 10^{-3} \text{ m} = 1.2 \text{ mm}.$

Example 3:

Two sinusoidal waves of same frequency and having amplitudes A_1 and A_2 respectively superpose coherently. Write the expression for the maximum and minimum values of the intensity of the resultant wave.

Solution:

Given A_1 and A_2 are the amplitudes.

We know that $I \propto A^2$

So, the resultant intensity is given by

$$I = A_1^2 + A_2^2 + 2 A_1 A_2 \cos \phi$$

where ϕ is the phase difference.

$\therefore \qquad I_{max} = A_1^2 + A_2^2 + 2 A_1 A_2 \qquad (\because \phi = 0)$

$$I_{max} = (A_1 + A_2)^2 \qquad \qquad ...(1)$$

$$I_{min} = A_1^2 + A_2^2 - 2 A_1 A_2$$

$$I_{min} = (A_1 - A_2)^2 \qquad \qquad ...(2)$$

Hence eqns. (1) and (2) are the expressions for the maximum and minimum values of the intensity of the resultant wave.

Example 4:

Green light of wavelength 5100 Å from a narrow slit is incident on a double slit. If the overall separation of 10 fringes on a screen 200 cm away is 2 cm, find the slit separation.

Solution:

$$\lambda = 5100 \, \text{Å} = 5100 \times 10^{-8} \, \text{cm}$$

$$D = 200 \, \text{cm}$$

$$10 \, \beta = 2 \, \text{cm}$$

$$\beta = \frac{2}{10} \, \text{cm} = 0.2 \, \text{cm}$$

$$\beta = \frac{\lambda D}{d} \quad \Rightarrow \quad d = \frac{\lambda D}{\beta} = \frac{5100 \times 10^{-8} \times 200}{0.2}$$

$$= \frac{102 \times 10^{-3}}{2} = 51 \times 10^{-3} \, \text{cm} = 0.051 \, \text{cm}.$$

Example 5:

Calculate the fringe width of interference pattern produced in Young's double slits experiment with two slits 10^{-3} m apart on a screen 1 m away. Wavelength of light is 5893×10^{-8} cm.

Solution:

Here,

$$D = 1\text{m}, \quad d = 10^{-3} \text{ m}$$

$$\lambda = 5893 \times 10^{-8} \text{cm} = 5893 \times 10^{-10} \text{m}$$

$$\beta = \frac{\lambda D}{d} = \frac{5893 \times 10^{-10} \times 1}{10^{-3}} = 5893 \times 10^{-7} \text{m}.$$

Example 6:

Two coherent sources are 0.18 mm apart and the fringes are observed on a screen 80 cm away. It is found that with a certain monochromatic source of light, the fourth bright fringe is situated at a distance of 10.8 mm from the central fringe. Calculate the wavelength of light.

Solution:

Here,

$$d = 0.18 \text{ mm} = \frac{0.18}{10} \text{cm} = 0.018 \text{ cm}$$

$$D = 80 \text{ cm}, \quad n = 4, \quad y = 10.8 \text{ mm} = \frac{10.8}{10} \text{ cm} = 1.08 \text{ cm}$$

$$y = \frac{n \lambda D}{d}$$

$$1.08 = \frac{4 \times \lambda \times 80}{0.018} \quad \Rightarrow \quad \lambda = \frac{1.08 \times 0.018}{4 \times 80} = 6075 \times 10^{-8} \text{cm} = 6075 \text{ Å}.$$

Example 7:

In a Young's double slit experiment 62 fringes are seen for Na light. How many fringes are observed for a wavelength of 6500 Å?

Solution:

For Na light,   For some other light,

$$n_1 = 62, \qquad n_2 = ?$$

$$\lambda_1 = 5893 \text{ Å}, \quad \lambda_2 = 6500 \text{ Å}$$

$$n_1 \lambda_1 = n_2 \lambda_2 \qquad\qquad [\because \text{ velocity remains same, } v = n\lambda]$$

$$62 \times 5893 = n_2 \times 6500$$

$$\Rightarrow \qquad n_2 = \frac{62 \times 5893}{6500} = 56 \qquad \Rightarrow \qquad n_2 = 56.$$

Example 8:

In Young's double slit experiment fringe width is 3.4 mm in air. The whole apparatus is immersed in water of refractive index 4/3. Find the fringe width in water.

Solution:

In air, $\quad \beta = \dfrac{\lambda D}{d} \qquad \dots(1)$

In water, $\quad \beta' = \dfrac{\lambda' D}{d} \qquad \dots(2)$

$$\frac{\beta'}{\beta} = \frac{\lambda' D}{d} \times \frac{d}{\lambda D}$$

$$\frac{\beta'}{\beta} = \frac{\lambda'}{\lambda} \qquad \Rightarrow \qquad \beta' = \frac{\lambda' \beta}{\lambda}$$

Now, $\quad \mu = \dfrac{V_1}{V_2} = \dfrac{n\lambda_1}{n\lambda_2} = \dfrac{n\lambda}{n\lambda'} = \dfrac{\lambda}{\lambda'}$

$\Rightarrow \quad \mu = \dfrac{\lambda}{\lambda'} = \dfrac{4}{3}$

So, $\quad \beta' = \dfrac{\beta}{\lambda/\lambda'} = \dfrac{3.4}{4/3} = 3.4 \times \dfrac{3}{4} = 2.5 \text{ mm}.$

Example 9:

In Young's double slit experiment the separation of the slits is 1.9 mm and the fringe spacing is 0.31 mm at a distance of 1 metre. Find the wavelength of light used.

Solution:

Here, $\qquad \beta = 0.31 \text{ mm} = 0.031 \text{ cm}$

$\qquad d = 1.9 \text{ mm} = 0.19 \text{ cm}$

$\qquad D = 1 \text{ m} = 100 \text{ cm}$

$\qquad \beta = \dfrac{\lambda D}{d}$

$\Rightarrow \qquad \lambda = \dfrac{\beta d}{D} = \dfrac{0.031 \times 0.19}{100}$

$\Rightarrow \qquad \lambda = 589 \times 10^{-8} \text{ cm} = 5890 \text{ Å}.$

Example 10:

A light source emits light of two wavelengths $\lambda_1 = 4300$ Å and $\lambda_2 = 5100$ Å. The source is used in a double slit interference experiment. The distance between the sources and the screen is 1.5 m and the distance between the slits is 0.025 mm. Calculate the separation between the third order bright fringes due to these two wavelengths.

Solution:

$$D = 1.5 \text{ m} \qquad\qquad d = 0.025 \text{ mm} = 25 \times 10^{-6} \text{ m}$$

$$\lambda_1 = 4300 \text{ Å} = 4.3 \times 10^{-7} \text{ m} \qquad \lambda_2 = 5100 \text{ Å} = 5.1 \times 10^{-7} \text{ m}$$

$$n = 3$$

$$x_1 = \frac{n\lambda_1 D}{d} \qquad\qquad x_2 = \frac{n\lambda_2 D}{d}$$

$$x_2 - x_1 = \left(\frac{n\lambda_2 D}{d}\right) - \left(\frac{n\lambda_1 D}{d}\right) = \frac{nD}{d}[\lambda_2 - \lambda_1]$$

$$= \left(\frac{3 \times 1.5}{25 \times 10^{-6}}\right)[5.1 \times 10^{-7} - 4.3 \times 10^{-7}] = 0.0144 \text{ m} = 1.44 \text{ cm}.$$

Example 11:

Two coherent sources of monochromatic light of wavelength 6000 Å produce an interference pattern on a screen kept at a distance of 1 m from them. The distance between two consecutive bright fringes on the screen is 0.5 mm. Find the distance between the two coherent sources.

Solution:

Here, $\qquad\qquad \lambda = 600 \text{ Å} = 6 \times 10^{-7} \text{ m}$

$$D = 1 \text{ m}$$

$$\beta = 0.5 \text{ mm} = 5 \times 10^{-4} \text{ m}$$

$$d = ?$$

$$\beta = \frac{\lambda D}{d}$$

$$\Rightarrow \qquad d = \frac{\lambda D}{\beta} = \frac{6 \times 10^{-7} \times 1}{5 \times 10^{-4}} = 1.2 \times 10^{-3} \text{ m}$$

$$d = 1.2 \text{ mm}.$$

Example 12:

Light of wavelength 5500 Å from a narrow slit is incident on a double slit. The overall separation of 5 fringes on a screen 200 cm away is 1 cm. Calculate: (i) the slit separation, and (ii) the fringe width.

Solution:

Here,
$$y = \frac{n\lambda D}{d}$$

$$n = 5, \quad D = 200 \text{ cm} = 2 \text{ m}$$
$$\lambda = 5500 \text{ Å} = 5.5 \times 10^{-7} \text{ m}$$
$$y = 1 \text{ cm} = 10^{-2} \text{ m}$$
$$d = ?$$

(i)
$$d = \frac{n\lambda D}{y} = \frac{5 \times 5.5 \times 10^{-7} \times 2}{10^{-2}} = 5.5 \times 10^{-4} \text{ m}$$

$$\Rightarrow d = 0.055 \text{ cm}$$

(ii)
$$\beta = \frac{x}{n} = \frac{1}{5} \text{cm} = 0.2 \text{ cm}.$$

Example 13:

In Young's double slit experiment, the distance between two slits is 0.5 mm. If wavelength of light used is $\lambda = 5 \times 10^{-5}$ cm and distance of the screen from the slit $D = 50$ cm, calculate the fringe width.

Solution:

$$\beta = \frac{\lambda D}{d}$$

Here,
$$\lambda = 5 \times 10^{-5} \text{cm}$$
$$D = 50 \text{ cm}, \quad d = 0.5 \text{ mm} = 0.05 \text{ cm}$$

$$\therefore \beta = \frac{\lambda D}{d} = \frac{5 \times 10^{-5} \text{cm} \times 50 \text{ cm}}{0.05 \text{ cm}} = 0.05 \text{ cm}.$$

Example 14:

In Young's double slit experiment, the screen is 1 m away from the slits. The separation between the slits is 1 mm and the distance between the 1st and 4th fringe is 1.2 mm. Determine the wavelength of light used.

Solution:

$$D = 1 \text{ m} = 100 \text{ cm}$$

$$d = 1 \text{ mm} = \frac{1}{10} \text{ cm} = 0.1 \text{ cm}$$

Distance between 1st and 4th fringe = 1.2 mm

$$\therefore \quad \text{Fringe width} = \frac{1.2 \text{ mm}}{4-1} = \frac{1.2}{3} = 0.4 \text{ mm} = 0.04 \text{ cm}$$

$$\therefore \quad \text{Wavelength} (\lambda) = \frac{\beta d}{D} = \frac{0.04 \times 0.1}{100} = 4000 \text{ Å}.$$

Example 15:

In a Young's double slit experiment, a point P on the screen is equidistant from both the slits. If any one of the slits is closed, the intensity at P due to the other slit is 0.02 ω/m². What is the intensity at P when both the slits are open?

Solution:

Given intensity at P due to one slit = 0.02 ω/m².

i.e., $\quad I_0 = 0.02$ ω/m².

$$\therefore \quad I_{max} = 4I_0 = 4 \times 0.02 = 0.08 \text{ ω/m}^2.$$

Example 16:

In Young's double slit experiment, the intensities on the screen due to the individual slits are I_1 and I_2 respectively. What is the difference between the maximum intensity in fringe pattern?

Solution:

$$I_{max} = I_1 + I_2 = 2\sqrt{I_1 I_2}$$

$$I_{min} = I_1 + I_2 - 2\sqrt{I_1 I_2}$$

$$I_{max} - I_{min} = 4\sqrt{I_1 I_2}$$

Example 17:

In a Young's double slit interference pattern, the fringe width is 0.1 mm. What would be the fringe width if the wavelength of light is increased by 10% and the distance between the source and the screen is reduced by 10% keeping other factors constant?

Solution:

The fringe width (β) in a Young's double slit experiment is given by,

$$\beta = \frac{\lambda D}{d} = 0.1 \text{ mm}$$

where λ is the wavelength of the light.

According to question, new wavelength,

$$\lambda' = \lambda + \frac{10}{100}\lambda = \lambda + \frac{\lambda}{10} = \frac{11\lambda}{10}$$

New distance is
$$D' = D - \frac{10}{100} D = D - \frac{D}{10} = \frac{9D}{10}$$

and
$$d' = d$$

Hence,
$$\beta' = \frac{\lambda' D'}{d'} = \frac{\dfrac{11\lambda}{10} \times \dfrac{9D}{10}}{d}$$

$$= \frac{99 \lambda D}{100 \, d} = \frac{99}{100} \times 0.1 \, \text{mm} = 0.099 \, \text{mm}.$$

Example 18:

The fringe width in a Young's double slit arrangement is 0.008 cm. What would be the fringe width if the slit separation is halved and slit-screen distance is doubled?

Solution:

Given,
$$\beta = \frac{\lambda D}{d} = 0.008 \, \text{cm}$$

The new distance between the slits, $d = d/2$.
The new slit-screen distance, $D' = 2D$

∴ The fringe width is given by, $\beta' = \dfrac{\lambda D'}{d'} = \dfrac{2\lambda D}{d/4}$

or, $\beta' = 8\left(\dfrac{\lambda D}{d}\right) = 8 \times 0.008 \, \text{cm} = 0.064 \, \text{cm}.$

Example 19:

A parallel beam of light $\lambda = 5890 \times 10^{-8}$ cm is incident on a thin glass plate having refractive index of 1.5, such that the angle of refraction into the plate is 60°. Calculate the smallest thickness of the glass plate which will appear dark by reflection.

Solution:

Here, $2\mu t \cos r = n\lambda$

$\mu = 1.5$, $r = 60°$, $\cos 60° = 0.5$

For smallest thickness, $n = 1$ and $\lambda = 5890 \times 10^{-8}$ cm

∴ $t = \dfrac{n\lambda}{2\mu \cos r} = \dfrac{1 \times 5890 \times 10^{-8}}{2 \times 1.5 \times 0.5}$

$t = 3.926 \times 10^{-5}$ cm.

Example 20:

A soap film 5×10^{-5} cm thick is viewed at an angle of 35° to the normal. Find the wavelengths of light in the visible spectrum which will be absent from the reflected light. ($\mu = 1.33$).

Solution:

Let i be the angle of incidence and r be the angle of refraction.

So, $\mu = \dfrac{\sin i}{\sin r} = \dfrac{\sin 35°}{\sin r}$ $\Rightarrow$ $1.33 = \dfrac{\sin 35°}{\sin r}$

$\Rightarrow$ $r = 25.55°$

and $\cos r = 0.90$

Again $t = 5 \times 10^{-5}$ cm

(i) For the 1st order, $n = 1$

$$2\mu t \cos r = n\lambda$$

$$2 \times 1.33 \times 5 \times 10^{-5} \times 0.90 = 1 \times \lambda_1$$

$\Rightarrow$ $\lambda_1 = 12.0 \times 10^{-5}$ cm

which lies in the infrared (invisible) region.

(ii) For the 2nd order, $n = 2$

$$\lambda_2 = \frac{2\mu t \cos r}{n} = \frac{2 \times 1.33 \times 5 \times 10^{-5} \times 0.90}{2}$$

$$\lambda_2 = 6.0 \times 10^{-5} \text{ cm}$$

which lies in the visible region.

(iii) Similarly, taking $n = 3$

$$\lambda_3 = 4.0 \times 10^{-5} \text{ cm}$$

which lies in the visible region.

(iv) If $n = 4$

$$\lambda_4 = 3.0 \times 10^{-5} \text{ cm}$$

which lies in the ultraviolet (invisible) region.

Hence, absent wavelengths in the reflected light are 6.0×10^{-5} cm and 4.0×10^{-5} cm.

Example 21:

A soap film of refractive index $\dfrac{4}{3}$ and of thickness 1.5×10^{-4} cm is illuminated by white light incident at an angle of 60°. The light reflected by it is examined by a spectroscope in which a dark band is found corresponding to a wavelength of 5×10^{-5} cm. Calculate the order of interference of dark band.

Solution:

Here, $2\mu t \cos r = n\lambda$ $\qquad\qquad \mu = \dfrac{4}{3}; \quad \lambda = 5 \times 10^{-5}\,\text{cm}$

$t = 1.5 \times 10^{-4}\,\text{cm}; \quad i = 60° \qquad \mu = \dfrac{\sin i}{\sin r} \Rightarrow \dfrac{4}{3} = \dfrac{\sin 60°}{\sin r}$

$\Rightarrow \quad \sin r = \dfrac{0.866}{4/3} = 0.6495 \qquad\qquad r = 40.5° \quad \text{and} \quad \cos r = 0.7604$

$\therefore \quad n = \dfrac{2\mu t \cos r}{\lambda} \qquad\qquad \Rightarrow \qquad n = \dfrac{2 \times 4/3 \times 1.5 \times 10^{-4} \times 0.7604}{5 \times 10^{-5}}$

$\Rightarrow \quad n = 6.0832$

Hence, the order, $n = 6$.

Example 22:

A thin film of soap solution is illuminated by white light at an angle of incidence, $i = \sin^{-1}\left(\dfrac{4}{5}\right)$. In reflected light, two dark consecutive overlapping fringes are observed corresponding to wavelengths 6.1×10^{-7} m and 6.0×10^{-7} m. μ for the soap solution is $\dfrac{4}{3}$. Calculate the thickness of the film.

Solution:

Here, $n\lambda_1 = (n+1)\lambda_2 \qquad n(6.1 \times 10^{-7}) = (n+1) \times 6 \times 10^{-7}$

$\Rightarrow \quad 6.1n = 6(n+1) \qquad\qquad \Rightarrow \quad 6.1n = 6n + 6$

$\Rightarrow \quad \dfrac{61}{10}n - 6n = 6 \qquad\qquad \Rightarrow \quad \dfrac{1n}{10} = 6$

$\Rightarrow \quad n = 60$

$\sin i = \dfrac{4}{5}, \quad \mu = \dfrac{4}{3}$

So, $\mu = \dfrac{\sin i}{\sin r} \qquad\qquad \Rightarrow \quad \dfrac{4}{3} = \dfrac{4/5}{\sin r}$

$\Rightarrow \quad \sin r = \dfrac{4}{5} \times \dfrac{3}{4} \qquad \Rightarrow \quad \sin r = \dfrac{3}{5} = 0.6$

$$\cos r = \left(1 - \sin^2 r\right)^{1/2} = \left[1 - (0.6)^2\right]^{1/2} = 0.8$$

Also, $2\mu t \cos r = n\lambda_1$

$$t = \frac{n\lambda_1}{2\mu \cos r} = \frac{60 \times 6.1 \times 10^{-7}}{2 \times (4/3) \times 0.8}$$

$$t = 1.72 \times 10^{-5}\,\text{m} \qquad \text{or} \qquad t = 1.72 \times 10^{-2}\,\text{mm}.$$

Example 23:

A soap film of refractive index 1.43 is illuminated by white light incident at an angle of 30°. The refracted light is examined by a spectroscope in which the dark band corresponding to the wavelength 6×10^{-7} m is observed. Calculate the thickness of the film.

Solution:

The condition for destructive interference (dark band) in refracted light in a thin film of uniform thickness t and refractive index μ is

$$2\mu t \cos r = (2n + 1)\frac{\lambda}{2}$$

Again, $\mu = \dfrac{\sin i}{\sin r}$

$$1.43 = \frac{\sin 30°}{\sin r}$$

$\Rightarrow \qquad \sin r = \dfrac{0.5}{1.43} = 0.349$

$$\cos r = \left[1 - \sin^2 r\right]^{\frac{1}{2}} = \left[1 - (0.349)^2\right]^{\frac{1}{2}}$$

$$= \left[1 - 0.1218\right]^{\frac{1}{2}} = (0.8782)^{\frac{1}{2}} = 0.937 \approx 0.94$$

For $n = 0$, $\lambda = 6 \times 10^{-7}$ cm

$$t = \frac{\lambda}{4 \times \mu \times t \times \cos r} = \frac{6 \times 10^{-7}}{4 \times 1.43 \times 0.94} = 0.11159 \times 10^{-7} \approx 1.12 \times 10^{-8}\,\text{m}.$$

Example 24:

A thin equiconvex lens of focal length 4 metres and refractive index 1.50 rests on and in contact with an optical flat, and using light of wavelength 5400 Å, Newton's rings are viewed normally by reflection. What is the diameter of the 5th bright ring?

Solution:

The diameter of the nth bright ring is given by

$$D_n = \sqrt{2(2n-1)\lambda R}$$

Here, $\qquad n = 5, \quad \lambda = 5460 \times 10^{-8}$ cm

$\qquad\qquad\quad \mu = 1.50, \quad f = 400$ cm

$$\frac{1}{f} = (\mu - 1)\left(\frac{1}{R_1} - \frac{1}{R_2}\right)$$

Here, $\qquad R_1 = R, \qquad\qquad R_2 = -R$

$\therefore \qquad\qquad \dfrac{1}{f} = (\mu - 1)\dfrac{2}{R}$

$$\frac{1}{400} = (1.50 - 1)\frac{2}{R}$$

$$R = .50 \times 2 \times 400$$

$$R = 400$$

$$D_5 = \sqrt{2(2n-1)\lambda R}$$

$$= \sqrt{2(2 \times 5 - 1) \times 5460 \times 10^{-8} \times 400}$$

$$= \sqrt{18 \times 5460 \times 4 \times 10^{-5}}$$

$$= \sqrt{0.39312} = 0.6269 \text{ cm}$$

$$D_5 \approx 0.627 \text{ cm}.$$

Example 25:

The radii of the 10th and 50th rings in Newton's rings experiment are 0.2 cm and 0.8 cm respectively. Find the wavelength of light if the radius of curvature of the convex surface of the plano-convex lens is 100 cm.

Solution:

Given, $\qquad r_{10} = 0.2$ cm $\qquad\qquad\qquad r_{50} = 0.5$ cm

$\therefore \quad D_{10} = 2 \times 0.2 = 0.4$ cm $\qquad\quad D_{50} = 2 \times 0.8 = 1.6$ cm

$\qquad\quad R = 100$ cm, $\qquad\qquad\qquad P = 50 - 10 = 40$

$\therefore \quad \lambda = \dfrac{D_{50}^2 - D_{10}^2}{4 \times R \times P}$

$$\lambda = \frac{(1.6)^2 - (0.4)^2}{4 \times 100 \times 40} = \frac{2.56 - 0.16}{16000} = \frac{2.4}{16000}$$

$$= \frac{24}{16 \times 10^4} = 1.5 \times 10^{-4} \, \text{cm}.$$

Example 26:

In Newton's rings experiment, the diameter of *n*th and (*n* + 8)th bright rings are 4.6 mm and 7.4 mm respectively. The radius of curvature of the lower surface of the lens is 2 m. Determine the wavelength of light.

Solution:

$$\lambda = \frac{D_{n+p}^2 - D_n^2}{4 \, RP}$$

Here, D_n = 4.6 mm = 0.46 cm

D_{n+p} = 7.4 mm = 0.74 cm

P = 8   R = 2 m = 200 cm

$$\lambda = \frac{(0.74)^2 - (0.46)^2}{4 \times 200 \times 8} = \frac{0.5476 - 0.2116}{6400}$$

$$= \frac{0.336}{6400} = \frac{336}{64 \times 10^5} = 5.25 \times 10^{-5} \, \text{cm.}$$

Example 27:

In a Newton's ring system, the diameters of the 5th and 10th dark rings are 0.122 cm and 0.150 cm respectively. What is the diameter of the 15th dark ring?

Solution:

Given,   Diameter of 5th dark ring, D_5 = 0.122 cm

Diameter of 10th dark ring, D_{10} = 0.150 cm

$$D_{10}^2 - D_5^2 = 4 \, RP \qquad\qquad \text{...(1)}$$

$$D_{15}^2 - D_{10}^2 = 4 \, RP \qquad\qquad \text{...(2)}$$

$$D_{10}^2 - D_5^2 = 4 \times R \times 5$$

$$D_{15}^2 - D_{10}^2 = 4 \times R \times 5$$

Comparing above two eqns.

$$D_{10}^2 - D_5^2 = D_{15}^2 - D_{10}^2$$

$$(0.150)^2 - (0.122)^2 = D_{15}^2 - (0.150)^2$$

$$0.0225 - 0.014884 = D_{15}^2 - 0.0225$$

$$D_{15}^2 = 0.007616 + 0.0225 = 0.030116$$

$$D_{15} = \sqrt{0.030116} = 0.1735 \text{ cm}.$$

Example 28:

In Newton's rings experiment in laboratory source of light having two wavelengths 6000 Å and 4500 Å is used. It is found that nth dark rings due to 6000 Å coincide with $(n+1)$th dark rings due to 4500 Å. Calculate the radii of nth dark rings due to 6000 Å and 4500 Å if radii of curvature of the plano-convex lens is 100 cm.

Solution:

$$D_n^2 = 4n\lambda R$$

So, $$\left[D_n^2 \right]_{6000} = 4nR \times 6000$$

$$\left[D_{n+1}^2 \right]_{4500} = 4(n+1)R \times 6000$$

According to question,

$$\left(D_n^2 \right)_{6000} = \left(D_{n+1}^2 \right)_{4500}$$

$$\Rightarrow \quad 4nR \times 6000 = 4(n+1)R \times 4500 \qquad \Rightarrow \quad 6000\,n = 4500\,n + 4500$$

$$\Rightarrow \quad (6000 - 4500)n = 4500$$

$$1500\,n = 4500$$

$$n = \frac{4500}{1500} = 3$$

Now, $$r_1 = \sqrt{n\lambda_1 R} = \sqrt{3 \times 6000 \times 10^{-8} \times 100}$$

$$\sqrt{18 \times 10^{-3}} = 0.134 \text{ cm}$$

$$r_2 = \sqrt{n\lambda_2 R} = \sqrt{3 \times 4500 \times 100 \times 10^{-8}} = 0.116 \text{ cm}.$$

Example 29:

Newton's rings experiment was conducted first in air medium then in water medium (i.e., water is inserted in between the plano-convex lens and glass plate). What happens to the diameter of a particular ring?

Solution:

Diameter of Newton's ring in air medium, $D_n^2 = 4Rn\lambda$

The diameter of that ring in water medium, $D_n'^2 = \dfrac{4Rn\lambda}{\mu}$

So, the diameter of a particular ring decreases $\dfrac{1}{\mu}$ times when water is introduced.

Example 30:

In Newton's rings experiment in laboratory sodium vapour lamp having two wavelengths 5890 Å and 5896 Å is used. If it is found that nth dark ring due to 5896 Å coincides with $(n+2)$th dark ring due to 5890 Å. Then calculate the radii of nth dark rings due to 5890 Å and 5896 Å. The radius of curvature of the plano-convex lens used is 200 cm.

Solution:

Here nth dark ring due to wavelength 5896 Å is given by,

$$D_n^2 = 4Rn\lambda \qquad\qquad D_n^2 = 4Rn(5890) \qquad ...(1)$$

and $(n+2)$ dark ring due to wavelength 5890 Å is given by,

$$D_{n+2}'^2 = 4R(n+2)(5890) \qquad\qquad ...(2)$$

According to the question, $\qquad D_n^2 = D_{n+2}'^2$

$$\Rightarrow \qquad 5896n = 5870n + 11780$$
$$\Rightarrow \qquad 5896n - 5890n = 11780 \qquad\qquad \Rightarrow \quad 6n = 11780$$

$$\therefore \qquad n = \frac{11780}{6} = \frac{5890}{3}$$

$\therefore$ The radius of nth dark ring due to wavelength 5896 Å is

$$D_n = \sqrt{4Rn\lambda} = \sqrt{4 \times 200 \times \frac{5890}{3} \times 5896 \times 10^{-8}}$$

$$= \sqrt{8 \times \frac{589}{3} \times 5896 \times 10^{-5}} = 92.60\ \text{cm} = 0.926\ \text{cm}$$

The radius of nth dark ring due to wavelength 5890 Å is

$$D_n' = \sqrt{4Rn\lambda} = \sqrt{4 \times 200 \times \frac{5890}{3} \times 5890 \times 10^{-8}}$$

$$= \sqrt{8 \times \frac{589}{3} \times 5890 \times 10^{-5}} = 92.51\ \text{cm} = 0.925\ \text{cm}.$$

Example 31:

A plano-convex lens of radius 300 cm is placed on an optically flat glass and is illuminated by monochromatic light. The diameter of the 8th dark ring in the transmitted system is 0.72 cm. Calculate the wavelength of light used.

Solution:

For the transmitted system, $\qquad r^2 = \dfrac{(2n-1)\lambda R}{2}$

Here, $\quad n = 8, \quad D = 0.72$ cm, $\quad R = 300$ cm, $\quad r = 0.36$ cm

So, $\qquad \lambda = \dfrac{2r^2}{(2n-1)R} = \dfrac{2 \times 0.36 \times 0.36}{(2 \times 8 - 1) \times 300}$

$\qquad\qquad = \dfrac{2 \times 0.36 \times 0.36}{15 \times 300} = 5700 \times 10^{-8}$ cm

$\qquad \therefore \ \lambda = 5760.$

Example 32:

In a Newton's rings experiment the diameter of the 15th ring was found to be 0.590 cm and that of the 5th ring was 0.336 cm. If the radius of the plano-convex lens is 100 cm, calculate the wavelength of light used.

Solution:

Here, $\qquad\qquad D_5 = 0.336$ cm $\qquad\qquad D_{15} = 0.590$ cm

$\qquad\qquad\qquad R = 100$ cm, $\qquad\qquad P = 15 - 10 = 10$

$\qquad \therefore \ \lambda = \dfrac{(D_{n+p})^2 - (D_n)^2}{4PR} = \dfrac{(0.590)^2 - (0.336)^2}{4 \times 10 \times 100}$

$\qquad\qquad = \dfrac{0.3481 - 0.112896}{4000} = \dfrac{0.235204}{4000} = 0.0000588$

$\qquad\qquad = 5880 \times 10^{-8}$ cm

$\qquad \lambda = 5880$ Å.

Example 33:

In a Newton's rings experiment, find the radius of curvature of the lens surface in contact with the glass plate when with a light of wavelength 5890×10^{-8} cm, the diameter of the third dark ring is 3.2 mm. The light is falling at such an angle that it passes through the air film at an angle of zero degree to the normal.

Solution:

For dark rings,

$$r^2 = n\lambda R; \qquad R = \frac{r^2}{n\lambda}$$

Here, $r = \frac{3.2}{2}$ mm $= 1.6$ mm $= 0.16$ cm

$n = 3,$ $\qquad \lambda = 5890 \times 10^{-8}$ cm

$\therefore \quad R = \frac{(0.16)^2}{3 \times 5890 \times 10^{-8}};$ $\quad R = 144.9$ cm.

Example 34:

In a Newton's rings experiment the diameter of the 10th ring changes from 1.40 cm to 1.27 cm when a liquid is introduced between the lens and the plate. Calculate the refractive index of the liquid.

Solution:

For liquid medium $D_1^2 = \frac{4n\lambda R}{\mu}$ $\qquad$ ...(i)

For air medium $D_2^2 = 4n\lambda R$ $\qquad$ ...(ii)

Dividing (ii) by (i) $\qquad \mu = \left(\frac{D_2}{D_1}\right)^2$

Here, $D_1 = 1.27$ cm $\qquad D_2 = 1.40$ cm

$\therefore \quad \mu = \left(\frac{1.40}{1.27}\right)^2 = 1.215.$

Example 35:

Newton's rings are observed in reflected light of $\lambda = 5.9 \times 10^{-5}$ cm. The diameter of the 10th dark ring is 0.5 cm. Find the radius of curvature of the lens and the thickness of the air film.

Solution:

$r^2 = n\lambda R$ $\qquad \lambda = 5.9 \times 10^{-5}$ cm $= 5.9 \times 10^{-7}$ m

$n = 10,$ $\qquad D = 0.5$ cm $\Rightarrow r = \frac{0.5}{2} = 0.25$ cm $= 0.25 \times 10^{-2}$ m

(i) For dark rings, $\qquad r^2 = n\lambda r$

$\Rightarrow \qquad R = \frac{r^2}{n\lambda} = \frac{(0.25 \times 10^{-2})^2}{10 \times 5.9 \times 10^{-7}}$

$\Rightarrow \qquad R = \frac{0.0625 \times 10^{-4}}{5.9 \times 10^{-6}} = \frac{6.25 \times 10^{-6}}{5.9 \times 10^{-6}} = 1.059$ m.

(ii) Thickness of the air film $= t$

For dark rings, $2t = n\lambda$

$$t = \frac{n\lambda}{2} = \frac{10 \times 5.9 \times 10^{-7}\,\text{m}}{2} = 5 \times 5.9 \times 10^{-7}\,\text{m}$$

$$= 29.5 \times 10^{-7}\,\text{m} = 2.95 \times 10^{-6}\,\text{m}.$$

QUESTIONS
(Short Type Questions)

1. Explain Huygens principle. Where does it find application?

2. What is wavefront? What is the angle between the normal to a wavefront and the direction of propagation of the wave at a given point?

3. What is interference? Can longitudinal waves exhibit interference?

4. What do you mean by coherent sources of light? How can coherent sources of light be realised in practice?

5. What is the relation between the path difference and phase difference between two waves at a point?

6. In a Newton's rings experiment, the diameter of bright rings are proportional to the square root of natural numbers. Are the rings formed by reflected light or transmitted light? Explain.

7. Mention the dependence of resolving power of a telescope on the wavelength of light and diameter of the aperture of the objective.

8. A thin liquid film of uniform thickness t and refractive index μ produces interference fringes due to reflection when illuminated by a parallel monochromatic beam of light of wavelength λ. If θ is the angle of refraction into the film, write the condition for production into the film, write the condition for production of bright fringes.

9. In a Young's double slit arrangement, the separation between the two slits is reduced to half of its original value. How will the fringe width be affected?

10. In Young's double slit experiment, the distance between two slits is 0.5 mm. If wavelength of light used is $\lambda = 5 \times 10^{-5}$ and distance of the screen from the slit D = 50 cm. Calculate the fringe width.

11. In a Newton's ring experiment performed in air, the diameter of a given ring is 0.2 mm. When a liquid is introduced between the lens and the plane glass plate below it, the diameter of the ring shrinks to 0.17 mm. Find the refractive index of the liquid.

12. In a Young's double slit experiment, a point P on the screen is equidistant from both the slits. If any one of the slits is closed, the intensity at P due to the other slit is 0.02 ω/m^2. What is the intensity at P when both slits are open?

13. The fringe width in a Young's double slit arrangement is 0.008 cm. What would be the fringe width if the slit separation is halved and the slit screen distance is doubled?

14. Optical interference phenomenon is broadly divided into two categories, namely:

    (i) Division of wavefront, and   (ii) Division of amplitude.

    Mention one example for each category of interference.

15. What is the colour of the central fringe in Newton's ring experiment as seen by transmitted monochromatic light when the space between curved surface of the planoconvex lens and plane glass plate contains ethyl alcohol?

Long Type Questions

1. Why does a soap bubble appear coloured when seen under the sunlight? What happens when the thickness of the bubble becomes extremely small?

2. Explain how the wavelength of monochromatic light is determined using Newton's ring arrangement.

3. The radii of the 10th and 50th rings in Newton ring experiment are 0.2 cm and 0.8 cm respectively. Find the wavelength of light if the radius of curvature of the convex surface of the plano-convex lens is 100 cm.

4. Mention the conditions for production of a sustained interference pattern. With the help of a suitable ray diagram, describe the production of Newton's rings.

5. In Newton's ring experiment, the diameter of nth and $(n+8)$th bright rings are 4.6 mm and 7.4 mm respectively. The radius of curvature of the lower surface of the lens is 2 m. Determine the wavelength of light.

6. In Young's double slit experiment, the screen is 1 m away from the slits. The separation between the slits is 1 mm and the distance between the 1st and 4th fringe is 1.2 mm. Determine the wavelength of light used.

7. In a Newton's ring system, the centre is bright. Is the ring observed is reflected or transmitted light? Justify your answer.

8. Draw a schematic diagram of the experimental arrangement to form Newton's rings. Also explain it.

9. In Newton's ring system, the diameters of the 5th and 10th dark ring are 0.122 cm and 0.150 cm respectively. What is the diameter of the 15th dark ring?

10. Obtain an expression for the fringe width in the interference pattern due to Young's double slit.

11. Explain with necessary theory how you will determine refractive index of water by using Newton's ring apparatus?

12. In a Young's double slit interference pattern, the fringe width is 0.1 mm. What would be the fringe width if the wavelength of light is increased by 10% and the distance between the source and the screen is reduced by 10% keeping other factors constant?

13. State Huygens principle. According to him what type of wave is light?

14. A plano-convex lens of radius of curvature 2.5 metre is placed on an optically plane glass plate in air medium and a parallel beam of monochromatic light is incident normally on the set up to observe the Newton's ring. The diameter of the 5th bright ring as seen by the reflected light is 0.70 cm. Calculate the wavelength of the light used.

15. What are constructive and destructive interference? Under which condition do they occur?

●●●

Chapter 8

Diffraction

The bending of light around an obstacle is known as diffraction of light. It shows deviation from the rectilinear propagation of light. In this case the light enters into the geometrical shadow of the obstacle that consists of maxima and minima. This is known as diffraction pattern. There are two classes of diffraction, one is Fraunhofer diffraction and the other is Fresnel diffraction. The amount of the bending of light depends upon the wavelength of the light and size of the obstacle 'a'.

If the size of the obstacle compared to the wavelength of the light wave is:

(i)   very small (i.e., $a \ll \lambda$), then the wave will undergo reflection and not diffraction.

(ii) very large (i.e., $a \gg \lambda$), then the wave will undergo less or minimum diffraction.

(iii) almost equal (i.e., $a = \lambda$), then the wave will undergo maximum diffraction.

Diffraction will not occur if the wave is not coherent and diffraction effects become weaker as the size of obstruction is made larger and larger compared to the wavelength.

Difference between Interference and Diffraction

Interference	Diffraction
1. This phenomenon results from the superposition of secondary wavelets originating from two coherent sources.	1. This phenomenon results from the superposition of secondary wavelets originating from a single coherent source.
2. Interference fringes are of equal width.	2. Diffraction fringes are never of equal width.
3. All bright fringes are of equal intensity.	3. The intensity of all bright fringes is not the same.
4. The intensity of all dark fringes is zero, i.e. all dark fringes are perfectly black.	4. The intensity of all dark fringes is not zero, i.e. all dark fringes are not perfectly black.

Types of Diffraction

There are two types of diffraction:

     (1) Fresnel diffraction, and

     (2) Fraunhofer diffraction.

Fresnel diffraction

Fresnel type of diffraction is observed when the source of light and the screen are situated at a finite distance from the diffracting aperture. Therefore, the wavefront incident in the obstacle is either cylindrical or spherical.

Fraunhofer diffraction

Fraunhofer type of diffraction is observed when the source of light and the screen are situated at infinite distances from the diffracting aperture. Therefore, the wavefront incident on the obstacle is plane. The infinite distance is created by placing a lens in between the light source and the obstacle so that the light source is at the focus of the lens.

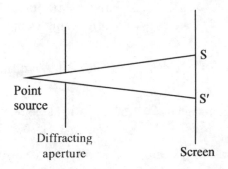

Fig. 8.1: Fresnel diffraction

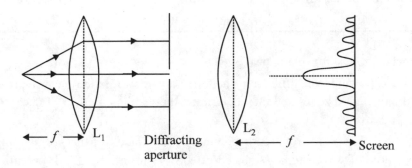

Fig. 8.2: Fraunhofer diffraction

Difference between Fresnel's diffraction and Fraunhofer diffraction

Fresnel's diffraction	*Fraunhofer's diffraction*
1. In Fresnel's diffraction, separation between light source and obstacle is finite.	1. In Fraunhofer's diffraction, separation between light source and obstacle is infinite.
2. The wavefront incident on the obstacle is either spherical or cylindrical in nature.	2. The wavefront incident on the obstacle is plane is nature.
3. No lens system is used in this class of diffraction.	3. Use of lens mainly convergent lens is made in this class of diffraction. One convex lens placed between the source and the obstacle, makes parallel beam of light. And another convex lens placed between the obstacle and the screen makes the rays to focus on the screen.
4. Diffraction in this case, the secondary wavelets originating from the obstacle having different phase.	4. However, diffraction in this case consists of maxima and minima which arises due to superposition of secondary wavelets originating from the obstacle having same phase.
5. The centre of diffraction pattern is either bright or dark.	5. The centre of diffraction pattern is always bright.
6. The diffraction pattern is the image of the obstacle or aperture.	6. The diffraction pattern is the image of the source itself.
7. In this type of diffraction, distance is important for analysis of diffraction.	7. In this type of diffraction, inclination are important for analysis of diffraction.
8. Examples of this class are diffraction at a straight edge, diffraction at a circular aperture, diffraction at a narrow wire, etc.	8. Examples of this class are diffraction at a single slit, diffraction at a double slit, diffraction due to 'n' slit (plane diffraction grating).

Fresnel's half period zones for plane wave

Fresnel's approach to the study of diffraction phenomenon was to divide the wavefront into small zones called half-period zones and then determine the combined effect of such zones at a point ahead of the wavefront.

Let ABCD be a plane wavefront (perpendicular to the plane of paper) originating from a monochromatic source of wavelength 'λ' placed effectively at infinity. Let P be an external point at which the effect of the entire wavefront is studied. PO is a perpendicular drawn from

P to the wavefront, where PO = b. O is called the pole of the wavefront with respect to P. With

P as the centre, construct concentric spheres with radii b, $b + \dfrac{\lambda}{2}$, $b = \dfrac{2\lambda}{2}$, ..., $b + \dfrac{n\lambda}{2}$ (where

$n = 0, 1, 2, 3, ...,$ etc.) on the wavefront thus cutting the wavefront into angular strips or zones. The sections of these spheres by the plane wavefront are concentric circles having common centre 'O' and radii OM_1, OM_2, ..., OM_{n-1}, OM_n, etc. as shown in (Fig. 8.3a and b).

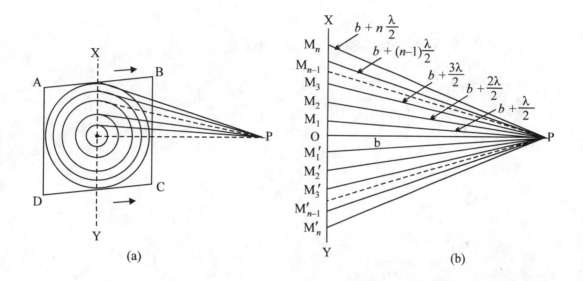

Fig. 8.3

The secondary wavelets from any two consecutive zones reach P with a path difference

$\lambda/2$ or time difference half period, i.e. $\dfrac{T}{2}$ where $T = \dfrac{\lambda}{C}$ is the time of the light wave. So the

zones are called half period zones. The area of the innermost or first circle is called first half period zone. The annular area between the first and second circle is called second half period zone and so on. Thus the annular area between the $(n-1)$th and nth circle is called nth half period zone. Thus this construction divides the entire wavefront into a large number of half period zones called Fresnel's half period zones.

'A Fresnel half period zone with respect to an external point P is a thin annular zone or a thin strip of the primary wavefront surrounding the point 'O' such that the distances of its outer and inner edges from 'O' differ by $\lambda/2$.'

Radii of half period zones:

The radius of first half period zone,

$$OM_1 = \sqrt{(M_1P)^2 - (OP)^2}$$

$$= \sqrt{\left(b + \frac{\lambda}{2}\right)^2 - b^2} = \sqrt{b^2 + \frac{\lambda^2}{4} + b\lambda - b^2} = \sqrt{b\lambda + \frac{\lambda^2}{4}}$$

$$r_1 = \sqrt{b\lambda} \qquad \text{as } b \gg \lambda$$

∵ Since $\dfrac{\lambda^2}{4} \ll b\lambda$, the second term is neglected.

The radius of second half period zone,

$$OM_2 = \sqrt{(M_2P)^2 - (OP)^2}$$

$$= \sqrt{\left(b + \frac{2\lambda}{2}\right)^2 - b^2} = \sqrt{b^2 + \lambda^2 + 2b\lambda - b^2}$$

$$r_2 = \sqrt{2b\lambda}$$

Similarly the radius of nth half period zone,

$$OM_n = \sqrt{(PM_n)^2 - (OP)^2}$$

$$= \sqrt{\left(b + \frac{n\lambda}{2}\right)^2 - b^2} = \sqrt{b^2 + \frac{n^2\lambda^2}{2} + bn\lambda - b^2}$$

$$r_n = \sqrt{nb\lambda} \qquad\qquad ...(1)$$

Eqn. (1) can be used to determine the area of different half period zones.

Area of a half period zone:

Area of the nth zone $= \pi(OM_n)^2 \, \pi(OM_{n-1})^2$

$$= \pi\left[(PM_n)^2 - (PO)^2\right] - \pi\left[(PM_{n-1})^2 - (PO)^2\right]$$

$$= \pi\left[\left(b + \frac{n\lambda}{2}\right)^2 - b^2\right] - \pi\left[\left(b + \frac{(n-1)\lambda}{2}\right)^2 - b^2\right]$$

$$= \pi \left\{ \left[b^2 + \frac{n^2\lambda^2}{4} + bn\lambda - b^2 \right] - \left[b^2 + (n-1)^2 \frac{\lambda^2}{4} + b(n-1)\lambda - b^2 \right] \right\}$$

$$= \pi \left\{ \frac{n^2\lambda^2}{4} + bn\lambda - \frac{(n-1)^2\lambda^2}{4} - b(n-1)\lambda \right\}$$

$$= \pi \left\{ \frac{n^2\lambda^2}{4} + bn\lambda - \frac{n^2\lambda^2}{4} + \frac{\lambda^2}{4} + \frac{2n\lambda^2}{4} - bn\lambda + b\lambda \right\}$$

$$= \pi \left\{ \frac{2n\lambda^2}{4} - \frac{\lambda^2}{4} + b\lambda \right\}$$

$$= \pi \left\{ b\lambda + \frac{\lambda^2}{4}(2n-1) \right\}$$

Area $= \pi b\lambda$ as $b \gg \lambda$ ...(2)

As area of nth zone is independent of n, thus the area of each half period zone is the same and is equal to $\pi b\lambda$.

Average distance:

The distance of inner boundary of nth zone, from P is $PM_{n-1} = b + (n-1)\frac{\lambda}{2}$

Similarly the distance of outer boundary of nth zone from P is $PM_n = b + \frac{n\lambda}{2}$

So, the average distance of nth zone from P is $= \dfrac{\left(b + \dfrac{n\lambda}{2}\right) + \left(b + (n-1)\dfrac{\lambda}{2}\right)}{2}$

$$= \frac{b + \dfrac{n\lambda}{2} + b + \dfrac{n\lambda}{2} - \dfrac{\lambda}{2}}{2} = \frac{2b + 2n\dfrac{\lambda}{2} - \dfrac{\lambda}{2}}{2}$$

$$= \frac{2b + n\lambda - \dfrac{\lambda}{2}}{2} = \frac{2\left(b + \dfrac{n\lambda}{2} - \dfrac{\lambda}{4}\right)}{2} = b + (2n-1)\frac{\lambda}{4} \qquad ...(3)$$

Amplitude at P due to an individual zone

The amplitude of the disturbance due to a given zone depends:

(i) directly on area of the zone.

(ii) inversely proportional to the distance of the point P from the given zone.

(iii) directly to the obliquity factor $(1 + \cos \theta)$, where θ is the angle between the normal to the zone and the line joining the zone to point P.

Thus amplitude at P due to nth zone is

$$R_n \propto \frac{\pi \left[b\lambda + \dfrac{\lambda^2}{4}(2n-1) \right]}{b + (2n-1)\dfrac{\lambda}{4}}(1+\cos\theta_n)$$

$$R_n \propto \frac{\pi\lambda \left[b + \dfrac{\lambda}{4}(2n-1) \right]}{\left[b + \dfrac{\lambda}{4}(2n-1) \right]}(1+\cos\theta_n)$$

$$R_n \propto \pi\lambda \left(1+\cos\theta_n\right)$$

As n increases, θ_n increases and $\cos \theta_n$ decreases. Thus the amplitude of the disturbance at P due to a given zone decreases as the order of the zone increases. This means that the amplitude due to first half period zone is maximum and it decreases regularly as we pass from the inner zone to the next outer zone.

The resultant amplitude due to the whole wavefront:

Let R_1, R_2, R_3, ..., R_n be the amplitude at P due to the first, second, third, ... nth half period zones respectively. As the path difference between the waves reaching P from any two consecutive half-period zones is $\lambda/2$, the wave from two consecutive zones reach P in the opposite phase. Therefore, if amplitude due to 1st zone is positive, that due to 2nd zone is negative, that due to 3rd zone is positive and so on, i.e. R_1, R_3, R_5, ..., etc. are positive and R_2, R_4, R_6, ..., etc. are negative.

Hence the resultant amplitude at p due to the entire wavefront is

$$R = R_1 - R_2 + R_3 - R_4 + ... (-1)^{n-1} R_n \qquad ...(4)$$

As the magnitudes of successive terms R_1, R_2, R_3, ... decreases gradually, R_2 is slightly less than R_1 and greater than R_3 so we can write,

$$R_2 = \frac{R_1 + R_3}{2} \qquad ...(5)$$

Similarly, $\qquad R_4 = \dfrac{R_3 + R_5}{2}$ $\qquad\qquad\qquad$ ...(6)

and so on,

Now eqn. (4) can be written as

$$R = \frac{R_1}{2} + \left(\frac{R_1}{2} - R_2 + \frac{R_3}{2}\right) + \left(\frac{R_3}{2} - R_4 + \frac{R_5}{2}\right) + \dots + \frac{R_n}{2}; \text{ if } n \text{ is odd}$$

and $\qquad R = \dfrac{R_1}{2} + \left(\dfrac{R_1}{2} - R_2 + \dfrac{R_3}{2}\right) + \left(\dfrac{R_3}{2} - R_4 + \dfrac{R_5}{2}\right) + \dots + \dfrac{R_{n-1}}{2} - R_n, \text{ if } n \text{ is even}$

Substituting the value of (5) and (6) in above eqn. we get,

$$R = \frac{R_1}{2} + \frac{R_n}{2}, \quad \text{if } n \text{ is odd} \qquad\qquad\qquad \text{...(7)}$$

$$R = \frac{R_1}{2} + \frac{R_{n-1}}{2} - R_n, \text{ if } n \text{ is even} \qquad\qquad \text{...(8)}$$

But usually n is large, so that we may write $R_{n-1} = R_n$ approximately. Then

$$\frac{R_{n-1}}{2} - R_n = \frac{-R_n}{2}$$

So, eqns. (7) and (8) become

$$R = \frac{R_1}{2} + \frac{R_n}{2}, \quad \text{when } n \text{ is odd} \qquad R = \frac{R_1}{2} - \frac{R_n}{2}, \quad \text{when } n \text{ is even}$$

Writing the above eqn. by a single eqn. we have,

$$R = \frac{R_1}{2} \pm \frac{R_n}{2}, \qquad\qquad\qquad\qquad \text{...(9)}$$

For a large wavefront n is very large, R_n nearly vanishes; so that we have

$$R = \frac{R_1}{2} \qquad\qquad\qquad\qquad\qquad \text{...(10)}$$

Thus, the amplitude due to a large wavefront at a point in front it is just half that due to the first half period zone acting alone.

The intensity at any point is proportional to the square of the amplitude, therefore the resultant intensity at P,

$$I \propto \frac{R_1^2}{4}$$

Thus, the intensity at P is only one-fourth of that due to the first half period zone alone.

Zone Plate: A zone plate is an interesting application of the division of a plane wavefront into Fresnel's half period zones which forms the basis of the proof of the rectilinear propagation of light. It is a specially constructed screen such that light is obstructed from every alternate zone. Its construction is based on the following three facts:

(i) The areas of half period zones are equal.

(ii) The radii of half period zones are proportional to the square roots of the natural numbers. The radius of nth half period zone is $\sqrt{(nb\lambda)}$; therefore the radii of successive half period zones are in the ratio $1:\sqrt{2}:\sqrt{3}:\sqrt{4}$, etc.

(iii) If the light is obstructed from alternate zones, the intensity of the image is sufficiently increased.

The resultant amplitude at image point P is

$$R = R_1 - R_2 + R_3 - R_4 + R_5 - \ldots\ldots$$

where $R_1, R_2, R_3, \ldots$, etc. are the amplitudes at image point P due to first, second, third, ..., etc. half period zones.

Now if the light is obstructed from even half period zones, then the resultant amplitude at image point P becomes

$$R = R_1 + R_2 + R_3 + \ldots\ldots$$

Further if the light is obstructed from odd half period zones, then the resultant amplitude at image point P becomes

$$R = \left(R_2 + R_4 + R_6 + \ldots\ldots \right)$$

In either case the amplitude and hence the intensity at P is enormously increased.

Construction: To construct a zone plate a large number of concentric circles with radii proportional to natural numbers are drawn on a sheet of white paper. Now there are two alternatives:

(1) The odd numbered zones are covered with black point. Then a reduced but accurate photograph of this is taken on a glass plate of uniform thickness. In the negative of the photograph the odd zones which were painted black appear transparent and the even zones appear black (Fig. 8.4a). The resulting glass negative is known as *positive zone plate*.

(2) The even numbered zones are covered with black ink. Then a reduced but accurate photograph of this is taken on a thin glass plate of uniform thickness. In the negative of the photograph the even zones which were painted black appear transparent and the

odd zones appear opaque (Fig. 8.4b). The resulting glass negative is known as negative zone plate.

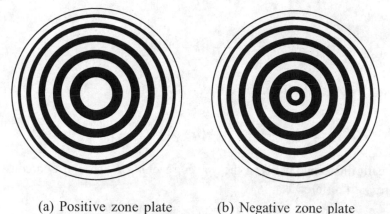

(a) Positive zone plate   (b) Negative zone plate

Fig. 8.4

Thus a zone plate is a specially constructed screen having alternate transparent and opaque concentric zones.

Theory: Fig. 8.5 represents the edgewise section of a zone plate. *S* is a point source emitting spherical waves of wavelength λ placed on the axis at a distance *a* from the centre *O* of the zone plate. Let *P* be the point on the screen at a distance *b* from *O* at which intensity is to be found.

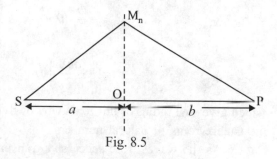

Fig. 8.5

Let $r_1, r_2, r_3, ..., r_n$ be the radii of first, second, third, ..., *n*th half period zones respectively.

In Fig. 8.5, the radius of *n*th half period zone, $r_n = OM_n$

Now
$$SM_n^2 = SO^2 + OM_n^2 = a^2 + r_n^2$$

$$\therefore \quad SM_n = \left(a^2 + r_n^2\right)^{1/2} = a\left(1 + \frac{r_n^2}{a^2}\right)^{1/2}$$

$$= a\left(1 + \frac{r_n^2}{2a^2}\right) \quad \text{by binomial theorem as } r_n \text{ is very small}$$

$$= a + \frac{r_n^2}{2a}$$

Similarly, $PM_n = \sqrt{\left(OP^2 + OM_n^2\right)} = b + \dfrac{r_n^2}{2b}$

$\therefore \qquad SM_n + PM_n = a + \dfrac{r_n^2}{2a} + b + \dfrac{r_n^2}{2b} = a + b + \dfrac{r_n^2}{2}\left(\dfrac{1}{a} + \dfrac{1}{b}\right);$

$\therefore \qquad$ Path difference, $SM_n P - SOP = \left(SM_n + PM_n\right) - \left(SO + OP\right)$

$$= a + b + \dfrac{r_n^2}{2}\left(\dfrac{1}{a} + \dfrac{1}{b}\right) - \left(a + b\right) = \dfrac{r_n^2}{2}\left(\dfrac{1}{a} + \dfrac{1}{b}\right) \qquad \text{...(1)}$$

As the path difference between the waves reaching P through two consecutive zones is $\lambda/2$; therefore path difference,

$$SM_n P - SOP = n.\lambda/2 \qquad \text{...(2)}$$

Therefore, from equations (1) and (2), we have

$$\dfrac{r_n^2}{2}\left(\dfrac{1}{a} + \dfrac{1}{b}\right) = n.\dfrac{\lambda}{2} \qquad \text{...(3)}$$

or, $$r_n^2 = \dfrac{ab}{a+b} n\lambda \qquad \text{...(4)}$$

or $$r_n = \sqrt{\left(\dfrac{ab\lambda}{a+b}\right)}\sqrt{n} \qquad \text{...(5)}$$

which gives the radius of nth zone. Obviously the radii of half period zones are proportional to the square roots of natural numbers.

As the waves from successive transparent zones differ in path by λ, the waves from them reach P in the same phase. The amplitude at P due to a zone depends on the area of the zone, the distance of the zone from P and the obliquity of the zone.

The area of nth zone $= \pi r_n^2 - \pi r_{n-1}^2$

$$= \pi \dfrac{ab}{a+b}.n\lambda - \pi.\dfrac{ab}{a+b}(n-1)\lambda \quad \text{using equation (4)}$$

$$= \dfrac{\pi ab\lambda}{a+b} \qquad \text{...(6)}$$

which is independent of order n. Therefore, the area of each zone is the same. But the distance of the zone from P and the obliquity increases as the order of the zone increases. Due to these two factors the amplitude of the disturbance due to a zone decreases with increase in the order of the zone. If R_1, R_2, R_3, ..., R_n are the amplitudes due to first, second, third, ..., nth zone respectively, the resultant amplitude at P, will be

$$R = R_1 + R_2 + R_3 + \dots \text{ for a positive zone plate} \qquad \dots(7a)$$

$$= -(R_2 + R_4 + R_6 + \dots) \text{ for a negative zone plate} \qquad \dots(7b)$$

which is enormously greater than $\dfrac{R_1}{2}$, the resultant due to all the zones. Hence the point

P will be sufficiently bright and can be said to be the image of S. This explains the focussing action of a zone plate.

Equation (3) may be written in the form

$$\frac{1}{a} + \frac{1}{b} = \frac{n\lambda}{r_n^2} \qquad \dots(8)$$

We see that this equation is similar to the lens formula

$$\frac{1}{u} + \frac{1}{v} = \frac{1}{f} \qquad \dots(9)$$

with a and b being the object and image distances respectively. Hence, comparing equations (8) and (9), we get

$$\frac{1}{f} = \frac{n\lambda}{r_n^2} \quad \text{i.e.,} \quad f = \frac{r_n^2}{n\lambda} \qquad \dots(10)$$

Here f is called the principal focal length of the zone plate. Thus, the zone plate acts as a convergent (convex) lens with multiple foci for a particular wavelength, depending on the values of n and r.

Multiple foci of zone plate: If the point source S is at infinity, i.e. if $a = \infty$, the area of the each zone $= nb\lambda$ [from equation (6)].

In this case the image P is formed at a distance $b = \dfrac{r_n^2}{n\lambda} =$ principal focal length.

Now consider a point P_3 along the axis of the zone plate at a distance $b/3$ from the zone

plate, then the area of each half period zone with respect to P_3 will be $\dfrac{1}{2}\pi b\lambda$, i.e., one third

of the area of each zone on the zone plate. Hence each zone on the zone plate will contain three half period elements corresponding to P_3. The resultant amplitude at P_3 will be

$$A = (m_1 - m_2 + m_3) + (m_2 - m_3 + m_2) + (m_{12} - m_{14} + m_{15}) + \dots$$

$$= \left(m_1 - \frac{m_1 + m_2}{2} + m_3\right) + \left(m_2 - \frac{m_1 + m_2}{2} + m_3\right) + \left(m_{13} - \frac{m_{13} + m_{15}}{2} + m_{15}\right)$$

$$= \frac{1}{2}(m_1 + m_3 + m_2 + m_2 + m_{12} + m_{15} + \dots)$$

where m_1, m_2, m_3, etc. are roughly one-third of R_1, R_2, R_3, etc.

Thus point P_3 is sufficiently bright. P_3 represents the position of the second focal point. Its distance from O is the second focal length and is given by

$$f_3 = \frac{r_n^2}{3n\lambda}$$

The resultant amplitude at P_3 is less than that P. Therefore, its intensity is less than that at P. Similarly other foci occur on the axis at distances $b/5$, $b/7$, etc. from the zone plate; but the intensity of successive foci decreases gradually.

It may be noted here that we get maximum intensity at those points for which the number of central exposed elements is odd, i.e.,

$$(2p-1) \text{ where } p = 1, 2, 3, \ldots.$$

If a clear zone is occupied by an even number of half period zones, they cancel in pairs and produce zero illumination together. On the other hand, if a clear space contains an odd number of half period zones, the complete deduction of light from all the zones together cannot take place, and hence there will be maximum intensity at P.

Generalising the results, we see that the various focal lengths are given by

$$f_p = \frac{r_n^2}{(2p-1)n\lambda} \qquad \qquad \ldots(11)$$

where $p = 1, 2, 3, \ldots\ldots$.

Equation (11) gives

$$f_1 = \frac{r_n^2}{1.n\lambda}, \; f_3 = \frac{r_n^2}{3.n\lambda} = \frac{f_1}{3}, \; f_5 = \frac{r_n^2}{5.n\lambda} = \frac{f_1}{5'}\ldots$$

Thus we see that a zone plate has multiple foci.

Comparison of a zone plate with a convex lens:

Similarities:

(i) The relations connecting the conjugate distances are similar.

For real image the formula connecting conjugate distances in the case of convex lens is

$$\frac{1}{v} + \frac{1}{u} = \frac{1}{f}$$

The corresponding formula for zone plate is

$$\frac{1}{a} + \frac{1}{b} = \frac{n\lambda}{r_n^2}$$

(ii) Focal lengths of both depend upon wavelength of light used. Hence, both show chromatic aberration.

(iii) Both form real image of an object on the side opposite to that of the object.

(iv) The formula for linear magnification for a zone plate and convex lens is similar.

Difference:

(i) A convex lens has only one focus on one side of it for a given wavelength of light; while

a zone plate has a number of foci at distances $\dfrac{r_n^2}{n\lambda}, \dfrac{r_n^2}{3n\lambda}, \dfrac{r_n^2}{5n\lambda}, \dots\dots, \dfrac{r_n^2}{(2p-1)n\lambda}$ of

continuously decreasing intensity.

(ii) For a convex lens all the waves reaching the image point have the same optical path; while for a zone plate the path difference between the waves reaching the image point through two consecutive transparent zones is λ.

(iii) For a zone plate the focal length for red colour is greater than that for violet, i.e. $f_r < f_e$; while for a convex lens $f_r < f_e$.

(iv) The intensity of the image formed by a zone plate is less than that of the image formed by a convex lens.

Phase reversal zone plate: Rayleigh suggested that if the alternate zones of a zone plate, instead of being blocked, are coated with a thin film of some transparent material which introduces an additional optical path difference $\lambda/2$, then the secondary disturbances from all the zones, both odd or even, would reach the image point in the same phase and so the amplitudes from successive zones will help each other. Consequently the intensity of the image would be four-fold. Such zone plates are called phase reversal zone plate. R.W. Wood in 1898, prepared such a zone plate by the following method:

A chemically cleaned thin glass plate is coated with a thin layer of gelatine solution and dried. It is than immersed in a weak solution of potassium dichromate for a few seconds and dried again in dark. It is then placed in contact with an ordinary zone plate and exposed to sunlight for several minutes. Light passing through transparent zones acts on gelatine and makes it insoluble in water while the gelatine in contact with opaque zones remains soluble in water. Finally the glass plate is immersed in water where the gelatine of the unexposed parts is

dissolved to such a depth that an additional optical path difference of half a wavelength $\left(\dfrac{\lambda}{2}\right)$

is introduced between the waves from successive zones. Thus the phase reversal zone plate is ready.

Diffraction at a straight edge: Let S be a narrow slit illuminated by monochromatic light of wavelength λ. O is the sharp straight edge of an obstacle OA. The edge of O and the slit are both parallel and they are perpendicular to the plane of the Fig. 8.6. XY is a screen perpendicular to the plane of the paper. Join SO and produce it to meet the screen at C. According to the laws of geometrical optics, we should get uniform illuminate darkness in the region CY. But actually it is observed that there are a few unequally spaced diffraction fringes in the illuminated region close to C and in the region of darkness the intensity does not become zero at C; but it falls rapidly and becomes zero at a small finite distant point from C.

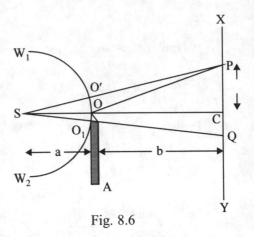

Fig. 8.6

Let $W_1 W_2$ be the section of the cylindrical wavefront diverging from S when it reaches the edge O of the obstacle.

Explanation of the diffraction fringes in the illuminated region: Let P be a point in the illuminated region. To find the intensity at P the wavefront is divided into half period strips with respect to P. The pole of the wavefront with respect to P is O. With P as centre and radii equal to

$$PO' + \frac{\lambda}{2}, \ PO' + \frac{2\lambda}{2}, PO' + \frac{3\lambda}{2}, \text{ etc.}$$

the points of wavefront are marked and then through these points lines are drawn parallel to the length of this slit. This divides the wavefront into half period strips. These are similar to halves of the wavefront lying on either side of O'. The effect of the entire wavefront $W_1 W_2$ on the point P is equal to the effect of the entire upper half of the wavefront plus the effect of the exposed portion of the lower half.

The amplitude at P due to entire upper half of the wavefront

$$= \frac{R_1}{2}$$

If the exposed portion $O'O$ of the lower half wavefront contains one half period strip, the amplitude at P

$$= \frac{R_1}{2} + R_1 = \frac{3R_1}{2} \text{ (maximum)}$$

If the exposed portion $O'O$ of the lower half wavefront contains to half period strips, the resultant amplitude at P_1 will be

$$= \frac{R_1}{2} + R_1 - R_2 = \frac{3R_1}{2} - R_2 \text{ (minimum)}$$

Similarly, we get maximum at P if the lower half wavefront exposes three, five, seven...half period strips and minimum at P if the lower half wavefront exposes four, six....half period strips. Thus, the point P will be of maximum or minimum intensity according as the exposed portion of the lower half wavefront contains odd or even number of half period strips.

Thus if we move away from C in the illuminated region, we get alternate maxima and minima; thus forming the diffraction fringes parallel to the edge of the obstacle. Moreover, the amplitudes and hence the intensities of maxima goes on decreasing and those of minima goes on increasing as we move away from C in the illuminated region.

When we reach sufficiently away from C in the illuminated region, the lower half wavefront exposes all the effective half period strips and then the resultant amplitude at P will be

$$\frac{R_1}{2} + \frac{R_1}{2} = R_1$$

i.e., the amplitude and hence the intensity is equal to that of the completely exposed wavefront. Thus there will be uniform illumination on the screen beyond the diffraction fringes in the illuminated region.

Mathematical theory: It is seen from above that the point P will be maximum and minimum intensity accordings as the exposed portion OO' of the lower half wavefront contains odd or even number of half period strips. This implies that the point P will be maximum or minimum intensity if the path difference $(PO–PO')$ is equal to the multiple of odd or even number of $\lambda/2$.

Thus, $\quad PO - PO' = (2n-1)\lambda/2 \ $ (for maxima), $n = 1, 2, 3...$ ...(1)

and $\quad PO - PO' = 2n \cdot \lambda/2$ (for minima), $n = 0, 1, 2...$ ...(2)

Let $\quad SO = a, \ OC = b$ and $CP = y.$

We have $\ PO = \sqrt{[(OC)^2 + (CP)^2]} = \sqrt{(b^2 + y^2)} - b\left(1 + \dfrac{y^2}{b^2}\right)^{1/2}$

$$= b\left(1 + \frac{y^2}{2b^2}\right) \text{ approx} \qquad\qquad \text{(by binomial theorem)}$$

or $\quad OP = b + \dfrac{y^2}{2b}$

$$PO' = SP - SO' = \sqrt{[(SC)^2 + (CP)^2]} - SO'$$

$$= \sqrt{[(a+b)^2 + y^2]} - a$$

$$= (a+b)\left(1 + \frac{v^2}{(a+b)^2}\right)^{1/2} - a$$

$$= (a+b)\left[1 + \frac{y^2}{2(a+b)^2}\right]^{1/2} - a$$

$$= b + \frac{y^2}{(a+b)}$$

Thus
$$PO - PO' = \left(b + \frac{y^2}{2b}\right) - \left(b + \frac{y^2}{2(a+b)}\right)$$

$$= \frac{y^2}{2b} - \frac{y^2}{2(a+b)}$$

or
$$PO - PO' = \frac{y^2}{2} \cdot \frac{a}{b(a+b)} \qquad \qquad ...(3)$$

Thus using equations (1) and (3) the condition of maxima is given

by
$$\frac{y^2}{2} \cdot \frac{a}{b(a+b)} = (2n-1)\frac{y}{2}$$

or
$$y_n = \sqrt{\left[\frac{b(a+b)}{a} \cdot (2n-1)\lambda\right]}; \qquad n = 1, 2, 3 \qquad \qquad ...(4)$$

This gives the distance of nth maxima from C.

Using equations (2) and (3), the condition of minimum is given by

$$\frac{y^2}{2} \cdot \frac{a}{(a+b)} = 2n \cdot \frac{\lambda}{2}$$

$$y_n = \sqrt{\left[\frac{2n(a+b)b\lambda}{a}\right]} \qquad \qquad ...(5)$$

This equation gives the distance nth minima from C.

The distance of successive maxima from C is given by

$$y_1 = \sqrt{\left[\frac{b(a+b)b\lambda}{a}\right]} 1 = K(n=1)$$

$$y_2 = \sqrt{\left[\frac{b(a+b)\lambda}{a}\right]} \sqrt{3} = \sqrt{3}K(n=2)$$

$$y_3 = \sqrt{\left[\frac{b(a+b)\lambda}{a}\right]} \sqrt{5} = \sqrt{5}K(n=3)$$

$$y_4 = \sqrt{\left[\frac{b(a+b)\lambda}{a}\right]}\sqrt{7} = \sqrt{7}K(n=4)$$

where $\qquad K = \sqrt{\left[\frac{b(a+b)\lambda}{a}\right]}$

Thus the separation between successive maxima are

$$y_2 - y_1 = \left(\sqrt{3}-1\right)K = 0.73\,K$$
$$y_3 - y_2 = \left(\sqrt{5}-\sqrt{3}\right)K = 0.50\,K$$
$$y_4 - y_2 = \left(\sqrt{7}-\sqrt{5}\right)K = 0.43\,K$$

where $\qquad K = \sqrt{\left[\frac{b(a+b)\lambda}{a}\right]}$ $\qquad\qquad$ ...(6)

Obviously the separation between successive maxima go on decreasing.

Similarly the separations between successive minima are

$$K\left(\sqrt{4}-\sqrt{2}\right), K\left(\sqrt{6}-\sqrt{4}\right), K\left(\sqrt{8}-\sqrt{6}\right)...$$

obviously they are also of decreasing magnitude.

Thus we see that the spacing between successive bright and dark rings decreases as the order increases.

Intensity at the edge of the geometrical shadow: Let us first find the intensity at the edge C of the geometrical shadow. The pole of the wavefront with respect to C is O. The wavefront is divided into half period strips with respect to C. As the lower half of the wavefront is completely blocked by the obstacle, the intensity at C is entirely due to the upper half of the wavefront OW_1. Thus the resultant amplitude at C is $\dfrac{R_1}{2}$ and hence the resultant intensity at

C is $\dfrac{R_1^2}{4}$. If the entire wavefront were exposed, the resultant amplitude at C had been

$\dfrac{R_1}{2} + \dfrac{R_1}{2} = R_1$ and hence the resultant intensity at C had been R_1^2. Thus the intensity at C is

one-quarter of the intensity of that produced by the complete exposed wavefront.

Intensity within geometrical shadow: Now consider a point Q within the geometrical shadow. The pole of the wavefront with respect to Q is O_1. Thus the lower half wavefront O_1W_2 is completely blocked and the portion O_1O of the upper half is also blocked. Thus the intensity at Q is due to only the portion OW_1 of the upper half wavefront O_1W_1. Let the wavefront be divided into half period strips with respect to Q. If point Q is such that O_1O cuts-off only first half period strip, the amplitude at Q will be

$$R_2 - R_3 + R_4 - R_5 + ... = \frac{R_2}{2}$$

If point Q move further into the geometrical shadow (away from C) gradually, first two, three, four etc. half period strips are successively blocked and so the amplitude at Q becomes $R_3/2$, $R_4/2$, $R_5/2$, etc. As R_1, R_2, R_3, R_4, etc. decrease in magnitude rapidly at first and then slowly, the amplitude and hence the intensity decreases rapidly at first and than slowly as we move from C inside the geometrical shadow. For a point, sufficiently way from C inside the geometrical shadow. For a point, sufficiently away of the upper half wavefront are blocked and hence the intensity at such a point becomes zero.

Thus we see that the intensity inside the geometrical shadow decreases continuously to zero as we move away from C.

The distribution of intensity of light on the screen is represented in Fig. 8.7. Inside the geometrical shadow the intensity decreases rapidly while outside it we get alternate maxima and minima. The intensity of maxima goes on decreasing while that of minima goes on increasing till uniform illumination results.

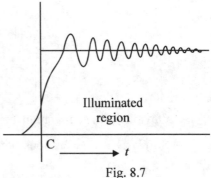

Fig. 8.7

Diffraction at a narrow wire: Let S be a narrow slit source, perpendicular to the plane of the paper, illuminated with monochromatic light of wavefront λ. Let AB be a narrow wire placed with its sides parallel to the slit S. Let XY be the trace of the screen perpendicular to the plane of the paper. Let $W_1 W_2$ be the cylindrical wavefront diverging from S incident on the wire AB. Let SA and SB produced meet the screen at M and N respectively. According to the laws of geometrical optics, we should get uniform illumination above M and below N and complete darkness within MN. But this does not happen and unequally spaced bonds with poor contrast and running parallel to the length of the slit are observed in the region above M and below N while equally spaced bands running parallel to the length of the slit are observed in the geometrical shadow MN.

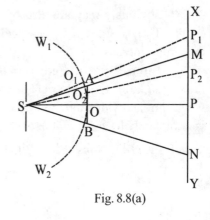

Fig. 8.8(a)

Explanation: Consider a point P_1 on the screen in the illuminated region. Join S to P_1 meeting the wavefront at O_1. Thus O_1 is the pole of the wavefront with respect to P_1. If we consider the wavefront to be divided into Fresnel's half period zones, then the intensity at P_1 will be partly due to the upper part of the wavefront (i.e., P_1W) and partly due to the part P_1A.

The intensity at P_1 due to the upper half wavefront P_1W_1 is same at all points and the effect due to the lower part of the wavefront BW_2 is negligible. Thus if O_1A contains an odd number of half period zones, the intensity at P_1 will be maximum and if O_1A contains an even number of half period zones, the intensity at P_1 will be minimum. Thus in the illuminated region above M and below N there will be diffraction bands similar to bands with a sharp edge. The intensity of the diffraction bands will go on decreasing and hence the distinction between maxima and minima will become less and less, as point P_1 is moved away from the edge M or N of the geometrical shadow. If the wire is extremely thin, the portion BW, of the wavefront also produces effect at P_1 and hence the maxima and minima on either side (above M and below N) cannot be distinguished.

Next consider a point P_2 in the region of geometrical shadow. Let O_2 be the pole of the wavefront relative to P_2. From the considerations of section 4.10 we know that the effect of the portion AW_1 of the cylindrical wavefront at any point P_2 within the geometrical shadow may be regarded as entirely due to a few half period elements at A. Thus for determining the intensity at P_2 the part AW_1 of the wavefront may be considered as a small linear source through A. Similarly the lower part BW_2 of the wavefront may be considered as a small linear source through B. Thus the edges A and B of the wire act as two coherent sources and give rise to equally spaced interference bands in the region of geometrical shadow similar to Young's double slit bands. These fringes have sharp contrast between maxima and minima. According to condition of interference, the point P_2 will be bright or dark depending on whether the path difference $(BP_2 - AP_2)$ is equal to even or odd multiple of half wavelength $\dfrac{\lambda}{2}$.

i.e., $BP_2 \sim AP_2 = 2n \cdot \dfrac{\lambda}{2}$ for maximum intensity at P_2, $n = 0, 1, 2, 3, ...$

$BP_2 \sim AP_2 = (2n-1)\dfrac{\lambda}{2}$ for minimum intensity at P_2, $n = 1, 2, 3,......$ obviously the centre of the shadow will be bright.

Further if d is the thickness (diameter) of the wire and D is its distance from the screen, then the fringe width ω is given by

$$\omega = \frac{D\lambda}{d}$$

knowing D and d, and measuring the fringe width ω with a micrometer eyepiece, we can determine the wavelength λ of the incident light.

If one edge of the wire is covered, the corresponding straight edge fringes in the illuminated region disappear and interference fringes in the geometrical shadow are replaced by the straight edge fringes on the other side. This proves conclusively that the fringes within the geometrical shadow are due to the combined effect of the two parts of the wavefront, viz. AW_1 and BW_2.

The intensity distribution due to a narrow wire is shown in (Fig. 8.8a).

Effect of increasing the thickness of the wire: If the thickness of the wire is gradually increased, the width of interference fringes decreases while the diffraction pattern in the illuminated region remains unaffected.

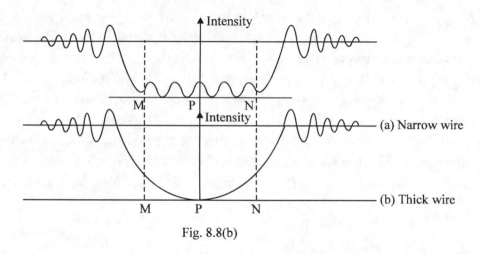

Fig. 8.8(b)

If the wire is sufficiently thick (of the order of mm) the interference fringes completely disappear. The intensity, however, falls off rapidly as we enter the geometrical shadow from either side (Fig. 8.8b).

Fraunhofer diffraction at a single slit: In Fig. 8.9, a slit source illuminated by monochromatic light of wavelength λ is placed perpendicular to the plane of the paper. L_1 is the collimating lens which is placed at a distance equal to the focal length of the lens from S. Thus the light emerges from lens L_1 as a parallel beam and the wavefront falling on the slit AB of width a is plane. Thus the source S is effectively at infinite distance from the slit. AB is normal to the plane of the paper. The diffracted light is focussed by another convex lens L_2 on the screen XY placed in the focal plane of lens L_2 perpendicular to the plane of the paper. According to the laws of geometrical optics, we should get a sharp image of the slit on the screen. But this is not the case and a diffraction pattern is obtained on the screen. This diffraction pattern consists of a central bright band having alternate dark and bright bands of decreasing intensity on both the sides.

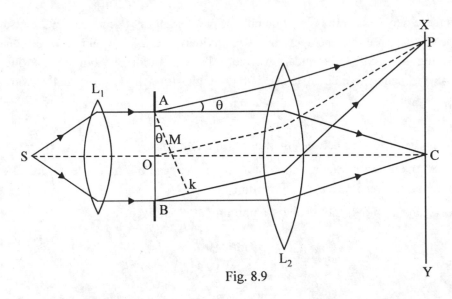

Fig. 8.9

Analytical Explanation: A plane wavefront is incident normally on the slit AB and according to Huygens theory each point of *AB* sends out secondary wavelets in all directions. In other words the diffracted rays start from every point in AB in every direction. Evidently all the wavelets start from various points in AB in the same plane. The rays diffracted along the direction of incident rays are focussed at *C*; while those diffracted through angle θ are focussed at *P*. The point *C* is optically equidistant from all points on the slit AB. Therefore all the secondary wavelets from *AB* reach *C* in the same phase. Hence there is maximum intensity at *C*.

Now let us find out the resultant intensity at *P*. Let *AK* be perpendicular to *BK*. The optical lengths of all the rays from the plane *AK* to the focal point *P* are the same. The path difference between the rays originating from extreme points A and B is given by

$$p = BK = AB\sin\theta = a\sin\theta \qquad \text{...(1)}$$

where *a* is the width of *AB*.

The path of the ray starting from A is less and the path of the ray straight from *B* is greatest. Hence the path difference of different rays originating from points in AB vary from 0 to $a\sin\theta$. Hence the rays arrive at *P* in succession and not simultaneously, i.e. the rays originating from points in *AB* do not arrive in the same phase; but the phase difference increases regularly for the rays from *A* to that from *B*.

∴ The *phase difference* between extreme rays

$$= \frac{2\pi}{\lambda} \times p = \frac{2\pi}{\lambda} \cdot a\sin\theta \qquad \text{...(2)}$$

As there may be infinitely large number of point sources of secondary wavelets between A and B, let us suppose that the aperture AB is divided into a large number n of equal parts, each part being the source of secondary wavelets. The amplitude of vibration at P due to each part will be the same*(say equal to a); but their phases will vary gradually from zero to $\frac{2\pi}{\lambda} \cdot a \sin\theta$.

The phase difference between the waves from two consecutive parts is $\frac{1}{n} \cdot \frac{2\pi}{\lambda} \cdot a \sin\theta = \delta$ (say). Thus we have to find the resultant of a vibrations each of amplitude a, having common phase difference between successive vibrations.

According to δ 5.2, the resultant amplitude at P is given by

$$R = \frac{\sin\dfrac{n\delta}{2}}{\sin\dfrac{\delta}{2}} = a. \frac{\sin\left(\dfrac{\pi a \sin\theta}{\lambda}\right)}{\sin\left(\dfrac{\pi a \sin\theta}{n\lambda}\right)} \qquad \text{...(3)}$$

Let us substitute

$$\alpha = \frac{\pi a \sin\theta}{\lambda} \qquad \text{...(4)}$$

$$\therefore \qquad R = a \frac{\sin\alpha}{\sin\alpha/n} = a \frac{\sin\alpha}{\alpha/n}$$

because n is very large so that αn is very small.

$$\therefore \qquad R = na \cdot \frac{\sin\alpha}{\alpha} = \frac{A\sin\alpha}{a} \qquad \text{...(5)}$$

where A = na = amplitude when all the vibrations are in the same phase. ...(6)

The resultant intensity at P is proportional to the square of the amplitude. Taking the constant of proportionality equal to unity for simplicity, the resultant intensity at P is given by

$$I = R^2 = A^2 \cdot \frac{\sin^2\alpha}{\alpha^2} \qquad \text{...(7)}$$

This equation gives the variation of intensity as a function of the direction. Maximum and minimum values of I can be found by putting $\frac{dI}{d\alpha} = 0$.

---

* As each part has the same width and the screen is effectively at infinite distance from the slit, the obliquity and the distance factors do not change the amplitude.

∴   Differentiating equation (7) and equating to zero, we get

$$\frac{dI}{d\alpha} = \frac{d}{d\alpha}\left(A^2 \cdot \frac{\sin^2 \alpha}{\alpha^2}\right) = A^2 \cdot \frac{2\sin\alpha}{\alpha} \cdot \frac{\alpha\cos\alpha - \sin\alpha}{\alpha^2} = 0$$

∴   This gives either

$$\frac{\sin\alpha}{\alpha} = 0 \text{ or } \frac{\alpha\cos\alpha - \sin\alpha}{\alpha^2} = 0$$

i.e., either $\dfrac{\sin\alpha}{\alpha} = 0$ or $\alpha\cos\alpha - \sin\alpha = 0$

i.e., either (i) $\dfrac{\sin\alpha}{\alpha} = 0$ or (ii) $\alpha = \tan\alpha.$ ...(8)

Directions of minimum intensity: When $\dfrac{\sin\alpha}{\alpha} = 0,$ it is clear from equation (7), that the intensity is zero.

Thus for minimum intensity,

$$\frac{\sin\alpha}{\alpha} = 0, \text{ or } \sin\alpha = 0$$

i.e., $\alpha = \pm m\pi,$ where $m = 1, 2, 3,....$

or $\dfrac{\pi a \sin\theta}{\lambda} = \pm m\pi$ [using equation (4)]

or $a\sin\theta = \pm m\lambda.$ ...(9)

Here $m = 1, 2, 3,...$ It is to be noted that $m \pm 0$, because when m = 0, $a \sin \theta = 0$ and this is the condition of maximum intensity. Equation (9) gives the directions of first, second, third,... minima by putting m = 1, 2, 3,... respectively.

Directions of maximum intensity: For maximum intensity,

we have $\alpha = \tan\alpha$ [see equation (8)]   ...(10)

This equation can be solved graphically by plotting the curves

$$y = \alpha \text{ and } y = \tan\alpha$$

The equation $y = \alpha$ represents a straight line passing through the origin and making an angle of 45° (Fig. 8.10).

The abscissae of the points of intersection of the two curves give the required values of α for which the intensity is maximum. The first value of α is zero while the remaining values are approximately

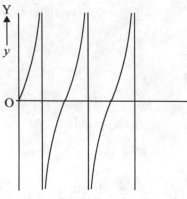

$$\frac{3\pi}{2}, \frac{5\pi}{2}, \frac{7\pi}{2}, \dots$$

The exact values of α are

$$\alpha = 0, \ 1.430\pi, \ 2.46\pi, \ 3.47\pi, \ 4.471\pi, \ 5.482 \ \pi, \dots$$

The value $\alpha = 0$ corresponds to the position of the central maximum. Thus for central maximum

Fig. 8.10

$$\alpha = 0 \ \text{ or } \ \frac{\pi a \sin\theta}{\lambda} = 0 \ \text{ or } \ \theta = 0$$

The intensity of the central principal maximum is

$$I = R^2 = A^2 \ \frac{\sin^2\alpha}{\lambda} = 0 \ \text{ or } \ \theta = 0 \qquad \dots(11)$$

The directions of secondary maxima are approximately given by

$$a = \pm\frac{(2m+1)\pi}{2}, m = 1, 2, 3, \dots \ \text{ or } \ a\sin\theta = \pm\frac{(2m+1)\lambda}{2} \qquad \dots(12)$$

If we put $m = 1, 2, 3, \dots$ we get the position of first, second, third,... secondary maxima respectively.

The intensity of the first secondary maxima,

$$I_1 = A^2 \frac{\left(\sin\dfrac{3\pi}{2}\right)^2}{\left(\dfrac{3\pi}{2}\right)^2} = \frac{A^2}{22} = \frac{I_0}{22}$$

The intensity of the second secondary maxima

$$I_2 = A^2 \frac{\left(\sin\dfrac{5\pi}{2}\right)^2}{\left(\dfrac{5\pi}{2}\right)^2} = \frac{A^2}{62} = \frac{I_0}{62}$$

Similarly the intensity of third secondary maxima,

$$I_2 = \frac{I_0}{120}$$

Evidently most of the light is concentrated in the central maximum. The intensity of the first secondary maxima is about $\frac{1}{22}$ th of the intensity of the central maximum. Moreover, the intensity of secondary maxima decreases very rapidly. The central maxima lies at $\alpha = 0$, the minima lie at $\alpha = \pm\pi, \pm2\pi, \pm3\pi, ...$ and the secondary maxima lie at $\alpha = \pm\frac{3\pi}{2}, \pm\frac{5\pi}{2},...$

The minima are equally spaced and their intensity is zero.

Thus the diffraction pattern consists of a bright central maximum surrounded alternatively by minima of zero intensity and feeble secondary maxima of rapidly decreasing intensities. The intensity distribution in Fraunhofer diffraction pattern is represented graphically in Fig. 8.11.

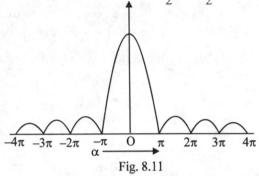

Fig. 8.11

It is obvious that the central maximum has double the width of the secondary maxima. Moreover, the secondary maxima do not fall exactly mid-way between the two minima; but are displaced towards the centre of the pattern which decreases as the order increases.

Spread of central diffraction maximum:

The direction of first minima is given by

$$a\sin\theta = \pm\lambda \qquad (\text{since } m = 1)$$

or
$$\sin\theta = \pm\frac{\lambda}{a} \qquad\qquad\qquad ...(13)$$

or
$$\theta = \pm\sin^{-1}\frac{\lambda}{a} \qquad\qquad\qquad ...(14)$$

Thus the central maximum extends between $\theta = \pm\sin^{-1}\lambda/a$.

Thus θ is the *angular half width of the central maximum.*

If the lens L_2 is very near to the slit AB or the screen is far away from the lens L_2, we have

$$\sin\theta = \pm y/f \qquad\qquad\qquad ...(15)$$

where y is the linear separation between the first minima on either side and the central maximum of the diffraction pattern, i.e. y is the linear half width of the central maximum and f is the focal length of lens L_2.

Comparing eqns. (14) and (15), we get

$$\therefore \qquad \frac{y}{f} = \frac{\lambda}{a} \quad \text{or} \quad y = \frac{f\lambda}{a}$$

Thus the width of the central maximum

$$= 2y = \frac{2f\lambda}{a} \qquad \qquad ...(16)$$

Obviously the width of the central maximum is proportional to the wavelength λ of light. As wavelength for red light is longer than that of violet light, the width of the central maximum with red light is more than with violet light.

Obviously the width of central maximum is inversely proportional to the width of the slit. Thus the width of central maximum will be greater for narrow slits.

Special Case (a = λ): The direction of first minima is given by

$$a \sin\theta = \pm\lambda$$

or $\qquad \sin\theta = \pm\dfrac{\lambda}{a} = \pm 1 \qquad$ (when $\lambda = a$)

or $\qquad \theta = \pm\dfrac{\pi}{2}.$

Thus the central maximum extends between $\theta = \pm\pi/2$, i.e. the light on the other side of the slit spreads in all directions. The intensity of light decreases steadily from the centre as θ increases and no other maxima is obtained.

Difference between Interference and Diffraction:

1. In the phenomenon of interference the interference occurs between two separate wavefronts originating from the two coherent sources; while in the phenomenon of diffraction the interference occurs between the secondary wavelets originating from the different points of the exposed part of the same wavefront.

It is clear from the double slit diffraction pattern that interference takes place between the secondary waves originating from the corresponding points of the two slits but the intensity of interference fringes is controlled by the amount of light reaching the screen due to diffraction occurring at each slit. Thus the resultant intensity at each point on the screen is due to joint effect of interference and of diffraction. Hence the pattern obtained on the screen may be called an interference pattern or a diffraction pattern. It will thus be realised that although the two phenomenon are similar, yet it is customary to reserve the term interference for those cases in which the amplitude is modified by the superposition of a small number of beams originating from coherent sources; while the term diffraction for those cases in which the amplitude is modified by summing the contributions of all the infinitesimal elements of the exposed part of the same wavefront.

2. In an interference pattern all the maxima are of same intensity but in diffraction pattern the intensity of central maximum is maximum and goes on decreasing on either side as the order of maxima increases.

3. In an interference pattern the minima are usually almost perfectly dark while it is not so in a diffraction pattern.

4. The interference fringes are usually equally-spaced while the diffraction fringes are never equally-spaced.

Fraunhofer diffraction due to N parallel equidistant slits:

Theory of diffraction grating: Let a plane wavefront of monochromatic light be incident normally on N parallel slits each of width a and separated by opaque space b. The light diffracted through N slits is focussed by lens L on the screen XY placed in the focal plane of lens L. The pattern obtained on the screen is called the Fraunhofer diffraction pattern due to N parallel equidistant slits.

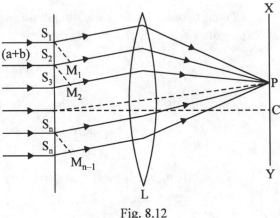

Fig. 8.12

When the wavefront reaches the plane of the slits, each point in the slits sends out secondary wavelets in all directions. From the theory of Fraunhofer diffraction at a single slit, the wavelets preceding from all

points in a slit in direction θ are equivalent to a single wave of amplitude R $\left(= \dfrac{A \sin \alpha}{\alpha} \right)$ starting

from the centre of the slit where $\alpha = \dfrac{\pi a \sin \theta}{\lambda}$. Thus the waves diffracted from all the slits in

direction θ are equivalent to N parallel waves, each wave starting from the middle points S_1, S_2, S_3,......,S_{n-1}. S_n of the slits.

Let $S_1 M_1$, $S_2 M_2$,......,$S_{n-1} M_{n-1}$ be the perpendiculars dropped from S_1, S_2,......,S_{n-1} on $S_2 M_1$, $S_3 M_2$,......,$S_n M_{n-1}$ respectively.

The path difference between the waves from S_1 and S_2 is

$$S_1 M_1 = (a+b) \sin \theta$$

The path difference between the waves S_2 and S_3 is

$$S_2 M_2 = (a+b) \sin \theta$$

The path difference between the waves from S_{n-1} and S_n is

$$S_{n-1} M_{n-1} = (a+b) \sin \theta$$

Thus as we pass from one vibration to another, the path goes on increasing by the same amount $(a + b)\sin\theta$.

The corresponding phase difference

$$= \frac{2\pi}{\lambda}(a+b)\sin\theta$$

Thus as we pass from one vibration to another, the phase goes on increasing by the same amount

$$\frac{2\pi}{\lambda}(a+b)\sin\theta$$

Thus in order to find the amplitude in a direction θ, we have to find the resultant amplitude of N wave, each having amplitude R and common phase difference.

$$\frac{2\pi}{\lambda}(a+b)\sin\theta = 2\beta \text{ (say)}$$

Applying the result of δ 5.1 the resultant amplitude in a direction θ is given by

$$R' = R\frac{\sin N\beta}{\sin \beta} = \frac{A\sin\alpha}{\alpha}\cdot\frac{\sin\alpha}{\beta}$$

The resultant intensity at P is given by

$$I = R'^2 = \frac{A^2 \sin^2\alpha}{\alpha^2}\cdot\frac{\sin^2 N\beta}{\sin^2 \beta} \qquad \text{...(1)}$$

The factor $\dfrac{A^2 \sin^2\alpha}{\alpha^2}$ gives the intensity distribution in the diffraction pattern due to a

single slit and has already been discussed earlier. While the factor $\dfrac{\sin^2 N\beta}{\sin^2 \beta}$ gives the distribution

of intensity in the pattern due to interference in the waves from all the slits.

Principal Maxima:

When $\qquad\qquad \sin\beta = 0, \ i.e. \ \beta = \pm n\pi, \ n = 0, 1, 2, 3, ...$

We have $\sin N\beta = 0$, so $\dfrac{\sin N\beta}{\sin \beta}$, let us differentiate the numerator and denominator and

find its ratio; thus

$$\underset{\beta \to \pm n\pi}{\text{Lim}} \frac{\sin N\beta}{\sin \beta} = \underset{\beta \to \pm n\pi}{\text{Lim}} \frac{\dfrac{d}{d\beta}(\sin N\beta)}{\dfrac{d}{d\beta}(\sin \beta)}$$

$$= \frac{\text{Lim}}{\beta \to \pm n\pi} \frac{N \cos N\beta}{\cos \beta} = N$$

Substituting this value of $\dfrac{\sin N\beta}{\sin \beta}$ in eqn. (1) we get

$$I = \frac{A^2 \sin^2 \alpha}{\alpha^2} \cdot N^2 \qquad \qquad ...(2)$$

which is maximum. Then the resultant intensity of maxima is $\dfrac{A^2 \sin^2 \alpha}{\alpha^2} \cdot N^2$. These maxima

are called the *principal maxima*. Thus in order to find the resultant intensity of any of the

principal maxima in the diffraction pattern, we have to multiply N^2 by the factor $= \dfrac{A^2 \sin^2 \alpha}{\alpha^2}$.

Hence if we increase the number of slits, the intensity of principal maxima increases.

$$\sin \beta = 0, \ i.e., \ \beta = \pm n\pi$$

or $\qquad \dfrac{\pi}{\lambda}(a+b)\sin\theta = \pm n\pi$

or $\qquad (a+b)\sin\theta = \pm n\lambda \qquad \qquad ...(3)$

If we put $n = 0$, we get $\theta = 0$. This is the direction in which all the waves arrive in the same phase and so we get bright central image. This maxima is called the zero order principal maxima. If we put $n = 1, 2, 3,.....$ we obtain first, second, third, ... order principal maxima respectively.

Minima: From equation (1), it is obvious that for minima, we must have

$$\sin N\beta = 0; \ \text{but} \ \sin \beta \neq 0$$

or $\qquad N\beta = \pm m\pi$

or $\qquad N \cdot \dfrac{\pi}{\lambda}(a+b)\sin\theta = \pm m\pi$

or $\qquad N \cdot (a+b)\sin\theta = \pm m\lambda \qquad \qquad ...(4)$

Here m can have all integral values except 0, N, $2N$, $3N$, ... for these values give $\sin \beta = 0$ which gives the positions of principal maxima. The positive and negative signs indicate that the minima of a given order lie symmetrically on both sides of the central principal maximum.

It is obvious from eqn. (4), that $m = 0$ gives principal maximum of zero order $m = 1$, 2, 3, ..., $(N-1)$ give minima and then $m = N$ gives principal maximum of first order. Thus there are $(N-1)$ equispaced minima between zero order and first order principal maxima. Similarly it can be shown that there are $(N-1)$ maxima between first and second order principal maxima and so on. Thus there are $(N-1)$ equispaced maxima between any two consecutive principal maxima.

Secondary Maxima: As there are $(N-1)$ minima between two consecutive principal maxima, there must be $(N-2)$ other maxima coming alternatively with the minima between two consecutive principal maxima. These maxima are called secondary maxima. The positions of the secondary maxima are obtained by differentiating eqn. (1) with respect to β and equating it to zero. Thus

$$\frac{dI}{d\beta} = \frac{A^2 \sin^2 \alpha}{\alpha^2} 2\left[\frac{\sin N\beta}{\sin \beta}\right] \frac{N \cos N\beta \sin \beta - \sin N\beta \cos \beta}{\sin^2 \beta} = 0$$

or $\qquad N \cos N\beta \sin \beta - \sin N\beta \cos \beta = 0$

or $\qquad \tan N\beta = N \tan \beta$. $\hspace{3cm}$ ...(5)

In order to find the intensity of secondary maxima, consider Fig. 8.13.

We have $\quad \sin N\beta = \dfrac{N \tan \beta}{\sqrt{\left(1 + N^2 \tan^2 \beta\right)}}$

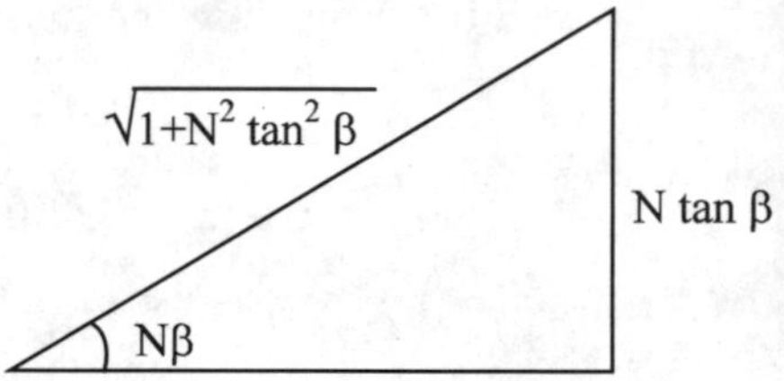

$\therefore \quad \dfrac{\sin^2 N\beta}{\sin^2 \beta} = \dfrac{N^2 \tan^2 \beta / (1 + N^2 \tan^2 \beta)}{\sin^2 \beta}$

$\qquad = \dfrac{N^2 \tan^2 \beta}{\left(1 + N^2 \tan^2 \beta\right)\sin^2 \beta} = \dfrac{N^2}{1 + \left(N^2 - 1\right)\sin^2 \beta}$

Fig. 8.13

Thus the intensity of secondary maxima from eqn. (1), is given by

$$I' = A^2 \frac{\sin^2 \alpha}{\alpha^2} \cdot \frac{N^2}{1 + \left(N^2 - 1\right)\sin^2 \beta} \hspace{2cm} ...(6)$$

From equations (2) and (6), we have

$$\frac{\text{Intensity of secondary maxima}}{\text{Intensity of principal maxima}} = \frac{1}{1 + \left(N^2 - 1\right)\sin^2 \beta} \hspace{1cm} ...(7)$$

Hence as N increases, the intensity of secondary maxima decreases. When N is very large, as in the case of diffraction grating, the secondary maxima are not visible in the spectrum. In such cases there is uniform darkness between any two consecutive principal maxima.

The intensity distribution on the screen is represented in Fig. 8.14.

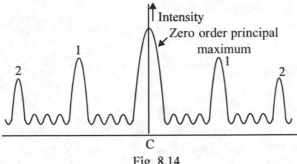

Fig. 8.14

The numbers 1, 2 on the two sides of zero order principal maximum represent, first, second order principal maxima respectively.

Width of the principal maxima: The angular width of principal maximum of any order is defined as the angular separation between the first two minima lying on its either side.

The direction of nth order principal maxima is given by

$$(a+b)\sin\theta_n = n\lambda \qquad \qquad ...(1)$$

If $(\theta_n + d\theta_n)$ and $(\theta_n - d\theta_n)$ represent the direction of first outer inner sided minima adjacent to the nth maximum, then $d\theta_n$ will be the angular half width of nth principal maximum.

The directions of minima are given by

$$N(a+b)\sin\theta = m\lambda \qquad \qquad ...(2)$$

where N is the total number of parallel slits. Here m takes all integral values except 0, N, $2N$..., nN, because these values of m, give the directions of zero, first, second, ..., nth order principal maxima respectively. As the first outer and inner sided minima adjacent to nth order principal maximum is obtained in the direction. $\theta_n \pm d\theta_n$, this corresponds to $m = (nN \pm 1)$.

Hence from equation (2), we have

$$N(a+b)\sin(\theta_n \pm d\theta) = (nN \pm 1)\lambda$$

or $\qquad N(a+b)(\sin\theta_n \cos d\theta_n \pm \cos\theta_n \sin d\theta_n) = (nN \pm 1)\lambda \qquad \qquad ...(3)$

For small values of $d\theta_n$, $\cos d\theta_n = 1$ and $\sin d\theta_n = d\theta_n$

So that equation (3) gives

$$N(a+b)\sin\theta_n \pm N(a+b)\cos\theta_n d\theta_n = nN\lambda \pm \lambda \qquad \qquad ...(4)$$

From equation (1), we have

$$N(a+b)\sin\theta_n = Nn\lambda$$

Substituting this in equation (4), we get

$$N(a+b)\cos\theta_n d\theta_n = \lambda$$

or $$d\theta_n = \frac{\lambda}{N(a+b)\cos\theta_n}$$ ...(5)

Thus the angular half width is inversely proportional to N, i.e. the width of principal maxima decreases as N increase. If N, i.e. the number of slits is sufficiently high, the half width will be small and hence the principal maxima will be very sharp. This fact is used in the construction of diffraction grating.

The width of nth order principal maxima $= 2d\theta_n$.

$$= \frac{2\lambda}{N(a+b)\cos\theta_n}$$ ...(6)

Plane diffraction grating

An arrangement consisting of a large number of parallel slits of equal width and separated from one another by equal opaque spaces is called a diffraction grating.

It may be made by ruling a large number of fine, equidistant and parallel lines with a diamond point on an optically plane glass plate. The ruled widths are opaque to the light, while the space between any two lines is transparent. Such a grating is called transmission grating. There are about 15,000 lines per inch in such a grating. On account of many difficulties faced in the construction of such a grating, replicas of these are made for practical purpose. The technique of preparing these replicas is as follows:

A thin layer of collodion solution is passed on the surface of a ruled grating. The liquid evaporates and the thin collodion film is stripped off from the grating surface. This film, which retains the impressions of the original grating, is fixed between two glass plates. This serves the purpose of a plane transmission grating.

Formation of spectrum with a grating: The direction of principal maxima are given by

$$(a+b)\sin\theta_n = \pm n\lambda$$ ...(1)

where a is the width of transparent portion and b is that of opaque portion. The width $(a+b)$ is called the grating element. This equation shows:

(i)   For a particular value of λ, the direction of principal maxima of different orders are different.

(ii)   For a given value of n, the angle of diffraction θ_n varies with the wavelength. The angle of diffraction increases as wavelength increases. As $\lambda_r > \lambda_v$, the angle of diffraction for red colour is greater than that for violet colour. Hence, if white light is incident normally on a grating, each order will contain principal maxima of different wavelengths in different directions (Fig. 8.15).

From equation (1) it is clear that for $n = 0$, $\theta = 0$ for all values of λ, i.e. zero order principle maxima for all wavelengths lie in the same direction. Thus the zero order principal maxima will be while. The first order principal maxima of all wavelengths from the first order spectrum. Similarly the second order principal maxima of all wavelengths from the second order spectrum and so on. Most of the light is concentrated in the principal maximum of zero order and the intensity goes on diminishing gradually as we go to higher orders.

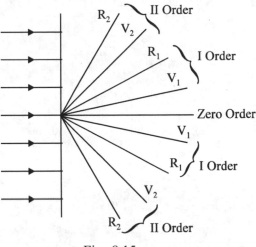

Fig. 8.15

Thus the spectrum consists of white maximum of zero order having on either side of it the first order spectra, the second order spectra and so on. The spectra of each other consists of spectral colours in the order from violet to red.

Conditions for absent spectra with a diffraction grating: The direction of principal maxima with a diffraction grating is given by

$$(a + b) \sin \theta = n\lambda \qquad \qquad ...(1)$$

where n is the order of the principal maximum.

The directions of minima in the diffraction pattern due to a single slit is

$$a \sin \theta = n\lambda \qquad \qquad ...(2)$$

where $m = 1, 2, 3, ...$

Now the intensity in the direction θ is given by

$$I = A^2 \frac{\sin^2 \alpha}{a^2} \cdot \frac{\sin^2 N\beta}{\sin^2 \beta}$$

where $\alpha = \dfrac{\pi a \sin \theta}{\lambda}$ and $\beta = \dfrac{\pi(a + b)\sin \theta}{\lambda}$

The term $\dfrac{A^2 \sin^2 \alpha}{\alpha^2}$ represents intensity due to diffraction at a single slit and $\dfrac{\sin^2 N\beta}{\sin^2 \beta}$ represents the intensity due to interference of waves from all the N slits.

Therefore if in any direction $\dfrac{\sin^2 N\beta}{\sin^2 \beta}$ is maximum and $\dfrac{A^2 \sin^2 \alpha}{\alpha^2}$ is zero, the principal maxima will not be present in the direction.

Thus if equations (1) and (2) are simultaneously satisfied, the principal maxima of order n will not be present in the grating spectrum.

Now dividing eqn. (1) by (2), we get

$$\frac{a+b}{a} = \frac{n}{m}$$

i.e., $$n = \frac{a+b}{a}m$$

This is the condition for nth order to be absent in the grating spectrum.

When $b = a$, then $n = \dfrac{a+b}{a}m = 2m = 2, 4, 6,...$ (since $m = 1, 2, 3,...$)

i.e., when $b = a$, 2nd, 4th, 6th,... order spectra will be absent.

When $b = 2a$, then $n = 3m = 3, 6, 9,...$

i.e., when $b = 2a$, 3rd, 6th, 9th .... order spectra will be absent and so on.

Oblique Incidence: Let a parallel beam of light be incident obliquely on the surface of the grating at an angle of incidence i (Fig. 8.16).

In this case the path difference between the resultant waves diffracted at angle θ through any two consecutive slits is given by

$$p = Ms_2 + S_2K$$

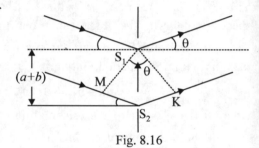

Fig. 8.16

If a is the width of the slit and b that of the opaque portion, we have from $\Delta s\ S_1MS_2$ and S_1KS_2, $S_2M = S_1S_2 \sin i$

$$= (a + b)i$$

and $$S_2K = S_1S_2 \sin\theta$$

$$= (a+b)\sin\theta$$

$\therefore$ $$p = MS_2 + S_2K = (a+b)\sin i + (a+b)\sin\theta$$

$$= (a+b)(\sin\theta + \sin i) \qquad ...(1)$$

This is the path difference for the beam diffracted upwards (Fig. 8.16).

If the beam is diffracted downwards, the path difference will be

$$p = (a+b)(\sin\theta - \sin i) \qquad ...(2)$$

For nth principal maximum, we have

$$(a+b)(\sin\theta_n + \sin i) = n\lambda \qquad ...(3)$$

or $\quad (a+b)\left[2\sin\dfrac{\theta_n+i}{2}\cdot\cos\dfrac{\theta_n-i}{2}\right]=n\lambda$

or $\quad \sin\dfrac{\theta_n+i}{2}=\dfrac{n\lambda}{2(a+b)\cos\dfrac{\theta_n-i}{2}}$ $\qquad$ ...(4)

The deviation of the diffracted beam form its original path

$\quad = \theta_n + i$

For the deviation $(\theta_n + i)$ to be minimum, $\sin\dfrac{\theta_n+i}{2}$ must be minimum. This will be so

if $\cos\dfrac{\theta_n-i}{2}$ will be maximum, i.e., $\dfrac{\theta_n-i}{2}=0$.

$\quad$ or $\quad \theta_n = i$

Thus the deviation produced by the grating in the diffracted beam is minimum when angle of incidence is equal to the angle of diffraction.

If D is the angle of minimum deviation.

$\quad \theta_n = \theta_n + i = 2\theta_n = 2i \quad$ (since $\theta_n = i$ for minimum deviation)

$\therefore \quad \theta_n = D/2$ and $i = D/2$

$\therefore$ For minimum deviation position, the equation for nth order principal maximum for wavelength λ becomes

$$2(a+b)\sin\dfrac{D}{2}=n\lambda \qquad ...(5)$$

Maximum number of orders with a diffraction grating:

The direction of nth order principal maxima for wavelength λ is given by

$$(a+b)\sin\theta_n = n\lambda \qquad \text{(for normal incidence)}$$

where a is the width of the slit and b is that of the opaque portion.

The above equation may be written as

$$n=\dfrac{(a+b)\sin\theta_n}{\lambda}$$

The maximum possible value of angle of diffraction $\theta_n = 90°$, therefore the maximum number of possible order is given by

$$n_{max}=\dfrac{(a+b)\sin 90°}{\lambda}=\dfrac{a+b}{\lambda} \quad \text{(for normal incidence)} \qquad ...(6)$$

When angle of incidence is i, we have

$$n = \frac{(a+b)(\sin \theta_n + \sin i)}{\lambda}$$

$$n_{max} = \frac{(a+b)(1+\sin i)}{\lambda} \quad (\text{since } \theta_n = 90°) \qquad \qquad ...(7)$$

When grating element $(a + b) < 2\lambda$, for normal incidence we have

$$n_{max} < \frac{2\lambda}{\lambda} < 2$$

This means that for normal incidence when $(a + b) < 2\lambda$, only first order is obtained.

Overlapping of spectral lines: The equation of nth principal maxima for wavelength λ is given by

$$(a+b)\sin \theta = n\lambda$$

Thus when the incident light has a large number of wavelengths $\lambda_1, \lambda_2, \lambda_3, ..., \lambda_n$ in decreasing order, the spectral lines will have the same angle of diffraction, i.e. they will overlap if

$$(a+b)\sin \theta = 1.\lambda_1 = 2.\lambda_2 = 3.\lambda_3 = ... = n.\lambda_n$$

where $l_1, \lambda_2, \lambda_3, ..., \lambda_n$ are wavelength in first, second, third, ..., nth order respectively.

When a line of wavelength λ in nth order coincides with a line of unknown wavelength λ' in order n', then

$$n'\lambda' = n\lambda \quad \text{or} \quad \lambda' = n\lambda/n'$$

The red light of wavelength 7000 Å in 3rd order, the green light of wavelength 5250 Å in 4th order and violet light of wavelength 4200 Å coincide because

$$(a+b)\sin \theta = 3 \times 7000 \times 10^{-8} = 4 \times 5250 \times 10^{-8} = 5 \times 42000 \times 10^{-3}.$$

Here $(a + b)$ is in cm.

For visible region of the spectrum there is no overlapping of the spectral lines. The visible region extends from 4000 Å to 7800 Å. Consequently the first and second order spectra can not overlap. The second order of violet 4000 Å will coincide with the first order 8000 Å which lies beyond the visible region and hence there is no such overlapping. If however we consider the higher order spectra, the overlapping is more pronounced. If the observations are made with a photographic plate, the spectrum may extend up to 2000 Å in the ultraviolet region. In such a case the spectral line corresponding to 4000 Å in the first order and a spectral line corresponding to 2000 Å in the second order overlap. Further it can be easily seen that the entire visible spectrum in the second order will overlap the whole of the infrared spectrum from

8000 Å to 15600 Å in the first order. When overlapping of spectra occurs, it is not possible to study any individual spectrum. This difficulty can be overcome by the use of suitable filters to absorb the undesired wavelengths of incident light which may overlap with the spectral lines under investigation.

Determination of wavelength of light with a plane transmission diffraction grating: The equation of nth order principal maxima for normal incidence is given by

$$(a+b)\sin\theta = n\lambda \qquad\qquad ...(1)$$

where a is the width of the slit, b is the width of the opaque portion. $(a + b)$ is called grating element. θ is the angle of diffraction for a particular order n for wavelength λ.

From equation (1) it is evident that if the grating element $(a + b)$, the angle of diffraction θ for order n are determined, the wavelength of light λ can be obtained.

Determination of (a + b): On every grating the number of rulings per inch is written. Therefore, if N is the number of rulings per inch then

$$N(a+b) = 1 = 2.54 \text{ cm}$$

or $\qquad (a+b) = \dfrac{2.54}{N} \text{ cm}$

Determination of angle of diffraction θ: The angle of diffraction θ is measured by using a spectrometer. Before using the spectrometer following adjustments are made:

(i) The eye-piece of the telescope is focussed on the cross-wires.

(ii) The collimator and the telescope are adjusted for parallel rays by Schuster's method.

(iii) The grating is adjusted on the prism table normal to the collimator. If this is so the light coming from the collimator is incident normally on the grating. The adjustment is very important and is achieved as follows:

(a) The collimator and the telescope are arranged in a line so that the direct image of the slit falls on the vertical cross-wire. This position of the telescope is noted.

(b) The telescope is rotated through 90° and then clamped. Now the axis of the telescope is perpendicular to that of the collimator.

(c) Now the grating mounted on the prism table and the prism table is rotated so that the image of the slit obtained by reflection from the surface of the grating is formed on the vertical cross-wire. The levelling screws are adjusted till this image lies equally above and below the point of the intersection of the cross-wires.

(d) The prism table is rotated from this position through 45° or 135° so that the ruled surface faces the incident light. In this position the grating is normal to the incident light.

(iv) The rulings are adjusted parallel to the slit. To do this the slit is rotated in its own plane till the diffracted images (i.e., the spectral lines) become very sharp and bright.

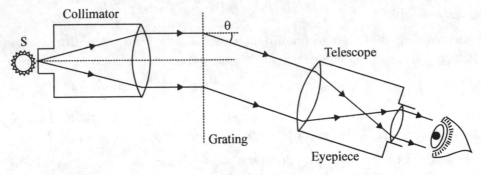

Fig. 8.17

Now the telescope is rotated to the left of the direct image and is adjusted so that the spectral line, whose wavelength is to be determined, falls on the vertical cross-wire for first order and the readings of both the verniers are taken. The telescope is then turned to the right of direct image and the readings of both the verniers for the same line in first order are taken again. The difference between the two readings of the same vernier gives 2θ for that line for first order, from which θ is found.

The same process is repeated for the second order.

Thus knowing $(a + b)$, θ, and n, the values of λ for a given spectral line is calculated by equation (1).

Dispersive power of a diffraction grating: The dispersive power of a diffraction grating is defined as the rate of change of the angle of diffraction with the wavelength of light.

It is denoted by $\dfrac{d\theta}{d\lambda}$.

For plane transmission grating, we have

$$(a+b)\sin\theta = n\lambda \qquad \qquad ...(1)$$

where $(a + b)$ is the grating element and θ is the angle of diffraction for nth order spectrum.

Differentiating equation (1) wrt λ, we get

$$(a+b)\cos\theta \cdot \frac{d\theta}{d\lambda} = n$$

$\therefore$ The dispersive power $= \dfrac{d\theta}{d\lambda} = \dfrac{n}{(a+b)\cos\theta}$ $\qquad \qquad ...(2)$

Here $d\theta$ is the angular separation between two lines having wavelength difference $d\lambda$. Equation (2) shows that:

(i) The dispersive power is directly proportional to n, i.e. the order of the spectrum. Thus higher is the order, greater is the dispersive power. Thus the angular separation of two spectral lines is double in the second order spectrum in comparison to the first order.

(ii) The dispersive power is inversely proportional to the grating element $(a + b)$. This means that the dispersive power is directly proportional to the number of lines per cm of the grating. Thus the angular dispersion power of two given lines is greater with a grating having larger number of lines per cm.

(iii) The dispersive power is inversely proportional to cos θ. Thus if the angle of diffraction $\theta = 0$, cos $\theta = 1$ and so the angular dispersion is minimum. Therefore if θ is small, the value of cos θ may be taken as unity, so the influence of cos θ may be neglected.

If we neglect the influence of the factor cos θ, then ($d\theta \propto d\lambda$ for a given order), i.e. the angular dispersion of two spectral lines in a particular order is directly proportional to the difference in wavelength between the two spectral lines. Such a spectrum is called a normal spectrum.

Linear dispersive power: If dx is the linear separation of two spectral lines differing in wavelength by $d\lambda$ in the focal plane of a lens of focal length f, we have

$dx = f d\theta.$

∴ The linear dispersive power

$$= \frac{dx}{d\lambda} = f \frac{d\theta}{d\lambda} = \frac{f \cdot n}{(a+b)\cos\theta} \qquad \qquad ...(3)$$

SOLVED EXAMPLES

Q.1. **A zone plate has the radius of the first right 0.05 cm. If plane waves ($\lambda = 5000$ AU) fall on the plate, where should the screen be placed so that the light is focussed to a bright spot?**

Sol. The focal length of the zone plate is given by

$$f = \frac{r_n^2}{n\lambda}$$

Given, radius of first ring, $r_1 = 0.05$ cm, for first ring, $n = 1$

and $\qquad\qquad \lambda = 5000 \text{ AU} = 5000 \times 10^{-8} \text{cm}$

∴ $\qquad\qquad f = \frac{(0.05)^2}{1 \times 5000 \times 10^{-8}} = 50 \text{ cm}.$

Q.2. **Find the radii of the first three transparent zones of a zone plate behaving like a convex lens of focal length 1 m, for a light of wavelength $\lambda = 5893$ Å.**

Sol. If r_n is the radius of nth half period zone, then we have

$$f = \frac{r_n^2}{n\lambda} \quad i.e. \quad r_n = \sqrt{n\lambda f}$$

Given $f = 1$ m and $\lambda = 5893$ Å $= 5893 \times 10^{-10}$ m

If we suppose that the even number of zones are opaque, then the first three transparent zones will correspond to $n = 1$, 3 and 5 respectively. Hence the radii of first three transparent zones are given by

$$r_1 = \sqrt{1 \times 5893 \times 10^{-10} \times 1} = 7.67 \times 10^{-4} \text{ m} = 0.0767 \text{ cm.}$$

$$r_2 = \sqrt{3 \times 5893 \times 10^{-10} \times 1} = 1.339 \times 10^{-3} \text{ m} = 0.1339 \text{ cm.}$$

$$r_3 = \sqrt{5 \times 5893 \times 10^{-10} \times 1} = 1.716 \times 10^{-3} \text{ m} = 0.1716 \text{ cm.}$$

Q.3. **The central circle of a zone plate has a radius of 0.07 cm. Light of wavelength 5000 Å coming from: (i) an object at infinity, (ii) an object 147 cm away from the plate falls on it. Find the position of the principal image in each case.**

Sol. If a and b are the distances of the object and image from the zone plate, then to pth focal length of the zone plate is given by

$$\frac{1}{a} + \frac{1}{b} = \frac{1}{f_p} \quad \dots (1) \text{ where } f_n = \frac{r_n^2}{(2p-1)n\lambda}$$

r_n being radius of nth zone.

The principal image corresponds to the principal (first) focal length (i.e., $p = 1$). Therefore, the focal length f of the zone plate is given by

$$f_p = f = \frac{r_n^2}{n\lambda}$$

For central zone $n = 1$;

$$\therefore \qquad r_n = r_1 = 0.07 \text{ cm and } \lambda = 5000 \times 10^{-5} \text{ cm}$$

$$\therefore \qquad f_p = \frac{(0.07)^2}{1 \times 5000 \times 10^{-3}} = 98 \text{ cm}$$

(i) Here $a = \infty$, $f_p = 98$ cm.   Hence from eqn. (1)

$$\therefore \qquad \frac{1}{\infty} + \frac{1}{b} = \frac{1}{98} \qquad \therefore \ b = 98 \text{ cm.}$$

(ii) Here $a = 147$ cm, $f_p = 98$ cm.   Hence from eqn. (1)

$$\frac{1}{147} + \frac{1}{b} = \frac{1}{98} \qquad i.e. \ \frac{1}{b} = \frac{1}{98} - \frac{1}{147} = \frac{1}{294} \qquad \therefore \quad b = 294 \text{ cm.}$$

Q.4. **A zone plate gives a series of images of a point source on its axis. If the first strongest and second strongest images are at distances of 30 cm and 6 cm respectively from the zone plate, both on the same side remote from the source, calculate the distance of the source from the zone plate.**

Sol. If a and b are the distances of the object and the image from the zone plate, we have

$$\frac{1}{a} + \frac{1}{b} = \frac{(2p-1)\,n\lambda}{r_n^2} \qquad \qquad ...(1)$$

where r_n is the radius of the nth zone.

For first image, $p = 1$ and $b = 30$ cm.

$$\therefore \qquad \frac{1}{a} + \frac{1}{30} = \frac{n\lambda}{r_n^2} \qquad \qquad ...(2)$$

For second image, $p = 2$ and $b = 6$ cm.

$$\therefore \qquad \frac{1}{a} + \frac{1}{6} = \frac{3n\lambda}{r_n^2} \qquad \qquad ...(3)$$

Dividing eqn. (3) by eqn. (2), we get

$$\frac{\dfrac{1}{a} + \dfrac{1}{6}}{\dfrac{1}{a} + \dfrac{1}{30}} = 3 \ \text{ or } \ \frac{1}{u} + \frac{1}{6} = \frac{3}{a} + \frac{3}{30}$$

or $\qquad \dfrac{3}{a} - \dfrac{1}{a} = \dfrac{1}{6} - \dfrac{3}{30}$ or $\dfrac{2}{a} = \dfrac{2}{30}$

$$\therefore \qquad a = 30 \text{ cm.}$$

Q.5. **A zone plate is made by arranging that the radii of the circles which define the zones are the same as the radii of Newton's rings formed between a plane surface and the surface having radius of curvature 200 cm. Find the principal focal length of the zone plate.**

Sol. In Newton's ring experiment the diameters of nth dark ring is given by

$$D_n^2 = 4n\lambda R$$

where R is the radius of curvature of the curved surface.

If r_n is the radius of nth dark ring, $\quad r_n = D_n/2$

$$\therefore \qquad r_n^2 = \lambda n R \qquad \qquad ...(1)$$

The principal focal length of the zone plate is

$$f = \frac{r_n^2}{n\lambda}, \quad r_n \text{ being the radius of } n\text{th zone.}$$

According to given problem the radii of the half period zones are same as the radii of Newton's rings, therefore

$$f = \frac{n\lambda R}{n\lambda} = R = 200 \text{ cm}.$$

Q.6. **In a particular experiment $\lambda = 6000$ Å, the distance between the slit source and the edge is 6 metres and between the edge and the eye-piece is 4 metres. Calculate the position of the first three maxima and their separation.**

Sol. The position of nth maxima relative to the edge of the geometrical shadow is given by

$$\therefore \qquad y_n = \sqrt{\left[\frac{b(a+b)}{a}(2n-1)\lambda\right]} = K\sqrt{(2n-1)}$$

where

$$K = \sqrt{\left[\frac{b(a+b)\lambda}{a}\right]}$$

Here, $a = 6$ m $= 600$ cm, $b = 4$ m $= 400$ cm, $\lambda = 6000$ Å $= 6 \times 10^{-5}$ cm

$$\therefore \qquad K = \sqrt{\frac{400(600+400) \times 6 \times 10^{-5}}{600}} = 0.2 \text{ cm}$$

The first three maxima correspond to $n = 1, 2, 3$

$\therefore$ The position of first maxima, $y_1 = K\sqrt{2 \times 1 - 1} = K = 0.2$ cm

The position of second maxima, $y_2 = K\sqrt{2 \times 2 - 1} = K\sqrt{3} = 0.2 \times \sqrt{3} = 0.346$ cm

The position of third maxima, $y_2 = K\sqrt{2 \times 3 - 1} = K\sqrt{3} = 0.2 \times \sqrt{5} = 0.447$ cm

Therefore, the separation between these maxima are

$$y_2 - y_1 = 0.346 - 0.200 = 0.146 \text{ cm}$$

and

$$y_3 - y_2 = 0.447 - 0.346 = 0.101 \text{ cm}.$$

Q.7. **A narrow slit illuminated by light of wavelength 6000 Å is placed at a distance of 10 cm from a straight edge. If measurements are done at a distance of 100 cm from the straight edge, calculate the distance between the first and second dark bands.**

Sol. The position of *n*th minima relative to the edge of geometrical shadow is given by

$$y_n = \sqrt{\left[\frac{2b(a+b)}{a}\lambda n\right]} = k\sqrt{2n}$$

where

$$k = \sqrt{\left[\frac{b(a+b)\lambda}{a}\right]}$$

The first and second dark bands (minima) correspond to $n = 1$ and $n = 2$ respectively. For the first dark band, $n = 1$

$$y_1 = k\sqrt{2 \times 1} = k\sqrt{2}$$

For second dark band, $n = 2$

∴ $$y_2 = k\sqrt{2 \times 2} = 2k$$

∴ The distance between the first and second dark band is

$$y_2 - y_1 = 2k - \sqrt{2} = k(2 - \sqrt{2})$$

Here $a = 10$ cm, $b = 100$ cm and $\lambda = 6000$ Å $= 6 \times 10^{-5}$ cm

∴ $$K = \sqrt{\left[\frac{b(a+b)\lambda}{a}\right]} = \sqrt{\left[\frac{100(10+100) \times 6 \times 10^{-5}}{10}\right]} = 0.257 \text{ cm}$$

∴ $$y_2 - y_1 = 0.257\left(2 - \sqrt{2}\right) = 0.257(2 - 1.414) = 0.257 \times 0.586 = 0.151 \text{ cm} = 1.51 \text{ mm}.$$

Q.8. Light of wavelength 5500 Å falls normally on a slit of width 22.0 × 10⁻⁸cm. Calculate the angular position of the first two minima on either side of the central maximum and show the intensity distribution graphically.

Sol. The angular position of minima in the diffraction pattern produced by a single slit are given by

$$a \sin\theta = m\lambda \quad \text{or} \quad \sin\theta = \frac{m\lambda}{a}$$

where *m* is the order of minima and *a* is the width of the slit.

Here $a = 22.0 \times 10^{-5}$ cm and $\lambda = 5500 \times 10^{-8}$ cm

For first order $m = 1$,

$$\sin\theta_1 = \frac{\lambda}{a} = \frac{5500 \times 10^{-8}}{22.0 \times 10^{-5}} = 0.25$$

$$\theta_1 = \sin^{-1}(0.25) = 14°29'$$

For second order $m = 2$,

$$\therefore \qquad \sin\theta_2 = \frac{2\lambda}{a} = \frac{2 \times 5500 \times 10^{-5}}{22.0 \times 10^{-5}} = \frac{1}{2}$$

$$\therefore \qquad \theta_2 = 30°$$

Thus the first two minima will occur at the angles 14°29′ and 30° on either side of the central maximum.

Q.9. **Microwaves of wavelength $\lambda = 2.0$ cm are incident normally on a slit 5.0 cm wide. Deduce the angular speed of the central maximum.**

Sol. The angular half width θ of the central maximum is given by

$$\sin\theta = \frac{\lambda}{a}$$

Here $\lambda = 2.0$ cm, $a = 5.0$ cm

$$\therefore \qquad \sin\theta = \frac{2.0}{5.0} = 0.4$$

$$\therefore \qquad \theta = \sin^{-1}(0.4) = 23° \, 35'$$

$\therefore$ Total angular width of central maximum $= 2\theta = 2 \times 23°35' = 47°10'$.

Q.10. **Plane waves of $\lambda = 6.0 \times 10^{-5}$ cm, falls normally on a straight slit of width 0.20 mm. Calculate the total angular width of the central maximum and also the linear width as observed on a screen placed 2 metres away.**

Sol. The angular half width θ of the central maximum is given by

$$\sin\theta = \frac{\lambda}{a}$$

Here $\lambda = 6.0 \times 10^{-5}$ cm and $a = 0.20$ mm $= 0.02$ cm

$$\sin\theta = \left(\frac{6.0 \times 10^{-5}}{0.02}\right) = (3 \times 10^{-3})$$

As sin θ is very small, we may write $\theta = \sin\theta = 3 \times 10^{-3}$ radians

$\therefore$ The total angular width of the central maximum $= 2\theta$

$$= 2 \times 3 \times 10^{-3} = 6 \times 10^{-3} \text{ radians}$$

If y is the linear half width and d is the distance of the screen from the slit, we have

$$y = d\theta \quad \text{as } \theta \text{ is small}$$

Here $d = 2$ metres $= 200$ cm and $\theta = 3 \times 10^{-3}$ radians.

$$\therefore \qquad y = 200 \times 3 \times 10^{-3} = 0.6 \text{ cm}$$

$\therefore \qquad$ The linear width of the central maximum on the screen

$$= 2y = 2 \times 0.6 = 1.2 \text{ cm}.$$

Q.11. **A screen is placed 200 cm away from a narrow slit which is illuminated with light of wavelength 6×10^{-5} cm. If the first minima lie 5 mm on either side of central maximum, calculate the slit width.**

Sol. The distance of the first minima from the central maxima, $y = 5$ mm $= 0.5$

The angular position θ of the first minima is given by $\sin\theta = \dfrac{\lambda}{a}$ ...(1)

where λ is the wavelength of light used and a is the width of the slit.

As θ is very small, we have

$$\theta = \sin\theta = \frac{\lambda}{a} \qquad ...(2)$$

Also from fig. for small θ, we have

$$\theta = \tan\theta = \frac{y}{d} \qquad ...(3)$$

Comparing eqns. (2) and (3) we have

$$\frac{y}{d} = \frac{\lambda}{a} \text{ or } a = \frac{d\lambda}{y}$$

Here $d = 200$ cm, $\lambda = 6 \times 10^{-5}$ cm and $y = 0.5$ cm

$$a = \frac{200 \times 6 \times 10^{-5}}{0.5} = 0.024 \text{ cm}.$$

Q.12. **A lens whose focal length is 40 cm forms a Fraunhofer diffraction pattern of a slit 0.3 mm width. Calculate the distances of the first dark band and of the next bright band from the axis (wavelength of light used is 5890 AU).**

Sol. The angular position of mth minima in the Fraunhofer diffraction pattern due to a slit is given by

$$a\sin\theta = m\lambda \qquad ...(1)$$

where $m = 1, 2, 3, ...$

        For first minima $m = 1$, $\therefore$ $a\sin\theta = \lambda$

·or $\quad \sin \theta = \dfrac{\lambda}{a}$

Here $\quad \lambda = 5890 \text{ AU} = 5890 \times 10^{-8} \text{cm}, \ a = 0.3 \text{ mm} = 0.03 \text{ cm}$

$$\sin \theta = \frac{5890 \times 10^{-8}}{0.03} = 1.963 \times 10^{-8}$$

As θ is very small, we may write $\sin \theta = \theta$.

$$\theta = \sin \theta = 1.963 \times 10^{-3} \text{ radians}$$

If y is the linear distance of first dark band from the axis, we have

$$\theta = \frac{y}{f} \text{ or } y = f\,\theta$$

where f is the focal length of the lens.

Here $\quad f = 40 \text{ cm and } \theta = 1.963 \times 10^{-3} \text{ radians}$

$\therefore \quad y = 40 \times 1.963 \times 10^{-3} = 0.0785 \text{ cm}$

The angular position θ' of the secondary maxima of either side is approximately given by

$$a \sin \theta' = \frac{(2m+1)\lambda}{2}$$

For first secondary maxima, $m = 1$

$\therefore \quad a\sin \theta' = \dfrac{3\lambda}{2a}$

or $\quad \sin \theta' = \dfrac{3\lambda}{2a} = \dfrac{3 \times 5890 \times 10^{-3}}{2 \times 0.03} = 2.945 \times 10^{-2}$

As θ' is small, we have $\theta' = \sin \theta' = 2.945 \times 10^{-3}$ radians

If the linear distance of first secondary maxima from the axis is y', we have

$$y' = f\,\theta' = 40 \times 2.945 \times 10^{-3} = 0.1178 \text{ cm} .$$

Q.13. Light of wavelength 5×10^{-3} cm is incident normally on a plane transmission grating of width 3 cm and 15000 lines. Find the angle of diffraction in first order.

Sol. The directions of principal maxima for normal incidence are given by

$$(a+b)\sin \theta = n\lambda$$

where $(a + b)$ is the grating element and n is the order of maxima.

Here $\quad (a+b) = \dfrac{\text{Width of grating}}{\text{Total number of lines on the grating}} = \dfrac{3}{15000} \text{ cm}$

$$n = 1 \text{ and } \lambda = 5 \times 10^{-5} \text{cm}$$

$$\therefore \qquad \frac{3}{15000} \sin \theta = 1 \times 5 \times 10^{-5}$$

$$\text{or} \qquad \frac{3}{15000} \sin \theta = 1 \times 5 \times 10^{-5}$$

$$\therefore \qquad \sin \theta = \sin^{-1}(0.25) = 14°29'.$$

QUESTIONS

1. What do you mean by diffraction of light waves?
2. Distinguish between Fresnel and Fraunhofer diffraction.
3. Differentiate between interference and diffraction of light waves.
4. What is a zone plate? Explain the formation of images by a zone plate. Compare its working with that of a converging lens.
5. Bring out the similarities and differences between zone plate and a convex lens.
6. Describe how the wavelength of monochromatic light can be determined using diffraction grating.
7. Obtain expression for intensity of the diffracted beam by a single slit. Discuss the position of minima and maxima.
8. Give the construction of a zone plate. How does the primary focal length of a zone plate depend on the wavelength of light used?
9. A monochromatic parallel beam of light is incident on a single slit. Graphically show the intensity distribution about the principal and secondary maxima in the diffracted beam.
10. In a plane diffraction grating, the width of each slit is equal to the width of the opaque space between two adjacent slits. Find the missing orders of spectra.
11. Show that the radii of the Fresnel half period zones are proportional to the square root of natural numbers.
12. What are the characteristics of grating spectra?
13. Derive an expression for the linear width of principal maxima in case of single slit Fraunhofer diffraction.
14. In single slit diffraction intensity of first secondary maximum is less than that of principal maximum by how many times?
15. What is the difference between Fresnel's first half period zone and Fresnel's second half period zone as envisioned by him to explain the optical diffraction phenomenon?

Problems

1. The second order maximum for a wavelength of 6360 Å in a transmission grating coincides with 3rd order maximum of an unknown light. Determine the wavelength of the unknown light. **(Ans. $\lambda = 4260$ Å)**

2. A transmission grating has 8000 rulings per cm. The first order principal maximum due to monochromatic source of light occurs at an angle of 30°. Determine the wavelength of light. **(Ans. $\lambda = 6.25 \times 10^{-5}$ cm)**

3. A monochromatic light of wavelength 6000 Å is incident on a plane diffraction grating with grating element of 6×10^{-5} cm. What is the maximum order of spectrum that can be observed? **(Ans. $n_{max} = 1$)**

4. A plane diffraction grating of width 2.5 cm has 12500 rulings on it. What is the maximum order of grating spectrum observable for incident light of wavelength 5500 Å? **(Ans. $n_{max} = 4$)**

5. In Fraunhofer diffraction due to single slit, the first order minimum is observed at an angle 30° with the incident beam. What is the width of the slit if wavelength of incident light is 6000 Å? **(Ans. $d = 12 \times 10^{-7}$ m)**

6. How many orders of diffraction bands will be visible theoretically if wavelength of incident radiation is 5893 Å and the number of lines per cm on the grating is 6000? **(Ans. $n = 3$)**

7. A zone plate is to have a principal focal length of 50 cm corresponding to wavelength of light $\lambda = 6 \times 10^{-5}$ cm. Find the radii. **(Ans. $r_1 = 0.05$ cm, $r_2 = 0.094$ cm)**

8. The primary focal length of a present zone plate is 8 m with monochromatic red light ($\lambda = 6000$ Å). What would be the focal length if a monochromatic light with $\lambda = 4800$ Å is used? **(Ans. $f_2 = 10$ cm)**

9. The bright image due to a Fresnel zone plate is formed at a distance of 45 cm from it. At what distance is the next bright image formed? **(Ans. $f_2 = 15$ cm)**

10. A zone plate is illuminated with sodium light ($\lambda = 5896$ Å) placed at distance of 100 cm. If the image of the point source is obtained at a distance of 2 m on the other side, what will be the power of the equivalent lens which may replace the zone plate without disturbing the set up? Also calculate the radius of the first zone of the plate. **(Ans. $P = +1.5$ Diopter, $r = 0.0627$ cm)**

● ● ●

Chapter 9

Polarisation of Light

Introduction

The phenomenon of interference and diffraction show that light travels in the form of waves. But they do not tell us about the type of light waves, i.e. whether the light waves are longitudinal or transverse or whether the vibrations are linear, circular, elliptical or torsional. Such important enquiries constitute the subject matter of polarisation of light. The phenomenon of polarisation can be explained only by assuming transverse character of waves of light. Now consider a mechanical experiment which explains what polarisation is.

Consider a string through two parallel slits S_1 and S_2. Consider that this string is held by two persons A and B in their hands and standing at some distance from each other. If the person A moves his hand up and down parallel to the slit S_1, the transverse waves will be produced in the string which travel towards the person B. In this case the particles of the string

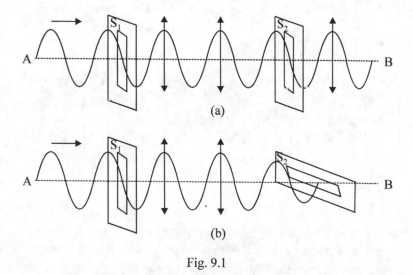

(a)

(b)

Fig. 9.1

vibrate along the direction parallel to S_1. As the slit S_2 is parallel to S_1, these transverse vibrations pass through S_2 and reach the person B as such. Such a wave is said to be plane polarised.

If A moves his hand in all possible directions instead of moving up and down, the particles of the string will vibrate in all directions. When these vibrations reach the vertical slit S_1, they are restricted to the vertical plane only. The waves between A and S_1 are unpolarised, while the waves passing through S_1 are polarised and thus S_1 act as a polariser. The polarised light reach the person B if S_2 is parallel to S_1 (Fig. 9.1a). If the slit S_2 is rotated to make it horizontal, i.e. perpendicular to S_1, the vibrations of the string are completely stopped by S_2 as the displacement of the particles is now at right angles to the length of S_2. Thus the string does not vibrate between S_2 and B. (Fig. 9.1b). In the intermediate positions of S_2, the waves will be partly stopped and partly transmitted through it, the waves will reach B with varying amplitudes. Thus we see that as the slit S_2 is rotated the amplitude of vibrations varies, being maximum when S_1 and S_2 are parallel and minimum when S_1 and S_2 are perpendicular.

If longitudinal waves are produced by moving the string forward and backward along its length, the waves will freely pass through S_1 and S_2 irrespective of their positions.

Thus we see that the amplitude of vibrations varies only in the case of transverse vibrations.

Now let us describe an optical experiment to illustrate the polarisation of light and to show that the light waves are transverse.

Light from a source S is allowed to fall normally on the flat surface of a thin plate of a tourmaline crystal, cut parallel to its axis (Fig. 9.2). Only a part of this light is transmitted through A.

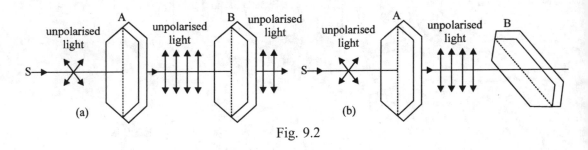

Fig. 9.2

If now the plate A is rotated, the character of transmitted light remains unchanged. Now another similar plate B is placed at some distance from A such that the axis of B is parallel to that of A. If the light transmitted through A is passed through B, the light is almost completely transmitted through B and no change is observed in the light coming out of B.

If now the crystal A is kept fixed and B is gradually rotated in its own plane, the intensity of light emerging out of B decreases and becomes zero when the axis of B is perpendicular to that of A. If B is further rotated, the intensity increases and becomes maximum when the axes of A and B are again parallel.

Thus we see that the intensity of light transmitted through B is maximum when axes of A and B are parallel and minimum when they are at right angles.

From this experiment it is obvious that light waves are transverse and not longitudinal; because if they were, the rotation of crystal B would not produce any change in the intensity of light.

We can thus consider that ordinary light (from source S) consists of transverse waves in which the vibrations take place in all directions in a plane perpendicular to the direction of propagation of light. When such a beam passes through first crystal A, its vibrations are confined in the direction of its axis alone. Thus the light coming out of crystal A is not symmetrical about the direction of propagation of light and its vibrations are confined only to a single line in a plane perpendicular to the direction of propagation. Such a light is said to be plane polarised or linearly polarised. This light will pass completely through B if its axis is parallel to the vibrations of polarised light (i.e., parallel to axis of A) and will be stopped completely if the axis of B is perpendicular to the vibrations of polarised light. For intermediate positions of B the light is partly stopped and partly transmitted. Therefore when B is rotated, the intensity of light coming out of B varies, being maximum when the axes of A and B are parallel and minimum when at right angles.

Polarised and unpolarised light and their representation

Unpolarised light: The light having vibrations along all possible straight lines perpendicular to the direction of propagation of light, is said to be unpolarised. It is symmetrical about its direction of propagation. It can be considered to consist of an infinite number of waves (Fig. 9.3) each having its own direction of vibration. Therefore the probability of occurrence of vibration along any direction is same, i.e. the probability of occurrence of vibration along the axis is same in all positions of the crystal. Therefore when such a light is passed through a single rotating tourmaline crystal; there is no change in the intensity of emergent light. Ordinary or natural light has such a property. Therefore ordinary light is unpolarised.

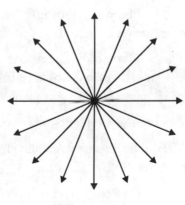

Fig. 9.3

Polarised light: The light having vibrations only along a single straight line perpendicular to the direction of propagation of light is said to be polarised. Its vibrations are one-sided, therefore there is lack of symmetry about the direction of propagation of light. When ordinary or unpolarised light is passed though a tourmaline crystal; the vibrations of emergent light are along a single straight line perpendicular to the direction of propagation of light. Therefore the emergent light is plane polarised or linearly polarised and this phenomenon is called polarisation. When polarised light is passed through a single rotating crystal, a change in the intensity of emergent light is observed.

According to Maxwell's electromagnetic theory of light a train of light wave consists of mutually perpendicular electric and magnetic waves; both being perpendicular to the direction of propagation of light. According to this idea the polarised light may be defined as follows:

A light wave is said to be plane or linearly polarised if the electric vector ($\vec{E}$) or magnetic vector ($\vec{H}$) at any point goes on vibrating along the same line perpendicular to direction of propagation of light wave.

Representation: In an unpolarised beam the vibrations take place along all possible directions at right angles to the direction of propagation of light. Therefore it is represented by a star (Fig. 9.4a).

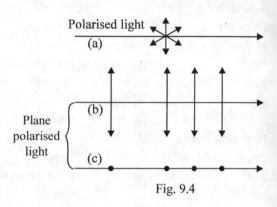

Fig. 9.4

In the polarised beam of light the vibrations are along a single straight line. If they are parallel to the plane of the paper, they are represented by arrows (Fig. 9.4b); while that perpendicular to the plane of the paper are represented by dots (Fig. 9.4c).

Plane of vibration: The plane containing the direction of vibration and direction of propagation of light its called the plane of vibration.

Plane of polarisation: The plane containing the direction of propagation of light, but containing no vibrations is called the plane of polarisation or the plane perpendicular to the plane of vibration is called the plane of polarisation.

Production of plane-polarised light

The different methods commonly used for the production of plane-polarised light may be classified under the following heads:

(1) Polarisation by reflection

(2) Polarisation by refraction

(3) Polarisation by double refraction

(4) Polarisation by selective absorption

(5) Polarisation by scattering

Polarisation by reflection: In 1808, Malus discovered the polarisation of light by reflection from the surface of glass. He found that when ordinary (unpolarised) light is incident on the surface of any transparent material, the reflected beam is partially plane polarised. The vibrations of the reflected beam are parallel to the reflecting surface. When this reflected light is passed through a tourmaline crystal and the latter is rotated, the intensity of the reflected beam is minimum (not necessarily zero) for one particular position of the crystal. If now the angle of incidence is changed, we see that intensity of the reflected beam is zero for one particular position of the crystal. This means that the reflected beam is completely plane polarised for a particular angle of incidence. This angle of incidence is called the polarising angle.

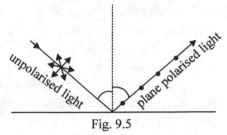

Fig. 9.5

This polarising angle depends upon the reflecting substance and slightly upon the wavelength of light. The value of polarising angle is 57° for air-glass reflection, 33° for glass-air reflection and 53° for air-water (at 0°C) reflection.

Here the reflecting surface acts as polariser. The production of polarised light by reflecting surface is explained as follows:

The vibrations of incident unpolarised light may be resolved in two directions; namely parallel and perpendicular to the reflecting surface.

The reflected light is only due to the components parallel to the reflection surface; whereas transmitted light is due to components perpendicular to the reflecting surface.

Demonstration of polarisation by reflection; Biot's polariscope:

Fig. 9.6 represents the arrangement and working of simple apparatus (Biot's polariscope) for demonstrating the polarisation by reflection and of analysing (or detecting) the plane polarised light.

PQ and *RS* are two glass plates with their back surfaces blackened to absorb refracted light. Let a beam of ordinary light *AB*, be incident on the polished face of *PQ* at the polarising angle (57°). This is reflected along *BC*. This ray *BC* is incident on polished face of another glass plate *RS* also at 57° and is reflected along *CD*. When the glass plates PQ and RS are parallel, all the rays *AB*, *BC* and *CD* lie in the same plane and the intensity of *CD* is found to

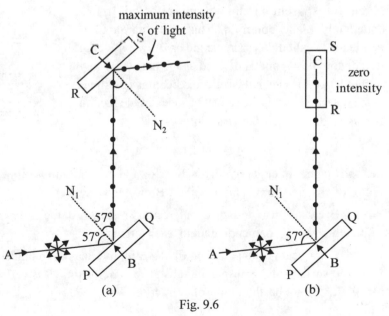

Fig. 9.6

be maximum. When the plate RS is gradually rotated about *BC* as axis, so as still to preserve the same angle of incidence on it, the intensity of the reflected beam decreases and becomes zero for a rotation of 90° (Fig. 9.6b). This means that no reflection takes place when the plate *PQ* is perpendicular to *RS*.

As the beam *AB* is incident on *PQ* at polarising angle, the reflected beam *BC* is completely plane polarised with its vibration perpendicular to the plane of incidence. When *RS* is parallel to *PQ*, the vibrations of beam BC are perpendicular to the plane of incidence and hence it is completely reflected along *CD*. But when RS is rotated through 90°, the vibrations of *BC* are parallel to the plane of incidence (with respect to plate RS); therefore it is not reflected.

If the plate *RS* is further rotated, the intensity again increases and becomes maximum for a rotation of 180° when again the condition of maximum intensity is satisfied. By further rotating the intensity decreases and becomes zero for a rotation of 273°. It again increases and becomes maximum for a rotation of 360°.

The plate *PQ* causes the beam *BC* to be polarised and hence is called polariser. The plane *RS* detects the polarised light and so it is called analyser. The two glass plates arranged in the above manner constitute a Biot's polariscope. By this method we get the polarised light by reflection.

Brewster's Law: In 1881, Sir David Brewster performed a series of experiments to study the polarisation of light by reflection. He observed that when a beam of ordinary light is incident on the surface of a given transparent medium, the reflected light is completely plane polarised. This particular angle of incidence is called polarising angle or Brewster's angle. It is denoted by i_v and is defined as the particular angle of incidence for which the reflected beam is completely plane polarised. If μ is the refractive index of the medium, Brewster discovered the relation

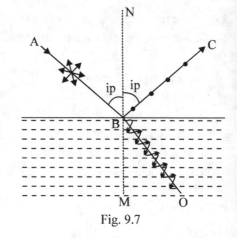

Fig. 9.7

$$\mu = \tan i_v$$

i.e., the tangent of angle of polarisation for a given medium is numerically equal to the refractive index of that medium. This is called Brewster's law.

Brewster also found an interesting result, 'At the polarising angle the reflected and refracted rays are perpendicular to each other.'

Let a beam of unpolarised light *AB* be incident at an angle equal to the polarising angle i_p on the surface of a transparent substance. The beam *AB* is reflected along BC and refracted along *BD*. Let *r* be the angle of refraction.

From Brewster's law we have $\mu = \tan i_p = \dfrac{\sin i_p}{\cos i_p}$ ...(1)

From Snell's law we have $\mu = \dfrac{\sin i_p}{\sin r}$ ...(2)

Comparing equations (1) and (2), we get $\cos i_p = \sin r = \cos\left(\dfrac{\pi}{2} - r\right)$

$\therefore \qquad i_p = \dfrac{\pi}{2} - r$

or $\qquad i_p + r = \dfrac{\pi}{2}$ ...(3)

We know $\qquad \angle NBM = \pi$

$\qquad\qquad i_p + \angle CBD + r = \pi$

or $\qquad \angle CBD = \pi - \left(i_p + r\right) = \pi - \dfrac{\pi}{2}$ from equation (3)

$\therefore \qquad \angle CBD = \dfrac{\pi}{2}$ ...(4)

Thus when the ray is incident at polarising angle, the reflected and refracted rays are perpendicular to each other.

Polarisation by refraction through pile of plates:

When ordinary unpolarised light is incident on the upper surface of a single glass plate at polarising angle i_p only a small fraction of the light is reflected, e.g. for glass ($\mu = 1.5$) out of incident light, 100% of the light having vibrations perpendicular to the plane of incidence is transmitted; whereas only 15% of light having vibrations perpendicular to the plane of incidence is reflected. Thus the reflected light is completely plane polarised with its vibrations perpendicular to the plane of incidence. While the transmitted or refracted light is partially polarised having its vibrations both in the plane of incidence as well as perpendicular to the plane of incidence. The refracted light is incident at the lower surface of the glass plate at an angle r where r is the angle of incidence at the upper face of the plate, we have

$$\tan r = \frac{\sin r}{\cos r} = \frac{\sin r}{\sin(90^\circ - r)} = \frac{\sin r}{\sin i_p}$$

(since $i_p + r = 90^\circ$)

or $\qquad \tan r = {}_a\mu_a$

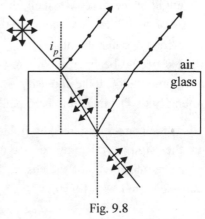

Fig. 9.8

Thus from Brewster's law, r is the angle of polarisation for a beam incident from glass to air. Therefore the light reflected at the lower face of the plate is completely plane polarised; while the light refracted into air is partially polarised. Thus the light refracted through a single glass plate is partially polarised.

In order to obtain completely polarised light, it is refracted through *pile of plates*. The pile of plates consists of adequate number of glass plates separated by air gaps. When a beam of ordinary light is incident at the polarising angle on the pile of plates, some of the vibrations

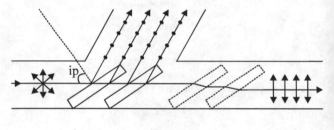

Fig. 9.9

perpendicular to the plane of incidence are reflected by the first plate whereas the rest are transmitted through it. The transmitted beam have also vibrations parallel to the plane of incidence. Thus the transmitted light is partially polarised. When this beam of light is refracted by the second plate parallel to the first, again some of the vibrations perpendicular to the plane of incidence are reflected by it and the rest are transmitted. The process continues.

When a beam of light, polarised by reflection at one plane surface, is allowed to fall on another similar reflecting surface at polarising angle, of plates, its vibrations are parallel to the plane of incidence. Obviously the polarisation depends upon the number of plates. If I_p and I_n are the intensities of the parallel and normal (perpendicular) components in the refracted light, the degree of polarisation is given by

$$P = \frac{I_p - I_n}{I_p + I_n} = \frac{m}{m + \left(\dfrac{2\mu}{1 - \mu^2}\right)^2}$$

where m is the number of plates and μ is refractive index of their material.

Law of Malus: In Fig. 9.6, if the plate *PQ* is rotated about the incident beam, keeping the angle of incidence constant, the intensity of the reflected beam *BC* remains unchanged. But if the plate *RS* is rotated about its incident beam *BC*, the beam *CD* reflected from it changes in intensity. When the planes of incidence of plates *PQ* and *PS* are parallel, the intensity of the final reflected beam is maximum (say I_θ). When the planes of incidence of plates *PQ* and *RS* are perpendicular, the intensity of the final reflected beam is zero. The law of variation of intensity of the final reflected beam was given by Malus.

When a beam of light, polarised by reflection at one plane surface, is allowed to fall on another similar reflecting surface at polarising angle, the intensity of the final reflected beam varies as the square of the cosine of the angle between the two planes of incidence. This law may be stated as follows:

The intensity of the light emerging from the analyser is proportional to the square of the cosine of the angle between the planes of transmission of the analyser and the polariser.

The law holds for any combination of the polariser and the analyser. The law can be proved as follows:

Let A be the amplitude of the plane polarised light. Let θ be the angle between the planes of transmission of the analyser and the polariser. The amplitude A may be resolved into two components $A \cos \theta$ and $A \sin \theta$. The vibrations of the former are parallel to the plane of transmission of the analyser; while the vibrations of latter are perpendicular to this. Evidently the former component (*i.e.*, $A \cos \theta$) is transmitted by the analyser; while the latter is reflected. Therefore the intensity of the transmitted beam

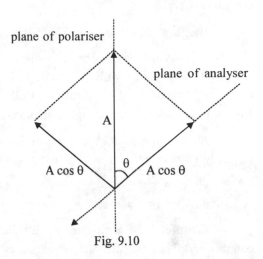

Fig. 9.10

$$I = A^2 \cos^2 \theta = I_\theta \cos^2 \theta \qquad \qquad ...(1)$$

where $I_\theta\ (= A^2)$ is the intensity of the incident plane polarised beam. Eqn. (1) represents law of Malus.

The Malus law may also be stated as:

When a completely plane polarised light beam is incident an a polarising sheet the intensity of the emergent light varies as the square of the cosine of the angle between the direction of electric field vector of the incident light and the polarising direction of the sheet, i.e.

$$I = I_0 \cos^2 \theta, \ I_0 \text{ being intensity of plane polarised light.}$$

This law does not hold when the incident light is unpolarised because in unpolarised light the angle θ made by the electric field vector with the polarising direction of the sheet is not constant but is a variable parameter; since in unpolarised light the electric field vector vibrates in all possible directions in a plane perpendicular to the direction of propagation. Hence in Malus law we have to put the average value of $\cos^2 \theta$, over all possible values of θ. Thus the intensity of the beam transmitted from the polarising sheet is given by

$$I = I_0 \cos^2 \theta = \frac{1}{2} I_0$$

Since average value of $\cos^2 \theta$ over all possible values of θ is $\frac{1}{2}$, i.e. $\cos^2 \theta = \frac{1}{2}$. Hence an ideal polariser is one that transmits 50% of the incident unpolarised light as plane polarised one.

Uniaxial and biaxial crystals: When a ray of unpolarised light is incident on a certain special crystal (known as doubly refracting crystal), there are two refracted rays. Such crystals are of two types:

(i) Uniaxial crystals, and (ii) biaxial crystals

In uniaxial crystals there is one direction, known as optic axis, along which the two refracted rays travel with the same velocity. Moreover in these crystals one of the two refracted rays obeys the ordinary laws of refraction; while the other refracted ray does not follow these laws. The examples of uniaxial crystals are calcite, tourmaline and quartz.

In biaxial crystals there are two optic axes and none of the refracted rays follows the ordinary laws of refraction. The examples of biaxial crystals are topaz and aragonite.

Here we shall consider uniaxial crystals. The most common example of uniaxial crystals is Iceland spar or Calcite.

Geometry of a calcite crystal: Iceland spar or calcite crystal is a colourless transparent crystal belonging to hexagonal system. In nature it is found in several forms. Particularly its fine specimens are found in Iceland, hence it is named as Iceland spar. Chemically it is crystallised Calcium carbonate ($CaCO_2$). A piece of Iceland spar of any shape readily breaks into simple rhombohedron as shown in Fig. 9.11. The rhombohedron or the rhomb is bounded by six faces, each of which is a parallelogram of angles 102° and 78° (more correctly 101°55' and 78°5'). At two diametrically opposite corners, all the angles of the faces meeting there are obtuse. These two corners A and B are termed as blunt corners of the crystal. At the rest of six corners, two angles are acute and the remaining one is obtuse.

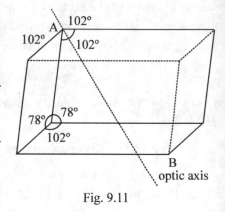

Fig. 9.11

Optic axis: A line passing through any one of the blunt corners and making equal angles with each of three edges which meet there gives the direction of the optic axis. In fact any line parallel to this line represents the optic axis. Thus optic axis is not a line, but it is a direction. It is to be noted that optic axis is not obtained by joining the two blunt corners. In a special case when the three edges of the crystal are equal, the line joining the two blunt corners coincides with the axis of the crystal and gives the direction of the optic axis.

Principal sections of the crystal: A plane containing the optic axis and perpendicular to the opposite faces of the crystal is called the principal section of the crystal for that pair of faces. As a crystal has six faces, therefore for every point inside there are three principal sections corresponding to each pair of opposite faces. A principal section also cuts the surfaces of a calcite crystal in a parallelogram with angles 109° and 71°.

Double refraction: When a ray of ordinary unpolarised light is passed through a uniaxial crystal (e.g., calcite) it is split up into two refracted rays. One of the refracted rays follows the ordinary laws of refraction and hence is called the ordinary ray (O-ray) whereas the other refracted ray does not follow the ordinary laws of refraction and is called the extraordinary ray (E-ray). Therefore if an object is viewed through such a crystal, two images of the object are observed. one image corresponds to O-ray and the other to E-ray. This phenomenon is called double refraction. It was discovered in 1669 by Dr Dane Erasmus Bartholinus. This phenomenon may be observed by the help of following simple experiment:

An ink dot is marked on a white paper and a calcite crystal is placed over it. The images of the ink dot (O and E) are observed (Fig. 9.12a). Now if the crystal is rotated slowly about a vertical axis and the eye is placed vertically above the crystal. It

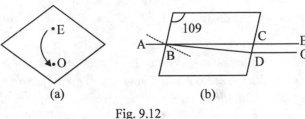

Fig. 9.12

is observed that one image remains fixed and the other image rotates with the rotation of the crystal. The fixed image is normal and is called ordinary image; while other image is abnormal and is called extraordinary image.

Let a ray *AB* be incident on a calcite crystal at an angle of incidence *i*. The ray *AB* is split up inside the crystal in two refracted rays along *BC* and *BD* such that angles of refraction are r_2 and r_1 respectively. The rays emerge from the crystal along *CE* and *DO* which are parallel (Fig. 9.12b).

The refractive index of ordinary ray, $\mu_0 = \dfrac{\sin i}{\sin r_1}$ and that of extraordinary ray,

$\mu_e = \dfrac{\sin i}{\sin r_2}$. It is observed that refractive index of ordinary ray is constant (since it obeys

ordinary laws of refraction), while refractive index of extraordinary ray varies with the angle of incidence *i*.

In calcite crystal $r_1 < r_2$, therefore $\mu_0 > \mu_e$. Hence the velocity of ordinary ray through calcite crystal is less than that of extraordinary ray. Such crystals are called uniaxial negative crystals. The other examples of negative crystals are tourmaline, sapphire, apatite, ruby and emerald. There are other crystals in which $\mu_e > \mu_0$, i.e. the velocity of ordinary ray is greater than that of extraordinary ray. Such crystals are called positive crystals. The examples of these crystals are quartz, iron oxide.

(i) When a ray of light is incident along the optic axis or in a direction parallel to the optic axis, the ray is not split up into ordinary and extraordinary components. In this case both the rays travel along the same direction with the same velocity.

(ii) When a ray of light is incident perpendicular to the optic axis, the ray of light is not split up into ordinary and extraordinary rays. In this case both the rays travel, along the same direction but with different velocities.

Principal plane: The plane containing the optic axis and the ordinary ray is called the principal plane of the ordinary ray. Similarly the plane containing optic axis and the extraordinary ray is called the principal plane of the extraordinary ray. In general the two principal planes do not coincide. Under the particular case, when the plane of incidence is the principal section of the crystal, then the principal planes of O and E rays and the principal section of the crystal coincide.

Polarisation by double refraction: It has been found that both the rays (ordinary and extraordinary) are polarised, into which an incident ray is split up by a doubly refracting calcite crystal are plane polarised, their vibration being at right angle to each other.

The ordinary ray is polarised in the principal section (i.e., it has vibrations perpendicular to the principal section), while extraordinary ray is polarised at right angles the principal section. This can be easily verified either by using a tourmaline plate or a calcite crystal as an analyser for the refracted rays.

Let us suppose that ordinary and extraordinary rays are coming out of a calcite crystal. Suppose a tourmaline plate is placed in the path of the two rays and the refracted light is received on a screen. As the tourmaline plate is rotated about the axis of the incident ray, the intensity of one of the images (say ordinary image) increases; while that of the other (extraordinary) decreases. In one position of the plate, the E-image disappears and the intensity of O-image becomes maximum. On further rotating the plate, the intensity of O-image decreases while that of E-image increases. When the glass plate has been rotated through 90° from the previous position (i.e., when E-image disappears) the intensity of E-image becomes maximum, while O-image disappears. This clearly reveals that both O and E rays are polarised and their vibrations are at right angles to each other.

Huygens in 1678, studied the polarisation of light by double refraction in calcite. He used two calcite crystals in his experiments and observed the following phenomenon:

(i) When a ray of light falls normally on a crystal surface, the O ray passes undeviated, while the E-ray emerges parallel to O-ray; but slightly displaced. The magnitude of the displacement depends upon the thickness of the calcite crystal. If another similar calcite crystal is placed in the path of the emergent rays from the first crystal and if the principal sections of the two crystals are parallel. It is observed that the two emergent rays O_1 and E_1 are parallel. The O ray passes undeviated through the second crystal and emerges as O_1 ray. The E ray from the first crystal passes through the second crystal along a path parallel to that inside the first and emerges as E_1 ray. The distance between O_1 and E_1 rays is equal to the sum of the two displacements found in each crystal if used separately (Fig. 9.13a).

(ii) If the second crystal is rotated about the incident light as axis each of the two rays O and E suffers double refraction in the second crystal, thus giving rise to four rays and hence four images. Thus along with O_1 and E_1, two new images O_2 and E_2 are observed. If the rotation is continued the images O_1 and O_2 remain fixed while E_1 and E_2 rotate about O_1 and O_2 respectively and the intensity of O_1 and E_1 goes on decreasing while that of O_2 and E_2 goes on increasing. When the principal section of the second crystal makes an angle of 45° with the principal section of the first, the intensity of all the four images is the same (Fig. 9.13b).

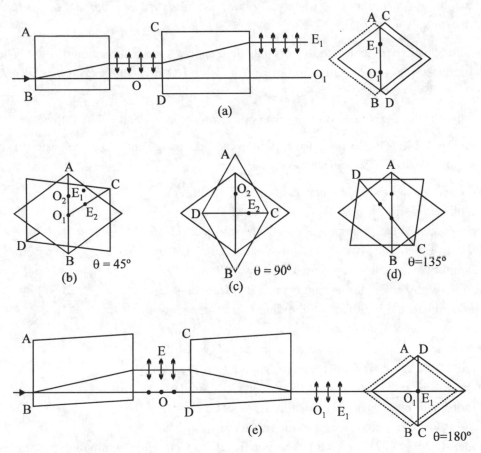

Fig. 9.13

(iii) At 90° rotation, the images O_1 and E_1 disappear and the new images O_2 and E_2 acquire maximum intensity (Fig. 9.13c).

(iv) If the rotation is continued further, the images O_1 and E_1 again appear and their intensity increases; while the intensity of images O_2 and E_2 decreases. At 135° rotation all the four images are once again equally intense (Fig. 9.13d).

(v) At 180° rotation, the principal sections of both the crystals are again parallel; but their optic axes oriented in opposite directions. In this position the images O_2 and E_2 vanish and the images O_1 and E_1 attain the initial intensities and overlap (Fig. 9.13e) provided the thickness of both the crystals is the same. The overlapping occurs because the displacement in the extraordinary ray due to first crystal is equal and opposite to that due to second crystal.

Thus this experiment with two crystals demonstrates the polarisation of light. The first crystal produces plane polarised vibrations in the E and O rays and second crystal analyses them.

The physical explanation of the phenomenon can be understood as follows:

In Fig. 9.14, AB is the principal section of the first crystal and CD is the principal section of the second crystal. The angle between AB and CD is θ.

When unpolarised light enters the first crystal, it breaks into two plane polarised O and E rays. The vibration of O ray are perpendicular to the principal section of the crystal; while that of E ray are in the plane of the principal section. Let a be the amplitude of each ray represented by PO and PE. When ordinary ray enters the second crystal, it suffers double refraction and hence it is further split up into two components O_1 and E_2. The amplitude of O_1 ray is $a\cos\theta$ perpendicular to the principal section of the second crystal CD, while amplitude of E_2 ray $a\sin\theta$ in the principal section. These amplitude are represented by PO_1 and PE_2 respectively. Similarly the extraordinary ray E, when passes through second crystal, splits up into ordinary ray O_2 and extraordinary ray E_1. The amplitude of O_2 ray is BO_2 ($= a\sin\theta$) perpendicular to CD; while amplitude of E_1 ray is PE_1 ($= a\cos\theta$) in the principal section. CD, of the second crystal. Thus the intensities of O_1 and E_1 are $a^2\cos^2\theta$ while those of O_2 and E_2 are $a^2\sin^2\theta$. Thus at $\theta = 0°$ or $180°$, we have $\cos^2\theta = 1$ and $\sin^2\theta = 0$. Hence the intensity of O_1 and E_1 is maximum; while that of O_2 and E_2 is zero. At $\theta = 45°$ or $135°$, we have $\cos^2\theta = \frac{1}{2}$ and $\sin^2\theta = \frac{1}{2}$.

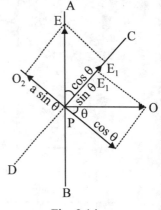

Fig. 9.14

Thus the intensity of O_1, E_1, O_2, E_2 is equal to $\dfrac{a^2}{2}$. Hence all the four images appear equally bright.

At $\theta = 90°$, we have $\cos^2\theta = 0$ and $\sin^2\theta = 1$.

Thus the intensity of O_1 and E_1 is zero, while that of O_2 and E_2 is maximum equal to a^2. As a result at $90°$ rotation O_2 and E_2 appear while O_1 and E_1 vanish.

Thus we see that all the observed results have been theoretically explained. For all positions the sum of the intensities of two components is

$$a^2\cos^2\theta = a^2\sin^2\theta = a^2$$

i.e., intensity of incident beam.

Nicol prism: In 1928, William Nicol invented a convenient optical device for producing and analysing plane polarised light, known as Nicol prism.

Principle: When a ordinary ray of light is passed through a calcite crystal it is broken up into two rays: (i) the ordinary ray which is polarised and has its vibrations perpendicular to the principal section of the crystal, and (ii) the extraordinary ray which is polarised and has its vibrations parallel to the principal section. If by some optical means one of the two rays is eliminated, the ray emerging through the crystal will be plane polarised. In Nicol prism, ordinary ray is eliminated by total internal reflection so that only the extraordinary ray, which is plane polarised, is transmitted through the prism.

Construction: A calcite crystal whose length is three times its breadth is taken. Let *ADFGBC* be such a crystal having *ABCD* as a principal section of the crystal with $\angle BAD=71°$ (Fig. 9.15a).

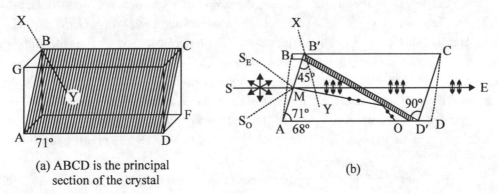

(a) ABCD is the principal
section of the crystal

(b)

Fig. 9.15

The end faces of the crystal are cut in such way that they make angles of $68°$ and $112°$ in the principal section instead of $71°$ and $109°$ (Fig. 9.15b). The crystal is then cut into two pieces from one blunt corner to the other along a plane perpendicular to the principal section.

The cut surfaces which are perpendicular to the end faces are ground, polished optically flat and then cemented together with a layer of Canada Balsam. Canada balsam is a transparent liquid and its refractive index lies midway between the refractive indices of calcite for the ordinary and the extraordinary rays.

Refractive index of calcite for O ray, $\mu_0 = 1.66$

Refractive index of Canada Balsam, $\mu = 1.55$

Refractive index of Calcite for E ray, $\mu_e = 1.49$

Thus we see that Canada Balsam is optically denser than calcite for *E* ray and rarer for *O* ray. Finally, the crystal is enclosed in a tube blackened inside.

Action: When a ray *SM* of unpolarised light (Fig. 9.15b) parallel to the face *AD'* is incident on the face *AB'* of the prism, it splits up into two refracted rays, the ordinary rays (*O* ray) and the extraordinary (*E* ray). Both of the *O* and *E* rays are plane polarised the vibrations of *O* ray being perpendicular to the principal section of the crystal; while that of *E* ray being in the principal section. The ordinary ray in going from calcite to Canada Balsam travels from optically denser medium ($\mu_0 = 1.66$) to a rarer medium ($\mu = 1.55$).

The refractive index of ordinary ray with respect to Canada Balsam.

$$= \frac{\mu_0}{\mu} = \frac{1.66}{1.55}$$

∴ If θ is critical angle, we have

$$\sin\theta = \frac{1.55}{1.66} = 0.933$$

∴ $\theta = \sin^{-1}(0.933) = 69°$

As the length of the crystal is large, the angle of incidence at Calcite-Balsam surface for the ordinary ray is greater than the critical angle. Therefore, when *O* ray is incident on Calcite-Balsam surface it is totally reflected and is finally absorbed by the side *AD'* which is blackened. The extraordinary ray travels from an optically rarer medium ($\mu_e = 1.49$) to a denser medium ($\mu = 1.55$), therefore it is not affected by the Calcite-Balsam surface and it is therefore transmitted through the prism. This *E* ray is plane polarised and has vibrations in the principal section parallel to the shorter diagonal of the end face of the crystal. Thus by Nicol prism we are able to get a single beam of plane polarised light. Thus Nicol prism can be used as a polariser.

Limitations: (i) When the angle of incidence at the crystal surface increases (e.g., for incident ray S_0M), the angle of incident at Calcite-Balsam surface decreases. When the angle S_0MS becomes greater than 14°, the angle of incidence at Calcite-Balsam surface becomes less than the critical angle (69°). In this position ordinary ray is also transmitted through the prism along with extraordinary ray so that the light emerging from the Nicol prism will not be plane polarised.

(ii) When the angle of incidence at the crystal surface is decreased (e.g., for incident rays $S_E M$), the extraordinary ray makes less angle with the optic axis, as a result its refractive index increases, because the refractive index of calcite crystal for E ray different in different directions through the crystal being maximum when the E ray travels at right angles to the optic axis and minimum when E ray travels along the optic axis. When $\angle S_E MS$ becomes greater than 14°, the refractive index of Calcite crystal for E ray becomes greater than the refractive index of Canada-Balsam and at the same time the angle of incidence for E rays at the Calcite-Balsam surface becomes greater than the corresponding critical angle. Hence in this position E ray is also totally reflected along with O ray and no light emerges from the prism.

Therefore in order to get plane polarised light, the incident light should be confined within an angle of 14° on either side of SM, i.e. in this case highly convergent or divergent light should not be used.

Nicol prism as an analyser: Consider two Nicol prisms arranged co-axially one after another. When a beam of unpolarised light is incident on the first prism P, the emergent beam is plane polarised with its vibrations in the principal section of prism P. This prism is called polariser. The beam emerging from the polariser falls on the second nicol prism A called the analyser. When the principal section of prism A is parallel to that of P, its vibrations will be parallel to the principal section of A. Thus the ray behaves as E ray in the prism A and is therefore completely transmitted. As a result the intensity of emergent light is maximum. This position and the position corresponding to an angle of 180° between the principal sections of the nicol is referred to as 'Parallel Nicols'.

When the prism A is rotated from this position, the intensity of the emergent beam decreases. When the principal section of A is at right angles to the principal section of P, then

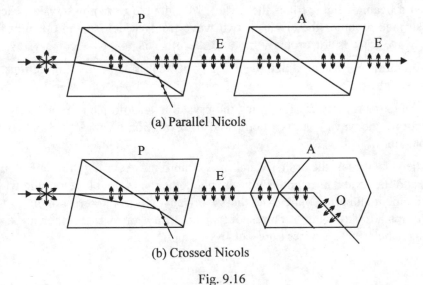

(a) Parallel Nicols

(b) Crossed Nicols

Fig. 9.16

no light emerges from the second prism A. In this position E ray has its vibrations in the principal section of P and hence normal to the principal section of A. Therefore E ray emerging from first prism P enters the prism A, it has no vibrations in the principal section of A and therefore it acts as ordinary ray (O ray) inside the prism A and is totally reflected at the Calcite-Balsam surface and therefore no light is transmitted through the prism A.

Here prism P produces plane polarised light and prism A detects or analyses it.

The above facts may be used to analyse plane polarised light with the help of a nicol prism.

If the light under examination on viewed through a rotating nicol shows variation in intensity and has minimum intensity equal to zero, the light is plane-polarised.

Huygens' Theory of Double Refraction: In order to explain double refraction by any theory the following experimental observations must be taken into account:

(i) The ordinary ray always obeys the ordinary laws of refraction (e.g., Snell's law of refraction).

(ii) The velocity of extraordinary ray is different along different directions inside the crystal.

(iii) Both the rays, ordinary and extraordinary, travel with the same velocity along the optic axis.

(iv) The refractive index, μ_0, for ordinary ray is constant for all directions inside the crystal; while the refractive index, μ_e, for extraordinary ray changes with the direction. The maximum value of μ_e in a calcite crystal is along the optic axis and is equal to μ_0; while the minimum value of μ_e is in a direction perpendicular to optic axis.

From these observations Huygens extended his principle of secondary wavelets to explain the phenomenon of double refraction. He assumed:

(i) When a beam of light strikes the surface of a double refracting crystal, each point on the surface becomes the origin of two wave-fronts which spread out into the crystal. One wave-front is for ordinary ray and the other is for extraordinary ray.

(ii) For the ordinary ray for which the velocity of light is same in all directions, the wave-front is spherical.

(iii) For the extraordinary ray for which the crystal is anisotropic (i.e., velocity varies with direction), the wave-front is an ellipsoid of revolution having optic axis as the axis of revolution.

(iv) As the velocities of the ordinary and the extraordinary rays are the same along the optic axis and the crystal does not exhibit the phenomenon of double refraction when the rays are incident along the optic axis, the spherical wave-front corresponding to O ray and the ellipsoid of revolution corresponding to E ray touch each other at any instant at points where these surfaces are cut by the optic axis.

Polarisation by scattering:

The sky is blue. Sunrises and sunsets are red. Skylight is largely linearly polarised, as can be readily verified by looking at the sky directly overhead through a polarising plate. It turns out that only the phenomenon of scattering of light is responsible for all the three effects mentioned above.

When ordinary (sun) light passes through the atmosphere containing extremely small suspended particles of dust, smoke, etc. and air molecules, it is scattered in all direction. Owing to the scattering the intensity of light is weakened.

Light scattered sideways from a gas can be wholly or partially polarised, even though the incident light is unpolarised. Fig. 9.17 shows an unpolarised beam moving upwards along Y-axis and striking a gas atom at point O. The electric field in unpolarised beam sets the electric charges in the gas atom is vibration. Since light is transverse wave, the directions of electric field in any component of incident unpolarised light lies in the XZ plane and the motion of the charges takes place in this plane. There is no field and hence no vibration in the direction of Y-axis.

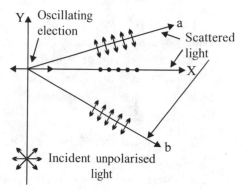

Fig. 9.17

An arbitrary component of the incident light, vibrating at any angle with the Z-axis, sets the electric charges in the atoms vibrating in the same direction. In the usual way we can resolve this vibration into two, one along the X and the other along Z-axis. Thus the result is that each component in the incident light produces the equivalent of two oscillating dipoles, oscillating with the frequency of incident light and lying along X and Z axes, represented in figure by the arrow at O along X-axis and the dot at O along Z-axis. An oscillating dipole does not radiate in the direction of its own length. Thus an observer at X would receive no radiation from the dipole along X-axis represented by arrow at O. The radiation reaching him would come entirely from the dipole along Z-axis, represented by dot at O; thus this radiation is plane-polarised, with the electric field parallel to dipole (i.e., along Z-axis). Observers at a and b would detect partially polarised light, since the dipole along X-axis would radiate somewhat in these directions. Observers viewing the transmitted or the back-scattered light would not detect any polarising effects because both dipoles at O would radiate equally in these two directions.

Blue colour of sky and red colour of sunrises and sunsets: Lord Rayleigh showed in 1817 that the intensity of light scattered from fine particles (dimension $\leq$ wavelength of light) varies inversely as the fourth power of wavelength of light, $\left(i.e., I \propto \dfrac{I}{\lambda^2} \right)$. Since the wavelength of violet light is approximately half that of red in the scattered light.

The fact that the scattering cross-section for blue light is higher than that for red light can be understood as follows. An electron in an atom or molecule is bound there by strong restoring forces. It has a definite natural frequency, like a small mass suspended in space by an assembly of springs. The natural frequency of electrons in atoms and molecules is usually in a region corresponding to violet or ultraviolet light. When light is allowed to fall on such bound electrons, it sets up forced oscillations at the frequency of the incident light beam. As the blue light is closer to the natural resonant frequency of the bound electron than is the red light. Therefore, we would expect the blue light to be more effective in causing the electron to oscillate, and thus it will be more effectively scattered.

The sun rays reach us after passing obliquely through the atmosphere. The presence of minute particles in the atmosphere scatter violet and blue light more strongly than the red light. During the passage of light from the scatterer to the eye, a part of violet is lost by further scattering and hence the light received by the observer is rich in blue. Thus the bright blue colour of sky is due to a considerable amount of light of shorter wavelength being scattered by the molecules of air.

The fact that the scattering of light is due to the presence of atmosphere or of minute particle is also clear from the fact that to observers in aeroplanes, rockets and satellites at high altitudes, where there is less or no atmosphere, the sky appears black. If the earth had no atmosphere, we would receive no skylight on earth's surface and the sky would appear black in day time too except when we looked directly at the sun. If the atmospheric particles had been larger then the wavelength of light, the scattered light would be white.

The path of the light through the atmosphere at sunrise and sunset is greatest. Since the violet and blue light is largely scattered, the transmitted light will have the colour of normal sunlight with the violet and blue largely removed; it is therefore more reddish in appearance. That is why sunrises and sunsets appear red.

SOLVED EXAMPLES

Q.1. **A ray of light is incident on the surface of a glass plate of refractive index 1.732 at the polarising angle. Calculate the angle of refraction of the ray.**

Sol. According to Brewster's law

$$\mu = \tan i_p$$

Here $\qquad \mu = 1.732 \qquad\qquad \therefore \quad 1.732 = \tan i_p$

or $\qquad i_p = \tan^{-1} 1.732 = \tan^{-1}\sqrt{3} = 60°$

If r is the angle of refraction, we have from equation (3) of section 7.5.

$$r + i_p = 90°$$

$$r = 90° - i_p = 90° - 60° = 30°.$$

Q.2. **A beam of light travelling in water strikes a glass plate which is also immersed in water. When the angle of incidence is 51°, the reflected beam found to be plane polarised. Calculate the refractive index of glass.**

Sol. By Brewster's law, $\mu = \tan i_p$

Here $i_p = 51°$ and the beam of light is travelling from water to glass.

$$\therefore \qquad _{10}\mu_9 = \tan i_p = \tan 51° = 1.235.$$

This is the refractive index of glass with respect to water.

Q.3. **At a certain temperature the critical angle of incidence of water for total internal reflection is 48° for a certain wavelength. What is the polarising angle and the angle of refraction for light incident on the water at an angle that gives maximum polarisation of the reflected light? (Given: Sin 48° = .7431)**

Sol. The refractive index μ and the critical angle C are related by

$$\mu = \frac{1}{\sin C} = \frac{1}{\sin 48°} = \frac{1}{.7431} = 1.345 \qquad \{\text{since } C = 48° \text{ (given)}\}$$

For maximum polarisation, the light must be incident at polarising angle. From Brewster's law we have $\mu = \tan i_p$.

$$\therefore \qquad \text{polarising angle, } i_p = \tan^{-1}(\mu) = \tan^{-1}(1.345) = 53°22'$$

If r is the angle of refraction we have $i_p + r = 90$

$$\text{or} \qquad r = 90° - i_p$$

$$\therefore \qquad r = 90° \quad 53°22' - 36°38'.$$

Q.4. **At a particular angle the coefficient of intensity reflection of glass plate is 4% for vibrations in the plane of incidence and 10% for vibrations perpendicular to the plane of incidence. Calculate the degree of polarisation of a beam passing through a pile consisting of 10 plates. The absorption of light is negligible.**

Sol. There are 10 plates in the pile. This mean reflection will take place at 20 surfaces, since at each plate the beam is reflected twice (i.e., from the upper and lower surface). At each surface 96% of parallel vibrations and 90% of perpendicular vibrations are transmitted. If I is the initial intensity in each plane, the intensities of the emergent beams will be given by

$$I_p = I \times (\cdot 96)^{20} = \cdot 4426\, I$$

and

$$I_n = I \times (\cdot 90)^{20} = 0 \cdot 1213\, I$$

$$\therefore \qquad P = \frac{I_p - I_n}{I_p + I_n} = \frac{0.4426\, I - 0.1213\, I}{0.4426\, I + 0.1213\, I} = \frac{0.4426 - 0.1213}{0.4426 + 0.1213} = \frac{0.3213}{0.5639} = 0.57.$$

Q.5. **Two polarising plates have parallel polarising directions so as to transmit maximum intensity of light. Through what angle must either plate be turned if the intensity of the transmitted beam is to drop by one-third?**

Sol. According to Malus, we have $\qquad I = I_\theta \cos^2 \theta$

Here $\qquad\qquad I = \dfrac{I_\theta}{3}, \qquad \therefore \quad \dfrac{I_0}{3} = I_0 \cos^2 \theta$

$\qquad$ or $\qquad \cos^2 \theta = \tfrac{1}{2} \qquad$ or $\quad \cos \theta = \pm \dfrac{1}{\sqrt{3}}$

$\qquad \therefore \qquad\qquad \theta = \cos^{-1} \pm (.5773) = \pm 55^\circ 18' \qquad$ or $\quad \pm 145^\circ 18'.$

Q.6. **A beam of light is passed through two nicols in series. In a particular setting maximum light is passed by the system and it is 500 units. If one the nicols are now rotated by 20°, calculate the intensity of transmitted light.**

Sol. $\qquad$ We have $\qquad I = I_0 \cos^2 \theta$

$\qquad\qquad$ Here $\qquad\qquad I_0 = 500$ units, $\theta = 20^\circ$

$$I = 500 \times \cos^2 20^\circ = 500 \times (0.9397)^2 = 500 \times 0.883 = 442 \text{ units}.$$

Q.7. **Two nicols are oriented with their principal planes making an angle of 30°. What percentage of incident unpolarised light will pass through the system?**

Sol. Let the intensity of the incident unpolarised light be $2 I_0$. On entering the first nicol it is broken up into O and E-components each of intensity I_0. The O-component is totally reflected and only E-component of intensity I_0 is transmitted through this (first) nicol. According to Malus law, the intensity of finally transmitted beam is given by

$$I = I_0 \cos^2 \theta = I_0 \cos^2 30^\circ = I_0 \left(\frac{\sqrt{3}}{2} \right)^2 = \frac{3}{4} I_0$$

$\therefore$ Percentage of incident unpolarised light transmitted the system

$$= \frac{I}{2I_0} \times 100 = \frac{\left(\dfrac{3}{4} I_0 \right)}{2 I_0} \times 100 = \frac{3}{5} \times 100 = 37.5\%.$$

Q.8. **Two nicols are first crossed and then one of them is rotated through 60°. Calculate the percentage of incident light transmitted.**

Sol. Let the intensity of incident unpolarised light be $2 I_0$. On entering the first nicol it is broken up into O and E-components each of intensity I_0 is transmitted through the first nicol.

When the nicols are crossed, the angle between their principal planes is 90°. Now if one of them is rotated through 60°, the angle between their planes of transmission is either 30° or 150°.

According to Malus law, the intensity of finally transmitted beam is given by

$$I = I_0 \cos^2 \theta = I_0 \cos^2 30^\circ = I_0 \cos^2 150^\circ = I_0 \times \left(\frac{\sqrt{3}}{2} \right)^2 = \frac{3}{4} I_0$$

∴ Percentage of incident light transmitted through the system

$$= \frac{I}{2I_0} \times 100 = \frac{3 I_0/4}{2 I_0} \times 100 = \frac{300}{8} = 37.5\%.$$

Q.9. **Two light sources are observed, one after the other with two nicol prisms mounted one behind the other as polariser and analyser. The intensity of beams transmitted by the prisms are observed equal from two sources when the angle between the principal section of the nicols are 45° and 70° respectively. Calculate the relative intensities of the two sources.**

Sol. Let the intensities of the given sources be $2I_0$ and $2I_0{}'$ respectively. On entering the first nicol (polarised), in each case the beams get broken up into O and E-components each of intensity I_0 in first case and I_0' in second case. The O-component is totally reflected and only E-component of intensity I_0 in first case and I_0' in second case is transmitted through the first nicol. According to Malus law, the intensities of finally transmitted beams are given by

$$I = I_0 \cos^2 0 \text{ for first source}$$

$$I' = I_0' \cos^2 \theta' \text{ for second source}$$

But $\qquad I = I'$ (given), therefore

$$I_0 \cos^2 \theta = I_0{}' \cos^2 \theta'$$

i.e., $\qquad \dfrac{I_0}{I_0'} = \dfrac{\cos^2 \theta'}{\cos^2 \theta}$

Hence the relative intensities of two sources

$$\frac{2I_0}{2I_0'} = \frac{\cos^2 \theta'}{\cos^2 \theta} = \frac{\cos^2 70}{\cos^2 45} = \frac{(0.3420)^2}{(1/\sqrt{2})^2} = \times (0.3420)^2 = 0.2339.$$

QUESTIONS

1. (i) What do you understand by the term polarisation of light?

   (ii) Distinguish between polarised and unpolarised light.

   (iii) Comment on the statement, 'Polarisation requires that the vibrations are transverse.'

   (iv) Comment on the statement, 'Light waves are longitudinal in character.'

   (v) Can sound waves be polarised? Give reasons for your answer.

2. (i) State clearly what do you understand by polarisation of light? Define plane of polarisation and plane of vibration. How would you show that light waves are transverse?

   (ii) Mention various methods of producing plane polarised light and describe the one which you consider to be the best.

3. (i) What are plane of polarisation and plane of vibration?

   (ii) State Brewster's law and use it to prove that when light is incident on a transparent substance at the polarising angle the reflected and refracted rays are at right angles to each other.

4. Explain how light is polarised by reflection. State Brewster's law and prove that the angles of incidence and refraction are complementary when complete polarisation is obtained by reflection at a plane glass plate.

5. Explain how would you obtain and detect experimentally a beam of plane polarised light by reflection. How does this experiment prove the transverse character of light vibration?

6. What do you mean by polarisation of light waves? Which character of light wave can explain polarisation? Explain by diagrams.

7. Distinguish between plane polarised and circularly polarised light.

8. Distinguish between polarised and unpolarised light.

9. Explain the following:

   (i) Polarisation by reflection   (ii) Polarisation by refraction

   (iii) Polarisation by double refraction   (iv) polarisation by selective absorption

   (v) polarisation by scattering.

10. State and explain Brewster's law.

11. Describe the working of a Biot's polariscope.

12. Explain how polarisation by refraction occurs through pile of plates.

13. State and explain the law of Malus.

14. Distinguish between uniaxial and biaxial crystals.

15. Describe the principle, construction and action of a Nicol Prism.

16. How does a Nicol Prism acts as an analyser?

17. Explain Huygens' theory of double refraction.

18. How does the sky appear blue?

19. How does the sun appear red at sunrise and sunset?

20. Distinguish between ordinary and extraordinary rays in a double refracting crystal.

21. What is the angle between plane of vibration and plane of polarisation?

22. What do you mean by polarised and unpolarised light?

23. Show that when light travelling in one transparent medium meets another transparent medium, the polarising angle, the reflected and transmitted rays are perpendicular to each other.

Problems

1. Calculate the Brewster's angle for a glass slab ($n = 1.5$) immersed in water ($n = 1.33$).

   (**Ans.** $i_p = 41.56°$)

2. The critical angle in certain substance is given to be $42°$. Calculate the polarising angle.

   (**Ans.** $i_p = 56.21°$)

3. Calculate the polarisation angle for crown glass ($\mu = 1.52$), flint glass ($\mu = 1.65$) and water ($\mu = 1.33$). (**Ans.** $56°40'$, $58°47'$, $53°4'$)

4. The angle of polarisation for rock salt is $47°4'$. What is the index of refraction?

   (**Ans.** $\mu = 1.544$)

5. With a slab of flint glass, the angle of polarisation is found to be $62°24'$. Calculate the refractive index of flint glass. (**Ans.** $\mu = 1.918$)

6. A ray of light is incident on the surface of a glass plate of refractive index 1.55 at the polarising angle. Calculate the angle of refraction. (**Ans.** $r = 32°50'$)

7. A plate of glass ($\mu = 1.5$) is used as a polariser. What is the polarising angle? What is the angle of refraction? (**Ans.** $i_p = 56.3°$, $r = 33.7°$)

8. Two Nicols are oriented with their principal planes making an angle of $60°$. What percentage of incident unpolarised light will pass through the system? (**Ans.** 12.5%)

9. Calculate the critical angle for total reflection of the ordinary ray at the layer of Canada balsam in a Nicol Prism for wavelength of light 5890 Å. (**Ans.** $i_c = 67.3°$)

10. Water has index of refraction 1.33. At what angle must a light be incident on the water surface in order that the reflected beam be plane polarised? (**Ans.** $i = 53°4'$)

11. If the polarising angle of a piece of glass for green light is $60°$. Find the angle of minimum deviation for a $60°$ prism made of same glass. (**Ans.** $60°$)

●●●

Chapter 10

Production and Analysis of Polarised Light

Plane, Circularly and Elliptically Polarised Light

In the preceding chapter we have discussed the production and analysis of plane (linearly) polarised light which represents a special and relatively simpler type of polarisation. According to electromagnetic theory of light, a light wave consists of electric and magnetic vectors vibrating in mutually perpendicular directions; both being perpendicular to the direction of propagation of light. As already indicated only the electric vector is responsible for the optical effects of a wave and hence the electric vector is also called light vector.

In the plane (linearly) polarised light, the vector vibrates simple harmonically along a fixed straight line perpendicular to the direction of propagation of light, i.e. the orientation of light vector remains unchanged while its magnitude undergoes a change during vibrations. When two plane polarised waves are superimposed, then under certain conditions the resultant light vector may rotate. If the magnitude of resultant light vector remains constant while its orientation varies regularly, then the tip of the vector traces a circle and the light is said to be circularly polarised. If however, both the magnitude and orientation of the light vector vary, the tip of the vector traces ellipse and the light is said to be elliptically polarised.

We shall now show analytically that the polarised light is in general elliptically polarised and the plane polarised and circularly polarised light are its special cases.

When unpolarised monochromatic light is incident on a Nicol prism P (Fig. 10.1), it is plane polarised. If the plane polarised light is incident normally on a thin plate of a uniaxial

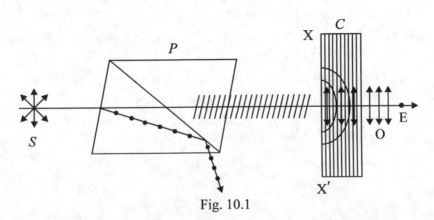

Fig. 10.1

doubly refracting crystal C (calcite, say) cut with faces parallel to the optic axis, it is split up into ordinary and extraordinary rays. Both of these rays are polarised at right angle and travel, in this case, along the same direction; but with different velocities (as $\mu_0 \neq \mu_e$). Hence there will be a phase difference δ between the two rays depending on the thickness t of the crystal.

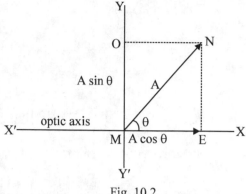

Fig. 10.2

Let the linear vibrations in the incident plane polarised light be along MN inclined at an angle θ with the optic axis. Let A be the amplitude of vibrations in the incident light. On entering the calcite crystal plate C, the incident polarised light is split up into two components:

(i) E-wave having vibrations parallel to optic axis of amplitude $A\cos\theta$.

(ii) O-wave having vibrations perpendicular to the optic axis of amplitude $A\sin\theta$.

As both these vibrations travel in crystal plate along the same direction; but with different velocities; a phase difference δ is produced between them. In calcite crystal plate, velocity of extraordinary wave is greater than that of ordinary wave (i.e., $v_e > v_0$); hence ordinary wave falls behind the extraordinary wave. Thus the rays emerging from the crystal plate may be represented by two simple harmonic vibrations of different amplitude and having a phase difference δ between them.

The displacement of the extraordinary ray (E-ray) along the optic axis is given by

$$x = A\cos\theta \cdot \sin(\omega t + \delta) \qquad \qquad ...(1)$$

and that perpendicular to optic axis is given by

$$y = A\sin\theta \cdot \sin\omega t \qquad \qquad ...(2)$$

Let us put $A\cos\theta = a$ and $A\sin\theta = b$ in Eqns. (1) and (2), so that we have

$$x = a\sin(\omega t + \delta) \qquad \qquad ...(3)$$

$$y = b\sin\omega t \qquad \qquad ...(4)$$

From equation (3), we have

$$\frac{x}{a} = \sin\omega t \cos\delta = \cos\omega t \sin\delta$$

$$= \sin\omega t \cos\delta + \sqrt{(1 - \sin^2\omega t)}\sin\delta \qquad \qquad ...(5)$$

From equation (4), we have;

$$\sin\omega t = \frac{y}{b} \qquad \qquad ...(6)$$

Equation (5) gives,

$$\frac{x}{a} = \frac{y}{b}\cos\delta + \sqrt{\left(1 - \frac{y^2}{b^2}\right)}\sin\delta$$

or

$$\left(\frac{x}{a} - \frac{y}{b}\cos\delta\right) - \sqrt{\left(1 - \frac{y^2}{b^2}\right)}\sin\delta$$

Squaring both the sides, we get

$$\left(\frac{x}{a} - \frac{y}{b}\cos\delta\right)^2 = \left(1 - \frac{y^2}{b^2}\right)\sin^2\delta$$

or

$$\frac{x^2}{a^2} + \frac{y^2}{b^2} - \frac{2xy}{ab}\cos\delta = \sin^2\delta \qquad \qquad ...(7)$$

This is the general equation of an ellipse inclined to the coordinate axes. Hence, the light emerging from the crystal plate is, in general, elliptically polarised. However, the exact nature of resultant depends on the value of δ which depends on the thickness t of the crystal plate.

Special cases:

(i) Plane polarised light: When the thickness of the plate is such that $\delta = n\pi$, equation (7) for even values of n gives

$$\frac{x^2}{a^2} + \frac{y^2}{b^2} - \frac{2xy}{ab} = 0$$

i.e.

$$\left(\frac{x}{a} - \frac{y}{b}\right) = 0$$

or

$$\pm\left(\frac{x}{a} - \frac{y}{b}\right) = 0$$

or

$$\pm y = \pm\frac{b}{a}x$$

which is the equation of a pair of coincident straight lines having a positive slope $\dfrac{b}{a}$ (Fig. 10.3a)

which is same as that of incident plane polarised ray. Thus, when $\delta = 0$, 2π, 4π, 6π, ..., the emergent light is plane polarised with the same direction of vibration as the incident light.

When n is odd, i.e. $\delta = \pi$, 3π, 5π, ..., equation (7) gives

$$\frac{x^2}{a^2} + \frac{y^2}{b^2} + \frac{2xy}{ab} = 0$$

or $$\left(\frac{x}{a}+\frac{y}{b}\right)^2 = 0$$ or $$\pm y = \mp \frac{b}{a}x$$

This is again the equation of a pair of coincident straight lines but with a slope $\left(-\dfrac{b}{a}\right)$.

Thus in this case the emergent light is again plane polarised; but the direction of vibration now makes an angle $\left(2\tan^{-1}\dfrac{b}{a}\right)$ with that of the incident light (Fig. 8.3b). This case forms the basis of half-wave plate.

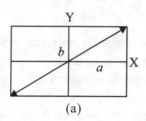

(a)

Fig. 10.3

(b)

$$\delta = 2n\pi, \quad n = 1, 2, 3, \dots \qquad\qquad \delta = (2n-1)\pi, \quad n = 1, 2, 3, \dots$$

(ii) Elliptically polarised light: When the thickness of the plate is such that,

$$\delta = (2n-1)\frac{\pi}{2}, \quad n = 1, 2, 3, \dots$$

then $\cos\delta = 0$ and $\sin\delta = 1$.

so that equation (7) gives $\dfrac{x^2}{a^2}+\dfrac{y^2}{b^2} = 1$

This is the equation of the ellipse with its axes coinciding with the coordinate axes. Thus under this condition the emergent light is elliptically polarised (Fig. 10.4).

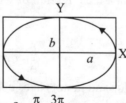

$$\delta = \frac{\pi}{2}, \frac{3\pi}{2}, \dots$$

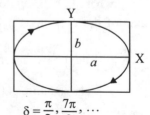

$$\delta = \frac{\pi}{2}, \frac{7\pi}{2}, \dots$$

Fig. 10.4

(iii) Circularly polarised light: When $a = b$ (i.e., $\theta = 45°$) and the thickness of plate is such that

$$\delta = (2n-1)\frac{\pi}{2}, \quad n = 1, 2, 3, \dots$$

then equation (7) gives, $x^2 + y^2 = a^2$

This is the equation of a circle of radius '*a*'. Thus under these conditions the emergent light is circularly polarised (Fig. 10.5).

This case forms the basis of *quarter-wave plate*. Thus we come to the following conclusion:

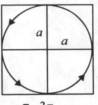 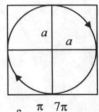

$$\delta = \frac{\pi}{2}, \frac{3\pi}{2}, \cdots \qquad \delta = \frac{\pi}{2}, \frac{7\pi}{2}, \cdots$$

Fig. 10.5

When plane polarised light passes through a thin doubly refracting crystal, cut with faces parallel to the optic axis, the incident vibration is split up into ordinary and extraordinary components. Both of these vibrations, polarised at right angles, travel along the same direction with different velocities and recombine into a vibration which is elliptical, circular or linear according to the phase difference within the crystal plate and the amplitude of the two components.

The light emerging from the crystal plate is generally elliptically polarised. It is to be noted that for $0 < \delta < \pi$, the rotation of the electric vector is related to the direction of propagation in the same sense as the rotation of a right-handed screw is rotated to this direction of translation. To the observer the rotation is anticlockwise and light is said to be right handed elliptically polarised light. For $\pi < \delta < 2\pi$ the rotation of electric vector appears clockwise and the light is said to be left handed elliptically polarised light.

However the exact nature of resultant vibration depends on value of phase difference δ and we have the following cases:

(i) When the phase difference $\delta = 0, 2\pi, \ldots$ or any even multiple of π, the emergent light is plane polarised with the same direction of vibration as the incident light.

When the phase difference $\delta = \pi, 3\pi, \ldots$ or any odd multiple of π, the emergent light is again plane polarised but the direction of vibration makes an angle 2θ with that of incident light, where θ is the angle which the direction of vibration of incident light makes with the optic axis of the crystal.

(ii) If the phase difference $\delta = (2n - 1)\dfrac{\pi}{2}$, $n = 1, 2, 3, \ldots$, the emergent light is elliptically polarised and its (major and minor) axes coincide with the axes of the component linear vibrations.

When $\delta = \dfrac{\pi}{2}, \dfrac{5\pi}{2}, \dfrac{9\pi}{2}, \ldots$, the rotation of light vector while describing the ellipse is anticlockwise and the light is right handed elliptically polarised light.

When $\delta = \dfrac{3\pi}{2}, \dfrac{7\pi}{2}, \ldots$, the light vector rotates clockwise while describing the ellipse and the light is left handed elliptically polarised light.

(iii) If the phase difference $\delta = (2n-1)\dfrac{\pi}{2}$ and $\theta = 45°$ (i.e., $a = b$), the emergent light is circularly polarised.

when $\delta = \dfrac{\pi}{2}, \dfrac{5\pi}{2}, \dfrac{9\pi}{2}, \ldots$ the light vector describes the circular vibration in the anticlockwise sense and the light is right handed circularly polarised light.

When $\delta = \dfrac{3\pi}{2}, \dfrac{7\pi}{2}, \ldots$ the light vector describes the circular vibration in the clockwise sense and the light is left handed circularly polarised light.

(iv) When $\delta \ne$ integral multiple of $\dfrac{\pi}{2}$, the resulting vibration is a slanted ellipse.

Thus it is obvious that the plane polarised and circularly polarised light are the special cases of elliptically polarised light.

Quarter-wave plate: A plate of a doubly refracting uniaxial crystal (e.g., quartz or calcite), whose refracting faces are cut parallel to the direction of optic axis and whose thickness is such as to produce a phase difference of $\dfrac{\pi}{2}$ or a path difference of $\dfrac{\lambda}{4}$ between the ordinary and the extraordinary waves, is called a quarter-wave plate.

If t is the thickness of quarter-wave plate, μ_0 and μ_E are the refractive indices for the ordinary and extraordinary rays respectively, then for normal incidence the path difference introduced between the O and E rays is given by:

For negative crystals like calcite ($\mu_0 > \mu_E$), so that the path difference $= (\mu_0 - \mu_E)\, t$.

For positive crystals like quartz ($\mu_E > \mu_0$), so that the path difference $= (\mu_E > \mu_0)\, t$.

But for a quarter-wave plate, path difference $= \dfrac{\lambda}{4}$

∴   For negative crystals like calcite, $(\mu_0 - \mu_E)\, t = \dfrac{\lambda}{4}$

or $\qquad\qquad t = \dfrac{\lambda}{4(\mu_0 - \mu_E)}$ ...(1)

For positive crystals like quartz $(\mu_E - \mu_0)\, t = \dfrac{\lambda}{4}$

$$\therefore \qquad t = \frac{\lambda}{4(\mu_E - \mu_0)} \qquad \qquad ...(2)$$

A quarter-wave plate is used to produce circularly and elliptically polarised light. When a plane polarised light is incident on a quarter-wave plate with its vibrations making an angle 45° with the optic axis the emergent light is circularly polarised. But if the vibration of the incident plane polarised light do not make an angle of 45° with the optic axis, the emergent light is elliptically polarised.

Half-wave plate: A plate of a doubly refraction uniaxial crystal (e.g., quartz or calcite), whose refracting faces are cut parallel to the direction of optic axis and whose thickness is such as to produce a phase difference of π or a path difference of $\frac{\lambda}{2}$ between the ordinary and the extraordinary waves is called a half-wave plate.

If t is the thickness of a half-wave plate, μ_0 and μ_E are the refractive indices for the ordinary and the extraordinary rays, then for normal incidence the path difference introduced between the O and E-rays is given by :

For negative crystals like calcite, path difference $= (\mu_0 - \mu_E)t$

For positive crystals like quartz, path difference $= (\mu_E - \mu_0)t$

But for a half-wave plate, path difference $= \frac{\lambda}{2}$

$\therefore$ For negative crystals like calcite, $(\mu_0 - \mu_E)t = \frac{\lambda}{2}$

or $\qquad t = \frac{\lambda}{2(\mu_0 - \mu_E)} \qquad \qquad ...(1)$

For positive crystals like quartz, $(\mu_E - \mu_0)t = \frac{\lambda}{2}$

or $\qquad t = \frac{\lambda}{2(\mu_E - \mu_0)} \qquad \qquad ...(2)$

When a plane polarised light is passed through a half-wave plate, the emergent light is also plane polarised for all the orientation of the plate with respect to the plane of vibration of the incident light; but now the direction of vibration is inclined at an angle 2θ with that in the incident light where θ is the angle which the direction of vibration in the incident light makes with the optic axis.

The quarter and the half-wave plates are also called retardation plates because they retard the motion of one of the beams. They are either made of quartz by cutting it parallel to optic axis or by splitting thin sheets of mica cleavage planes. Mica is a negative biaxial crystal, but the angle between the two axes is very small. Quartz is a positive crystal. It is cut with its faces parallel to optic axis and then its faces are polished to make them optically plane.

Production of plane, circularly and elliptically polarised light:

(1) Plane polarised light: In order to produce plane polarised light, a beam of unpolarised monochromatic light is passed through a Nicol prism. As the beam enters the Nicol prism, it is split up into ordinary and extraordinary components. The ordinary components is totally internally reflected at the Canada-balsam layer and is absorbed; while the extraordinary component passes through the Nicol prism. The emergent light is plane polarised having its vibrations parallel to the shorter diagonal of end face of the Nicol.

Detection: To detect plane polarised light, it is passed through another Nicol prism which is rotated gradually about the direction of propagation of light. If the intensity of the light emerging from the rotating Nicol varies with zero minimum, the light is plane polarised.

(2) Circularly polarised light: Circularly polarised light is produced if the amplitudes of the ordinary and extraordinary rays are equal and there is a phase difference of $\pi/2$ or a path difference of $\lambda/4$ between them

For the purpose the ordinary monochromatic light is passed through a Nicol prism. The light emerging from Nicol prism is plane polarised. This plane polarised light is then allowed to fall normally on a quarter-wave plate such that the vibrations in the incident plane polarised light make an angle of 45° with the optic axis of the plate. When this condition is satisfied the plane polarised light on entering this quarter-wave plate is split up into ordinary and extraordinary components having equal amplitude and period. The light emerging from the quarter-wave plate is circularly polarised.

Detection: The circularly polarised light, when observed through a rotating Nicol, shows no variation in intensity. The same is observed when ordinary unpolarised light is viewed through a rotating Nicol. In this respect circularly polarised light resembles the unpolarised light. hence in order to detect circularly polarised light, it is first passed through a quarter-wave plate which converts the circularly polarised light into plane polarised light. When this emergent light is viewed through a rotating Nicol, it shows a variation in intensity with zero minimum (Fig. 10.6a).

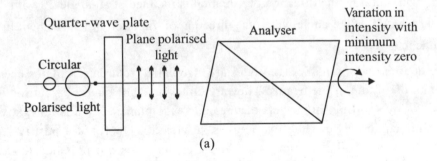

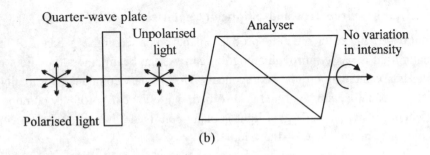

Fig. 10.6

If the given light were unpolarised, it would have remained unchanged on passage through the quarter-wave plate. Hence on passing through the rotating Nicol the emergent light would have shown no variation in intensity (Fig. 10.6b).

(3) Elliptically polarised light: Elliptically polarised light is produced if the amplitudes of the ordinary and extraordinary rays are unequal and there is a phase difference of $\pi/2$ or a path difference of $\pi/4$ between them.

For the purpose the ordinary monochromatic light is passed through a Nicol prism. The light emerging from the Nicol prism is plane polarised. This plane polarised light is then allowed to fall normally on a quarter-wave plate such that the vibrations in the plane polarised incident light make an angle θ ($\theta \neq 0°, 45°, 90°$) with the optic axis of the plate. When this condition is satisfied the plane polarised light on entering the quarter-wave plate is split up into ordinary and extraordinary components having unequal amplitudes and equal periods. The light emerging from quarter-wave plate is elliptically polarised.

Detection: When elliptically polarised light is observed through a rotating Nicol, the variation in intensity is observed with minimum intensity not zero. [The intensity is maximum when the principal section of the analysing Nicol is parallel to the major axis of the elliptical vibration and minimum when parallel to the minor axis.] The same is observed when partially plane polarised light is passed through a rotating Nicol. In this respect elliptically polarised light

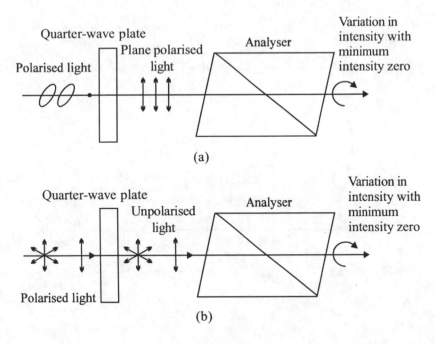

Fig. 10.7

resembles the partially plane polarised light. In order to detect the elliptically polarised light, it is passed through a quarter-wave plate before passing through the analysing Nicol. The quarter-wave plate converts the elliptically polarised light into plane polarised light. So when the light emerging from the quarter-wave plate is passed through a rotating Nicol, the finally emergent light shows a variation in intensity with minimum intensity zero (Fig. 10.7a).

However, if the light were partially plane polarised, it would have remained unchanged by the quarter-wave plate. Hence on passing through the rotating Nicol, the emergent light would have shown the variation in intensity with minimum intensity not zero (Fig. 10.7b).

Distinction between a quarter-wave and half-wave plates:

In order to distinguish between a quarter-wave and half-wave plates, plane polarised light is allowed to fall normally at each of them. The emerging light is then viewed with a Nicol prism which is rotated about the direction of propagating of light as axis.

In the case of a quarter-wave plate (Fig. 10.8a), the light emerging from the quarter-wave plate will be in general elliptically polarised. It will be circularly polarised when the vibrations in the incident plane polarised light make an angle 45° with the optic axis of the plate and it will be plane polarised if the vibrations in the incident plane polarised light make an angle to 0° or 90° with the optic axis. Hence if light emerging from quarter-wave plate is viewed through the rotating Nicol, it will be observed.

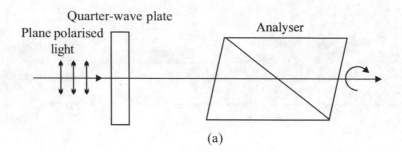

(a)

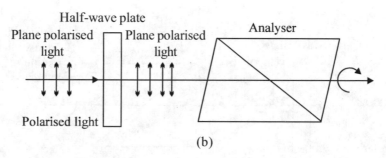

(b)

Fig. 10.8

Either

  (i) Variation in intensity with minimum intensity not zero.

  (ii) No variation in intensity.

 (iii) Variation in intensity with minimum intensity zero.

Conditions for Variation in Intensity:

  (i) When the light emerging from the quarter-wave plate is elliptically polarised, there will be variation of intensity with minimum intensity not zero.

  (ii) When the light emerging from the quarter-wave plate is circularly polarised, there will be no variation in intensity.

 (iii) When the emergent light is plane polarised, there will be variation in intensity with minimum intensity zero.

SOLVED EXAMPLES

Q.1. **Calculate the thickness of a quarter-wave plate of quartz for sodium light of wavelength 5893 AU. The refractive indices of quartz for E-ray and O-ray are equal to 1.5533 and 1.5442 respectively.**

Sol. The thickness of the quarter-wave plate of quartz is given by

$$t = \frac{\lambda}{4(\mu_E - \mu_0)}$$

Here, $\lambda = 5893$ AU $= 5893 \times 10^{-3}$ cm, $\mu_E = 1.5533$, $\mu_0 = 1.5442$

$$i = \frac{5893 \times 10^{-2}}{4 \times (1.5533 - 1.5442)} = \frac{5893 \times 10^{-8}}{4 \times 0.0091} = 1.62 \times 10^{-3} \text{ cm.}$$

Q.2. **Calculate the thickness of a calcite plate which would convert plane polarised light into circularly polarised light. The principal refractive indices are $\mu_0 = 1.658$ and $\mu_E = 1.486$ at the wavelength of light 5890 AU.**

Sol. The plane polarised light will be converted into circularly polarised light if the thickness of the plate introduces a phase difference of $\pi/2$ or an odd multiple of $\pi/2$, i.e., a path difference of $\lambda/4$ or an odd multiple of $\lambda/4$ between ordinary and extraordinary rays. Hence the thickness 't' of the calcite plate is given by

$$(\mu_0 - \mu_E)t = \frac{n\lambda}{4}, \quad n = 1, 3, 5, \ldots$$

or $$t = \frac{n\lambda}{4(\mu_0 - \mu_E)}$$

Here $\lambda = 5890$ AU $= 5890 \times 10^{-8}$ cm, $\mu_E = 1.658$

$\therefore$ $$t = \frac{n \times 5890 \times 10^{-3}}{4(1.658 - 1.486)} = \frac{n \times 5890 \times 10^{-2}}{4 \times 0.172} = 8.56 \times 10^{-2} n \text{ cm,} \quad n = 1, 3, 5, \ldots$$

$\therefore$ The thickness of the calcite plate is given by

$$t = 8.56 \times 10^{-2} \times 1, 8.56 \times 10^{-2} \times 3, 8.56 \times 10^{-5} \times 5, \ldots$$

$$= 8.56 \times 10^{-5} \text{ cm}, 25.68 \times 10^{-5} \text{ cm}, 42.80 \times 10^{-5} \text{ cm}, \ldots$$

The least thickness of the plate $= 8.56 \times 10^{-5}$ cm.

Q.3. **A given calcite plate behaves as a half-wave plate for a particular wavelength λ. Assuming variation of refractive index with λ to be negligible, how would the above plate behaves for another light of wavelength 2λ?**

Sol. The thickness of a half-wave plate of calcite crystal for wavelength λ is given by

$$t = \frac{\lambda}{2(\mu_0 - \mu_E)} = \frac{2\lambda}{4(\mu_0 - \mu_E)} = \frac{\lambda'}{4(\mu_0 - \mu_E)} \qquad \text{where } \lambda' = 2\lambda.$$

Obviously the half-wave plate for λ will behave as a quarter-wave plate for $\lambda'(= 2\lambda)$ provided the variation of refractive index with wavelength is negligible.

Q.4. **The faces of a quartz plate are parallel to the optic axis of the crystal: (a) What is the thinnest possible plate that would serve to put the ordinary and extraordinary rays of $\lambda = 5890$ AU a half-wave apart an their exit? (b) What multiples of this thickness would give the same result? The indices of refraction of quartz are $\mu_E = 1.553$ and $\mu_0 = 1.554$.**

Sol. (a) The thinnest possible plate that will produce a path difference of $\lambda/2$ (half-wave), is a half-wave plate. The thickness of a half-wave plate of quartz is given by

$$t = \frac{\lambda}{2(\mu_E - \mu_0)}$$

Here $\qquad \lambda = 5890 \times 10^{-8}$ cm, $\mu_E = 1.553$ and $\lambda_0 = 1.544$

$$\therefore \qquad t = \frac{5890 \times 10^{-8}}{2(1.553 - 1.544)} = \frac{5890 \times 10^{-3}}{2 \times .009} = 3.27 \times 10^{-3} \text{ cm}$$

(b) The thickness which would give the same result are t, $3t$, $5t$, ...

$$= 3.27 \times 10^{-3} \text{ cm}, \quad 3 \times 3.27 \times 10^{-3} \text{ cm}, \quad 5 \times 3.27 \times 10^{-3} \text{ cm}, ...$$

Q.5. **The value of μ_E and μ_0 for quartz are 1.5508 and 1.5418 respectively. Calculate the phase retardation for $\lambda = 5000$ Å when the plate thickness is 0.032 mm.**

Sol. The path difference between the ordinary and extraordinary rays for quartz plate of thickness t is given by:

$$\text{Path difference} = (\mu_E - \mu_0)\, t.$$

The phase retardation $= \dfrac{2\pi}{\lambda} \times$ path difference $= \dfrac{2\pi}{\lambda} \times (\mu_E - \mu_0)\, t$

Here $\qquad \mu_E = 1.5508, \quad \mu_0 = 1.5418,$

$$t = 0.032 \text{ mm} = 0.0032 \text{ cm}$$

and $\qquad \lambda = 5000 \times 10^{-3}$

$$= \frac{2\pi}{5000 \times 10^{-3}} \times (1.5508 - 1.5418) \times 0.0032 = \frac{2\pi \times 0.009 \times 0.0032}{5000 \times 10^{-3}} = 1.152\,\pi \text{ radians.}$$

Q.6. **A plate of thickness 0.020 mm is cut from calcite with optic axis parallel to the face. Given $\mu_0 = 1.648$ and $\mu_E = 1.481$ (ignoring variations with wavelength), find out those wavelengths in the range 4000 Å to 7800 Å (visible range) for which the plate behaves as a half-wave plate and also those for which the plate behaves as a quarter-wave plate.**

Sol. The given calcite plate will behave as a half-wave plate for wavelength λ if

$$\left(\mu_0 - \mu_E\right)t = \left(2n - 1\right)\frac{\lambda}{2} \qquad \text{where } n = 1, 2, 3, \ldots$$

i.e.,
$$\lambda = \frac{2\left(\mu_0 - \mu_E\right)t}{2n - 1} = \frac{2\left(1.648 - 1.481\right) \times 0.0020}{2n - 1}\,\text{cm}$$

$$= \frac{66800}{2n - 1} \times 10^{-2}\,\text{cm} = \frac{66800}{2n - 1}\,\text{Å}.$$

Substituting $n = 5, 6, 7, 8$, we get $\lambda = 7422$ Å, 6073 Å, 5138°, 4453 Å in the visible region. These are the required wavelengths for which the given calcite plate behaves as a half-wave plate.

The given calcite plate will behave as a quarter-wave plate for wavelength λ if

$$\left(\mu_0 - \mu_E\right)t = \left(2n - 1\right)\frac{\lambda}{4} \quad \text{where } n = 1, 2, 3, \ldots$$

i.e.,
$$\lambda = \frac{4\left(\mu_0 - \mu_E\right)t}{2n - 1} = \frac{4\left(1.648 - 1.481\right) \times 0.0020}{2n - 1}$$

$$= \frac{133600}{2n - 1} \times 10^{-3}\,\text{cm} = \frac{133600}{2n - 1}\,\text{Å}.$$

Substituting $n = 10, 11, 12, 13, 14, 15, 16, 17$, respectively, we get

$\lambda = 7032$ Å, 6362 Å, 5807 Å, 5344 Å, 4948 Å, 4607 Å, 4310 Å, 4048 Å, in the visible region. These are the required wavelengths for which the given calcite plate behaves as a quarter-wave plate.

Q.7. **Plane polarised light falls normally on a quarter-wave plate. Explain what will be the nature of emergent light if the plane of polarisation of the incident light makes the following angles with the principal plane of the quarter-wave plate:**

0°, 30°, 45°, 90°

Sol. If plane polarised light falls normally on a quarter-wave plate of uniaxial crystal cut with its optic axis parallel to the face of the plate, its vibrations are broken at the first face into one parallel to optic axis (extraordinary or E-vibrations) and the other perpendicular to optic axis (ordinary or O-vibrations). Let A be amplitude of the incident plane polarised light and O the angle which the plane of polarisation of incident light makes with the principal plane of quarter-wave plate, then referring to (Fig. 10.2), the amplitudes 'a' and 'b' of E and O-vibrations are given by

$$a = A \cos \theta, \quad b = A \sin \theta \qquad \qquad ...(1)$$

These E-and O-vibrations travel through the plate along the same direction (along the normal) but with different velocities V_E and V_0 since $\mu_E \neq \mu_0$. Consequently on emergence from the opposite face of the quarter-wave plate, a path difference of $\dfrac{\lambda}{4}$ and a phase difference of $\dfrac{\pi}{2}$ is introduced between the E- and O-vibrations.

For $\theta = 0°$, $a = A$ and $b = 0$, i.e. O-vibration becomes absent and only E-vibration passes through the plate. Therefore the emergent light remains plane polarised having its vibrations along the optic axis.

For $\theta = 30°$, the two rectangular vibrations of amplitude $A \cos 30°$ and $A \sin 30°$ $\left(\text{i.e., } A\sqrt{3}/2 \text{ and } A/2\right)$ and having a phase difference of $\dfrac{\pi}{2}$ combine to form elliptical vibrations of eccentricity.

$$\epsilon = \sqrt{1 - \frac{b^2}{a^2}} = \sqrt{1 - \frac{\left(A/2\right)^2}{\left(A\sqrt{3}/2\right)^2}} = \sqrt{1 - \frac{1}{3}} = \sqrt{\frac{2}{3}}$$

For $\theta = 45°$, the amplitude of O- and E-vibrations are equal (refer eqn. (1) $a = A \cos 45°$, $b = A \sin 45°$ so that $a = b = A/\sqrt{2}$) and therefore ellipse reduces to a circle. The emergent light is, therefore, circularly polarised.

For $\theta = 90°$, $a = 0$ and $b = A$, i.e. E-vibration becomes absent and O-vibrations passes through the plate. Therefore, the emergent light remains plane polarised having its vibrations perpendicular to the optic axis.

QUESTIONS

1. Give the pictorial representation of the end-on view of unpolarised light and plane polarised light.
2. Describe how elliptical and circularly polarised light is generated. Explain the process for realisation of plane polarised light from elliptically polarised light.
3. How can you distinguish between circularly polarised light and unpolarised light?
4. Give the construction and use of a half-wave plate. Can it be used for light of all wavelengths?
5. Distinguish between half-wave plate and quarter-wave plate.
6. Give the general mathematical treatment for production of plane, circularly and elliptically polarised light.

Problems

1. An analysing Nicol examines two adjacent plane polarised beams A and B whose plane of polarisation are mutually perpendicular. In one position of the analyser, a beam B shows zero intensity. From this position a rotation of $30°$ shows the two beams as matched (i.e., of equal intensity). Deduce the intensity ratio I_A/I_B of the two beams.

   (Ans. 1/3)

2. A quarter-wave plate is meant for $\lambda = 5893$ Å. Calculate the phase retardation which this plate will show for $\lambda = 4358$ Å. The variation of refractive indices with λ may be assumed to be negligible.

   (Ans. 2.12 radian)

3. Calculate the thickness of a quartz-half-wave plate for the Fraunhofer C-line (λ for C-line is 6563 AU) for which the extraordinary and ordinary refractive indices of quartz are 1.55085 and 1.54181 respectively.

4. A quartz wedge of refracting angle $\left(\dfrac{1}{2}\right)^°$ is cut with its optic axis parallel to the refracting edge. It is placed between crossed Nicols with its principal section inclined at $45°$ with the vibration plane of incidence plane polarised light from the polariser. If the wavelength of light used is 6000 AU, calculate the distance between successive fringes. Given:

   $$\mu_E = 1.556 \quad \text{and} \quad \mu_0 = 1.546 \qquad \textbf{(Ans. 1.03 cm)}$$

5. A doubly refracting crystal plate is placed between two parallel Nicols. If the vibration plane of polarised light from the polarising Nicol makes an angle $60°$ with the optic axis of the plate, calculate the amplitude ratio of vibration components transmitted through the crystal plate and the analyser. **(Ans. I: $\sqrt{3}$; 1:1)**

6. Plane polarised light is incident normally on a calcite plate, cut with its optic axis in the face, the angle between the principal section of the plate and the polariser being $30°$. The light emerging from the plate is found to be right handed elliptically polarised. Evaluate the possible thickness of the plate and the ratio between the major and minor axes.

   (Given: $\mu_0 = 1.658$, $\mu E = 1.486$ for $\lambda = 6000$ Å)

   (Ans. (i) $8.722\,(4n-3) \times 10^{-8}$ cm, for $n = 1, 2, 3, ...,$ **(ii)** $\sqrt{3}$)

7. It is desired to produce, by the use of a thin quartz plate, a beam of elliptically polarised light (sodium) having a ratio of major axis to minor axis as 2:1 and rotating in the counter clockwise direction as viewed against the direction of light. How thick plate should be chosen and how should it be used? (Given: $\mu_e = 1.5443$, $\mu_e = 1.5533$).

   (Ans. $t_{min} = 1.636 \times 10^{-3}$ cm, optic axis at $63°26'$)

●●●

Chapter 11

Optical Rotation

Introduction

When monochromatic light is passed through two Nicol prisms (polarising Nicol P and analysing Nicol A) placed in crossed position, no light emerges out of the analyser (Fig. 11.1a). The polarising Nicol P renders the beam plane polarised with vibrations in its principal plane. These plane polarised vibrations are not transmitted by the analysing Nicol A because its principal plane is perpendicular to the direction of vibrations of plane polarised light. If now a thin plate of calcite cut with its faces perpendicular to the optic axis is placed between two crossed Nicols, still no light emerges from the analysing Nicol. If, however, a plate of quartz cut with its faces perpendicular to the optic axis is placed between two crossed Nicols, it is observed that some light emerges from the analyser (Fig. 11.1b). If now the analysing Nicol is rotated slowly about the direction of propagation of light, it is found that there is again complete extinction of light (i.e., no light again emerges) for some different position of the analysing Nicol. This indicates that light after passing through the quartz plate is still plane polarised; but the plane of polarisation is rotated (due to the passage in the quartz plate) though a certain angle which is equal to the angle through which analysing Nicol has to be rotated to restore the original condition of complete extinction of light. This property of the plane of polarisation is possessed not only by quartz; but also by other

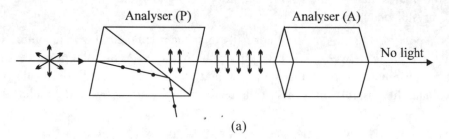

(a)

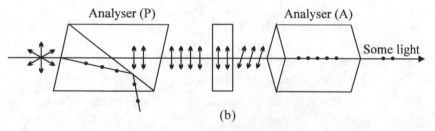

(b)

Fig. 11.1

crystals like sodium chlorate, cinnabar, sugar crystals and solutions like turpentine, sugar solution, quinine sulphate solution.

The property or the phenomenon of rotating the plane of polarisation about the direction of propagation of light by certain crystals or substance is called optical rotation.

There are two types of optically active substances:

(i) **Right-handed or dextrorotatory:** The substances which rotate the plane of polarisation in the clockwise direction with respect to an observer looking towards light travelling towards him are called right-handed or dextrorotatory.

(ii) **Left-handed or laevorotatory:** The substances which rotate the plane of polarisation in the anti-clockwise direction with respect to an observer looking towards light travelling towards him are called left-handed or laevorotatory.

Some crystals of quartz are dextrorotatory; while others are laevorotatory. Cane sugar is dextrorotatory; while fruit sugar is laevorotatory.

Biot's laws for rotatory polarisation:

Biot, in 1815, studied the phenomenon of optical rotation in detail and gave the following laws:

1. The amount of rotation of the plane of polarisation is directly proportional to the length of the optically active substance traversed.

$$(i.e., \ \theta \propto l)$$

2. In the case of solutions the amount of rotation is directly proportional to the concentration of the solution.

$$(i.e., \ 0 \propto c)$$

3. The amount of rotation is approximately inversely proportional to the square of the wavelength. Thus angle of rotation is least for red and greater for violet.

$$\left(i.e., \ \theta \propto \frac{1}{\lambda^2}\right)$$

4. The rotation produced by number of optically active substances is equal to the algebraic sum of the individual rotations. The rotation in the anti-clockwise direction is taken as positive while that in clockwise direction is taken as negative.

$$\theta = \theta_1 + \theta_2 + \theta_3 + \ldots = \Sigma_i \ \theta_i$$

5. The amount of rotation also depends on the nature and temperature of the substance.

The phenomenon may be used to find the percentage of the optically-active material present in a solution. By measuring the angle of rotation of the plane of polarisation, the amount of sugar present in the urine of a diabetic patient may be readily determined.

Specific rotation and molecular rotation: Optically active solutions, e.g. sugar solution rotate the plane of polarisation of light. The amount of the rotation depends upon:

(i) the thickness of the medium, (ii) concentration of the solution or the density of optically active substance in the solvent, (iii) wavelength of light, and (iv) temperature.

The optical activity of a substance is measured by its specific rotation or specific rotatory power. The specific rotation of a solution at a given temperature and for a given wavelength of light is defined as the rotation (in degrees) produced by 1 decimetre (10 cm) length of the solution when its concentration is 1 gm per cc, i.e.

$$\text{specific rotation } \alpha\frac{t}{\lambda} = \frac{\theta}{lc} \qquad \qquad ...(1)$$

where $\alpha\frac{t}{\lambda}$ represents specific rotation at temperature $t°$ for wavelength λ, l is the length of the solution in decimetres and c is the concentration of the active substance in gm per cc in the solution. The unit of specific rotation is dig (decimetre)$^{-1}$ (gm cm^3)$^{-1}$.

If length of the solution is in cm, the specific rotation is given by

$$\alpha\frac{t}{\lambda} = \frac{100\theta}{lc} \qquad \qquad ...(2)$$

Molecular rotation is the product of specific rotation and the molecular weight.

Rotatory Dispersion: The angle of rotation of the plane of polarisation depends on:

(i) the thickness of the medium traversed, i.e. path length,

(ii) the concentration of the solution,

(iii) the wavelength of light, and

(iv) temperature.

For a solution of given concentration, given path length and at a given temperature, the angle of rotation of plane of polarisation varies approximately as $\frac{1}{\lambda^2}$. Therefore, if white light is passed through an optically active solution, the angle of rotation of the plane of polarisation for different wavelengths will be different, being minimum for red and maximum for violet. This phenomenon is called rotatory dispersion.

Polarimeter: The device which measures the angle through which the plane of polarisation of a plane polarised beam is rotated by a given medium is called a polarimeter. By measuring this angle θ the specific rotation $\alpha\frac{t}{\lambda}\left(=\frac{100\theta}{lc}\right)$ can be evaluated if concentration and the length of the solution are known. There are two types of polarimeters in use:

(i) Laurent's half-shade polarimeter,

(ii) Bi-quartz polarimeter.

When these polarimeters are used to determine the quantity of sugar in a solution, they are called saccharimeter.

Laurent's Half-Shade Polarimeter:

Apparatus: The experimental arrangement is shown in Fig. 11.2. *S* is a source of monochromatic light (usually a radium) the focus of convex lens *L*. The beam, rendered parallel by lens *L*, falls on a Nicol prism (called polariser) *P*. After passing through polariser *P* the light becomes plane polarised. This polarised light beam passes through a half-shade device *H* (called Laurent's plate) and then through a tube *T* containing optically active solution. The transmitted light passes through another Nicol (called analyser) *A* which can be rotated about the direction of propagation of light as axis and its rotation can be read on a circular scale graduated in degrees, with the help of a vernier. The light emerging from the analyser is observed through a telescope.

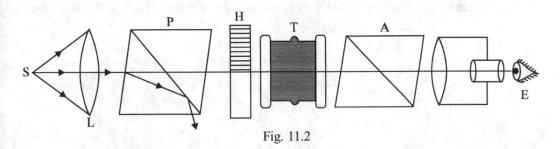

Fig. 11.2

Action of Laurent's half-shade plate: Half-shade plate H is combination of two semi-circular plates XBY and XDY (Fig. 11.3). The plate DXY is of quartz and is cut parallel to optic axis while the plate XBY is of glass. Both the plates are joined along the diameter XY. The thickness of the quartz plate is such that it introduces a path difference $\lambda/2$ or a phase difference π between ordinary and extraordinary vibrations, i.e. it is a half-wave plate. The thickness of the glass plate is such that it absorbs the same amount of light as the quartz plate.

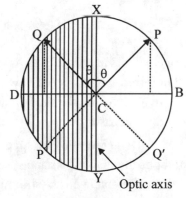

Fig. 11.3

The light emerging from the polariser (*P*) is plane polarised and falls normally on the half-shade plate. Let CP be the direction of vibrations in the plane polarised light. The light passing through the glass plate remains unaffected. But the light falling on the quartz plate is broken up into two components parallel and perpendicular to the optic axis XCY. The ordinary component is perpendicular to the optic axis, i.e. along CB; while the extraordinary component is parallel to optic axis, i.e. along

CX. As in quartz ordinary (O) component travels faster; hence passing through the quartz plate it gains a path of $\lambda/2$ or a phase of π over extraordinary (*E*-component). Hence on emergence from the quartz plate *O*-component has vibrations along *CD*. The *E*-component has vibrations still along CX. Thus the light emerging from the quartz plate has resultant vibrations along *CQ* where $\angle PCD = \angle QCX = \theta$ (say). Thus the vibration plane of light emerging from quartz is inclined to the vibration plane of light emerging from glass at an angle 2θ.

Thus on emergence from the combination we have two plane polarised beams, one passing out of the glass plate and the other passing out of the quartz plate.

When the principal plane of the analyser is parallel to *QCQ*; light from the quartz plate will pass through the analysing Nicol and the light from the glass plate will be partly stopped by the analyser. Therefore left half will be brighter as compared to right half (Fig. 11.4a). If, however, the principal section of the analyser is parallel to PCP, the light from the glass plate will pass through the analyser; while the light from the quartz plate will be partly stopped. Therefore, the right-half will appear brighter as compared to left-half (Fig. 11.4b). But when the

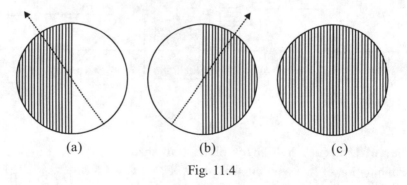

(a)   (b)   (c)

Fig. 11.4

principal plane of analyser is parallel to XCY; the two halves of the field appear equally illuminated (Fig. 11.4c). If now the analyser is rotated from this position slightly, the intensity of the two halves of the field of view changes rapidly; because the transmitted portion for one rapidly increases and for the other rapidly decreases.

Determination of specific rotation of sugar: In order to find the specific rotation of an optically active substance (say, sugar), the tube is first filled with water and the analyser is adjusted to obtain equally illuminated position of the field of view. The position of the analyser is read on the scale. Now the sugar solution of known concentration is filled in the tube T (without any air gap). The solution rotates the planes of vibration CP and CQ (i.e., of light emerging from glass half and the quartz half) through the same angle in the same direction. Sugar solution rotates the plane of vibration in clockwise direction. Now the analyser is rotated in the clockwise direction to obtain equally-illuminated position of the field of view again. This position of the analyser is again read on the scale. As the angle through which the analyser has been rotated gives the angle through which the plane of vibration of incident beam has been rotated by sugar solution, hence the difference between the two positions of the analyser gives the angle of rotation θ. In actual experiment the solutions of various concentrations are prepared and the corresponding angles of rotation are measured. Then a graph is plotted between concentration C and the angle of rotation θ. The graph is a straight line (Fig. 11.5).

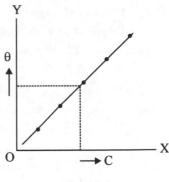

Fig. 11.5

From the graph, a set of θ and C is found and the specific rotation of sugar is then calculated from the following relation:

$$\alpha = \frac{10\theta}{lC}$$

where l is the length of the tube in cm, C is the concentration of the solution in gm/cc and θ is the rotation in degrees.

SOLVED EXAMPLES

Q.1. **For a given wavelength 1 mm of quartz cut perpendicular to optic axis rotator the plane of polarisation by 18°. Find for what thickness will no light of this wavelength be transmitted when the quartz piece is interposed between a pair of parallel nicols?**

Sol. The polariser and analyser are parallel in this case, so the light is completely transmitted through the analyser. If no light is allowed to be transmitted through the analyser, the plane of vibration (or plane of polarisation) of light passing through quartz must be rotated through 90°.

It is given that 1 mm of quartz produces a rotation of 18°.

As the rotation is directly proportional to the thickness of the quartz, therefore thickness of

the quartz required to produce a rotation of $90° = \frac{1}{18} \times 90 = 5$ mm.

Q.2. **A 20 gm of cane sugar is dissolved in water to make 50 cc of solution. A 20 cm length of this solution causes +53°30′ optical rotation. Calculate the specific rotation (α).**

Sol. The specific rotation is given by $\alpha = \dfrac{10\theta}{lC}$

Here $\qquad \theta = 53°\,30' = 53.5°, \quad C = \dfrac{20}{50} = 0.40$ gm/cc, $\quad l = 20$ cm

∴ $\qquad \alpha = \dfrac{10 \times 53°\,30'}{0.40 \times 20} = 66.9°$ deg dm^{-1} (gm cc)$^{-1}$

Q.3. A tube of sugar solution 20 cm long is placed between crossed nicols and illuminated with light of wavelength 6×10^{-5} cm. If the optical rotation produced is 13° and the specific rotation is 65°, determine the strength of the solution.

Sol. The specific rotation (α) is given by $\alpha = \dfrac{10\,\theta}{lC}$

Here $\qquad \alpha = 65°, \quad \theta = 13°, \quad l = 20 \, cm$

$\therefore \qquad 65° = \dfrac{10 \times 13°}{20\,C}$

$\therefore \qquad C = \dfrac{10 \times 13°}{20 \times 65°} = \dfrac{1}{10} = 0.1 \, gm/cm^2.$

Q.4. On introducing a polarimeter tube of 35 cm long containing a sugar solution of unknown strength it is found that the plane of polarisation is rotated through 10°. Find the strength of the solution in gm/cm². Given specific rotation of sugar solution 60° per decimetre per unit concentration.

Sol. The specific rotation (α) is given by

$$\alpha = \frac{10\,\theta}{lC} \quad \text{where } l \text{ is in cm}$$

$\therefore \qquad C = \dfrac{10\,\theta}{l\alpha}$

Here $\qquad \theta = 10°, \quad l = 25 \, cm$

$\qquad\qquad \alpha = 60 \, deg \, (decimetre)^{-1} \, (gm/cm^2)^{-1}$

$\therefore \qquad C = \dfrac{10 \times 10}{25 \times 60} = \dfrac{1}{15} = 0.0667 \, gm/cm^2.$

Q.5. The optical rotation produced by a particular material is found to be 30 per mm at $\lambda = 5000$ Å and 50° per mm at $\lambda = 4000$ Å. In the visible region of the spectrum the rotation of the plane of polarisation is given by $\theta = a + b\lambda^{-2}$. Find the values of the constants a and b for the given material. Calculate the possible thickness of the given material which when interposed between two crossed Nicols will produce maximum transmission for $\lambda = 5500$ Å.

Sol. For wavelength λ, the rotation of the plane of polarisation is given to be

$$\theta = a + \frac{b}{\lambda^2} \qquad\qquad\qquad ...(1)$$

where a and b are constants,

When $\lambda = 5000° = 5000 \times 10^{-3}$ cm $= 5 \times 10^{-5}$ cm, $\theta = 30°$ per mm $= 300°$ per cm.

$\therefore$ From equation (1), $300 = a + \dfrac{b}{\left(5 \times 10^{-5}\right)^2}$ ...(2)

When $\lambda = 4000$ Å $= 4 \times 10^{-5}$ cm, $\theta = 50°$ per mm $= 500°$ per cm.

$\therefore$ From eqn. (1), $500 = a + \dfrac{b}{\left(4 \times 10^{-5}\right)^2}$ ...(3)

Subtracting equation (2) from (3), we get

$$200 = \frac{b}{\left(4 \times 10^{-5}\right)^2} - \frac{b}{\left(5 \times 10^{-5}\right)^2}$$

or $$200 = \frac{b}{10^{-10}}\left[\frac{1}{16} - \frac{1}{25}\right] - \frac{b}{10^{-10}} \times \frac{9}{16 \times 25}$$

or $$b = \frac{200 \times 10^{-10} \times 16 \times 25}{9} = \frac{80}{9} \times 10^{-7} = 8.88 \times 10^{-7} \text{ deg/cm.}$$

Substituting the value of b in equation (2), we get

$$300 = a + \frac{(80/9) \times 10^{-7}}{\left(5 \times 10^{-5}\right)^2} = a + 355.55$$

$\therefore$ $a = 300 - 355.55 = 55.55$ deg. cm.

when $\lambda = 5500$ Å $= 5.5 \times 10^{-5}$ cm.

The optical rotation is given by $\theta = a + \dfrac{b}{\lambda^2} = -55.55 + \dfrac{\left(80 \times 10^{-7}/9\right)}{\left(5.5 \times 10^{-5}\right)^2}$

$$= -55.55 + \frac{80 \times 10^{-5}}{9 \times 5.5 \times 5.5 \times 10^{-10}} = -55.55 + 293.85 = 238.30 \text{ deg/cm.}$$

As the analyser and polariser are crossed in this case, therefore no light is transmitted through the analyser. In order to produce maximum transmission the plane of polarisation must be rotated through 90° or any odd multiple of it.

Here the rotation produced is 238.30 deg/cm. As the rotation is directly proportional to thickness, the thickness of the given material to produce maximum transmission (for wavelength 5500 Å) $= \dfrac{90}{238.30} = .3776$ cm $= 3.776$ mm.

QUESTIONS

1. What is meant by optical rotation? For a given material how does the rotation generally vary with wavelength?

2. Give a qualitative explanation of optical rotation in molecular substances.

3. Explain the optical rotation observed in concentrated sugar solution.

4. Describe the principle of half-shade device used in a polarimeter.

5. Define specific rotation. Describe the construction and working of Laurent's half-shade polarimeter.

6. How can you determine the specific rotation of sugar?

7. How can you differentiate between rotatory polarisation and rotatory dispersion?

8. Explain Biot's laws for rotatory polarisation.

9. What is the difference between polarimeter and saccharimeter?

10. Define specific rotation of an optically active substance and give its units. How is it related to molecular rotation? Describe in detail, an experiment to determine the specific rotation of an optically active substance like glucose or sugar solution.

Problems

1. A 15 cm tube containing cane sugar solution (specific rotation 66°) shows optical rotation 7°. Calculate the strength of the solution. **(Ans. 0.07 gm/cc)**

2. Calculate the specific rotation of the plane of polarisation turned through 26.4°, in traversing 20 cm length of 20% sugar solution. **(Ans. 66°)**

3. Calculate the specific rotation of sugar solution from the following data:
   Length of the tube containing solution = 22 cm
   Volume of solution = 88 cc
   Amount of sugar in solution = 6 gm
   Angle of rotation = 9°54'. **(Ans. 66° (decimetre) (gm/cc)$^{-1}$)**

4. A 20 cm long tube containing sugar solution rotates the plane of polarisation by 11°. If the specific rotation of sugar is 66°. Calculate the strength of the solution.
   (Ans. 0.0833 gm/cm^3)

5. A tube 20 cm long filled with an aqueous solution containing 15 gm of cane sugar per 100 cc of solution, is placed in the path of plane polarised light. Find the angle of rotation of the plane of polarisation, if the specific rotation of sugar is 66°.
   (Ans. 19.8°)

PART II

Differentiation of Vectors

1.1 GRADIENT (DEL) (∇)

Let us suppose that ϕ (x, y, z) is a scalar point function, i.e. a function whose value depends on the values of the coordinates (x, y, z). As a scalar it must have the same value at a given fixed point in space, independent of the radiation of our coordinate system or

$$\phi^i\left[x_1^1,\ x_2^1,\ x_3^1\right] = \phi(x_1, x_2, x_3) \qquad \qquad ...(1.1)$$

By differentiating w.r.t. x_i^1 we obtain

$$\frac{\partial x^i\left(x_1^1, x_2^1, x_3^1\right)}{\partial x_i^1} = \frac{\partial\phi(x_1, \dot{x}_2, x_3)}{\partial x_i^1}$$

$$\sum_j \frac{\partial x}{\partial x_j} \cdot \frac{\partial x_j}{\partial x_i^1} = \sum_j a_{ij} \cdot \frac{\partial x_j}{\partial xj} \qquad \qquad ...(1.2)$$

By comparing this with the vector transformation law, we get a vector with components

$\left(\dfrac{\partial\phi}{\partial x_j}\right)$. This vector we label the gradient of ϕ. It is given by

$$\nabla\phi = \hat{i}\frac{\partial\phi}{\partial x} + \hat{j}\frac{\partial\phi}{\partial y} + \hat{k}\frac{\partial\phi}{\partial z} \qquad \qquad ...(1.3)$$

or,
$$\nabla = \hat{i}\frac{\partial}{\partial x} + \hat{j}\frac{\partial}{\partial y} + \hat{k}\frac{\partial}{\partial z} \qquad \qquad ...(1.4)$$

$\nabla\phi$ (or, del ϕ) is our gradient of the scalar ϕ. ∇ itself is a vector differential operator. It operates on a scalar 'ϕ' or it is available to differentiate a scalar ϕ. This operator is a hybrid operator that must satisfy both the laws for handling vectors and the laws of partial differentiation.

Geometrical Interpretation of Gradient

One immediate application of $\nabla\phi$ is to dot it into an increment of length

$$d\vec{r} = \hat{i}\,dx + \hat{j}\,dy + \hat{k}\,dz \quad ...(1.5)$$

Thus, we obtain

$$\nabla\phi \cdot d\vec{r} = \frac{\partial\phi}{\partial x}dx + \frac{\partial\phi}{\partial y}dy + \frac{\partial\phi}{\partial z}dz = d\phi \quad ...(1.6)$$

This equation shows how a scalar function ϕ changes corresponding to a change in position $d\vec{r}$.

Now let us consider P and Q to be two points on a surface $\phi\,(x,\,y,\,z) = C$, a **constant.** These points are chosen so that ϕ is a distance $d\vec{r}$ from P. Then moving from P to Q the change in $\phi\,(x,\,y,\,z) = C$ is given by:

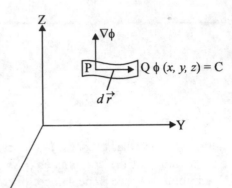

Fig. 1.1

$$d\phi = (\nabla\phi)d\vec{r} = 0 \qquad\qquad ...(1.7)$$

Since, we stay on the surface $\phi\,(x,\,y,\,z) = C$

This shows that $\nabla\phi$ is $\perp$ or $d\vec{r}$. Since $d\vec{r}$ may have any direction from P as long as it stays in the surface ϕ, point Q being restricted to the surface, but having arbitrary direction, $\nabla\phi$ is seen as normal to the surface ϕ = constant.

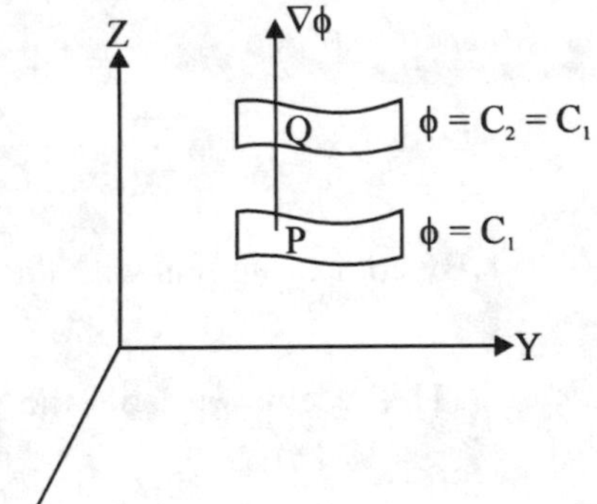

If we now permit $d\vec{r}$ to take us from one surface $\phi = C_1$ to an adjacent surface $\phi = C_2$.

$$d\phi = C_2 - C_1 = \Delta C = (\nabla\phi)\cdot d\vec{r} \quad ...(1.8)$$

For a given $d\phi$, $|d\vec{r}|$ is minimum when it is chosen parallel to $\nabla\phi$ (cos $\theta = 1$) or for a given $d\vec{r}$ the change in the scalar function ϕ is maximised by choosing $d\vec{r}$ parallel to $\nabla\phi$.

Fig. 1.2

This identifies $\nabla\phi$ as a vector having the direction of the maximum space rate of change of ϕ.

1.2 DIVERGENCE ($\nabla\cdot$)

We have already defined ∇ as a vector operator. Now paying careful attention to both its vector and its differential properties we let it operate on a vector. First as a vector we dot it in to a second vector.

Thus, $\quad \nabla\cdot\vec{V} = \left(\hat{i}\dfrac{\partial}{\partial x} + \hat{j}\dfrac{\partial}{\partial y} + \hat{k}\dfrac{\partial}{\partial z}\right)\cdot\left(\hat{i}\,Vx + \hat{j}\,Vy + \hat{k}\,Vz\right)$

$$= \left| \frac{\partial Vx}{\partial x} + \frac{\partial Vy}{\partial y} + \frac{\partial Vz}{\partial z} \right| \qquad \dots(1.9)$$

We read it as divergence of V. This is a scalar.

Geometrical Interpretation: The name divergence is well chosen, for $\nabla \cdot \vec{V}$ is a measure of how much the vector $\vec{V}$ spreads out (diverges) from the point in question. The vector function in Fig. 1.3 has a large (+ ve) divergence at the point P; it is spreading out. On the other hand, the function in Fig. 1.4 has zero divergence at P; it is not spreading out at all.

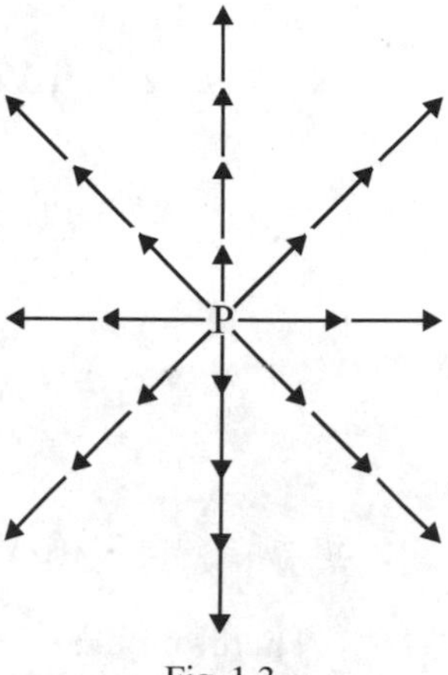

Physical Interpretation: Let us consider $\nabla \cdot (\rho \vec{V})$ with $\vec{V}$ (x, y, z), the velocity of a compressible fluid and ρ (x, y, z) its density at a point (x, y, z). If we consider a small volume *dx.dy.dz* (*see* Fig. 1.5), the fluid flowing into this volume per unit time (+ve x-direction) through the face EFGH is

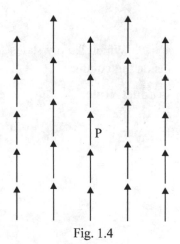

Fig. 1.3

$$\left(\text{Rate of flow out}\right)_{ABCD} = \rho V_x\big|_{x=dx} dy \cdot dz$$

The components of the flow ρVy and ρVz tangential to this face contribute nothing to the flow through this face. The rate of flow through the face EFGH is $\rho Vxdydz$.

The net flow out is given by

$$\left(\text{Rate of flow out}\right)_{ABCD} = \rho V_x\big|_{x=dx} dy \cdot dz$$

$$= \left[\rho V_x + \frac{\partial}{\partial x}(\rho Vx)dx \right]_{x=0} dy \cdot dz$$

Net rate of flow out $\big|_x = \dfrac{\partial}{\partial x}(\rho Vx)dx \cdot dy \cdot dz$

Fig. 1.4

Similarly we can have for y and z directions.

Net rate flow out $\big|_y = \dfrac{\partial}{\partial y}(\rho Vy)dx \cdot dy \cdot dz$

Net rate flow out $\big|_y = \dfrac{\partial}{\partial z}(\rho Vz)dx \cdot dy \cdot dz$

Adding all the 3 terms we get,

(Net flow out) i (unit volume) over the entire volume is

$$\frac{\partial}{\partial x}(\rho\, Vz)dx \cdot dy \cdot dz$$

$$+\frac{\partial}{\partial x}(\rho Vy)dx \cdot dy \cdot dz$$

$$+\frac{\partial}{\partial z}(\rho Vz)dx \cdot dy \cdot dz$$

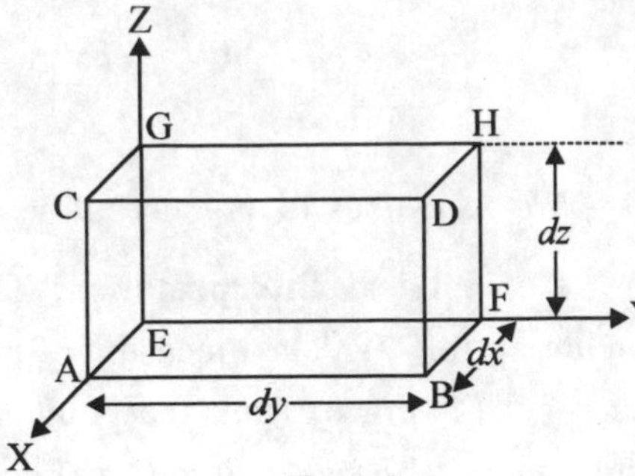

Fig. 1.5

$$=\left[\frac{\partial}{\partial x}(\rho Vx)+\frac{\partial}{\partial y}(\rho Vy)+\frac{\partial}{\partial z}(\rho Vz)\right]dx \cdot dy \cdot dz$$

$$=\nabla \cdot (\rho \vec{V})dx \cdot dy \cdot dz \qquad \qquad ...(1.10)$$

Therefore the net flow of our compressible fluid out of the volume element $dx.dy.dz$ per unit volume per unit time is $\nabla.\,(\rho \vec{V})$.

Hence, the name divergence. A direct application is the continuity equation.

$$\frac{\partial \rho}{\partial t}+\nabla.\left(\rho \vec{V}\right)=0 \qquad \qquad ...(1.11)$$

Which simply states that the net flow out of the volume results in a decreased density inside the volume.

Product Rule:

In general

$$\nabla \cdot (f \vec{V}) = \frac{\partial}{\partial x}(f\, Vx)+\frac{\partial}{\partial y}(f\, Vy)+\frac{\partial}{\partial z}(f\, Vz)$$

$$=\frac{\partial f}{\partial x}Vx + f\frac{\partial f}{\partial x}+Vy\frac{\partial f}{\partial y}+f\frac{\partial Vy}{\partial y}+Vz\frac{\partial f}{\partial z}+f\frac{\partial f}{\partial z}$$

$$=(\nabla f)\cdot \vec{V}+f\,\nabla \cdot \vec{V} \qquad \qquad ...(1.12)$$

If we have the special case of the divergence of the vector vanishing,

$$\nabla \cdot \vec{B} = 0, \qquad \qquad ...(1.13)$$

where the vector $\vec{B}$ is solenoidal. $\vec{B}$ is the magnetic induction.

When a vector is solenoidal it may be written as the curl of another vector known as the vector potential.

1.3 CURL (∇X)

When the vector operator ∇ is crossed into a vector then we say it as the curl of the vector.

$$\nabla \times \vec{V} = \begin{vmatrix} \hat{i} & \hat{j} & \hat{k} \\ \dfrac{\partial}{\partial x} & \dfrac{\partial}{\partial y} & \dfrac{\partial}{\partial z} \\ Vx & Vy & Vz \end{vmatrix} = \hat{i}\left(\frac{\partial Vz}{\partial y} - \frac{\partial Vy}{\partial z}\right) + \hat{j}\left(\frac{\partial Vx}{\partial z} - \frac{\partial Vz}{\partial x}\right) + \hat{k}\left(\frac{\partial Vy}{\partial x} - \frac{\partial Vx}{\partial y}\right) \quad \ldots(1.14)$$

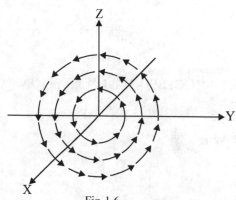

Fig. 1.6

The curl of a vector function is a vector. We can't have the curl of a scalar that is meaningless.

Geometrical Interpretation: The name curl is also well chosen for $\nabla x \cdot \vec{V}$ is a measure of how much the vector $\vec{V}$ curls around the point in question. The function in Fig. 1.6 has a large curl, in fact this curl points in the z-direction in accordance with the right hand rule.

Product Rule:

If ∇ is crossed into the product of a scalar and a vector, we can have

$$\nabla \times (f\,\vec{V})|_x = \left[\frac{\partial}{\partial y}(f\,Vz) - \frac{\partial}{\partial z}(f\,Vy)\right] = f\frac{\partial Vz}{\partial y} + \frac{\partial f}{\partial y}Vz - f\frac{\partial Vy}{\partial z} - \frac{\partial f}{\partial z}Vy$$

$$\Rightarrow \qquad \nabla \times (f\,\vec{V}) = f\,\nabla \times \vec{V}\,|_x - (\nabla f) \times \vec{V}\,|_x$$

In general, we have $\qquad \nabla \times (f\,\vec{V}) = f\,\nabla \times \vec{V} + (\nabla f) \times \vec{V} \qquad\qquad \ldots(1.15)$

1.4 SUMMARY OF PRODUCT RULES

(i) $\nabla(f\,g) = f\,\nabla g + g\nabla f$

(ii) $\nabla(\vec{A}\cdot\vec{B}) = \vec{A}\times(\nabla\times\vec{B}) + \vec{B}\times(\nabla\times\vec{A}) + (\vec{A}\cdot\nabla)\vec{B} + (\vec{B}\cdot\nabla)\vec{A}$

(iii) $\nabla\cdot(f\,\vec{A}) = f(\nabla\cdot\vec{A}) + \vec{A}\cdot(\nabla f)$

(iv) $\nabla\cdot(\vec{A}\times\vec{B}) = \vec{B}\cdot(\nabla\times\vec{A}) - \vec{A}\cdot(\nabla\times\vec{B})$

(v) $\nabla\times(f\,\vec{A}) = f(\nabla\times\vec{A}) - \vec{A}\times(\nabla f)$

(vi) $\nabla\times(\vec{A}\times\vec{B}) = (\vec{B}\cdot\nabla)\vec{A} - (\vec{A}\cdot\nabla)\vec{B} + \vec{A}(\nabla\cdot\vec{B}) - \vec{B}(\nabla\cdot\vec{A})$

(vii) $\nabla\left(\dfrac{f}{g}\right) = \dfrac{g\nabla f - f\nabla g}{g^2}$

(viii) $\nabla \times \left(\dfrac{\vec{A}}{g}\right) = \dfrac{g(\nabla\vec{A}) + \vec{A}(\nabla g)}{g^2}$

(ix) $\nabla \times \left(\dfrac{\vec{A}}{g}\right) = \dfrac{g(\nabla \times \vec{A}) + \vec{A} \times (\nabla g)}{g^2}$

1.5 SUCCESSIVE APPLICATIONS OF ∇

(i) $\nabla \cdot \nabla \phi$ \qquad (ii) $\nabla \times \nabla \phi$ \qquad (iii) $\nabla\nabla \cdot \vec{V}$ \qquad (iv) $\nabla \cdot \nabla \times \vec{V}$ \qquad (v) $\nabla \times \nabla \times \vec{V}$

Proof:

(i) $\nabla \cdot \nabla \phi = \left(\hat{i}\dfrac{\partial}{\partial x} + \hat{j}\dfrac{\partial}{\partial y} + \hat{k}\dfrac{\partial}{\partial z}\right) \cdot \left(\hat{i}\dfrac{\partial \phi}{\partial x} + \hat{j}\dfrac{\partial \phi}{\partial y} + \hat{k}\dfrac{\partial \phi}{\partial z}\right) = \dfrac{\partial^2 \phi}{\partial x^2} + \dfrac{\partial^2 \phi}{\partial y^2} + \dfrac{\partial^2 \phi}{\partial z^2}$ \qquad ...(1.16)

$\Rightarrow \qquad \nabla^2 \phi = \dfrac{\partial^2 \phi}{\partial x^2} + \dfrac{\partial^2 \phi}{\partial y^2} + \dfrac{\partial^2 \phi}{\partial z^2}$ \qquad ...(1.17)

∇^2 is often known as Laplacian operator. It has applications in electrostatics. When ϕ is the electrostatic potential

we have $\qquad \nabla \cdot \nabla \phi = 0$

$\Rightarrow \qquad \nabla^2 \phi = 0$ \qquad ...(1.18)

(ii) $\qquad \nabla \times \nabla \phi = \left(\hat{i}\dfrac{\partial}{\partial x} + \hat{j}\dfrac{\partial}{\partial y} + \hat{k}\dfrac{\partial}{\partial z}\right) \times \left(\hat{i}\dfrac{\partial \phi}{\partial x} + \hat{j}\dfrac{\partial \phi}{\partial y} + \hat{k}\dfrac{\partial \phi}{\partial z}\right)$

$$= \begin{vmatrix} \hat{i} & \hat{j} & \hat{k} \\[4pt] \dfrac{\partial}{\partial x} & \dfrac{\partial}{\partial y} & \dfrac{\partial}{\partial z} \\[8pt] \dfrac{\partial \phi}{\partial x} & \dfrac{\partial \phi}{\partial y} & \dfrac{\partial \phi}{\partial z} \end{vmatrix}$$

$$= \hat{i}\left(\dfrac{\partial^2 \phi}{\partial y \partial z} - \dfrac{\partial^2 \phi}{\partial z \partial y}\right) + \hat{j}\left(\dfrac{\partial^2 \phi}{\partial z \partial x} - \dfrac{\partial^2 \phi}{\partial x \partial z}\right) + \hat{k}\left(\dfrac{\partial^2 \phi}{\partial x \partial y} - \dfrac{\partial^2 \phi}{\partial y \partial z}\right)$$

$$= \nabla \times \nabla \phi = 0 \qquad ...(1.19)$$

Assuming that the order of partial differentiation may be interchanged. This is true as long as these second partial derivatives of 'ϕ' are continuous functions.

(iii) $\qquad \nabla \nabla \cdot \vec{V} = \left(\hat{i} \dfrac{\partial}{\partial x} + \hat{j} \dfrac{\partial}{\partial y} + \hat{k} \dfrac{\partial}{\partial z} \right) \cdot \left(\dfrac{\partial Vx}{\partial x} + \dfrac{\partial Vy}{\partial y} + \dfrac{\partial Vz}{\partial z} \right)$

$= \hat{i} \dfrac{\partial}{\partial x} \left(\dfrac{\partial Vx}{\partial x} + \dfrac{\partial Vy}{\partial y} + \dfrac{\partial Vz}{\partial z} \right) + \hat{j} \dfrac{\partial}{\partial y} \left(\dfrac{\partial Vx}{\partial x} + \dfrac{\partial Vy}{\partial y} + \dfrac{\partial Vz}{\partial z} \right) + \hat{k} \dfrac{\partial}{\partial z} \left(\dfrac{\partial Vx}{\partial x} + \dfrac{\partial Vy}{\partial y} + \dfrac{\partial Vz}{\partial z} \right)$

$$...(1.20)$$

(iv) $\nabla \cdot \nabla \times \vec{V} = \left(\hat{i} \dfrac{\partial}{\partial x} + \hat{j} \dfrac{\partial}{\partial y} + \hat{k} \dfrac{\partial}{\partial z} \right) \cdot \begin{vmatrix} \hat{i} & \hat{j} & \hat{k} \\ \dfrac{\partial}{\partial x} & \dfrac{\partial}{\partial y} & \dfrac{\partial}{\partial z} \\ Vx & Vy & Vz \end{vmatrix}$

$\left(\hat{i} \dfrac{\partial}{\partial x} + \hat{j} \dfrac{\partial}{\partial y} + \hat{k} \dfrac{\partial}{\partial z} \right) \cdot \left[\hat{i} \left(\dfrac{\partial Vz}{\partial y} - \dfrac{\partial Vy}{\partial z} \right) + \hat{j} \left(\dfrac{\partial Vx}{\partial z} - \dfrac{\partial Vz}{\partial x} \right) + \hat{k} \left(\dfrac{\partial Vy}{\partial x} - \dfrac{\partial Vx}{\partial y} \right) \right]$

$= \dfrac{\partial^2 Vz}{\partial x \partial y} - \dfrac{\partial^2 Vy}{\partial x \partial z} + \dfrac{\partial^2 Vx}{\partial y \partial z} - \dfrac{\partial^2 Vz}{\partial y \partial x} + \dfrac{\partial^2 Vy}{\partial z \partial x} - \dfrac{\partial^2 Vx}{\partial z \partial y} = 0$

$\Rightarrow \qquad \nabla \cdot \nabla \times \vec{V} = 0 \qquad\qquad ...(1.21)$

The divergence of a curl vanishes or, all curls are solenoidal.

(v) $\nabla \times \nabla \times \vec{V} = \nabla (\nabla \cdot \vec{V}) - \vec{V}(\nabla \cdot \nabla) = \nabla \nabla \cdot \vec{V} - \nabla \cdot \nabla \vec{V} = \nabla \nabla \cdot \vec{V} - \nabla^2 \vec{V}$

(By applying BAC-CAB rule)

$\nabla \times \nabla \times \vec{V} = \left(\hat{i} \dfrac{\partial}{\partial x} + \hat{j} \dfrac{\partial}{\partial y} + \hat{k} \dfrac{\partial}{\partial z} \right) \times \begin{vmatrix} \hat{i} & \hat{j} & \hat{k} \\ \dfrac{\partial}{\partial x} & \dfrac{\partial}{\partial y} & \dfrac{\partial}{\partial z} \\ Vx & Vy & Vz \end{vmatrix}$

$= \left(\hat{i} \dfrac{\partial}{\partial x} + \hat{j} \dfrac{\partial}{\partial y} + \hat{k} \dfrac{\partial}{\partial z} \right) \times \left[\hat{i} \left(\dfrac{\partial Vz}{\partial y} - \dfrac{\partial Vy}{\partial z} \right) + \hat{j} \left(\dfrac{\partial Vx}{\partial z} - \dfrac{\partial Vz}{\partial x} \right) + \hat{k} \left(\dfrac{\partial Vy}{\partial x} - \dfrac{\partial Vx}{\partial y} \right) \right]$

So the *x*-component of this quantity is

$\text{Curl}_x \ \text{Curl} \ \vec{V} = \hat{i} \left[\dfrac{\partial}{\partial y} \left(\dfrac{\partial Vy}{\partial x} - \dfrac{\partial Vx}{\partial y} \right) - \dfrac{\partial}{\partial z} \left(\dfrac{\partial Vx}{\partial z} - \dfrac{\partial Vz}{\partial x} \right) \right]$

Since $\quad \hat{j} \times \hat{k} = -\hat{k} \times \hat{j} = \hat{i}$

$\therefore \qquad$ Curl$_x$ Curl $\vec{V} = \hat{i}\left[\dfrac{\partial^2 Vy}{\partial x \partial y} + \dfrac{\partial^2 Vz}{\partial z \partial x}\right] - \left[\dfrac{\partial^2 Vx}{\partial y^2} + \dfrac{\partial^2 Vx}{\partial z^2}\right]\hat{i}$

$$= \hat{i}\left[\dfrac{\partial^2 Vx}{\partial x^2} + \dfrac{\partial^2 Vy}{\partial x \partial y} + \dfrac{\partial^2 Vz}{\partial x \partial z}\right] - \hat{i}\left[\dfrac{\partial^2 Vx}{\partial x^2} + \dfrac{\partial^2 Vx}{\partial y^2} + \dfrac{\partial^2 Vx}{\partial z^2}\right]$$

$$\left[\text{on adding and subtracting } \dfrac{\partial^2 Vx}{\partial x^2}\right]$$

The first term on right hand side of this equation is x-component of $\nabla \nabla . \vec{V}$ according to equation (1.20) and second term is x-component of $\nabla^2 \vec{V}$ according to equation (1.17).

So, $\qquad$ Curl$_x$ Curl $\vec{V} = \nabla x(\nabla \cdot \vec{V}) - \nabla^2 Vx$

$\qquad\qquad$ Curl$_y$ Curl $\vec{V} = \nabla y(\nabla \cdot \vec{V}) - \nabla^2 Vy$

$\qquad\qquad$ Curl$_z$ Curl $\vec{V} = \nabla z(\nabla \cdot \vec{V}) - \nabla^2 Vz$

Adding all these terms we get

$\qquad$ Curl Curl $\vec{V} = \nabla \times \nabla \times \vec{V} = \nabla \nabla \cdot \vec{V} - \nabla^2 V$ $\qquad$ (Proved)

(vi) Vector Integration: There are 3 types of integrals, namely

$\qquad$ (a) Line integrals, (b) Surface integrals, and (c) Volume integrals.

(a) Line integrals: Let us take

$$d\vec{r} = \hat{i}\,dx + \hat{j}\,dy + \hat{k}\,dz$$

where $d\vec{r}$ is an increment of length. Then we may encounter the line integrals

$$\int_c \phi\, d\vec{r} \qquad\qquad\qquad\qquad ...(1.22)$$

$$\int_c \vec{V} \cdot d\vec{r} \qquad\qquad\qquad\qquad ...(1.23)$$

$$\int_c \vec{V} \times d\vec{r} \qquad\qquad\qquad\qquad ...(1.24)$$

In each of which the integral is over some contour C that may be open or closed. Because of its physical interpretation that follows, the second form is by far the most important of the three. With ϕ, a scalar, the first integral reduces immediately to

$$\int_c \phi \, d\vec{r}$$

$$= \hat{i} \int_c \phi\,(x,y,z)dx + \hat{j} \int_c \phi\,(x,y,z)dy + \hat{k} \int_c \phi\,(x,y,z)dz \qquad \ldots(1.25)$$

This separation has employed the relation

$$\int_c \hat{i} \, \phi \, dx = \hat{i} \int_c \phi \, dx \qquad \ldots(1.26)$$

which is permissible because Cartesian unit vectors $\hat{i}, \hat{j}, \hat{k}$ are constant in both magnitude and direction. This relation is only applicable in case of Cartesian system of coordinates, but it will not be true in non-Cartesian system of coordinates.

The three integrals on the right hand side of equation (1.25) are ordinary scalar integrals and, to avoid complications, we assume that they are Riemann integrals. It should be noted that the integral *w.r.t.* x can't be evaluated unless y and z are known in terms of x and similarly for the integrals *w.r.t.* y and z. This simply means that the path of integration C must be specified unless the integrand has special properties that lead the integral to depend only on the value of the end points, the value will depend on the particular choice of contour 'C'. For instance, if we choose the very special case $\phi = 1$, equation (1.22) is just the vector distance from the start of contour 'C' to the end point, in this case, end point of the choice of path connecting fixed end points.

With $d\vec{r} = \hat{i}\,dx + \hat{j}\,dy + \hat{k}\,dz$, the 2nd and 3rd forms given in equations (1.23) and (1.24) also reduce to scalar integrals and like equation (1.22), are dependent, in general, on the choice of path. The equation (1.23) is exactly the same as that encountered when we calculate the work done by a force that varies along the path.

$$\omega = \int \vec{F} \cdot d\vec{r}$$

$$= \int Fx(x,y,z)dx + \int Fy(x,y,z)dy + \int Fz(x,y,z)dz \qquad \ldots(1.27)$$

In this expression $\vec{F}$ is the force exerted on a particle.

For example, if the force exerted on a body is $\vec{F} = -\hat{i}\,y + \hat{j}\,x$, then the work done in going from the origin to the point (1, 1) is given by

$$\omega = \int_{(0,0)}^{(1,1)} \vec{F}\,d\vec{r} = \int_{(0,0)}^{(1,1)} (-ydx + xdy) \qquad \ldots(1.28)$$

Separating the two integrals we obtain $\omega = -\int_0^1 y\,dx + \int_0^1 x\,dy$ ...(1.29)

The first integral can't be evaluated unless we specify the values of y as x ranges from 0 to 1. Likewise, the second integral requires x as a function of y.

Let us consider first the path shown in Fig. 1.7.

Then $\omega = -\int_0^1 0\,dx + \int_0^1 1\,dy = 1$ ...(1.30)

Since $y = 0$ along the first segment of the path and $x = 1$ along the second.

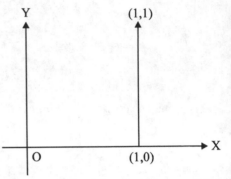

Fig. 1.7: A path of integration

If we select the path $[x = 0, 0 \le y \le 1]$ and $[0 \le x \le 1, y = 1]$, then equation (1.29) gives $\omega = 1$. For this force, the work done depends on the choice of the path.

Because of its physical interpretation explained about the integral given in equation (1.23) is by far the most important of the three vector fields which satisfy this integral are derived from the scalar field by taking the gradient of the latter and can be written as

$$\vec{V} = \nabla\phi$$

The curl of such a vector field is zero. The line integrals are frequently occur in physics. As shown above work is the line integral of $\vec{F}.d\vec{r}$, similarly if $\vec{V}$ is the electric field strength, then the line integral expresses the potential difference between the end points. If $\int \vec{V}.d\vec{r}$ is independent of the path, then the vector field $\vec{V}$ (x, y, z) is said to be conservative or non-curl or Lamellar vector. The integration of a vector along a curve is called line integral.

(b) Surface integrals: Analogous to the line integrals, surface integrals may appear in the forms:

$$\int_\sigma \phi \cdot d\vec{\sigma}$$ ...(1.31)

$$\int_\sigma \vec{V} \cdot d\vec{\sigma}$$ ...(1.32)

$$\int_\sigma \vec{V} \times d\vec{\sigma}$$ ...(1.33)

In the above integrals the element of area is also taken as a vector, i.e. $d\vec{\sigma}$ often this area element is written $\hat{n}\,dA$ in which $\hat{n}$ is a unit vector to indicate the +ve direction.

There are two conventions for choosing the +ve direction. First, if the surface is a close surface, we agree to take the outward normal as +ve. Second, if the surface is an open surface, the +ve normal depends on the direction in which the perimeter of the open surface is transversed. If the right hand fingers are placed in the direction of travel around the perimeter, the +ve normal is indicated by the thumb of the right hand.

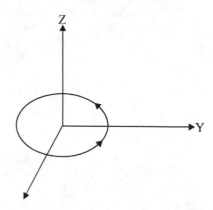

As an illustration a circle in the *xy*-plane (Fig. 1.8), mapped out from x to y to $-x$ to $-y$ and back to x will have its +ve normal parallel to the +ve *z*-axis.

Fig. 1.8: Right hand rule for the +ve normal

Physical Interpretation

The surface integral $\int \vec{V} \cdot d\vec{r}$ may be interpreted as a flow or flux over a given surface. This is identified with Gauss's theorem discussed later. The total flux through a surface σ is given by

$$\int_{\sigma} \vec{V} \cdot \hat{n}\,d\sigma = \int_{\sigma} V\cos\theta\,d\sigma = \int_{\sigma}\left(V_x d\sigma_x + V_y d\sigma_y + V_z d\sigma_z\right) \qquad \dots(1.34)$$

Another example of surface integral gives the amount of fluid passing normally through the surface element $d\vec{\sigma}$ in unit time, i.e.

$$\int_{\sigma} \vec{V} \cdot d\vec{\sigma} = \text{Amount of fluid passing normally through the surface element } d\vec{\sigma} \text{ in unit time.}$$

Here, $\vec{V}$ denotes the velocity of the moving fluid in which a fixed surface σ is drawn.

Definition: The integration of a vector over an area of a surface (closed or open) in known as the surface integral.

(c) Volume integrals: Volume integrals are somewhat simpler, for the volume element $d\tau$ is a scalar quantity. We have,

$$\int_{\sigma} \vec{V} \cdot d\tau = \hat{i}\int_{v} \vec{V}_x\,d\tau + \hat{j}\int_{v} \vec{V}_y\,d\tau + \hat{k}\int_{v} \vec{V}_z\,d\tau \qquad \dots(1.35)$$

Thus, we have reduced the vector integral to a vector sum of scalar integrals.

Definition: The integration of a vector over a volume $\vec{V}$ is called the volume integral.

(vii) Integral Definitions of Gradient, Divergence and Curl: The integrals of gradient, divergence and curl are given by

$$\nabla\phi = \lim_{\int d\tau \to 0} \frac{\int \phi\, d\vec{\sigma}}{\int d\tau} \qquad\qquad ...(1.36)$$

$$\nabla \cdot \vec{V} = \lim_{\int d\tau \to 0} \cdot \frac{\int \vec{V} \cdot d\vec{\sigma}}{\int d\tau} \qquad\qquad ...(1.37)$$

$$\nabla \times \vec{V} = \lim_{\int d\tau \to 0} \cdot \frac{\int d\vec{\sigma} \times \vec{V}}{\int d\tau} \qquad\qquad ...(1.38)$$

In these three equations $\int d\tau$ is the volume of a small region of space and $d\vec{\sigma}$ is the vector area element of this volume.

Proof: To verify equation (1.36) we will follow as under. Let us choose ' $\int d\tau$ ' to be the differential volume $dxdydz$, i.e.

This is shown in Fig. 1.9.

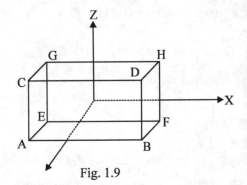

Fig. 1.9

This time we place the origin at the geometric centre of our volume element. The area integral leads to integrals, one for each of the six faces.

$$\int \phi\, d\vec{\sigma} = -\hat{i} \int_{EFHG}\left(\phi - \frac{\partial\phi}{\partial x}\frac{dx}{2}\right)dy\, dz + \hat{i} \int_{ABDC}\left(\phi + \frac{\partial\phi}{\partial x}\frac{dx}{2}\right)dy\, dz$$

$$-\hat{j} \int_{AEGC}\left(\phi - \frac{\partial\phi}{\partial y}\frac{dy}{2}\right)dy\, dz + \hat{j} \int_{BFHD}\left(\phi + \frac{\partial\phi}{\partial y}\frac{dy}{2}\right)dx\, dz$$

$$-\hat{k} \int_{ABFE}\left(\phi - \frac{\partial\phi}{\partial z}\frac{dz}{2}\right)dx\, dy + \hat{k} \int_{CDHG}\left(\phi + \frac{\partial\phi}{\partial z}\frac{dz}{2}\right)dx\, dy$$

$$= \left(\hat{i} \frac{\partial \phi}{\partial x} + \hat{j} \frac{\partial \phi}{\partial y} + \hat{k} \frac{\partial \phi}{\partial z} \right) dxdydz \qquad \qquad ...(1.39)$$

Dividing this by $\int d\tau = dxdydz$ we get

$$\lim_{\int d\tau = 0} \frac{\int \phi\, d\vec{\sigma}}{\int d\tau} = \hat{i} \frac{\partial \phi}{\partial x} + \hat{j} \frac{\partial \phi}{\partial y} + \hat{k} \frac{\partial \phi}{\partial z} = \nabla \phi \qquad \qquad ...(1.40)$$

Thus, we have verified the integral (1.36). Similarly the integrals (1.37) and (1.38) can be verified.

SOLVED EXAMPLES

Q.1. Find the gradient of a function of $\vec{r}$. [f ($\vec{r}$)].

Sol. Let us calculate the gradient of $f(r) = f\left(\sqrt{x^2 + y^2 + z^2}\right)$

$$\nabla f(r) = \hat{i} \frac{\partial f(r)}{\partial x} + \hat{j} \frac{\partial f(r)}{\partial y} + \hat{k} \frac{\partial f(r)}{\partial z}$$

Now $f(r)$ depends on x through the dependence of r on x. Therefore,

$$\frac{\partial f(r)}{\partial x} = \frac{\partial f(r)}{\partial r} \cdot \frac{\partial r}{\partial x}$$

From r as a function of (x, y, z)

$$\frac{\partial r}{\partial x} = \frac{\partial \left(x^2 + y^2 + z^2\right)^{\frac{1}{2}}}{\partial x} = \frac{1}{2}\left(x^2 + y^2 + z^2\right)^{-\frac{1}{2}} \cdot 2x = \frac{x}{\left(x^2 + y^2 + z^2\right)^{\frac{1}{2}}} = \frac{x}{r}$$

Therefore, $\qquad \dfrac{\partial f(r)}{\partial x} = \dfrac{\partial f(r)}{\partial r} \cdot \dfrac{\dot{x}}{r}$

Permuting coordinates $(x \to y, y \to z, z \to x)$ to obtain the y and z derivatives, we get

$$\nabla f(r) = \left(\hat{i}x + \hat{j}y + \hat{k}z\right)\frac{1}{r}\frac{df}{dr} \quad \left[\because \Delta f(r) = \left(\hat{i}\frac{\partial f(r)}{\partial r} \cdot \frac{x}{r} + \hat{j}\frac{\partial f(r)}{\partial r} \cdot \frac{y}{r} + \hat{k}\frac{\partial f(r)}{\partial r} \cdot \frac{z}{r}\right) \right]$$

$$= \frac{\vec{r}}{r} \cdot \frac{df}{dr} = \hat{r}_0 \frac{df}{dr} = \hat{r}_0 \frac{df}{d\vec{r}}$$

Here $\hat{r}_0$ is a unit vector $\left(\vec{r}/r\right)$ in the +ve radial direction. *The gradient of a function of $\vec{r}$ is a vector in the (+ve or negative) radial direction.*

Q.2. **What is the meaning of gradient $\vec{r}$ where $\vec{r}$ is position vector?**

Sol. Grad $r = \nabla\left(x^2 + y^2 + z^2\right)^{1/2}$

$$= \frac{x\hat{i}}{\sqrt{x^2 + y^2 + z^2}} + \frac{y\hat{j}}{\sqrt{x^2 + y^2 + z^2}} + \frac{z\hat{k}}{\sqrt{x^2 + y^2 + z^2}} = \frac{\vec{r}}{\sqrt{r^2}} = \vec{r}_0$$

where $\vec{r}_0$ is unit vector.

Q.3. **Calculate div $\vec{r}$, where $\vec{r}$ is the position vector.**

Sol. $\vec{r} = x\hat{i} + y\hat{j} + z\hat{k}$

$$\text{div } \vec{r} = \nabla \cdot \vec{r} = \frac{\partial x}{\partial x} + \frac{\partial y}{\partial y} + \frac{\partial z}{\partial z} = 3.$$

Q.4. **Show that $\nabla(uv) = u\,\nabla v + v\,\nabla u$, where u and v are differentiable scalar functions.**

Sol. $\nabla(uv) = \hat{i}\dfrac{\partial(uv)}{\partial x} + \hat{j}\dfrac{\partial(uv)}{\partial y} + \hat{k}\dfrac{\partial(uv)}{\partial z}$

$$= \left(\hat{i}\frac{\partial u}{\partial x} + \hat{j}\frac{\partial u}{\partial y} + \hat{k}\frac{\partial u}{\partial z}\right)v + \left(\hat{i}\frac{\partial v}{\partial x} + \hat{j}\frac{\partial v}{\partial y} + \hat{k}\frac{\partial v}{\partial z}\right)u$$

$$= v\nabla u + u\nabla v$$
$$= u\nabla v + v\nabla u \qquad \text{(Proved.)}$$

Q.5. **Show that $\nabla(u + v) = \nabla u + \nabla v$.**

Sol. $\nabla(u+v) = \hat{i}\dfrac{\partial(u+v)}{\partial x} + \hat{j}\dfrac{\partial(u+v)}{\partial y} + \hat{k}\dfrac{\partial(u+v)}{\partial z}$

$$= \left(\hat{i}\frac{\partial u}{\partial x} + \hat{j}\frac{\partial u}{\partial y} + \hat{k}\frac{\partial u}{\partial z}\right) + \left(\hat{i}\frac{\partial v}{\partial x} + \hat{j}\frac{\partial v}{\partial y} + \hat{k}\frac{\partial v}{\partial z}\right)\frac{df}{dr}$$

$$= \nabla u + \nabla v \qquad \text{(Proved.)}$$

Q.6. **Prove $\Delta r^n = nr^{n-2}\,\vec{r}$.**

Sol. $\Delta r^n = \left(\hat{i}\dfrac{\partial}{\partial x} + \hat{j}\dfrac{\partial}{\partial y} + \hat{k}\dfrac{\partial}{\partial z}\right)r^n$

$$= \hat{i}\frac{\partial r^n}{\partial x} + \hat{j}\frac{\partial r^n}{\partial y} + \hat{k}\frac{\partial r^n}{\partial z} = \hat{i}nr^{n-1}\frac{\partial r}{\partial x} + \hat{j}nr^{n-1}\frac{\partial r}{\partial y} + \hat{k}nr^{n-1}\frac{\partial r}{\partial z}$$

$$= nr^{n-1}\left(\hat{i}\frac{\partial r}{\partial x} + \hat{j}\frac{\partial r}{\partial y} + \hat{k}\frac{\partial r}{\partial z} \right)$$

Since $\qquad r^2 = x^2 + y^2 + z^2$

$$\Rightarrow \qquad 2r\frac{\partial r}{\partial x} = 2x \qquad\qquad \Rightarrow \qquad \frac{\partial r}{\partial x} = \frac{x}{r}$$

Similarly, $\qquad \dfrac{\partial r}{\partial y} = \dfrac{y}{r}, \quad \dfrac{\partial r}{\partial z} = \dfrac{z}{r}$

So, $\qquad \nabla r^n = nr^{n-1}\left(\hat{i}\dfrac{x}{r} + \hat{j}\dfrac{y}{r} + \hat{k}\dfrac{z}{r} \right)$

$$= \frac{nr^{n-1}}{r}(\hat{i}x + \hat{j}y + \hat{k}z) = nr^{n-2}\,\vec{r} \qquad\qquad \text{(Proved.)}$$

Q.7. **If** $S(x, y, z) = \left(x^2 + y^2 + z^2 \right)^{-3/2}$, **find**

   (i) ∇S **at the point (1, 2, 3)**

   (ii) The magnitude of the gradient of S, (∇S) at (1, 2, 3)

   (iii) The direction cosines of ∇s at (1, 2, 3).

Sol. (i) $\qquad S = \left(x^2 + y^2 + z^2 \right)^{-3/2}$

$$\nabla S = \hat{i}\frac{\partial S}{\partial x} + \hat{j}\frac{\partial S}{\partial y} + \hat{k}\frac{\partial S}{\partial z}$$

$$= \hat{i}\frac{\partial\left(x^2 + y^2 + z^2\right)^{-3/2}}{\partial x} + \hat{j}\frac{\partial\left(x^2 + y^2 + z^2\right)^{-3/2}}{\partial y} + \hat{k}\frac{\partial\left(x^2 + y^2 + z^2\right)^{-3/2}}{\partial z}$$

$$= \frac{-3}{2}\left(x^2 + y^2 + z^2\right)^{-5/2}\left[\hat{i}\,2x + \hat{j}2y + \hat{k}2z \right]$$

$$= -3\left(x^2 + y^2 + z^2\right)^{-5/2}\left[\hat{i}x + \hat{j}y + \hat{k}z \right]$$

$$= -3\left(1^2 + 2^2 + 3^2\right)^{-5/2}\left[\hat{i} + 2\hat{j} + 3\hat{k} \right]$$

$$= -3(14)^{-5/2}\left[\hat{i} + 2\hat{j} + 3\hat{k} \right]$$

(ii) Magnitude of ∇S is $|\nabla S| = \left| -3(14)^{-5/2}\sqrt{1+4+9} \right|$

$$= \left| -3(14)^{-5/2}(14)^{1/2} \right|$$

$$= \left| (-3)(14)^{-4/2} \right| = \left| (-3)(14)^{-2} \right| = \left| \frac{(-3)}{(14)^2} \right|$$

$$= \frac{-3}{14^2} = \frac{3}{196}$$

(iii) Direction cosines of ∇S are $\left\langle -3(14)^{-5/2},\ -6(14)^{-5/2},\ -9(15)^{-5/2} \right\rangle$

Q.8. Show that $\nabla(\vec{a}\cdot\vec{r})=\vec{a}$, **where** $\vec{a}$ **is a constant vector and** $\vec{r}$ **is the position vector.**

Sol.
$$\nabla(\vec{a}\cdot\vec{r}) = \left(\hat{i}\frac{\partial}{\partial x} + \hat{j}\frac{\partial}{\partial y} + \hat{k}\frac{\partial}{\partial z} \right)(a_x x + a_y y + a_z z)$$

$$= \hat{i}a_x + \hat{j}a_y + \hat{k}a_z = \vec{a}$$

Q.9. If $\vec{r}$ **is the position vector of a point. Find the value of** $\nabla\left(\dfrac{1}{\vec{r}}\right)$.

Sol.
$$\nabla = \left(\hat{i}\frac{\partial}{\partial x} + \hat{j}\frac{\partial}{\partial y} + \hat{k}\frac{\partial}{\partial z} \right)$$

$$\vec{r} = \left(\hat{i}x + \hat{j}y + \hat{k}z \right)$$

$$\frac{1}{\vec{r}} = \frac{1}{\left(x^2 + y^2 + z^2 \right)^{1/2}}$$

$$\nabla\left(\frac{1}{\vec{r}}\right) = \left(\hat{i}\frac{\partial}{\partial x} + \hat{j}\frac{\partial}{\partial y} + \hat{k}\frac{\partial}{\partial z} \right)\left[\frac{1}{\left(x^2 + y^2 + z^2 \right)^{1/2}} \right]$$

$$= \hat{i}\frac{\partial}{\partial x}\left[\frac{1}{\left(x^2 + y^2 + z^2 \right)^{1/2}} \right] + \hat{j}\frac{\partial}{\partial y}\left(x^2 + y^2 + z^2 \right)^{-1/2} + \hat{k}\left(x^2 + y^2 + z^2 \right)^{-1/2}$$

$$= \left| \hat{i} \frac{\partial}{\partial x} \left[\frac{1}{\left(x^2 + y^2 + z^2\right)^{1/2}} \right] + \hat{j} \frac{\partial}{\partial y} \left(x^2 + y^2 + z^2\right)^{1/2} + \hat{k} \left(x^2 + y^2 + z^2\right)^{-1/2} \right|$$

$$= -\left(x^2 + y^2 + z^2\right)^{-3/2} \left(\hat{i}x + \hat{j}y + \hat{k}z\right) = \frac{\vec{r}}{\left(x^2 + y^2 + z^2\right)^{3/2}} = \frac{-\vec{r}}{\left(r^2\right)^{3/2}} = \frac{-\vec{r}}{r^3}$$

Q.10. Find $\nabla\phi$ and $|\nabla\phi|$ for the function $\phi = 2xz^4 - x^2 y$ at the point (2, –2, –1).

Sol.
$$\nabla\phi = \left(\hat{i} \frac{\partial}{\partial x} + \hat{j} \frac{\partial}{\partial y} + \hat{k} \frac{\partial}{\partial z} \right)\left(2xz^4 - x^2 y\right)$$

$$= \hat{i} \frac{\partial}{\partial x}\left(2xz^4 - x^2 y\right) + \hat{j} \frac{\partial}{\partial y}\left(2xz^4 - x^2 y\right) + \hat{k} \frac{\partial}{\partial z}\left(2xz^4 - x^2 y\right)$$

$$= \hat{i}\left(2z^4 - 2xy\right) + \hat{j}\left(-x^2\right) + \hat{k}\left(8xz^3\right)$$

$$= \hat{i}\left(2 + 8\right) - 4\hat{j} - 16\hat{k}$$

$$= 10\hat{i} - 4\hat{j} - 16\hat{k}$$

$$\nabla\phi = \sqrt{100 + 16 + 256} = \sqrt{372} = 2\sqrt{93} \ .$$

Q.11. If $u = x^2 z + e^{(v/x)}$ and $v = 2z^2 y - xy^2$.

Find: (i) $\nabla(u + v)$, and (ii) $\nabla(uv)$ at the point (1, 0, –2).

Sol. (i) $\nabla(u + v) = \nabla u + \nabla v$

$$\nabla u = \hat{i} \frac{\partial}{\partial x}\left(x^2 z + e^{v/x}\right) + \hat{j} \frac{\partial}{\partial y}\left(x^2 z + e^{v/x}\right) + \hat{k} \frac{\partial}{\partial z}\left(x^2 z + e^{v/x}\right)$$

$$= \hat{i}\left(2xz - \frac{e^{v/x}}{x^2} y\right) + \hat{j}\left(\frac{e^{v/x}}{x}\right) + \hat{k}\left(x^2\right)$$

$$\nabla v = \hat{i} \frac{\partial}{\partial x}\left(2z^2 y - xy^2\right) + \hat{j} \frac{\partial}{\partial y}\left(2z^2 y - xy^2\right) + \hat{k} \frac{\partial}{\partial z}\left(2z^2 y - xy^2\right)$$

$$= \hat{i}\left(-y^2\right) + \hat{j}\left(2z^2 - 2xy\right) + \hat{k}\left(4yz\right)$$

$$\nabla(u + v) = \nabla u + \nabla v$$

$$= \left(2xz - \frac{e^{y/x}}{x^2}y - y^2\right)\hat{i} + \left(2z^2 - 2xy + \frac{e^{y/x}}{x}\right)\hat{j} + \left(4y^2 + x^2\right)\hat{k}$$

$$= (-4 - 0 - 0)\hat{i} + (8 + 1)\hat{j} + 1.\hat{k} = -4\hat{i} + 9\hat{j} + \hat{k}$$

$$\therefore \quad \nabla(u + v) \text{ at } (1, 0, -2) \text{ is } -4\hat{i} + 9\hat{j} + \hat{k}$$

(ii) $\nabla(uv) = u\nabla v + v\nabla u$

$$= u\left[-\hat{i}y^2 + \hat{j}(2z^2 - 2xy) + \hat{k}(4yz)\right] + v\left[\hat{i}\left(2xz - y\frac{e^{y/x}}{x^2}\right) + \hat{j}\left(\frac{e^{e/x}}{x}\right) + \hat{k}x^2\right]$$

$$= \left(x^2z + e^{y/x}\right)\left[-\hat{i}y^2 + \hat{j}(2z^2 - 2xy) + \hat{k}(4yz)\right] +$$

$$\left(2z^2y - xy^2\right)\left[-\hat{i}\left(2xz - y\frac{e^{y/x}}{x^2}\right) + \hat{j}\left(\frac{e^{y/x}}{x}\right) + \hat{k}\,x^2\right]$$

$$= \left(-2 + e^0\right)\left[-0\hat{i} + \hat{j}(8) + \hat{k}0\right] + (2.0 - 0)\left[\hat{i}\left(2xz - y\frac{e^{y/x}}{x^2}\right) + \hat{j}\left(\frac{e^{y/x}}{x}\right) + \hat{k}\,x^2\right]$$

$$= (-2 + 1)\left[0 + 8\hat{j} + 0\right] + 0$$

$$= (-1)\left(8\hat{j}\right)$$

$$= -8\hat{j}$$

Q.12. Prove $\nabla\left(\dfrac{u}{v}\right) = \dfrac{v\Delta u - u\Delta v}{v^2}$, **provided** $v \neq 0$.

Sol. $\phi = \dfrac{u}{v} = uv^{-1}$

$$\nabla\phi = \left(\hat{i}\frac{\partial}{\partial x} + \hat{j}\frac{\partial}{\partial y} + \hat{k}\frac{\partial}{\partial z}\right)\left(uv^{-1}\right)$$

$$= \hat{i}\frac{\partial}{\partial x}\left(uv^{-1}\right) + \hat{j}\frac{\partial}{\partial y}\left(uv^{-1}\right) + \hat{k}\frac{\partial}{\partial z}\left(uv^{-1}\right)$$

$$= \hat{i}\left[u\frac{\partial}{\partial x}v^{-1} + v^{-1}\frac{\partial}{\partial x}u\right] + \hat{j}\left[u\frac{\partial}{\partial y}v^{-1} + v^{-1}\frac{\partial}{\partial x}u\right] + \hat{k}\left[u\frac{\partial}{\partial z}v^{-1} + v^{-1}\frac{\partial}{\partial z}u\right]$$

$$= u\left[\hat{i}\frac{\partial v^{-1}}{\partial x} + \hat{j}\frac{\partial v^{-1}}{\partial y} + \hat{k}\frac{\partial v^{-1}}{\partial z}\right] + v^{-1}\left[\hat{i}\frac{\partial u}{\partial x} + \hat{j}\frac{\partial u}{\partial y} + \hat{k}\frac{\partial u}{\partial z}\right]$$

Multiplying and dividing by $-v^2$.

$$= u\left(-v^2\right)\left[\hat{i}\frac{\partial v}{\partial x} + \hat{j}\frac{\partial v}{\partial y} + \hat{k}\frac{\partial v}{\partial z}\right] + v^{-1}\nabla u$$

$$= \frac{-u}{v^2}\nabla v + v^{-1}\nabla u = \frac{\nabla u}{v}\frac{v\nabla v}{v^2} = \frac{v\nabla u - v\nabla u}{v^2}\sin^{-1}$$

$\Rightarrow \qquad \nabla\left(\dfrac{u}{v}\right) = \dfrac{v\nabla u - u\Delta v}{v^2}$ \qquad (Proved.)

Q.13. **If** $\vec{A} = 2x^2\,\hat{i} - 3y^2\hat{i} + xz^2\hat{k}$

and $\phi = 2z - x^3 y$

Find: **(i)** $\vec{A}\cdot\nabla\phi$

(ii) $\vec{A}\times\nabla\phi$ **at the point (1, –1, 1).**

Sol. $\qquad \nabla\phi = \left[\hat{i}\dfrac{\partial}{\partial x}\left(2z - x^3 y\right) + \hat{j}\dfrac{\partial}{\partial y}\left(2z - x^3 y\right) + \hat{k}\dfrac{\partial}{\partial z}\left(2z - x^3 y\right)\right]$

$$= \hat{i}\left(-3x^2 y\right) + \hat{j}\left(-x^3\right) + \hat{k}(2)$$

$$= -3x^2 y\hat{i} + -x^3\hat{j} + 2\hat{k}$$

at (1, –1, 1) $\nabla\phi = 3\hat{i} - \hat{j} + 2\hat{k}$

(i) $\qquad \vec{A}\cdot\nabla\phi = 6x^2 - 3yz + 2xz^2$

$$= 6 - 3 + 2 = 5$$

(ii) $\qquad \vec{A}\times\nabla\phi = \begin{vmatrix} \hat{i} & \hat{j} & \hat{k} \\ 2x^2 & -3yz & xz^2 \\ 3 & -1 & 2 \end{vmatrix}$

$$= \hat{i}\left(-6yz + xz^2\right) + \hat{j}\left(3xz^2 - 4x^2\right) + \hat{k}\left(-2x^2 + 9y^2\right)$$

at (1, –1, 1) $\vec{A}\cdot\nabla\phi = \hat{i}(6+1) + \hat{j}(3.1 - 4.1) + \hat{k}(-2.1 - 9.1)$

$$= 7\hat{i} - \hat{j} - 11\hat{k}$$

$$\vec{A} \times \nabla\phi = 7\hat{i} - \hat{j} - 11\hat{k}.$$

Q.14. If *u* is differentiable function of *x, y, z*, prove $\nabla u . d\vec{r} = du$.

Sol. $u = u \ (x, \ y, \ z)$

$$du = \frac{\partial u}{\partial x}dx + \frac{\partial u}{\partial y}dy + \frac{\partial u}{\partial z}dz$$

$$\vec{r} = \hat{i}x + \hat{j}y + \hat{k}z$$

$$d\vec{r} = dx\hat{i} + dy\hat{j} + dz\hat{k}$$

$$\nabla u = \hat{i}\frac{\partial u}{\partial x} + \hat{j}\frac{\partial u}{\partial y} + \hat{k}\frac{\partial u}{\partial z}$$

$$\nabla u \cdot d\vec{r} = \left(\hat{i}\frac{\partial u}{\partial x} + \hat{j}\frac{\partial u}{\partial y} + \frac{\partial u}{\partial z}\hat{k}\right) \cdot \left(dx\hat{i} + dy\hat{j} + dz\hat{k}\right)$$

$$= \frac{\partial u}{\partial x}dx + \frac{\partial u}{\partial y}dy + \frac{\partial u}{\partial z}dz$$

$$= du.$$

Q.15. If a vector function $\vec{F}$ depends on both space coordinates (*x, y, z*) and time *t*.

Show that $d\vec{F} = (d\vec{r}\cdot\nabla)\vec{F} + \frac{\partial F}{\partial t}dt$.

Sol. $\vec{F} = \vec{F} \ (x, \ y, \ z, \ t)$

$$d\vec{F} = d\vec{F} = \frac{\partial\vec{F}}{\partial x}dx + \frac{\partial\vec{F}}{\partial y}dy + \frac{\partial\vec{F}}{\partial z}dz + \frac{\partial\vec{F}}{\partial t}dt \qquad \text{...(1)}$$

$$d\vec{r} = dx\hat{i} - dy\hat{j} - dz\hat{k}$$

$$\nabla = \left(\hat{i}\frac{\partial}{\partial x} + \hat{j}\frac{\partial}{\partial y} + \hat{k}\frac{\partial}{\partial z}\right)$$

$$d\vec{r} \cdot \nabla = \frac{\partial}{\partial x}dx + \frac{\partial}{\partial y}dy + \frac{\partial}{\partial z}dz$$

$$(d\vec{r}\cdot\nabla)\vec{F} = \frac{\partial\vec{F}}{\partial x}dx + \frac{\partial\vec{F}}{\partial x}dy + \frac{\partial\vec{F}}{\partial x}dz \qquad \text{...(2)}$$

From equations (1) and (2) we get

$$d\vec{F} = (d\vec{r}\cdot\nabla)\vec{F}+\frac{\partial\vec{F}}{\partial t}dt \qquad \text{(Proved.)}$$

Q.16. Show that $\nabla\cdot(\vec{u}+\vec{v})=\Delta\vec{u}+\Delta\vec{v}$.

Sol. If $\vec{u}$ and $\vec{v}$ be two vector point functions expressed as

$$\vec{u} = u_x\hat{i}+u_y\hat{j}+u_z\hat{k}$$

$$\vec{v} = v_x\hat{i}+v_y\hat{j}+v_z\hat{k}$$

then $\nabla\cdot(\vec{u}+\vec{v}) = \left(\hat{i}\frac{\partial}{\partial x}+\hat{j}\frac{\partial}{\partial y}+\hat{k}\frac{\partial}{\partial x}\right)\cdot\left[(u_x+v_x)\hat{i}+(u_y+v_y)\hat{j}+(u_z+v_z)\hat{k}\right]$

$$= \frac{\partial}{\partial x}(u_x+v_x)+\frac{\partial}{\partial y}(u_y+v_y)+\frac{\partial}{\partial x}(u_z+v_z)$$

$$= \left(\frac{\partial u_x}{\partial x}+\frac{\partial u_y}{\partial y}+\frac{\partial u_z}{\partial z}\right)\cdot\left(u_x\hat{i}+u_y\hat{j}+u_z\hat{k}\right)+\left(\hat{i}\frac{\partial}{\partial x}+j\frac{\partial}{\partial y}+\hat{k}\frac{\partial}{\partial z}\right)\cdot\left(v_x\hat{i}+v_y\hat{j}+v_z\hat{k}\right)$$

$$= \nabla\vec{u}+\nabla\vec{v} \qquad \text{(Proved.)}$$

Q.17. Show that $\nabla\cdot(\vec{u}\,v)=(\nabla v)\cdot\vec{u}+v(\nabla\cdot\vec{u})$.

Sol. If the vector point function $\vec{u}$ is expressed as

$$= u_x\hat{i}+u_y\hat{j}+u_z\hat{k}$$

and v is a scalar point function, then

$$\nabla\cdot(\vec{u}\,v) = \left(\hat{i}\frac{\partial}{\partial x}+\hat{j}\frac{\partial}{\partial y}+\hat{k}\frac{\partial}{\partial z}\right)\cdot\left[(u_x\hat{i}+u_y\hat{j}+u_z\hat{k})v\right]$$

$$= \left(\hat{i}\frac{\partial}{\partial x}+\hat{j}\frac{\partial}{\partial y}+\hat{k}\frac{\partial}{\partial z}\right)\cdot\left[\hat{i}u_xv+\hat{j}u_yv+\hat{k}u_zv\right]$$

$$= \frac{\partial}{\partial x}u_xv+\frac{\partial}{\partial y}u_yv+\frac{\partial}{\partial z}u_zv$$

$$= u_x\frac{\partial v}{\partial x}+v\frac{\partial u_z}{\partial x}+u_y\frac{\partial v}{\partial y}+v\frac{\partial v_y}{\partial y}+u_z\frac{\partial v_z}{\partial z}+v\frac{\partial u_z}{\partial z}$$

$$= u_x \frac{\partial v}{\partial x} + v \frac{\partial u_x}{\partial x} + u_y \frac{\partial v}{\partial y} + v \frac{\partial v_y}{\partial y} + u_z \frac{\partial v_z}{\partial z} + v \frac{\partial u_z}{\partial z}$$

$$= \left(\hat{i} \frac{\partial v}{\partial x} + \hat{j} \frac{\partial v}{\partial y} \hat{k} + \frac{\partial v}{\partial z} \right) \cdot \left(u_x \hat{i} + u_y \hat{j} + u_z \hat{k} \right)$$

$$+ v \left(\hat{i} \frac{\partial}{\partial x} + \hat{j} \frac{\partial}{\partial y} \hat{k} + \frac{\partial}{\partial z} \right) \cdot \left(u_x \hat{i} + u_y \hat{j} + u_z \hat{k} \right)$$

$$= (\nabla v) \cdot \vec{u} + v (\nabla \cdot \vec{u}) \qquad \text{(Proved.)}$$

Q.18. Find the value of $\nabla \cdot \vec{r} f(r)$.

Sol. $\nabla \cdot \vec{r} f(r) = \frac{\partial}{\partial x} \left[x f(r) \right] + \frac{\partial}{\partial y} \left[y f(r) \right] + \frac{\partial}{\partial z} \left[z f(r) \right]$

$$= 3 f(r) + \frac{x^2}{r} \frac{df}{dr} + \frac{y^2}{r} \frac{df}{dr} + \frac{z^2}{r} \frac{df}{dr}$$

$$= 3 f(r) + r \frac{df}{dr}$$

Q.19. If $f(r) = r^{n-1}$, then find $\nabla \cdot (\vec{r} \, r^{n-1})$.

Sol. $\nabla \cdot \vec{r} \, r^{n-1} = 3 r^{n-1} + r \frac{dr^{n-1}}{dr}$

$$= 3 r^{n-1} = r(n-1) r^{n-2}$$

$$= 3 r^{n-1} + (n-1) r^{n-1}$$

$$= r^{n-1} (3 + n - 1)$$

$$= (2 + n) r^{n-1}$$

The divergence vanishes for $n = -2$.

Q.20. If $\phi = 2 x^3 y^2 z^4$, then find divergence of grad $\phi = \left[\nabla \cdot (\nabla \phi) \right]$.

Sol. $\nabla \cdot \nabla \phi = \left(\hat{i} \frac{\partial}{\partial x} \hat{j} + \frac{\partial}{\partial y} \hat{k} + \frac{\partial}{\partial z} \right) \cdot \left(\hat{i} \frac{\partial \phi}{\partial x} \hat{j} + \frac{\partial \phi}{\partial y} \hat{k} + \frac{\partial \phi}{\partial z} \right)$

$$= \left(\hat{i} \frac{\partial}{\partial x} \hat{j} + \frac{\partial}{\partial y} \hat{k} + \frac{\partial}{\partial z} \right) \cdot \left(\hat{i} 2 \frac{\partial}{\partial x} x^3 z^4 y^2 + \hat{j} 2 \frac{\partial}{\partial y} x^3 y^2 z^4 + \hat{k} 2 \frac{\partial \phi}{\partial y} x^3 y^2 z^4 \right)$$

$$= \left(\hat{i} \frac{\partial}{\partial x} \hat{j} + \frac{\partial}{\partial y} \hat{k} + \frac{\partial}{\partial z} \right) \cdot \left(\hat{i} 6x^2 y^2 z^4 + \hat{j} 4x^3 yz^4 + \hat{k} 8x^3 y^2 z^3 \right)$$

$$= \frac{\partial}{\partial x} 6x^2 y^2 z^4 + \frac{\partial}{\partial y} 4x^3 yz^4 + \frac{\partial}{\partial z} 8x^3 y^2 z^3$$

$$= 12xy^2 z^4 + 4x^3 z^4 + 24x^3 y^2 z^2$$

Q.21. Show that $\nabla \cdot \nabla \phi = \nabla^2 \phi$.

Sol. $\nabla \cdot \nabla \phi = \left(\hat{i} \frac{\partial}{\partial x} + \hat{j} \frac{\partial}{\partial y} + \hat{k} \frac{\partial}{\partial z} \right) \cdot \left(\hat{i} \frac{\partial \phi}{\partial x} + \hat{j} \frac{\partial \phi}{\partial y} + \hat{k} \frac{\partial \phi}{\partial z} \right)$

$$= \frac{\partial^2 \phi}{\partial x^2} + \frac{\partial^2 \phi}{\partial y^2} + \frac{\partial^2 \phi}{\partial z^2}$$

$$= \nabla^2 \phi \qquad \text{(Proved.)}$$

Q.22. If $\vec{V} = 3y^4 z^2 \hat{i} + 4x^3 z^2 \hat{j} - 3x^2 y^2 \hat{k}$, **then prove that** $\vec{V}$ **is a solenoidal vector.**

Sol. A vector is solenoidal if the divergence of the vector is zero.

$$\nabla \cdot \vec{V} = \frac{\partial}{\partial x} \left(3y^4 z^2 \right) + \frac{\partial}{\partial y} \left(4y^3 z^2 \right) + \frac{\partial}{\partial z} \left(3x^4 y^2 \right) = 0 + 0 - 0 = 0$$

So, vector $\vec{V}$ is solenoidal.

Q.23. Prove that $\nabla^2 \left(\dfrac{1}{r} \right) = 0$, **where** $r^2 = x^2 + y^2 + z^2$.

Sol. $\nabla^2 \left(\dfrac{1}{r} \right) = \dfrac{\partial^2}{\partial x^2} \left(\dfrac{1}{r} \right) + \dfrac{\partial^2}{\partial y^2} \left(\dfrac{1}{r} \right) + \dfrac{\partial^2}{\partial z^2} \left(\dfrac{1}{r} \right)$

$$= \frac{\partial^2}{\partial x^2} \left[\frac{1}{(x^2 + y^2 + z^2)^{1/2}} \right] + \frac{\partial^2}{\partial y^2} \left[\frac{1}{(x^2 + y^2 + z^2)^{1/2}} \right] + \frac{\partial^2}{\partial z^2} \left[\frac{1}{(x^2 + y^2 + z^2)^{1/2}} \right]$$

$$= \frac{\partial^2}{\partial x^2} \left(x^2 + y^2 + z^2 \right)^{-1/2} + \frac{\partial^2}{\partial y^2} \left(x^2 + y^2 + z^2 \right)^{-1/2} + \frac{\partial^2}{\partial z^2} \left(x^2 + y^2 + z^2 \right)^{-1/2}$$

$$= \frac{\partial}{\partial x} \left[\frac{\partial}{\partial x} \left(x^2 + y^2 + z^2 \right)^{-1/2} \right] + \frac{\partial}{\partial y} \left[\frac{\partial}{\partial y} \left(x^2 + y^2 + z^2 \right)^{-1/2} \right] + \frac{\partial}{\partial z} \left[\frac{\partial}{\partial z} \left(x^2 + y^2 + z^2 \right)^{-1/2} \right]$$

$$= \frac{\partial}{\partial x} \left[\left(-\frac{1}{2} \right) \left(x^2 + y^2 + z^2 \right)^{-3/2} \cdot 2x \right] + \frac{\partial}{\partial y} \left[\left(-\frac{1}{2} \right) \left(x^2 + y^2 + z^2 \right)^{-3/2} \cdot 2y \right]$$

$$= -\frac{\partial}{\partial x}\left[x\left(x^2 + y^2 + z^2\right)^{-3/2}\right] - \frac{\partial}{\partial y}\left[y\left(x^2 + y^2 + z^2\right)^{-3/2}\right] - \frac{\partial}{\partial z}\left[x\left(x^2 + y^2 + z^2\right)^{-3/2}\right]$$

Now, $\dfrac{\partial}{\partial x}\left[x\left(x^2 + y^2 + z^2\right)^{-3/2}\right]$

$$= x\frac{\partial}{\partial x}\left(x^2 + y^2 + z^2\right)^{-3/2} + \left(x^2 + y^2 + z^2\right)^{-3/2}\frac{\partial x}{\partial x}$$

$$= x \cdot \left(\frac{-3}{2}\right)\left(x^2 + y^2 + z^2\right)^{-5/2} \cdot 2x + \left(x^2 + y^2 + z^2\right)^{-3/2}$$

$$= -3x^2\left(x^2 + y^2 + z^2\right)^{-5/2} + \left(x^2 + y^2 + z^2\right)^{-3/2}$$

Similarly, $\dfrac{\partial}{\partial y}\left[y\left(x^2 + y^2 + z^2\right)^{-3/2}\right]$

$$= 3y^2\left(x^2 + y^2 + z^2\right)^{-5/2} + \left(x^2 + y^2 + z^2\right)^{-3/2}$$

and, $\quad = \dfrac{\partial}{\partial x}\left[z\left(x^2 + y^2 + z^2\right)^{-3/2}\right]$

$$= -3z^2\left(x^2 + y^2 + z^2\right)^{-5/2} + \left(x^2 + y^2 + z^2\right)^{-3/2}$$

Now, $\nabla^2\left(\dfrac{1}{r}\right) = \dfrac{\partial}{\partial x}\left[x\left(x^2 - y^2 + z^2\right)^{-3/2}\right] - \dfrac{\partial}{\partial y}\left[y\left(x^2 - y^2 + z^2\right)^{-3/2}\right]$

$$- \frac{\partial}{\partial z}\left[z\left(x^2 - y^2 + z^2\right)^{-3/2}\right]$$

$$= -\left[\begin{array}{l} -3x^2\left(x^2 + y^2 + z^2\right)^{-5/2} + \left(x^2 + y^2 + z^2\right)^{-3/2} - 3y^2\left(x^2 + y^2 + z^2\right)^{-5/2} \\ + \left(x^2 + y^2 + z^2\right)^{-3/2} - 3z^2\left(x^2 + y^2 + z^2\right)^{-5/2} + \left(x^2 + y^2 + z^2\right)^{-3/2} \end{array}\right]$$

$$= 3\left(x^2 + y^2 + z^2\right)^{-5/2} \cdot \left(x^2 + y^2 + z^2\right) - 3\left(x^2 + y^2 + z^2\right)^{-3/2}$$

$$= 3\left(x^2 + y^2 + z^2\right)^{\frac{-5}{2}+1} - 3\left(x^2 + y^2 + z^2\right)^{\frac{-3}{2}}$$

$$= 3\left(x^2 + y^2 + z^2\right)^{\frac{-3}{2}} - 3\left(x^2 + y^2 + z^2\right)^{\frac{-3}{2}} = 0 \quad \text{(Proved.)}$$

Q.24. Prove that $\nabla^2(uv) = u\nabla^2 v + 2\nabla u \cdot \nabla v + v\nabla^2 u.$

Sol. $\quad \nabla^2(uv) = \nabla(\nabla uv)$

Now, $\quad \nabla uv = \left(\hat{i}\dfrac{\partial uv}{\partial x} + \hat{j}\dfrac{\partial uv}{\partial y} + \hat{k}\dfrac{\partial uv}{\partial z} \right)$

$$= \hat{i}\left(u\dfrac{\partial v}{\partial x} + v\dfrac{\partial u}{\partial x} \right) + \hat{j}\left(u\dfrac{\partial v}{\partial y} + v\dfrac{\partial u}{\partial y} \right) + \hat{k}\left(u\dfrac{\partial v}{\partial z} + v\dfrac{\partial u}{\partial z} \right)$$

$$= u\left(\hat{i}\dfrac{\partial u}{\partial x} + \hat{j}\dfrac{\partial v}{\partial y} + \hat{k}\dfrac{\partial v}{\partial z} \right) + v\left(\hat{i}\dfrac{\partial u}{\partial x} + \hat{j}\dfrac{\partial u}{\partial y} + \hat{k}\dfrac{\partial u}{\partial z} \right)$$

$$= u\nabla v + v\nabla u$$

$$\nabla(\nabla uv) = \nabla(u\nabla v + v\nabla u)$$

$$= \nabla(u\nabla v) + \nabla(v\nabla u) \text{ (Here, Product rule is applied)}$$

$$= u\nabla^2 v + \nabla v\, \nabla u + \nabla v\, \nabla u + v\nabla^2 u$$

$$= u\nabla^2 v + 2\nabla u\, \nabla v + \nabla\, \nabla^2 u \qquad \text{(Proved.)}$$

Q.25. Prove $\nabla \cdot (u\nabla v - v\nabla u) = u\nabla^2 v - v\nabla^2 u.$

Sol. $\quad \nabla \cdot (u\nabla v - v\nabla u) = \nabla \cdot u\nabla v - \nabla \cdot u\nabla v$

Now, $\quad \nabla \cdot u\nabla v = \nabla u \cdot \nabla v + u\nabla^2 v$

$\qquad \nabla \cdot v\nabla u = \nabla v \cdot \nabla u + v\nabla^2 u$

$\therefore \qquad \nabla(u\nabla v - v\nabla u) = \nabla u \cdot \nabla v + u\nabla^2 v - \nabla v \cdot \nabla u - v\nabla^2 u$

$$[\because \nabla u \cdot \nabla v = \nabla v \cdot \nabla u, \ u \text{ and } v \text{ are both scalar}]$$

$\Rightarrow \qquad \nabla \cdot (u\nabla v - v\nabla u) = u\nabla^2 v - v\nabla^2 u. \qquad \text{(Proved.)}$

QUESTIONS

1. Define gradient divergence and curl of a physical quantity. Discuss on the physical significance of each.

2. Expand $\nabla \times \nabla \times \vec{V}$.

3. Discuss on the line integrals, surface integrals and volume integrals.

4. Write down the integral equation of gradient, divergence and curl.

5. Show that the $\nabla \cdot \vec{E}$ and $\nabla \times \vec{E}$ vanish for $\vec{E} = \dfrac{1}{4\pi \, \epsilon_0 \, r^3} q \vec{r}$. What is the nature of electrostatic field?

6. What is the meaning of a scalar function? How will you express gradient in Cartesian coordinates?

7. What is the curl of a vector field? Express it in Cartesian coordinates.

8. Evaluate $\nabla^2 r$ where $r = \sqrt{x^2 + y^2 + z^2}$.

Problems

1. Evaluate curl $\vec{A}$, where $\vec{A} = xy\hat{i} + yz\hat{j} + xz\hat{k}$.

2. Evaluate $\nabla \vec{A}$, where $\vec{F} = 2xy\hat{i} + x^2 y^2 \hat{j} + xyz\hat{k}$ and $\hat{i}, \hat{j}, \hat{k}$ are unit vectors along x, y, and z directions respectively.

3. A vector field is given by $\vec{F} = 2xy\hat{i} + 5y\hat{j}$. Evaluate the divergence of the vector.

4. Evaluate the divergence of the vector field $\vec{F} = 2xy\hat{i} + y^2 \hat{j} + 3xz\hat{k}$ at (1, 1, 0).

5. Evaluate the curl of the vector field $\vec{A} = 2x^2 y\hat{i} + y^3 \hat{j} + 3xyz\hat{k}$.

6. Evaluate the divergence of the vector $\vec{A} = xy\hat{i} + y^2 \hat{j} + 2xz\hat{k}$ at the point (2, 1, 0).

7. Evaluate the surface integral for the vector function $\vec{F} = 4xz\hat{x} - y^2 \hat{y} + yz\hat{z}$ over the surface S, where S is the surface of the unit cube bounded by $x = 0$, $x = 1$; $y = 0$, $y = 1$; $z = 0$, $z = 1$ planes.

8. Evaluate $\vec{\nabla} \times \vec{r}$ where $\vec{r}$ is the position vector.

9. Evaluate $\nabla \cdot \vec{r}$ where $\vec{r}$ is the position vector.

10. Evaluate ∇q, where $q = ax^2 - 2by + c^2 z^2$ where a, b and c are constants at (1, –2, 3).

●●●

Chapter 2

Electrostatics

2.1 COULOMB'S LAW OF ELECTROSTATICS

The quantity of electricity, or charge Q, possessed by a body is simply the aggregate of the amount by which the negative charge exceeds or is less than the positive charges in the body. The term point charge is used to indicate that the charge is not distributed over a large area but rather is concentrated at a specifically located point.

Charles-Augustin de Coulomb was the first investigator to place the law of force between electrostatic charges upon an experimental basis. His relatively rough experiments established in 1784 the law, now known as Coulomb's law of electrostatics, states that the force F between two point charges Q and Q' varies directly with each charge, inversely with the square of the distance S between the charges and is a function of the nature of the medium surrounding the charges. In symbols, Coulomb's law is

$$F = K \frac{QQ'}{S^2} \qquad \qquad ...(A)$$

where, the dimensional factor K is introduced to take care of the units of F, Q and S and also to provide for the properties of the medium around the charges, in so far as these properties affect the force between the charges. It must be noted that K is not a dimensionless proportionality constant. The dimensions of K are $\dfrac{\text{Force} \times \text{distance}^2}{\text{charge}^2}$

Since F is always a vector quantity, it must be noted that the relation (A) gives only its magnitude. The direction of the force is always along the line joining Q and Q'.

$$K = \frac{1}{4\pi \, \epsilon_0} \qquad \qquad ...(B)$$

where, 'ϵ_0' is the permittivity of empty space. In a medium other than empty space,

$$K = \frac{1}{4\pi \, \epsilon} \qquad \qquad ...(C)$$

where, ϵ is the permittivity of the medium surrounding the charges.

Units of Charge: Two families of units are useful in the areas of electrostatics, namely, the MKS system and the system of CGS electrostatic units.

CGS electrostatic unit: The CGS electrostatic unit of charge is statcoulomb (statC). It is defined as a point charge of such a magnitude that it is repelled by a force of one dyne if it is placed one centimetre away from an equal charge in empty space.

MKS practical unit: The MKS unit of charge is coulomb. It is defined as the point charge of such a magnitude that it is repelled by a force of one newton from an equal charge placed one metre apart in empty space.

However, the coulomb is an enormous charge in terms of those ordinarily met within electrostatic phenomena. For this reason, smaller units of charge are also in use. These are called microcoulomb and micromicrocoulomb (picocoulomb).

$$1\mu C = 10^{-6} C$$

$$1\mu\mu C = 1pC = 10^{-12} C$$

One Coulomb is approximately 3×10^9 statC.

2.2 ELECTRIC FIELD

A region surrounding a charge in which if another charge is kept, it experiences a force, is called an electric field.

2.3 ELECTRIC FIELD INTENSITY

The force experienced by the unit test charge when at rest relative to the observer at any point in the field is known as the electric field intensity or the electric field strength E at the point.

Unit: Newton per Coulomb or Dyne per statcoulomb.

The force f on a charge q place at a point where the electric field intensity E is

$$f = qE \qquad\qquad ...(2.1)$$

If q is positive f has the direction of E, otherwise it has the opposite direction.

The complete relationship between these quantities is expressed by the vector equation

$$\vec{f} = q\vec{E}$$

It is clear from the equation that if the electric intensity is known at every point in the field the force on a charge of any magnitude placed at any point can be computed at once; provided the charges producing the field are held rigidly in position, so that the field remains unaltered by the introduction of the charge on which the force is to be determined.

The electric field intensity at a distance 'r' from a point charge 'q' is obtained at once from Coulomb's law by making q' equal to unity.

It is $E = \dfrac{q}{4\pi \in_0 r^2}$ in the direction of the radius vector drawn from q.

2.4 ELECTRIC POTENTIAL (V)

The potential (V) at any point in an electrostatic field is the work necessary to bring a unit positive charge from infinity, i.e. from outside the field up to the point in question, the charges producing the field being held rigidly in position during the process.

2.5 ELECTRIC POTENTIAL DUE TO A POINT CHARGE

The potential due to a point charge q placed at 'O' can be calculated by the following way:

To find the potential at a point P distance r_p from q, we shall compute the work done by the field on a unit charge as it moves from 'p' to infinity.

Let PQ be the path followed.

The electric intensity at Q_1 is $E = \dfrac{q}{4\pi \in_0 r^2}$ in the direction of radius vector 'r'.

The work done by the field in moving the unit charge the distance dl from Q_1 to Q_2 is

$$\frac{q}{4\pi \in_0 r^2} dl \cos\alpha = \frac{q}{4\pi \in_0 r^2} dr$$

where α is the angle between E and dl.

Integrating from r_p to infinity, we get

$$V = \frac{q}{4\pi \in_0} \int_{r_p}^{\infty} \frac{dr}{r^2} = \frac{q}{4\pi \in_0 r_p} \qquad \qquad ...(2.2)$$

It shows that the value of the potential depends only upon the coordinates of the point 'p' and is independent of the path followed by the unit positive charge.

If the field is due to the point charge $q_1, q_2, \dots$ at distances $r_1, r_2, \dots$ from P, the potential at P is the scalar sum.

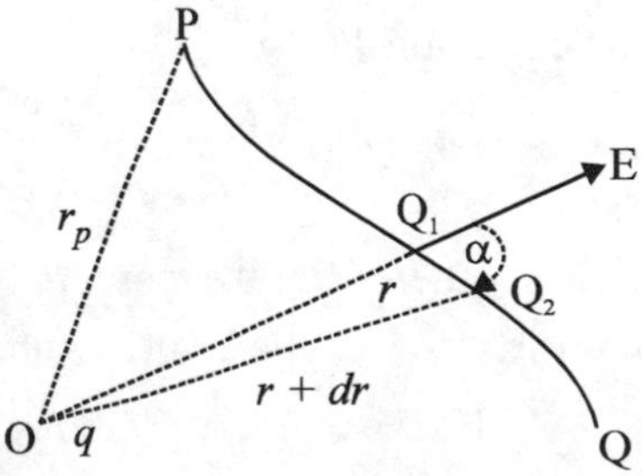

Fig. 2.1

$$V = \frac{q_1}{4\pi \in_0 r_1} + \frac{q_2}{4\pi \in_0 r_2} + \frac{q_3}{4\pi \in_0 r_3} + \dots = \frac{1}{4\pi \in_0} \sum \frac{q}{r} \qquad \qquad ...(2.3)$$

2.6 VOLUME DENSITY OF CHARGE

It is defined as the charge per unit volume of the conductor. It is denoted by ρ.

$$\rho = \frac{q}{\text{vol.}} \qquad \qquad ...(2.4)$$

2.7 SURFACE DENSITY OF CHARGE

It is defined as the charge per unit area distributed over a surface. It is denoted by σ.

$$\sigma = \frac{q}{\text{area}} \qquad \qquad ...(2.5)$$

The potential due to both volume and surface distribution of charges is given by

$$V = \frac{1}{4\pi \in_0} \int_V \frac{\rho dr}{r} + \frac{1}{4\pi \in_0} \int_S \frac{\sigma ds}{r} \qquad \qquad ...(2.6)$$

The distance r now being measured from P.

2.8 POTENTIAL GRADIENT

Consider two nearby points P and Q at a distance dl apart and denote the potentials at P and Q by V and V + dV.

As the force on a unit positive charge is the electric intensity E, the drop in potential in going from P to Q is the work done by the electric intensity.

Therefore, $\qquad V_1 - V_Q \equiv -dV = E\cos\alpha\, dl$

Hence, the component of E in the direction of the displacement PQ is

$$E_l = E \cos \alpha = -\frac{\partial V}{\partial l} \qquad ...(2.7)$$

equal to the space rate of decrease of potential in the direction PQ.

The LHS of this equation is greatest for α equal to zero.

Therefore, the potential decreases most rapidly in the direction of the electric intensity.

If dl is parallel to the X-axis, it becomes dx and E cos α becomes E_x; the X-component of the electric intensity.

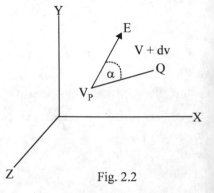

Fig. 2.2

Consequently, $\quad E_x = -\dfrac{\partial V}{\partial x}$ $\qquad \qquad ...(2.8)$

Similarly, expression for Y and Z-components of E.

So, if V is expressed as a function of coordinates x, y, z.

$$E_x = -\frac{\partial V}{\partial x}, \quad E_y = -\frac{\partial V}{\partial y}, \quad E_z = -\frac{\partial V}{\partial z} \qquad \qquad ...(2.9)$$

'V' as a Function of Spherical Coordinates: If potential V is expressed as a function of the spherical coordinates r, θ, ϕ where 'r' represents the radius vector.

θ = the polar angle

ϕ = the azimuth, the space rates of decrease of potential and therefore, the components of electric intensity in the direction of increasing r, θ, ϕ are respectively

$$E_r = -\frac{\partial V}{\partial r}, \quad E_\theta = -\frac{\partial V}{r\partial\theta}$$

$$E_\phi = -\frac{\partial V}{r \sin\theta\, \partial\phi} \qquad \qquad ...(2.10)$$

'V' as a Function of Cylindrical Coordinates: If V is expressed as a function of the cylindrical coordinates r, θ, z.

where r = the distance from the z-axis.

θ = the azimuth measured around the axis.

The components of electric intensity in the directions of increasing r, θ, z are respectively

$$E_r = -\frac{\partial V}{\partial r}, \quad E_\theta = -\frac{\partial V}{r\partial\theta}, \quad E_z = -\frac{\partial V}{\partial z} \qquad \qquad ...(2.11)$$

This electrostatic potential at all cases is a function only of the coordinates of the point P at which it is to be evaluated and is independent of the path along which the unit charge moves. It is this property which makes the concept of potential important. In fact, we say that a field of force possesses a potential only in regions where these conditions are fulfilled.

2.9 GAUSS' LAW

Physical Definition: Gauss' Law $\frac{\Sigma q}{\epsilon_0}$ states that the net outward electric flux through any closed surface is equal to $\frac{1}{\epsilon_0}$ times the total charge contained within that surface.

Mathematical Definition:

Mathematically we can write, $\phi = \int_S \vec{E} \cdot \vec{d}s = \int_S E \cos\alpha \, ds = \frac{\Sigma q}{\epsilon_0}$ $\qquad \qquad ...(2.12)$

and
$$\int_S E \cos \alpha \, ds = \frac{1}{\epsilon_0} \int_V \rho \, dV \qquad \qquad ...(2.13)$$

In the first of these Σq represents the sum of all the charges contained in the closed surface S, and in the second volume integral is to be taken over the entire volume 'V' surrounded by S.

Case I: A point charge 'q' inside a closed surface S

If $\hat{n}$ represents the normal to the surface at P, the electric flux dF through the element AB of area ds due to q is

$$dF = E \cos \alpha \, ds = \frac{q \, ds \cos \alpha}{4\pi \, \epsilon_0 \, r^2} \qquad \qquad ...(2.14)$$

where 'r' is the distance of P from the charge.

If we draw straight lines from all points of the periphery of the surface element AB to q, the cone so described is said to define a conical angle or solid angle. The solid angle is measured by the area intercepted on the surface of a sphere of unit radius having the vertex of the cone as centre.

Since the area of a sphere is proportional to square of its radius, the magnitude of a solid angle is also equal to the area intercepted on the surface of any sphere with centre at the vertex of the angle divided by the square of the radius furthermore as the superficial area of a sphere of radius r is $4\pi r^2$, the solid angle subtended at q by a surface such as 'S' entirely surrounding it is 4π.

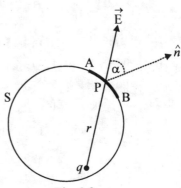

Fig. 2.3

Now $ds \cos \alpha$ is the projection of the area AB perpendicular to the radius vector r and this projection divided by r^2 is the solid angle $d\Omega$ subtended by AB at q.

Hence, $dF = q d\Omega / 4\pi \, \epsilon_0$ ...(2.15)

and if we integrate over the entire surface, the total outward flux is seen to be,

$$F = \frac{q}{\epsilon_0} \qquad \qquad ...(2.16)$$

Case II: Charge not enclosed by the surface

Consider now a charge 'q' lying outside the closed surface. With q as vertex, describe a cone of angular aperture $d\Omega$. Let ds_1 be the area of the surface intercepted at P_1 and ds_2 that at P_2. The projections of these surfaces perpendicular to the radius vector are $ds_1 \cos \alpha_1$ and $ds_2 \cos \alpha_2$ and

$$d\Omega = \frac{ds_1 \cos \alpha_1}{r_1^2} = \frac{ds_2 \cos \alpha_2}{r_2^2}$$ ...(2.17)

Since the angle $\pi - \alpha_1$ between the direction $\vec{E}_1$ and $\hat{n}_1$ is obtuse, the flux through ds_1 is negative, signifying that it is directed inward through the closed surface instead of outward. Taking ds_1 and ds_2 together, the outward flux is,

$$E_1 \cos \alpha \, ds_1 + E_2 \cos \alpha_2 \, ds_2$$

$$= \frac{q \, ds_1 \cos \alpha_1}{4\pi \in_0 r_1^2} + \frac{q \, ds_2 \cos \alpha_2}{4\pi \in_0 r_2^2}$$ ...(2.18)

But the geometrical relation above shows that this expression vanishes. Hence, as the whole surface S can be divided into pairs of elements subtending the same solid angle at q such that the inward flux through one annuls outward flux through the other, the net outward flux through the entire surface due to a charge outside is zero.

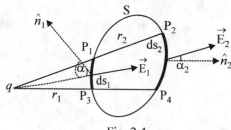

Fig. 2.4

Special Case: If a number of point charges are present, some inside and others outside the surface S, the normal component of the resultant electric intensity is equal to the sum of the normal components of the electric intensities due to the individual charges. Therefore, if $q_1, q_2, q_3, \ldots$ are the point charges lying inside the surface, the entire outward flux through the surface is

$$Q = \frac{q_1}{\in_0} + \frac{q_2}{\in_0} + \frac{q_3}{\in_0} + \ldots \frac{\sum q}{\in_0}$$ ...(2.19)

Since the charges outside make no contribution. If charge is distributed continuously with density ρ units of charge per unit volume.

$$Q = \frac{1}{\in_0} \int_V \rho dV$$ ...(2.20)

where the integral is taken through the volume 'V' enclosed by the surface S.

2.10 APPLICATIONS OF GAUSS' LAW

(i) Coulomb's Law from Gauss' Law:

Gauss' law constitutes a powerful tool for finding the electric intensity in fields of such symmetry that we can draw surface everywhere normal to $\vec{E}$ at all points of which the magnitude of electric intensity is same.

Let us consider, for instance, a sphere S of radius 'a' inside which electric charge is so distributed that the charge density ρ is a function of the radius vector 'r' alone.

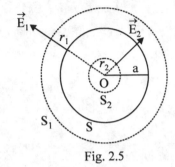

Describe a concentric spherical surface S_1 of radius r_1 greater than a.

It is clear from symmetry that the electric intensity is everywhere normal to S_1 and has the same magnitude E_1 at all points of this surface.

Fig. 2.5

Therefore, the total flux through S_1 is $4\pi r_1^2\, E_1$ and Gauss' law requires that

$4\pi r_1^2\, E_1 = Q/\epsilon_0$ where Q represents the entire charge inside S.

or,
$$E_1 = \frac{Q}{4\pi\, \epsilon_0\, r_1^2}$$
...(2.21)

This is the expression if all the charge had been concentrated at the centre 'O' of the sphere S.

Therefore, any distribution of charge in which the density is a function of the radius vector only, such as that on an isolated charged conducting sphere, produces the same field at exterior points as if the entire charge were located at the central point.

If we apply Gauss' law to the spherical surface S_2 of radius r_2 less than a, we find in the same way

$$E_2 = \frac{Q_2}{4\pi\, \epsilon_0\, r_2^2}$$
...(2.22)

where Q_2 is the portion of the charge inside S_2.

In this case, the electric intensity is that which would be produced by a charge Q_2 located at O.

The charge lying between the spherical surfaces S_2 and S is without effect.

If all the charge lay between these two surfaces, there would be no field at points on the surface S_2 or at points inside their surface.

Thus a charge spread uniformly over the surface of a sphere produces no field at points on its interior, although the field at exterior points is the same as if the entire charge were concentrated at the centre of the sphere.

On the other hand, if the charge is distributed uniformly through the volume of the sphere,

$$Q_2 = \frac{r_2^3}{a^3} Q, \qquad \qquad ...(2.23)$$

and $\quad E_2 = \frac{Q}{4\pi \in_0 a^3} r_2$ in the entire of the sphere. $\qquad ...(2.24)$

Suppose, we wish to find the force between two spherical charges Q_1 and Q_2 in each of which the charge density ρ is a function of the distance from the centre only. Denote the centres of the two spheres by O_1 and O_2 and the distance between centres by R.

Replace the second spherical distribution by a point charge Q_2 located at O_2.

The electric intensity at O_2 due to the first sphere is $E_1 = \frac{Q_1}{4\pi \in_0 R^2} \qquad ...(2.25)$

Therefore, the force on the point charge Q_2 at O_2 due to Q_1 is,

$$f = \frac{Q_1 Q_2}{4\pi \in_0 R^2} = Q_1 E_2 \qquad \qquad ...(2.26)$$

where, E_2 is the electric intensity at O_1 due to Q_2.

(ii) $\vec{E}$ near a Uniformly Charged Plane: Next consider a uniformly charged plane MN of very great extent. Let σ be the charge per unit area of the surface.

For symmetry it is clear that the electric intensity is perpendicular to the plane, being directed upward above the plane and downward below if the charge is positive.

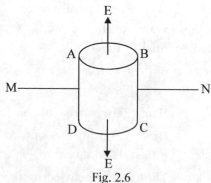

Fig. 2.6

Describe a pill box shaped surface ABCD.

The flat base AB and CD of the box lying parallel to and equidistance from the plane MN.

If ΔS = area of one of the bases, the flux through the surface of the pill box is = 2 EΔs and the charge enclosed is $\sigma\Delta S$.

Therefore, Gauss' law takes the form

$$2E\Delta S = \sigma\Delta S / \in_0$$

$$\Rightarrow \quad E = \frac{\sigma}{2\,\epsilon_0} \qquad\qquad ...(2.27)$$

Consequently, the field is uniform on each side of the plane, the magnitude of E being independent of the distance from the plane.

(iii) $\vec{E}$ near Two Parallel Conducting Plates: Consider the field due to two parallel conducting plates AB and CD of very great extent. The lower of which has a +ve charge density σ and the upper of which has an equal negative charge density.

At a point P between the plates the electric intensity due to AB is $\sigma/2\,\epsilon_0$ upward.

The electric intensity due to CD is $\sigma/2\,\epsilon_0$ upward.

Therefore, the total field is

$$E = \frac{\sigma}{\epsilon_0} \quad \text{upward, everywhere between the plates.}$$

The electric intensity at a point Q outside the plates is $\sigma/2\,\epsilon_0$ upward due to AB and $\sigma/2\,\epsilon_0$ downward due to CD.

Hence, the field outside the plates vanishes. These results apply regardless of the way in which the charge densities divide between the inside and outside surfaces of their respective plates.

However, there can be no field within the plates themselves and this requires that the charge densities be confined to the inside surfaces.

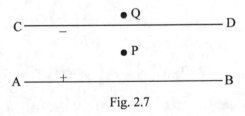

Fig. 2.7

(iv) $\vec{E}$ at a Distance 'r' from the Line: From symmetry, E due to a uniform linear charge can only be radially directed. As a Gaussian surface, we choose a circular cylinder of radius 'r' and length n, closed at each end by plane caps normal to the axis.

E is constant over the cylindrical surface and the flux of E through this surface is E $(2\pi rh)$.

Here, $2\pi rh$ = area of the surface.

Here E, lies in the surface at every point therefore there is no flux through the circular caps.

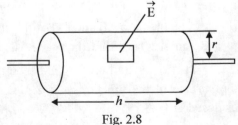

Fig. 2.8

The charge enclosed by the Gaussian surface is λh.

Gauss' law $\epsilon_0 \oint E \cdot ds = q$

then becomes $\epsilon_0 E(2\pi rh) = \lambda h$ $\qquad (\because q = \lambda h)$

Whence $\qquad E = \dfrac{\lambda}{2\pi \epsilon_0 \, r}$ $\qquad ...(2.28)$

The direction of E is radially outward for a line of positive charge.

(v) $\vec{E}$ near a Charged Conductor: The direction of $\vec{E}$ for points close to the surface is at right angles to the surface, pointing away from the surface if the charge is positive.

If $\vec{E}$ were not normal to the surface, it would have a component lying in the surface.

Such a component would act on the charge carries in the conductor and set up surface currents. Since, there are no such currents under the assumed electrostatic conditions; $\vec{E}$ must be normal to the surface.

The magnitude of $\vec{E}$ can be found out from Gauss's law using a small flat 'pill box' of cross section A or a Gaussian surface.

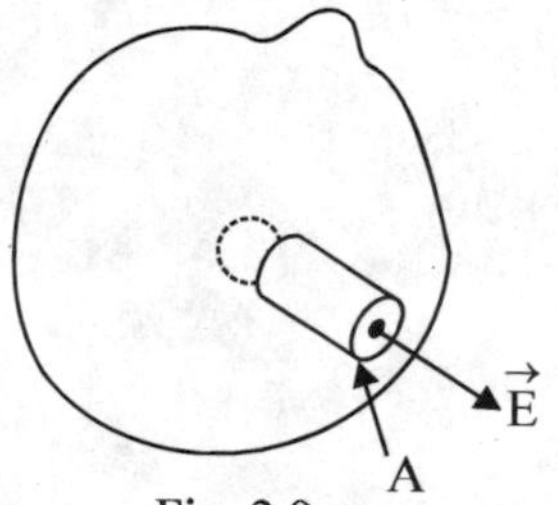

Gauss's law $\qquad \epsilon_0 \oint E \cdot ds = q$

becomes $\qquad \epsilon_0 (EA) = \sigma A$

where σA is the net charge within the Gaussian surface.

This yields, $\qquad E = \dfrac{\sigma}{\epsilon_0}$ $\qquad ...(2.29)$

Fig. 2.9

2.11 DIFFERENTIAL FORM OF GAUSS' LAW

We have so far represented Gauss' theorem by an integral equation. But, we can readily turn it into a differential form, for continuous charge distributions. By applying Gauss' divergence theorem, we have

$$\int_S \vec{E} \cdot d\vec{S} = \int_V \nabla \cdot \vec{E} \, dV$$

where $\vec{E}$ is the electric field intensity. Re-writing Q_{ene} in terms of the charge density ρ, we have

$$Q_{ene} = \int_V \rho \, dV$$

So, Gauss' law becomes $= \int_V \nabla \cdot \vec{E} \, dV = \int_V \left(\dfrac{\rho}{\epsilon_0} \right) dV$ $\qquad \left[\because \oint_S \vec{E} \cdot d\vec{S} = \dfrac{Q_{ene}}{\epsilon_0} \right]$

or, $\qquad \nabla \cdot \vec{E} = \dfrac{\rho}{\epsilon_0}$ $\hfill ...(2.30)$

This is the differential form of Gauss' law. ϵ_0 is the permittivity of empty space.

2.12 CURL OF ELECTROSTATIC FIELD

The electrostatic field $\vec{E}$ can be expressed as gradient of a scalar field (potential) V,

i.e., $\qquad \vec{E} = -\vec{\nabla} V$

Now, Curl $\vec{E} = -\vec{\nabla} \times \vec{E}$

$$= \vec{\nabla} \times (\vec{\nabla} V)$$

$$= \left(\hat{i} \frac{\partial}{\partial x} + \hat{j} \frac{\partial}{\partial y} + \hat{k} \frac{\partial}{\partial z} \right) \times \left(\hat{i} \frac{\partial V}{\partial x} + \hat{j} \frac{\partial V}{\partial y} + \hat{k} \frac{\partial V}{\partial z} \right)$$

$$= \begin{vmatrix} \hat{i} & \hat{j} & \hat{k} \\ \dfrac{\partial}{\partial x} & \dfrac{\partial}{\partial y} & \dfrac{\partial}{\partial z} \\ \dfrac{\partial V}{\partial x} & \dfrac{\partial V}{\partial y} & \dfrac{\partial V}{\partial z} \end{vmatrix}$$

$$= \hat{i} \left(\frac{\partial^2 V}{\partial y \partial z} - \frac{\partial^2 V}{\partial z \partial y} \right) + \hat{j} \left(\frac{\partial^2 V}{\partial z \partial x} - \frac{\partial^2 V}{\partial x \partial z} \right) + \hat{k} \left(\frac{\partial^2 V}{\partial x \partial y} - \frac{\partial^2 V}{\partial y \partial x} \right)$$

Curl $\vec{E} = 0$ $\hfill ...(2.31)$

Thus, the curl of an electrostatic field is zero.

However, for electric field generated by changing magnetic fields or for magnetic fields generated by electric currents, the curl is not zero.

SOLVED EXAMPLES

Q.1. What is the force between a point charge of –50 µC and another of +25 µC when they are 50 cm apart in vacuum?

Sol.

$Q = -50\,\mu C, \ Q' = +25\,\mu C, \ r = 50\,cm = 0.5\,m$

$$F = \frac{1}{4\pi \, \epsilon_0} \frac{QQ'}{r^2} = \frac{9 \times 10^9 \times 50 \times 10^{-6} \times 25 \times 10^{-6}}{(0.5)^2} = 45\,N.$$

Q.2. **Two charges are 60 cm apart in air. One charge Q_1 is 1.67×10^{-7}C; the other Q_2 is -1.67×10^{-7} C. What is the electric field intensity at P midway between the charges?**

Sol. $Q_1 = +1.67 \times 10^{-7}$C, $Q_2 = -1.67 \times 10^{-7}$C,

$$r = 60 \text{ cm} = 0.6 \text{ m}, r_1 = r/2 = 0.3 \text{ m}$$

$$E = E_1 + E_2$$

$$= \frac{KQ_1 q}{q r_1^2} + \frac{KQ_2 q}{r_1^2 q}$$

$$= \frac{Kq}{q r_1^2}\left[Q_1 + Q_2\right]$$

$$= \frac{\mu_o}{4\pi}\left[\frac{Q_1 + Q_2}{r_1^2}\right] = 3.34 \times 10^4 \text{ N/C}.$$

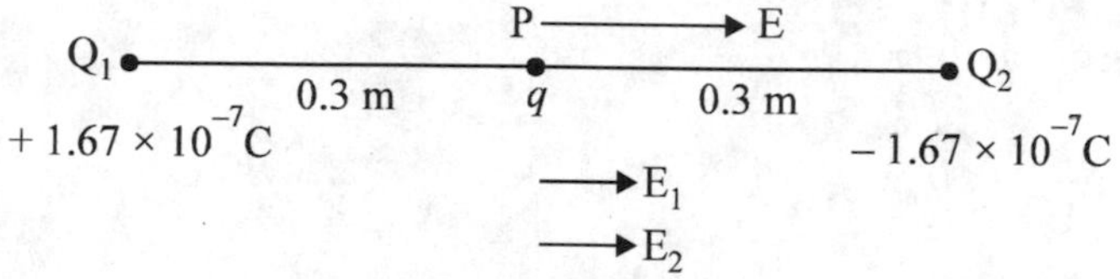

Fig. 2.10

Q.3. **4 grams of gold is beaten into a thin leaf of area 1 m². A small piece is cut out of it and placed on a conductor. Calculate the surface density of charge required by the conductor so that the gold leaf is just lifted up.**

Sol. Mass of gold leaf = 4 gm = 0.004 kg

Area of gold leaf = 1 m²

Mass per unit area of gold leaf $= \frac{0.004}{1} = 4 \times 10^{-3} \text{ kg/m}^2$

Suppose the area of piece = A m²

Then mass 'm' of the piece $= 4 \times 10^{-3} A$ kg

Outward pull experienced by the piece due to the charge $= \dfrac{\sigma^2 A}{2 \epsilon_0}$

As the weight of the piece is just balanced by the outward force.

$$mg = \frac{\sigma^2 A}{2 \epsilon_0}$$

or, $\quad 4 \times 10^{-3} A \times 9.8 = \dfrac{\sigma^2 A}{2 \epsilon_0}$

or, $\qquad \sigma^2 = 2 \epsilon_0 \times 4 \times 10^{-3} \times 9.8$

or, $\qquad \sigma = \sqrt{2 \times 8.85 \times 10^{-12} \times 4 \times 9.8 \times 10^{-3}}$

$\qquad = \sqrt{8 \times 8.85 \times 9.8 \times 10^{-15}}$

$\qquad = \sqrt{8 \times 885 \times 98 \times 10^{-18}} = 832.79 \times 10^{-9} \, \text{C}/\text{m}^2.$

Q.4. **Calculate the charge which must be placed on a sphere of radius 20 cm in order that the repulsion per square metre of the surface may just balance the atmospheric pressure. (Given: Atmospheric pressure = 0.76 m of Hg; Density of Hg = 13.6 × 10³ kg/m³)**

Sol. Force of repulsion per unit area of the charged surface is given by

$$P_E = \frac{\sigma^2}{2 \epsilon_0} \, \text{N}/\text{m}^2 \quad \text{in air.}$$

For a sphere of radius r,

$$\sigma = \frac{q}{4\pi r^2} \, \text{N}/\text{m}^2$$

$$P_E = \left(\frac{q}{4\pi r^2} \right) \text{N}/\text{m}^2$$

$$= \frac{q^2}{16\pi^2 r^4 \times 2 \epsilon_0} = \frac{q^2}{32 \epsilon_0 \pi^2 r^4}$$

$$= \frac{q^2}{32\pi^2 \times 8.85 \times 10^{-12} \times (0.2)^4} \quad (\because \quad r = 0.2 \text{ m})$$

As P_E is just balanced by atmospheric pressure P, we have

$$\frac{q^2}{32\pi^2 \times 8.85 \times 10^{-12} \times 16 \times 10^{-4}} = 0.76 \times 13.6 \times 10^3 \times 9.8$$

or $\qquad q^2 = 32 \times 16 \times 8.85 \times 9.87 \times 10^{-16} \times 0.76 \times 13.6 \times 9.8 \times 10^3$

or $\qquad q = 2128.4 \times \left(10^{-13}\right)^{1/2} = \frac{2128.4 \times 10^{-6}}{r\sqrt{10}}$

$$\frac{2128.4}{3.16} \times 10^{-6} = 673.5 \times 10^{-6} \text{C}.$$

Q.5. **A sphere of radius 10 cm is charged to a potential of 3000 volts. Calculate the outward pull per unit area.**

Sol. Here, $V = 3000$ volts, $r = 10$ cm $= 0.1$ m
Now, charge on the sphere $= 4\pi \in_0 Vr$

or, $q = 4 \times 3.14 \times 8.85 \times 10^{-12} \times 3 \times 10^3 \times 10^{-1}$

or, $q = 12 \times 3.14 \times 8.85 \times 10^{-10} C$

or, $q = 333.468 \times 10^{-10} C$

Charge density, $\sigma = \dfrac{q}{4\pi r^2} = \dfrac{333.468 \times 10^{-10}}{4 \times 3.14 \times 10^{-2}} = 26.55 \times 10^{-8} \; C/m^2$

Hence, outward pull per unit area in air $= \dfrac{\sigma^2}{2 \in_0}$

$$= \dfrac{\left(26.55 \times 10^{-8}\right)^2}{2 \times 8.85 \times 10^{-12}} = \dfrac{704.9025 \times 10^{-16} \times 10^{12}}{17.70} = 39.825 \times 10^{-4} \; N/m^2.$$

Q.6. **An insulated soap bubble of 6 cm radius is given a charge of $48 \times 10^{-9}C$. If the atmospheric pressure is 10^5 N/m^2, calculate the increase in radius due to the charge, neglecting the surface tension effect.**

Sol. Let P be the atmospheric pressure and V the original volume of the bubble. If r

is the original radius of the bubble, $V = \dfrac{4}{3}\pi r^3$. From Boyle's law, we have

$$P\left(\dfrac{4}{3}\pi r^3\right) = K \; \text{(a constant)}$$

or, $P = \dfrac{K}{\dfrac{4}{3}\pi r^3} = \dfrac{K'}{r^3}$, ...(i)

where $K' = \dfrac{K}{\dfrac{4}{3}\pi}$. Let dr be the increase in radius and dp be the change in pressure of

the bubble due to the charge. Differentiating (i) with respect to r, we have

$$\dfrac{dp}{dr} = -\dfrac{3K'}{r^4} = -\dfrac{3p}{r}$$

or, $\qquad dr = -\dfrac{r}{3p} dp$

But the change in pressure is given by

$$-dp = \frac{\sigma^2}{2 \in_0} = \left(\frac{q}{4\pi r^2}\right)^2 \times \frac{36\pi \times 10^9}{2}$$

$$dr = \frac{r}{3p} \times \frac{q^2 \times 18 \times 10^9}{16\pi \, r^4}$$

$$= \frac{3}{8} \frac{q^2 \times 10^9}{\pi p r^3} = \frac{3}{8} \times \frac{\left(48 \times 10^{-9}\right)^2 \times 10^9 \times 7}{22 \times 10^5 \left(0.06\right)^3} = 1.27 \times 10^{-8} \, \text{m}$$

where we have taken $\dfrac{1}{\in_0} = 36\pi \times 10^9 \ Nm^2/C^2$.

Q.7. **Calculate the charge on the soap bubble of radius 0.2 m, if the air pressure inside and outside the bubble is the same. Surface tension = 3 × 10^{-2} N/m.**

Sol. Here, $r = 0.2$ m, $T = 3 \times 10^{-2}$ N/m, as the pressure inside and outside the bubble is the same.

$$P_E = \frac{4T}{r}$$

or, $\qquad \dfrac{\sigma^2}{2 \in_0} = \dfrac{4T}{r} \quad \text{or} \quad \left(\dfrac{q}{4\pi r^2}\right)^2 \times \dfrac{1}{2 \in_0} = \dfrac{4T}{r}$

or, $\qquad q^2 = \dfrac{16\pi^2 r^4 \times 2 \in_0 \times 4T}{r}$

or, $\qquad q^2 = \dfrac{128\pi^2 r^3 \times T}{36\pi \times 10^9}$

or, $\qquad q^2 = \dfrac{32}{9} \dfrac{\pi r^3 T}{10^9}$

or, $\qquad q = \dfrac{4}{3 \times 10^4} \times \left(\dfrac{\pi r^3 T}{10}\right)^{\frac{1}{2}}$

$$= \frac{4r}{3 \times 10^4} \times \left(\frac{\pi r T}{10}\right)^{\frac{1}{2}}$$

$$= \frac{4 \times 0.2}{3 \times 10^4} \times \left(\frac{3.14 \times 0.2 \times 3 \times 10^{-2}}{10}\right)^{\frac{1}{2}}$$

$$= \frac{0.8 \times 10^{-4}}{3} \times 10^{-1} \times \left(\frac{3.14 \times 0.6}{10}\right)^{\frac{1}{2}}$$

$$= 0.266 \times 10^{-5} \times (3.14 \times 0.06)^{\frac{1}{2}}$$

$$= 0.266 \times 10^{-5} \times 0.434$$

$$= 0.1154 \times 10^{-5} = 115.4 \times 10^{-8} \text{C}.$$

Q.8. **The pressure of air inside an electrically charged soap bubble of radius 1.5 cm is the same as that of the air outside. If the surface tension of soap solution is 27 dyne/cm, find in CGS units the electric potential of the bubble.**

Sol. Here, r = 1.5 cm; T = 27 dyne/cm

As the pressure inside and outside the bubble is the same,

$$P_E = \frac{4T}{r}$$

or, $\qquad \dfrac{q^2}{8\pi r^4} = \dfrac{4T}{r} \qquad\qquad \left(\because \text{ In CGS units, } P_E = 2\pi r^2 = \dfrac{q^2}{8\pi r^4}\right)$

or, $\qquad q = \sqrt{32\pi r^3 T} = \sqrt{32 \times \pi \times (1.5)^3 \times 27} = 95.7 \text{ esu}$

Electric potential, $V = \dfrac{q}{r} = \dfrac{95.7}{1.5} = 63.8 \text{ esu}.$

Q.9. **A soap bubble of radius r and surface tension T is given a potential of V volt. Show that the new radius R of the bubble is related to its initial radius by the equation**

$$P\left(R^3 - r^3\right) + 4T\left(R^2 - r^2\right) - \frac{1}{2}\epsilon_0 V^2 R = 0$$

where P is the atmospheric pressure.

Sol. Before the bubble is charged, its radius is r. The pressure outside the bubble is atmospheric, P. As a result of surface tension, there is an excess pressure p (= 4T/r) inside the bubble over that outside. Therefore, the pressure inside the bubble would be given by

$$p_1 = P + p = P + \frac{4T}{r} \qquad\qquad\qquad\qquad \text{...(i)}$$

and the volume of the bubble would be

$$v_1 = \frac{4}{3} \pi r^3 \qquad \text{...(ii)}$$

When the bubble is given a charge q (say), its surface experiences an outward pull (electrostatic pressure) of magnitude $\sigma^2 / 2\epsilon_0$, where σ is the charge density. The bubble therefore expands to a new radius R until the excess pressure inside it falls to p', such that

$$p' + \frac{\sigma^2}{2\epsilon_0} = \frac{4T}{R}$$

The total pressure inside the bubble is now therefore

$$p_2 = P + p' = P + \frac{4T}{R} - \frac{\sigma^2}{2\epsilon_0}$$

But, $\qquad \sigma = \dfrac{q}{4\pi R^2}$

So, $\qquad p_2 = P + \dfrac{4T}{R} - \dfrac{q^2}{32\pi^2 R^4 \epsilon_0} \qquad \text{...(iii)}$

The new volume of the bubble is

$$v_2 = \frac{4}{3} \pi R^3 \qquad \text{...(iv)}$$

As the temperature has remained constant, we have by Boyle's law,

$$p_1 v_1 = p_2 v_2$$

or, $\qquad \left(P + \dfrac{4T}{r} \right)\left(\dfrac{4}{3}\pi r^3 \right) = \left(P + \dfrac{4T}{R} - \dfrac{q^2}{32\pi^2 R^4 \epsilon_0} \right)\left(\dfrac{4}{3}\pi R^3 \right)$

or, $\qquad \left(P + \dfrac{4T}{r} \right) r^3 = \left(P + \dfrac{4T}{R} - \dfrac{q^2}{32\pi^2 R^4 \epsilon_0} \right) R^3$

or, $\qquad P\left(R^3 - r^3 \right) + 4T\left(R^2 - r^2 \right) - \dfrac{q^2}{32\pi^2 R \epsilon_0} = 0 \qquad \text{...(v)}$

The capacity of a spherical body of radius R is given by

$$C = \frac{q}{V} = 4\pi \in_0 R$$

or, $\quad q = 4\pi \in_0 RV$

Substituting this value of q in equation (v), we have

$$P\left(R^3 - r^3\right) + 4T\left(R^2 - r^2\right) - \frac{1}{2} \in_0 V^2 R = 0.$$

Q.10. **If the radius and surface tension of a spherical soap bubble be r and T respectively, show that the charge required to double its radius would be $8\pi r \left[\in_0 r \left(7\,Pr + 12T \right) \right]^{1/2}$ Coulomb where P is the atmospheric pressure.**

Sol. Before the bubble is charged, the pressure p_1 inside the bubble is greater than the pressure P (atmospheric) outside by $4T/r$, where T is surface tension, i.e.,

$$p_1 = P + \frac{4T}{r} \qquad \qquad ...(i)$$

When the bubble is given a charge q (say), it experiences outward pull (electrostatic pressure) $\sigma^2 / 2 \in_0$ and expands. If $2r$ is the new radius, then its volume will increase to 8 times and so the pressure inside the bubble will decrease to $p_1/8$. This decrease is compensated by the electrostatic pressure. Thus,

$$\frac{p_1}{8} + \frac{\sigma^2}{2 \in_0} = P + \frac{4T}{2r}$$

or, $\qquad \dfrac{p_1}{8} = P + \dfrac{4T}{2r} - \dfrac{\sigma^2}{2 \in_0}$

But, $\qquad \sigma = \dfrac{q}{4\pi(2r)^2} = \dfrac{q}{16\pi r^2}$

Hence, $\qquad \dfrac{p_1}{8} = P + \dfrac{4T}{2r} - \dfrac{q^2}{512\pi^2 r^4 \in_0}$

Substituting the value of p_1 from equation (i), we get

$$\frac{P}{8} + \frac{4T}{8r} = P + \frac{4T}{2r} - \frac{q^2}{512\pi^2 r^4 \in_0}$$

or,
$$\frac{q^2}{512\pi^2 r^4 \, \epsilon_0} = \frac{7P}{8} + \frac{3T}{2r}$$

or,
$$q^2 = 64\pi^2 r^3 \, \epsilon_0 \, (7\,Pr + 12T)$$

or,
$$q = 8\pi r \left[r \, \epsilon_0 \, (7\,Pr + 12T) \right]^{1/2}$$

which is the required result.

Q.11. **An isolated and charged spherical soap bubble has a radius r and the pressure inside is atmospheric. If T is the surface tension of soap solution, show that the bubble must have a potential $2\sqrt{(2Tr/\epsilon_0)}$.**

Sol. The pressure inside the charged bubble is atmospheric, i.e. same as outside. Hence, it is the electrostatic pressure (outward pull) $\sigma^2/2\,\epsilon_0$ which balances the surface tension pull $4T/r$, and keeps the bubble in equilibrium. Thus, $\dfrac{\sigma^2}{2\,\epsilon_0} = \dfrac{4T}{r}$

If q is the charge on the bubble, then $\sigma = \dfrac{q}{4\pi r^2}$, so that

$$\frac{q^2}{32\pi^2 r^4 \, \epsilon_0} = \frac{4T}{r}$$

or,
$$q^2 = \frac{128\pi^2 r^4 \, \epsilon_0}{r}$$

or,
$$q = \sqrt{128\pi^2 r^3 \, \epsilon_0} = 8 \times \pi r \sqrt{2rT \, \epsilon_0} \qquad \qquad \text{...(i)}$$

The capacity of a spherical body of radius r is given by

$$C = \frac{q}{V} = 4\pi \, \epsilon_0 \, r$$

So that $q = 4\pi \, \epsilon_0 \, rV$

Substituting this value of q in equation (i), we get

$$4\pi \, \epsilon_0 \, rV = 8\pi r \sqrt{2rT \, \epsilon_0}$$

or,
$$V = 2\sqrt{(2rT/\epsilon_0)}, \text{ the required result.}$$

QUESTIONS

1. State and prove Gauss' law in electrostatics. Find the electric field intensity inside and just outside a charged sphere.

2. Show that Gauss' law vanishes for charges not enclosed by a surface.

3. Deduce Coulomb's law from Gauss' law.

4. Use Gauss' law to find the electric intensity at a point near an infinite plane sheet of charge.

5. Find the field intensity at a distance r greater than 'a' from the axis of an infinitely long straight cylindrical rod of radius 'a' whose charge per unit length is λ.

6. Using Gauss' law, find the electric field intensity near a plane sheet of charge.

7. State and prove Gauss' law. Show that the energy associated with unit volume of the medium is $\dfrac{\in E^2}{8\pi}$.

Problems

1. With the help of Gauss divergence theorem, show that volume of a sphere is $\dfrac{\pi d^3}{6}$, where 'd' is the diameter of the sphere.

2. An electron and a proton is placed inside a sphere of radius 15 cm. Calculate the total electric flux coming out of the surface of the sphere.

3. Two charges are 60 cm apart in air. One charge Q_1 is $+1.67 \times 10^{-7}$C and the other Q_2 is $+1.67 \times 10^{-7}$C. What is the electric field intensity at P midway between the charges? **(Ans.** 3.34×10^4 N/C)

4. Calculate the position of the point in the neighbourhood of two point charges of 1.67 µC apart where a third charge would experience no force. **(Ans.** 15 cm)

5. What is the force of the electric field on a proton if the field is set up between the plates of a discharge tube and has a magnitude of 5×10^4 N/C? **(Ans.** 8×10^{-15}C)

6. What is the intensity of the electric field which will just support a water droplet having a mass of 10 µg and a charge of 10^{-7} µC? **(Ans.** 9.8×10^5 N/C)

● ● ●

Dielectrics

3.1 DIELECTRICS

Definition: Dielectrics are actually insulators which become conducting by the application of external electric field.

Example: Air, Water, Gas (CO_2), Compound (CCl_4), etc.

Types of Dielectrics:

There are two types of dielectrics:

(1) Polar Dielectrics, (2) Non-polar Dielectrics

Polar Dielectrics

The dielectrics which have permanent electric dipole moment are known as Polar Dielectrics. Example: Water, HCl.

Non-polar Dielectrics

The dielectrics which have no permanent electric dipole moment are known as Non-polar Dielectrics. Example: CO_2, CCl_4.

Polarisation of Dielectric:

Whether or not the molecules have permanent electric dipole moments, they acquire them by induction when placed in an electric field. The external dielectric field tends to separate the negative and positive charges in the atom or molecule. This induced electric dipole moment is present only when the electric field is present. It is proportional to the electric field (for normal field strengths) and is lined up with the electric field as shown in Fig. 3.1.

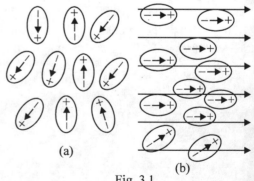

(a)

(b)

Fig. 3.1

(i) Molecules with a permanent electric dipole moment, showing their random orientation in the absence of an external electric field.

(ii) An electric field is applied, producing partial alignment of the dipoles. Thermal agitation prevents complete alignment.

Let us use a parallel plate capacitor, carrying a fixed charge q and not connected to a battery, to provide a uniform external electric field E_0 into which put a dielectric slab.

The overall effect of alignment and induction is to separate the centre of positive charge of the entire slab slightly from the centre of negative charge.

The slab as a whole, although remaining electrically neutral, becomes polarized as (Fig. 3.2b) suggests.

$E_0 = 0$

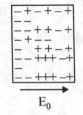

E_0

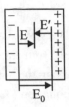

E_0

Fig. 3.2(a): No external electric field is applied.   Fig. 3.2(b): An external electric field E_0 is applied.   Fig. 3.2(c): The surface charges set up a field E'.

The net effect is a pile-up of positive charge on the right face of the slab and of negative charge on the left face; within the slab no excess charge appears in any given volume element.

Since the slab as a whole remains neutral, the positive induced surface charge must be equal in magnitude to the negative induced surface charge.

In this process, electrons in the dielectric are displaced from their equilibrium positions by distances that are considerably less than an atomic diameter. There is not transfer of charge over macroscopic distances such as occurs when a current is set up in a conductor.

The induced field E' which opposes the external field E_0 associated with the charges on the capacitor plates. The resultant field $(\vec{E} = \vec{E}_0 + \vec{E}' = E_0 + (-E'))$ in the dielectric is thus less than E_0.

Fig. (3.2c) shows that the induced surface charges will always appear in such a way that the electric field set up by them ($\vec{E}$) opposes the external electric field $\vec{E}_0$. The resultant field in the dielectric $\vec{E}$ is the vector sum of $\vec{E}_0$ and $\vec{E}'$. It ($\vec{E}$) points in the same direction as $\vec{E}_0$ but is smaller. If a dielectric is placed in an electric field, induced surface charges appear which tend to weaken the original field within the dielectric.

This weakening of the electric field reveals itself as a reduction in potential difference between the plates of a charged isolated capacitor when a dielectric is introduced between the plates.

The relation V = Ed for a parallel plate capacitor holds whether or not dielectric is present and shows the reduction on V is directly connected to the reduction in E described in Fig. 3.2.

So, if a dielectric slab is introduced into a charged parallel plate capacitor, then

$$\frac{E_0}{E} = \frac{V_0}{V_d} = \epsilon_r \text{ (dielectric constant or relative permittivity)} \qquad ...(3.1)$$

where,　　　　　　E_0 = Applied external electric field

　　　　　　　　E = Net electric field

　　　　　　　　V_0 = Potential without dielectric

　　　　　　　　V_d = Potential when dielectric is present

Dielectric Constant (ϵ_r)

As it is evident from equation (3.1) above, the dielectric constant ϵ_r is a dimensionless constant which depends on the nature of the dielectric. The value of 'ϵ_r' for all dielectrics is greater than unity. From equation (3.1) we can write,

$$E = \frac{E_0}{\epsilon_r} \qquad ...(3.2)$$

Therefore, it implies that the presence of a dielectric medium reduces the external electric field by a factor ϵ_r.

For all practical purposes, the value of ϵ_r for air or vacuum can be taken as unity. But, the actual value of ϵ_r for air at 1 atm pressure is 1.00059.

Complete Polarisation

When the external electric field applied to the dielectric is strong enough to produce electric dipole moment in each and every molecule of the dielectric, then *complete polarisation* of the dielectric occurs. In this case, there is complete alignment of dipoles along the direction of the field.

Partial Polarisation

When the external electric field strength is weak, it can't produce electric dipole moment in each and every molecules of the dielectric, the *partial polarisation* of the dielectric occurs. In this case, there is partial alignment of dipoles along the direction of the external electric field.

Dielectric Strength

If the applied electric field across the dielectric material is fairly high, the electrons may be pulled out of the molecules. The dielectric then starts conducting and this phenomenon is termed as *dielectric breakdown*. The maximum electric field that a dielectric can withstand without breakdown is called its *dielectric strength,* values of which for some materials are given below:

Medium	Dielectric constant	Dielectric strength V/m
Vacuum	1	∞
Mica	3 to 6	160×10^6
Wood	2.5 to 8	100×10^6
Benzene	2.3	
Water	81	
Air	1.00059	3×10^6
Paper	3.5	14×10^6

3.2 GAUSS'S LAW IN DIELECTRIC

Let us apply Gauss's law to a parallel plate capacitor filled with a dielectric of dielectric constant $\in_r$.

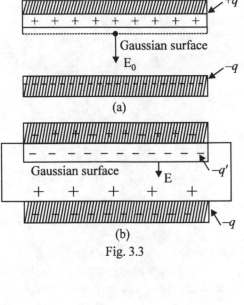

A parallel plate capacitor is shown in Fig. 3.3(a) without and (b) with a dielectric. Charge q on the plates is same in each case.

Fig. 3.3 shows the capacitor both with and without the dielectric. It is assumed that the charge q on the plates is the same in each case.

If no dielectric is present, Gauss's law gives

$$\in_0 \oint E \cdot ds = \in_0 E_0 A = q$$

since,

$$\oint E \cdot ds = \frac{q}{\in_0}$$

when $\theta = 0$, $\cos \theta = 1$ and

So,

$$E_0 = \frac{q}{\in_0 A}, \quad \int \vec{E} \cdot ds = \int E ds \cos \theta$$

$$= \int E ds$$

(as no dielectric is present $E = E_0$ and E is normal to the Gaussian surface).

Thus, we have $E_0 = \dfrac{q}{\in_0 A}$ ...(3.3)

If the dielectric is present, Gauss's law gives,

$$\in_0 \oint \vec{E} \cdot d\vec{s} = \in_0 \oint E ds \cos \theta = \in_0 \oint E ds$$

or, $\qquad\qquad \in_0 EA = q - q'$

or, $\qquad\qquad E = \dfrac{q}{\in_0 A} - \dfrac{q'}{\in_0 A},$ $\qquad\qquad\qquad$ ...(3.4)

Here $\qquad\qquad q' = $ the induced surface charge.

$\qquad\qquad\qquad q = $ free charge on the plates.

The two charges, both of which lie within the Gaussian surface, are opposite in sign.

$\qquad\qquad q - q' = $ the net charge within the Gaussian surface.

Equation (3.1) shows that

$$E = \frac{E_0}{\in_r}$$

Combining this with equation (3.3) we have,

$$E = \frac{E_0}{\in_r} = \frac{q}{\in_r \in_0 A} \qquad\qquad ...(3.5)$$

Inserting equation (3.5) in (3.4), it yields,

$$\frac{q}{\in_r \in_0 A} = \frac{q}{\in_0 A} - \frac{q'}{\in_0 A} \qquad\qquad ...(3.6a)$$

$$\Rightarrow \qquad \frac{q}{\in_r} = q - q'$$

$$\Rightarrow \qquad q' = q - \frac{q}{\in_r} = q\left(1 - \frac{1}{\in_r}\right) \qquad\qquad ...(3.6b)$$

This shows correctly that the induced surface charge q' is always less in magnitude than the free charge q and is equal to zero if no dielectric is present, i.e. if $\in_r = 1$.

Now we write Gauss's law for case of (Fig. 3.3b) in the form (i.e., the equation 3.4).

$$\in_0 \oint \vec{E} \cdot d\vec{s} = q - q' \qquad\qquad ...(3.7)$$

$q - q' = $ the net charge within the Gaussian surface.

Substituting from equation (3.6b) for q' leads, after some rearrangement, to

$$\in_0 \oint \vec{E} \cdot d\vec{s} = q - \left[q - \frac{q}{\in_r}\right] = \frac{q}{\in_r}$$

$$\Rightarrow \qquad \in_0 \oint \in_r \cdot \vec{E} \cdot d\vec{s} = q \qquad\qquad ...(3.8)$$

But the electric displacement vector $|\vec{D}| = \dfrac{q}{A}$

$$\Rightarrow \quad |\vec{D}| = \frac{q}{A} = \epsilon_r \, \epsilon_0 \left(\frac{q}{\epsilon_r \epsilon_0 \, A} \right) \qquad ...(3.9)$$

From equation (3.5) $|\vec{E}| = \dfrac{q}{\epsilon_r \epsilon_0 \, A}$

Putting this value of $|\vec{E}|$ in the above equation (3.9) we get

$$\vec{D} = \epsilon_r \, \epsilon_0 \, \vec{E} \text{ (taking direction into consideration)}$$

From equation (3.8) $\epsilon_0 \oint \epsilon_r \, \vec{E} \cdot d\vec{s} = q$

$$\Rightarrow \quad \oint \vec{D} \cdot d\vec{s} = q \qquad ...(3.10)$$

This is the form in which Gauss's law is usually written when dielectrics are present.

Note: (1) The flux integral now contains a factor ϵ_r.

(2) The charge q contained within the Gaussian surface ϵ_r taken to be the free charge only. Induced surface charge is deliberately ignored in the right side of this equation having been taken into account by the introduction of ϵ_r on the left side.

3.3 THREE ELECTRIC VECTORS

The three electric vectors are $\vec{E}$, $\vec{P}$ and $\vec{D}$.

    (i) Polarisation Vector ($\vec{P}$),

    (ii) Electric Field Intensity ($\vec{E}$), and

    (iii) Electric Displacement Vector ($\vec{D}$).

Polarisation Vector: It is the induced surface charge per unit area. Mathematically, it can be written as:

$$|\vec{P}| = P = \frac{q'}{A} \qquad ...(3.11)$$

where, q' = Induced charge

            A = Surface area

Electric Field Intensity: It is defined as the force per unit positive test charge. It is given by

$$E = \frac{F}{q} \qquad ...(3.12)$$

Electric Displacement Vector: It is defined as the free surface charge per unit area. It is given by

$$D = \frac{q}{A} \qquad \qquad ...(3.13)$$

Relation between $\vec{E}$, $\vec{P}$ and $\vec{D}$

Gauss's law in dielectric is given by:

$$\int_S \vec{E}, d\vec{s} = \frac{(q-q')}{\epsilon_0} \qquad \text{(from equation 3.7)}$$

$$\Rightarrow \qquad \epsilon_0 \int_S \vec{E} \cdot d\vec{s} = q - q'$$

$$\Rightarrow \qquad \epsilon_0 \cdot EA = q - q'$$

$$\Rightarrow \qquad \epsilon_0 \cdot E = \frac{q}{A} - \frac{q'}{A}$$

$$\Rightarrow \qquad \epsilon_0 \cdot E = D - P \quad \text{[from equations (3.11) and (3.12)]}$$

$$\Rightarrow \qquad D = \epsilon_0 E + P$$

In general, $\qquad \vec{D} = \epsilon_0 \vec{E} + \vec{P} \quad$ (Vector form) $\qquad\qquad ...(3.14)$

Relation between E and D

$$D = \frac{q}{A} = \frac{q \,\epsilon_0 \epsilon_r}{\epsilon_r \epsilon_0 A} = \epsilon_0 \epsilon_r \left(\frac{q}{\epsilon_r \epsilon_0 A} \right)$$

$$= \epsilon_0 \epsilon_r \, E \qquad\qquad ...(3.15)$$

$$\left[\text{from equation } (3.5) E = \frac{q}{\epsilon_r \epsilon_0 A} \right]$$

Relation between P and E

$$P = \frac{q'}{A} = \frac{q}{A} \left(1 - \frac{1}{\epsilon_r} \right) \qquad\qquad \left[\because q' = q \left(1 - \frac{1}{\epsilon_r} \right) \right]$$

$$= \epsilon_r \epsilon_0 \, E \left(1 - \frac{1}{\epsilon_r} \right)$$

$$= \epsilon_0 \left(\epsilon_r - 1 \right) E \qquad\qquad ...(3.16)$$

This shows that in a vacuum ($\epsilon_r = 1$). So, the polarisation vector $\vec{P}$ is zero.

3.4 ENERGY STORAGE IN AN ELECTRIC FIELD

We know that all charge configurations have a certain electric potential energy U, equal to the work W that must be done to assemble them from their individual components, originally assumed to be infinitely far apart and at rest. This potential energy reminds us of the potential energy stored in a compressed spring or the gravitational potential energy stored in, say, the earth-moon system.

For a simple example,

Work must be done to separate two equal and opposite charges. This energy is stored in the system and can be recovered if the charges are allowed to come together again.

Similarly, a charged capacitor has stored in it an electric potential energy U equal to the work W required to charge it. This energy can be recovered if the capacitor is allowed to discharge.

We can visualize the work of charging by imagining that an external agent pulls electrons from the positive plate and pushes them onto the negative plates, thus bringing about the charge separation; normally the work of charging is done by a battery, at the expense of its store of chemical energy.

Suppose that at a time 't' a charge $q'(t)$ has been transferred from one plate to the other.

The potential difference $V(t)$ between the plates at that moment will be $q'(t)/C$. If an extra increment of charge dq' is transferred, the small amount of additional work needed will be

$$dW = Vdq = \left(\frac{q'}{C}\right)dq'$$

If this process is continued until a total charge q has been transferred, the total work will be found from

$$W = \int dW = \int_0^{q'} \frac{q'}{c} dq' = \frac{1}{2}\frac{q^2}{c} \qquad ...(3.17)$$

from the relation $q = CV$, we can also write this as,

$$W(=U) = \frac{1}{2}CV^2 \qquad ...(3.18)$$

It is reasonable to suppose that the energy stored in a capacitor resides in the electric field.

As q or V in equations (3.17) and (3.18) increase, for example, then E also increases, when q and V are zero, then E is also zero.

In a parallel plate capacitor, the electric field has the same value for all points between the plates. Thus, the energy density 'u', which is the stored energy per unit volume, should also be uniform. So, u [from eqn. (3.13)] is given by

$$u = \frac{U}{Ad} = \frac{\frac{1}{2}CV^2}{Ad}$$

where Ad = the volume between the plates.

Substituting the relation $C = \epsilon_0 \, \epsilon_r \, A/d$, the above equation leads to

$$u = \frac{1}{2} \frac{\epsilon_0 \, \epsilon_r \, AV^2}{d \cdot Ad} = \frac{\epsilon_r \, \epsilon_0}{2} \left(\frac{V}{d} \right)^2$$

However, V/d is the electric field strenth E, $\left[\because \frac{V}{d} = E \right]$

So that   Energy density $u = \frac{1}{2} \epsilon_r \, \epsilon_0 \, E^2$,   In MKS system.   ...(3.19)

In CGS system, $u = \frac{1}{2} \frac{\epsilon_0 \, \epsilon_r \, E^2}{4\pi \, \epsilon_0}$

$\Rightarrow$ $u = \frac{\epsilon_r \, E^2}{8\pi}$ ...(3.20)

Although this equation was derived for the special case of a parallel plate capacitor, it is true in general.

If an electric field $\vec{E}$ exists at any point in space, we can think of that point as the site of stored energy in amount, per unit volume, of $\frac{1}{2} \epsilon_r \, \epsilon_0 \, E^2$.

3.5 ENERGY OF CHARGED SYSTEMS

To charge a system of conductors, requires the expenditure of energy. This energy is stored electrostatically and becomes available again when the system is discharged.

Let us build of a charge Q on an isolated conductor by bringing up infinitesimal charges dq through the surrounding medium. If q is the charge and v the potential at any time during this process, $q = Cv$.

Similarly $Q = CV$, where Q and V are the final charge and the final potential respectively.

Bringing up an infinitesimal charge from infinity does not charge the potential by any finite amount, so the amount of work done, that is, the increase in the potential energy of the system during this process is Vdq.

The total energy of the charged conductor is therefore given by,

$$U = \int_0^Q vdq = \frac{1}{2} \int_0^Q qdq = \left[\frac{q^2}{2C} \right]_0^Q$$

$$= \frac{1}{2}\frac{Q^2}{C} = \frac{1}{2}QV = \frac{1}{2}CV^2 \qquad \qquad ...(3.21)$$

With a capacitor, we may take the element of charge from the negative to the positive plate and,

$$U = \int_0^Q (v_1 - v_2)dq = \frac{1}{C}\int_0^Q qdq \; \frac{1}{2}\frac{Q^2}{C}$$

$$= \frac{1}{2}Q(V_1 - V_2) = \frac{1}{2}C(V_1 - V_2)^2 \qquad \qquad ...(3.22)$$

By an extension of the above method, we can calculate the energy of any number 'N' of charged conductors.

As before we build up the charge on each conductor in infinitesimal steps, but in such a way that every charge at any time is the same fraction of its final value.

Thus, at any instant, $q_1 = \alpha Q_1$, $q_2 = \alpha Q_2$, ..., $q_N = \alpha Q_N$

where α lies between zero and infinity one step in the charging process must then consist of bringing up from infinity to a charge $dq_1 = Q_1 d\alpha$, to a charge $dq_2 = Q_2 d\alpha$, and so on. When all charges are multiplied by a given factor, the potentials are multiplied by the same factor; so using the same notation as in equation (3.21).

$$v_1 = \alpha V_1, \; v_2 = \alpha V_2, ..., v_N = \alpha V_N$$

The work done per step is

$$v_1 dq_1 + v_2 dq_2 + ...+ v_N dq_N$$

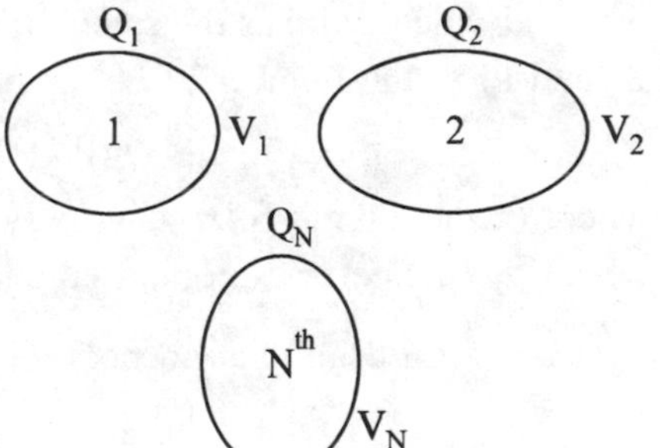

Therefore, $U = \int_0^Q (v_1 dq_1 + v_2 dq_2 + ... + v_N dq_N)$

$$= (Q_1 V_1 + Q_2 V_2 + ... + Q_N V_N)\int_0^1 \alpha d\alpha$$

Fig. 3.4

$$= \frac{1}{2}\sum_1^N Q_1 V_1 \qquad \qquad ...(3.23)$$

When a dielectric is present, at least some of the energy is in the region between the charges, that is, in the field.

$$U = \frac{1}{2}\epsilon_0 \int_\tau \epsilon_r E^2 \, d\tau \qquad \qquad ...(3.24)$$

3.6 FORCE: STRESS IN A DIELECTRIC

As the polarisation in an isotropic dielectric is parallel to the electric intensity, each atomic dipole may be considered to be fixed up with its axis parallel to the lines of force as in the figure given.

If, E is the electric intensity at $-q$,

that at q is $E + \dfrac{\partial E}{\partial l} . l$

The resultant force on the dipole is

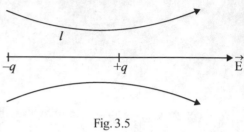

Fig. 3.5

$$-dE + q\left(E + \frac{\partial E}{\partial l} . l\right) = p\frac{\partial E}{\partial l}$$

that of the negative charge. Since $\rho'R$ is $\vec{P}$, the electric moment per unit volume.

$$E_2 = -\frac{P}{3\epsilon_0} \qquad\qquad ...(3.25)$$

Then $\qquad\qquad E_1 = E + \dfrac{1}{3\epsilon_0}P \qquad\qquad ...(3.26)$

which shows that polarising field in the dielectric is actually greater than the total field.

Evidently P has the same direction as E_1. In magnitude, it is found to be proportional to E_1 and to n, the number of atoms per unit volume. Accordingly,

$$P = n\alpha E_1 \qquad\qquad ...(3.27)$$

where α, the atomic polarizability is a constant characteristic of the type of atom (or atoms) of which the dielectric is composed and independent of the density.

To find the relation between α and ϵ_r, we note that

$$(\epsilon_r - 1)E = \frac{1}{\epsilon_0}P \qquad\qquad ...(3.28)$$

So that (3.26) may be written as,

$$\frac{P}{3\epsilon_0} = \frac{\epsilon_r - 1}{\epsilon_r + 2}E_1 \qquad\qquad ...(3.29)$$

Comparing this with eqn. (3.27), we get

$$\frac{\epsilon_r - 1}{\epsilon_r + 2} = \frac{n\alpha}{3\epsilon_0} \qquad\qquad ...(3.30)$$

This is called Clausius-Mossotti equation.

3.7 ELECTROSTATIC BOUNDARY CONDITIONS

Let us consider two dielectrics K_1 and K_2 in Fig. 3.7 in contact. Construct the pill-box shaped surface ABCD about an area ds of the surface of separation, the height BC of the pill-box being very small compared with the diameter of the bases.

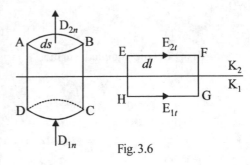

Fig. 3.6

Let D_{1n} and D_{2n} be the normal components of the displacement in the two media. Then, as there is no free charge inside the pill-box, Gauss's Law requires that

$$D_{2n}ds - D_{1n}ds = 0$$

or, $$D_{2n} = D_{1n} \qquad \qquad ...(3.31)$$

Therefore, the normal component of the displacement is the same on both sides of the surface of separation.

Next consider the rectangle EFGH of length dl and negligible height. Let E_{1t} and E_{2t} be the tangential components of the electric intensity in the two media. Then, the work done in taking a unit charge around the rectangle is

$$E_{2t}dl - E_{1t}dl$$

As this represents the drop in potential around a closed path, it must vanish. Hence,

$$E_{2t} = E_{1t} \qquad \qquad ...(3.32)$$

Hence, the tangential component of the electric intensity is the same on both sides of the surface.

SOLVED EXAMPLES

Q.1. **The parallel-plate capacitor of plate area 0.01 m^2 is filled with an insulating material of dielectric constant 6. Its capacitance is 10^{-10} farad and it has been charged to 50 volts. Find the electric field intensity in the dielectric. What is induced surface charge on the dielectric? Take $\epsilon_0 = 8.9 \times 10^{-12}$ Coul^2/N–m^2.**

Sol. The charge on the capacitor is $q = CV = 10^{-10} \times 50 = 5 \times 10^{-9}$ Coulomb.

The electric intensity in the dielectric is

$$E = \frac{q}{\epsilon_r \, \epsilon_0 \, A} = \frac{5 \times 10^{-9}}{6 \times \left(8.9 \times 10^{-12}\right) \times 0.01} = 9.4 \times 10^3 \text{ Volt/m}$$

Now, the electric polarisation is $P = \epsilon_0 \left(\epsilon_r - 1\right)E$.

$$= 8.9 \times 10^{-12} (6-1) \times (9.4 \times 10^3)$$

$$= 4.18 \times 10^{-7} \text{ Coul/m}^2.$$

If q' be the induced surface charge on the dielectric, we have

$$P = \frac{q'}{A}$$

or, $q' = PA = (4.18 \times 10^{-7}) \times (0.01) = 4.18 \times 10^{-9} \text{ Coul.}$

Q.2. A dielectric slab of thickness b = 0.5 cm and dielectric constant ϵ_r = 7 is placed between the plates of a parallel-plate capacitor of plate area A = 100 cm² and separation d = 1 cm. A potential difference V_0 = 100 volts is applied with no dielectric present. The battery is then disconnected and the dielectric slab inserted. Calculate the three electric vectors E, D and P in the dielectric.

(Given: ϵ_0 = 8.9 × 10⁻¹² C²/Nm²)

Sol. The electric intensity in air between the two places is

$$E_0 = \frac{V_0}{d} = \frac{100}{10^{-2}} = 10^4 \text{ Volt/m}$$

when the dielectric ($\epsilon_r = 7$) is introduced, the electric intensity in the dielectric reduces to

$$E = \frac{E_0}{\epsilon_r} = \frac{10^4}{7} = 1.43 \times 10^3 \text{ Volt/m}$$

The electric displacement in the dielectric is

$$D = \epsilon_r \epsilon_0 E$$

$$= 7 \times 8.9 \times 10^{-12} \times 1.43 \times 10^3 = 8.9 \times 10^{-8} \text{ Coul/m}^2$$

The electric polarisation in the dielectric is

$$P = (\epsilon_r - 1) \epsilon_0 E$$

$$= (7-1) \times 8.9 \times 10^{-12} \times 1.43 \times 10^3$$

$$= 6 \times 8.9 \times 1.43 \times 10^{-9} = 7.6 \times 10^{-8} \text{ Coul/m}^2$$

In air, the values are

$$E_0 = 10^4 \text{ Volt/m}, \ D_0 = D = 8.9 \times 10^{-8} \text{ Coul/m}^2 \text{ and } P_0 = 0.$$

Q.3. The electric susceptibility of a material is 35.4×10^{-12} Coul^2/N-m^2. What are the values of the dielectric constant and the permittivity of the material?

Sol. The dielectric coefficient ϵ_r of a material is related to its electric susceptibility χ_e by

$$\epsilon_r = 1 + \frac{\chi_e}{\epsilon_0} = 1 + \frac{35.4 \times 10^{-12}}{8.85 \times 10^{-12}} = 1 + 4 = 5$$

The permittivity is $\epsilon = \epsilon_r \epsilon_0$

$$= 5 \times 8.85 \times 10^{-12} = 44.3 \times 10^{-12} \text{ Coul}^2/\text{N-m}^2.$$

Q.4. The dielectric constant of helium at $0°$ C is 1.000074. Calculate its electric susceptibility at this temperature.

Sol.

We have $\quad \epsilon_r = 1 + \dfrac{\chi_e}{\epsilon_0}$

Here, $\quad \epsilon_r = 1.000074$

So, $\quad 1.000074 + 1 + \dfrac{\chi_e}{\epsilon_0}$

or, $\quad 0.000074 = \dfrac{\chi_e}{\epsilon_0}$

or, $\quad 74 \times 10^{-6} \times \epsilon_0 = \chi_e$

$\chi_e = 74 \times 10^{-6} \times 8.85 \times 10^{-12} = 654.9 \times 10^{-18}$

$= 6.549 \times 10^{-16} \text{ Coul}^2/\text{N-m}^2.$

Q.5. The dielectric constant of helium at $0°$ C and 1 atmospheric pressure is 1.000074. Calculate the dipole moment induced in each helium atom when the gas in an electric field of 1 volt/m. (Given: $\epsilon_0 = 8.85 \times 10^{-12}$ Coul^2/N-m^2 and molecular density of helium $= 2.69 \times 10^{25}$ molecules/m^2 at NTP)

Sol. The electric polarisation P of a dielectric placed in an electric field E is given by

$$P = (\epsilon_r - 1) \epsilon_0 E$$

Now, P is the induced electric dipole moment per unit volume in the dielectric. If P' is the dipole moment induced in each helium atom and 'n' be the number of helium atoms per m^3, then

$$p = np' = (\epsilon_r - 1)\, \epsilon_0\, E$$

or, $$p' = \frac{(\epsilon_r - 1)\, \epsilon_0\, E}{n}$$

Putting the given values, we have

$$p' = \frac{(1.000074 - 1) \times (8.85 \times 10^{-12}) \times 1}{2.69 \times 10^{25}} = 2.43 \times 10^{-41}\ \text{Coul-m.}$$

Q.6. **The dielectric constant of neon gas at NTP is 1.000134. Calculate the dipole moment induced in each atom of the gas when it is placed in an electric field of intensity 9×10^4 Volt/m. Also find the atomic polarisability of neon. (Given: $\epsilon_0 = 8.85 \times 10^{-12}$ Coul^2/N-m^2, Avogadro's number = 6.023×10^{26} atoms/kg-mole).**

Sol. Let p' be the dipole moment induced in each neon atom, and n the number of neon atoms per m^2. Then the electric polarisation of the gas is given by

$$P = np' = (\epsilon_r - 1)\, \epsilon_0\, E$$

or $$P' = \frac{(\epsilon_r - 1)\, \epsilon_0\, E}{n} \qquad\qquad ...(i)$$

Now, at NTP, one kg-atom of neon gas occupies a volume of 22.4 m^3, and contains 6.023×10^{26} atoms. Therefore, the number of neon atoms in 1 m^3 is given by

$$n = \frac{6.023 \times 10^{26}}{22.4} = 2.69 \times 10^{25}$$

Substituting $\quad \epsilon_r = 1.000134,\ E = 9 \times 10^4$ Volt/m

and $\quad n = 2.69 \times 10^{25}$ in equation (i), we get

$$p' = \frac{(1.000134 - 1)(8.85 \times 10^{-12})(9 \times 10^4)}{2.69 \times 10^{25}} = 3.97 \times 10^{-36}\ \text{Coul-m.}$$

The atomic polarisability is given by

$$\alpha = \frac{p'}{E} = \frac{3.97 \times 10^{-36}}{9 \times 10^4} = 4.4 \times 10^{-41}\ \text{Coul-m}^2\ \text{volt.}$$

Q.7. **An isolated metal sphere of diameter 10 cm has a potential of 8000 Volt. Compute the energy density at the surface of the sphere.**

Sol. The capacitance of the sphere is $C = 4\pi \in_0 R$

The charge on its surface is $q = CV = 4\pi \in_0 RV$

The field at the surface of the sphere is

$$E = \frac{1}{4\pi \in_0} \frac{q}{R^2} = \frac{1}{4\pi \in_0} \frac{4\pi \in_0 RV}{R^2} = \frac{V}{R} = \frac{8000}{5 \times 10^{-2}} = 1.6 \times 10^5 \text{ Volt/m}.$$

QUESTIONS

1. What do you mean by dielectrics?

2. What are different types of dielectrics? Give examples of each.

3. What are three electric vectors? Establish relations among them.

4. Deduce Gauss' law in dielectrics.

5. What do you mean by dielectric polarisation? Explain partial and complete polarisation.

6. Show that the energy density of an electric field in the presence of dielectric is $\frac{1}{2} \in_r \in_0 E^2$.

7. Compute the energy stored in a charged system with the presence of a dielectric.

Problems

1. Find the absolute permittivity of a dielectric medium having relative permittivity of 25.

   (Ans. 221.25×10^{-12} Coul^2/Nm^2)

2. The electric field intensity between the capacitor plates decreases by 25% due to introduction of a dielectric slab. Find the relative permittivity and absolute permittivity of the dielectric medium. **(Ans.** 35.4×10^{-12} Coul^2/Nm^2)

3. A parallel plate capacitor is made of 350 plates separated by paraffin paper 0.001 cm thick ($\in_r = 2.5$). The effective size of each plate is 15 by 30 cm. What is the capacitance of this capacitor? **(Ans.** 35 μf)

4. Find the permittivity of a dielectric medium of the relative permittivity 5.

   (Given: $t_0 = 8.87 \times 10^{-7}$ Coul^2/Nm^2)

5. Find the energy density in a dielectric medium ($\in_r = 2$) if the electric field intensity is 8 N/C. **(Ans.** 566.4×10^{-12} J/m^3)

●●●

Ampere's Law and Electromagnetic Induction

4.1 AMPERE'S LAW

In 1819, the Danish physicist Oersted discovered that a current exerts a force on a magnet held near it, which tends to turn the magnet until its axis is at right angles to the wire. In the following year Ampere showed that currents also exert forces on one another and succeeded in formulating analytically the force or torque exerted by one current circuit on another. Although Oersted's and Ampere's experiments were confined to conduction currents in wires, Rowland showed in 1876 that the convection current due to a moving electric field produces the same effect on a magnet as a conduction current.

We conclude, then, that a charge moving relative to observers inertial system produce a magnetic field as well as an electric field.

The magnitude of the magnetic field is proportional to the velocity of the charge relative to the observer, provided the velocity is small compared with the velocity of light as it usually the case. Using CGS electromagnetic units, we find that the magnetic intensity at a distance r_m from a charge q_m moving with a velocity V_m is given by

$$\vec{H}_m = \frac{q_m \vec{V}_m \times \vec{r}_m}{r_m^3} \qquad \qquad ...(4.1)$$

To express the law in MKS practical units it is only necessary to transform the equation.

$$q = 10\, q_m$$

$$\vec{H} = \frac{q \vec{V} \times \vec{r}}{4\pi r^3} \qquad \qquad ...(4.2)$$

$\vec{H}$ is perpendicular to the plane of $\vec{V}$ and $\vec{r}$.

If we take instead of P other points on the circumference of a circle passing through p and lying in a plane perpendicular to $\vec{V}$ with its centre on the X-axis, we note that the magnitude of $\vec{H}$ remains unaltered.

Its direction however changes so as to be everywhere tangent to the circle. The magnetic lines of force, therefore, are circles in planes perpendicular to the velocity with their centres on a line through the charge in the direction of the velocity. Moreover, the sense in which these lines of force are described is that of the rotation of a right handed screw advancing in the direction of the velocity. Finally, if we consider different

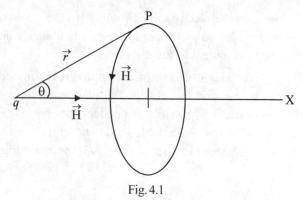

Fig. 4.1

points on the surface of a sphere with centre at q, we note that H vanishes a pole $\theta = 0$, and $\theta = \pi$ and is maximum along the equator.

$$\theta = \frac{\pi}{2}$$

Let us consider a conducting medium of cross-section A carrying a current i. We shall confine ourselves to the case where the moving charges responsible for the current of one sign only. If n is the number of the free charges per unit volume, e the charge of each and V their drift velocity, then

$$i = AneV \qquad \qquad ...(4.3)$$

When the conducting medium is a metallic wire, in which the moving charges are electrons, i has the direction opposite to $\vec{V}$ since e is negative.

The total moving charge in a length dl of the tube is $q = Anedl$ $\qquad ...(4.4)$

Substituting this in eqn. (4.2), the field due to a length dl of a current is found to be,

$$d\vec{H} = \frac{(Ane\,\vec{V}) \times \vec{r}}{4\pi r^3} dl = \frac{\vec{i} \times \vec{r}}{4\pi r^3} dl \qquad \qquad ...(4.5)$$

As $\vec{dl}$ may be considered to be a vector in the direction of the current this may be written equally well

$$d\vec{H} = i\frac{\vec{dl} \times \vec{r}}{4\pi r^3} \qquad ...(4.6)$$

This is Ampere's law for the magnetic field due to a current element idl. Although we have formulated it for a conduction current, it is equally valid for a convection current produced by the motion of charged particles in empty space. In the case of a conduction current in a wire, the electric field due

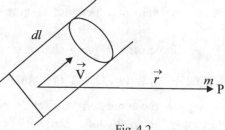

Fig. 4.2

to the moving electrons is annulled by the opposite electric field produced by the stationary atomic nuclei, so that the only electric field inside the wire is the impressed field along its length which is responsible for the current.

Although the law of action and reaction does not apply to electromagnetic forces in all cases, it is found to hold for the magnetic forces between a current element and a magnetic pole. Hence, it is possible to deduce from eqn. (4.5) the force exerted on a current element by an extended magnetic field. To do so, let us consider pole '*m*' placed at the point P in the fig. above. The force exerted on the pole by the current element is

$$m\,d\vec{H} = i \times \left(\frac{m\,\vec{r}}{4\pi r^3} \right) dl \qquad \qquad ...(4.7)$$

According to the law of action and reaction, the force exerted by *m* on the current element is equal and opposite to this. If now we use $\vec{H}$ to represent the magnetic intensity at the current element due to the pole, we have

$$\vec{H} = \frac{-m\,\vec{r}}{4\pi\mu_0 r^3}$$

Since, here the vector $\vec{r}$ is directed toward the pole instead of away from it. Consequently, the force $\vec{df} = -m\,d\vec{H}$ on the current element is

$$\vec{df} = \mu_0 i \times \vec{H}\,dl \qquad \qquad ...(4.8)$$

As this expression contains no reference to the pole *m*, it represents the force due to an external field $\vec{H}$ no matter what the origin of the field may be. Dividing by *dl*, the force per unit length of the circuit is

$$\vec{f_l} = \mu_0 i \times \vec{H} \qquad \qquad ...(4.9)$$

Again, if we replace *i* by *j*A $\quad d\vec{f} = \mu_0 i \times \vec{H}\,A\,dl$

and the force per unit volume of the current is $\vec{f_\tau} = \mu_0 j \times \vec{H}$ $\qquad ...(4.10)$

Force on a moving charge gives $\vec{f} = q\mu_0 \vec{V} \times \vec{H}$ $\qquad \qquad ...(4.11)$

4.2 CIRCUITAL FORM OF AMPERE'S LAW

Let us superpose the current circuit whose field we wish to investigate to be replaced by a very thin magnetic shell whose periphery coincides with the circuit and whose strength ϕ is μ_0 times equal to the current *i*.

$$W = \frac{\phi}{\mu_0}$$

The work done by the magnetic field produced by a current is

$$W = i \qquad \qquad ...(4.12)$$

In the field of a shell, the work done on a pole in passing through shell is equal and opposite to that done along the part of the path lying outside the shell, so that the net work done around a closed path is zero. The field in the interior of the equivalent shell, however is not the same as that due to the circuit. In the case of the latter, the field is continuous all the way around. So as the equivalent shell is extremely thin; the work done on a pole as it passes from a point on one side around to an opposite point on the other does not differ appreciably from the work done on it in completely encircling the current circuit.

If α is the angle between the magnetic intensity $\vec{H}$ produced by a current circuit and an element dl of a closed curve surrounding the current, the work done by the field when a unit positive pole describing a closed path around the current is

$$Mmf = \oint H \cos \alpha \, dl = \oint \vec{H} \cdot \vec{dl} \qquad \qquad ...(4.13)$$

$$emf = \oint E \cos \alpha \, dl = \oint \vec{E} \cdot \vec{dl}$$

The unit of M*mf* is ampere-turn.

Comparing equations (4.12) and (4.13), we have

$$\oint \vec{H} \cdot \vec{dl} = i \qquad \qquad ...(4.14)$$

The closed path around which the integral is taken may be any closed curve arbitrarily chosen. i being the current passing through the curve in the direction of advancing of a right-handed screw rotated in the sense in which the closed path is described. If the closed path surrounds no current, the right hand side is zero. If it encircles 'n' wires, each of which carries a current i, the right-hand member becomes ni.

$$\oint \vec{H} \cdot \vec{dl} = 0, \text{ for } i = 0 = ni, \text{ for } n \text{ turns.} \qquad \qquad ...(4.15)$$

4.3 ELECTROMAGNETIC INDUCTION

When there is a relative motion between a coil and a magnet, there is change in magnetic flux which causes a current in the coil. This phenomenon is called electromagnetic induction. This was discovered by Faraday in 1831.

The process of producing electric current or emf by the relative motion between a coil and a magnet is known as *electromagnetic induction*. In this process the electric current is produced by the magnetic field. The current so produced is called *induced current* and the emf which causes induced current is called *induced emf*.

4.4 DEMONSTRATION OF THE PHENOMENON OF ELECTROMAGNETIC INDUCTION

Michael Faraday of England and Joseph Henry of America independently demonstrated that the change of magnetic flux linked with a coil induces an emf in the coil. The induced emf is found to oppose the inducing emf by Lenz.

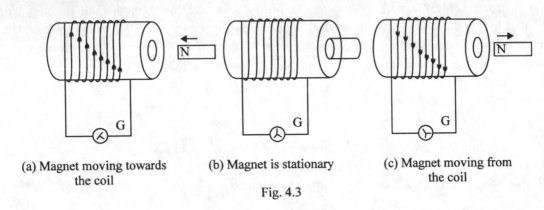

(a) Magnet moving towards   (b) Magnet is stationary   (c) Magnet moving from
      the coil   the coil

Fig. 4.3

The above Fig. 4.3 represents a coil of wire connected to a sensitive galvanometer 'G'. If the north pole of a bar magnet is moved towards the coil, there is deflection in the galvanometer indicating momentary current in the coil in the anticlockwise direction (Fig. 4.3a). This current which is set up in the coil due to the change of flux linked with the coil is known as the induced current and the process of generating the induced emf is known as electromagnetic induction.

When the bar magnet remains at rest within the coil, no current is induced, so there is no deflection by the galvanometer (Fig. 4.3b).

When the magnet is quickly removed from the coil, the galvanometer will indicate a current in the direction opposite to that observed in the 1st case. This current is in clockwise direction (Fig. 4.3c).

Conclusion: By conducting a series of such experiments, Faraday found that an emf is induced in a coil (conductor) when there is any change of magnetic flux linked with the conductor.

Note:
(i) In the above experiments, magnet is moved towards or away from the coil and induction effect is noticed. The same effect can also be produced by moving the coil (conductor) towards a fixed magnet.

(ii) An emf may also be induced in a coil by the change in the magnetic field associated with a change in current in a nearby circuit. This is due to the fact that the change in current in coil or in a circuit produces a magnetic field. In this magnetic field, if a conductor is moved, a current is set up in the conductor. This is nothing but the induced current.

4.5 FARADAY'S LAWS OF ELECTROMAGNETIC INDUCTION

First law: Whenever there is a change in the magnetic flux linked through the coil, an induced current flows through it.

Increase in magnetic flux produced an inverse (opposite sense) current and decrease in magnetic flux produces a direct (same sense) current.

Second law: The magnitude of the induced emf 'E' is directly proportional to the number of turns (N) of the coil and the rate of change of flux ($d\phi/dt$) linked through the coil.

Mathematically,

$$E \propto N$$

$$E \propto \frac{d\phi}{dt}$$

$$E \propto N\frac{d\phi}{dt}$$

or,

$$E = KN\frac{d\phi}{dt} = N\frac{d\phi}{dt} \qquad\qquad ...(4.16)$$

where K = constant of proportionality = 1.

Using suitable units, '$d\phi$' is the change in flux linked through the coil in time interval 'dt'. In MKS system, '$d\phi$' is expressed in webers, 'dt' in seconds and 'E' in volts.

4.6 LENZ'S LAW

Lenz's law: This law states that whenever an emf is induced the induced current is in such a direction as to oppose (by its magnetic action) the change inducing the current.

Explanation: Lenz's law is a particular example of the principle of conservation of energy. An induced current can produce heat to do chemical or mechanical work. The energy must come from the work done in inducing the current. When induction is due to motion of a magnet or a coil, work is done; therefore the motion must be resisted by a force. This opposing force comes from the action of the magnetic field of the induced current. When a change in current in a primary coil induces an emf in a neighbouring secondary coil, the current in the secondary will be in such a direction as to require the expenditure of additional energy in the primary to maintain the current.

4.7 EXPERIMENTAL VERIFICATION OF LENZ'S LAW

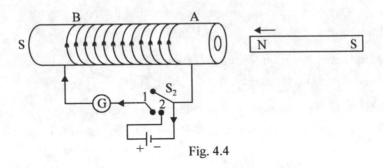

Fig. 4.4

Components:

(i) Coil AB, (ii) Hollow wooden cylinder, (iii) Galvanometer (G), (iv) Battery (E), (v) Switches (S_1, S_2).

Procedure: The circuit is completed as shown in Fig. 4.4. Gap 1 is closed by S_1 keeping gap 2 open. Consequently the source of emf E sends a current through the circuit in the direction as shown in the figure. The direction in which the pointer of galvanometer moves is noted. Let us say, it moves towards right. The direction of this current, when looked into face 'A' of the coil is anticlockwise. So, we conclude that a current which appears anticlockwise from face 'A' gives a deflection toward right in galvanometer. Now gap 2 is closed by S_2 keeping gap 1 open, so that battery goes out of circuit. A magnet N-S towards the end 'A' of the coil is moved. It will be observed that if north pole 'N' of the magnet approaches the coil, the current set up in the coil produces a deflection towards right indicating that the directional induced current must be anticlockwise as seen through the face 'A'. According to the rule of magnetic effects of currents, face A should develop a north polarity due to magnetic field of induced current. The approaching north pole of the magnet is opposed by this north polarity, i.e. the cause which produces induced emf. Hence, Lenz's law is verified.

Consequences of Lenz's law: The following results are observed from the electromagnetic induction and Lenz's law:

(i) If the magnetic flux linked with a coil increases, the induced current set up in the coil rises.

(ii) If the magnetic flux linked with a coil decreases, the induced current produced in the coil also decreases.

(iii) Any change of magnetic flux through the coil causes induced current in the coil.

(iv) The induced current or the induced emf is always in a direction opposite to the very cause from which it is due.

In the light of Lenz's law, the final expression for induced emf is modified as

$$E = -N\frac{d\phi}{dt}$$

...(4.17)

4.8 FLEMING'S RIGHT HAND RULE

Fleming's right hand rule gives the direction of induced current. This rule states that if the thumb, forefinger and middle finger of the right hand are stretched out making right angles to each other, then the forefinger points in the direction of the magnetic field, the thumb points in the direction of motion of the conductor. The middle finger will give the direction of the induced current.

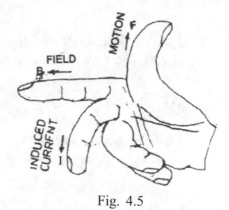

Fig. 4.5

4.9 ENERGY ASSOCIATED WITH AN INDUCTOR

Potential energy is stored by a magnetic field so that when a current grows to a steady value I_0, definite amount of potential energy is stored in the growing magnetic field. The potential energy stored is equal to the work done by the current in the inductor.

At time 't' the induced emf is,

$$\epsilon = -L\frac{dI}{dt} \text{ and the rate of working is}$$

$$\epsilon I = -L\frac{dI}{dt} \text{ watt}$$

$$dW = \epsilon I\, dt = -LI\frac{dI}{dt}, dt = -LI\, dI \text{ Joules}$$

In the time taken to establish the field, I changes from O to I_0, so that the total work done by the current is,

$$W = \int dW = \int_0^{I_0} -\left(LI\, dI\right)$$

$$W = -L\int_0^{I_0} I\, dI = \frac{1}{2}LI_0^2 \text{ Joule} \qquad \qquad ...(4.18)$$

In accordance with this equation L can be defined as twice the work done in establishing the magnetic flux associated with a steady current of one unit in the circuit.

SOLVED EXAMPLES

Q.1. **The magnetic flux through a coil changes from 6×10^{-3} weber to 12×10^{-3} weber in 0.01 sec. Calculate the induced emf.**

Sol. $|E| = \dfrac{\Delta\phi}{\Delta t} = \dfrac{12 \times 10^{-3} - 6 \times 10^{-3}}{0.01} = 0.6$ volt.

Q.2. **Explain whether an induced current will be produced in a conductor if it is moved parallel to the magnetic field.**

Sol. There is no induced emf because the magnetic flux linked with the conductor does not change.

$$\frac{d\phi}{dt} = 0. \quad \text{So,} \quad e = e\frac{d\phi}{dt} = 0$$

Hence, no induced current will be produced as there is no source of emf.

Q.3. **When current through a coil of inductance 0.03 H changes at the rate of 200 A/S. What is the induced emf at the ends of the coil?**

Sol. $e = -L\dfrac{di}{dt} = -0.03 \times 200 = -6$ volt.

Q.4. **The magnetic flux linked through the coil of 500 turns changes from 5×10^{-6} weber to 3×10^{-6} weber in 0.001 sec. What is the average induced emf in the coil?**

Sol. $e = -N\dfrac{d\phi}{dt} = -500 \times \dfrac{5 \times 10^{-6} - 3 \times 10^{-6}}{0.001}$

$$= -500 \times \frac{2 \times 10^{-6}}{10^{-3}}$$

$$= -1000 \times 10^{-3} = -10^{3} \times 10^{-3} = -1 \text{ V}.$$

Q.5. **Compute the time-rate of change of flux of the induced emf is –5 volt in a coil having 500 turns.**

Sol. $e = -N\dfrac{d\phi}{dt}$

or, $\dfrac{d\phi}{dt} = -\dfrac{e}{N} = -\dfrac{(-5)}{500} = \dfrac{1}{100} = 10^{-2}$ W/S.

QUESTIONS

1. What do you mean by electromagnetic induction? Illustrate Faraday's experiments on electromagnetic induction.

2. State and explain Faraday's laws of electromagnetic induction.

3. State and explain Ampere's law. Give circuital form of Ampere's law.

4. State Lenz's law. Design the experiment to verify this law.

5. Explain Fleming's right hand rule.

6. Derive the expression for the energy associated with an inductor.

Problems

1. The north pole of a bar magnet is moved away from a coil of 400 turns so that the flux linked through the coil changes from 4×10^{-6} weber to 2×10^{-6} weber in 1 millisecond. What is the average emf in the coil?

   (Ans. E = –0.8 volt)

2. A coil of 600 turns is threaded by a flux of 8×10^{-5} weber. If the flux is reduced to 3×10^{-5} weber in 0.15 sec, what is the average induced emf?

   (Ans. E = –2 volt)

3. A field of 0.0125 tesla is at right angle to a coil of area 50 cm^2 with 1000 turns. It is removed from the field in 1/20 sec. Find the emf produced.

   (Ans. E = 1250 volts)

4. A coil having 100 turns is placed in a magnetic field. If the change of flux linked with the coil is 50 weber in time 5 secs, then find the emf induced.

   (Ans. E = –1000 volts)

5. Find the emf induced in a coil of 200 turns if the time-rate of change of flux is 1/40 weber/sec.

   (Ans. E = –5 volts)

●●●

Chapter 5

Maxwell's Electromagnetic Field Equations and Electromagnetic Waves

5.1 GAUSS'S LAW FOR ELECTRIC FIELD

Gauss's law for the electric field of charge yields

div $$\vec{D} = \nabla \cdot \vec{D} = \rho$$

where $\vec{D}$ is the electric displacement in coul/m² and ρ is the free charge density in coul/m².

Let's consider a surface S bounding a volume τ contains no net charge but we allow the dielectric to be polarised say by placing it in an electric field. We also deliberate some charge on the dielectric body. Thus we have two types of charges:

(a) Real charge of density ρ, and

(b) Bound charge of density ρ'.

Gauss's law then can be written as

$$\oint_S \vec{E} \cdot \vec{ds} = \frac{1}{\epsilon_0} \int_\tau (\rho + \rho')d\tau$$

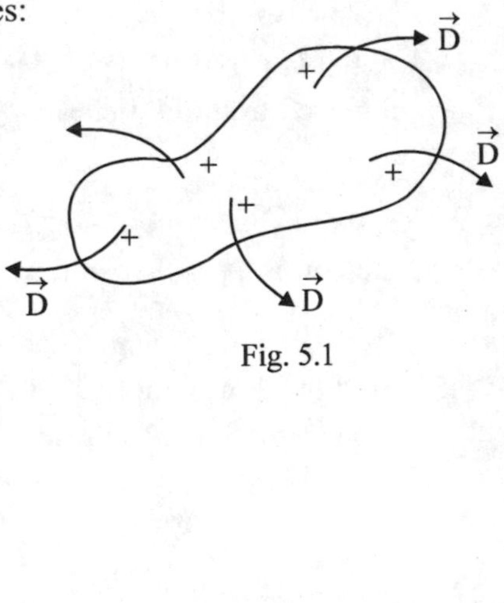

Fig. 5.1

i.e., $$\epsilon_0 \oint_S \vec{E} \cdot \vec{ds} = \int_\tau \rho \, d\tau + \int_\tau \rho' \, d\tau \qquad \ldots(5.1)$$

But as the bound charge density ρ' is defined as

$$\rho' = -\mathrm{div}\,\vec{P} \text{ and } \oint_S \vec{E} \cdot \vec{ds} = \int_\tau \mathrm{div}\,\vec{E}\,d\tau$$

So eqn. (5.1) becomes,

$$\epsilon_0 \int_\tau \mathrm{div}\,\vec{E}\,d\tau = \int_\tau \rho\,d\tau - \int_\tau \mathrm{div}\,\vec{P}\,d\tau$$

or, $$\int_\tau (\mathrm{div}\,\vec{D} - \rho)d\tau = 0$$

Since the equation is true for all volumes the integral in it must vanish.

Thus, we have, div $\vec{D} = \nabla \cdot \vec{D} = \rho$ ...(A)

5.2 GAUSS'S LAW FOR MAGNETIC FIELD

Experiments to date have shown that magnetic monopoles don't exist. This in turn implies that the magnetic lines of force are either closed group or go off to infinity. Hence the number of magnetic lines of force entering any arbitrary closed surface is exactly the same leaving it. Therefore the flux or magnetic induction $\vec{B}$ across any closed surface is always zero, i.e.,

$$\oint_S \vec{B} \cdot \vec{ds} = 0$$

Transforming this surface integral into volume integral by Gauss's theorem, we get

$$\int_\tau \text{div } \vec{B} \, d\tau = 0$$

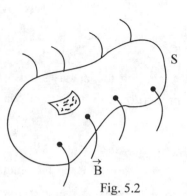

Fig. 5.2

But as the surface bounding the volume is quite arbitrary, the above eqn. will be true only when the integral vanishes, i.e.,

$$\text{div } \vec{B} = \nabla \cdot \vec{B} = 0$$ ...(B)

5.3 AMPERE'S CIRCUITAL LAW

From Ampere's circuital law the work done in carrying unit magnetic pole once around a closed arbitrary path linked with current I is expressed as,

$$\oint_C \vec{H} \cdot \vec{dl} = 1$$

i.e., $\quad \oint_C \vec{H} \cdot \vec{dl} = \int_S \vec{J} \cdot \vec{ds} \quad \left(\because I = \int \vec{J} \cdot \vec{ds} \right)$

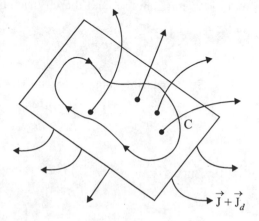

where S is the surface bounded by the closed path C. Now changing the line integral into surface integral by Stokes, theorem,

we get, $\quad \int_S \text{curl } \vec{H} \cdot \vec{ds} = \int_S \vec{J} \cdot \vec{ds}$

Fig. 5.3

i.e., $\quad \text{curl } \vec{H} = \vec{J},$ ...(5.2)

But Maxwell found it to be incomplete for changing electric fields and assumed that a

quantity $\vec{J}_d = \dfrac{\partial \vec{D}}{\partial t}$ called displacement current must also be included in it, so that it may satisfy

the continuity eqn., i.e. $\vec{J}$ must be replaced in eqn. (5.2) by $\vec{J} + \vec{J}_d$. Hence,

$$\text{curl } \vec{H} = \vec{J} + \vec{J}_d$$

or $\quad\quad\quad\quad \text{curl } \vec{H} = \vec{J} + \dfrac{\partial \vec{D}}{\partial t}$ ...(C)

5.4 FARADAY'S LAW OF ELECTROMAGNETIC INDUCTION

According to Faraday's law of electromagnetic induction, we know that the induced emf is proportional to the rate of change of flux, i.e.,

$$e = -\dfrac{d\phi_B}{dt} \quad\quad ...(5.3)$$

Now if $\vec{E}$ be the electric intensity at a point, the work done in moving a unit charge through a small distance dl is $\vec{E} . \vec{dl}$. So the work done in moving the unit charge once round the circuit is $\oint \vec{E} \cdot \vec{dl}$.

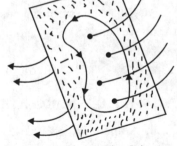

Now, as emf is defined as the amount of work done in moving a unit charge round the electric circuit.

$$e = \oint_C \vec{E} \cdot \vec{dl} \quad\quad ...(5.4)$$

Fig. 5.4

So comparing eqns. (5.3) and (5.4), we get,

$$\oint_C \vec{E} \cdot \vec{dl} = \dfrac{-d\,\phi_B}{dt} \quad\quad ...(5.5)$$

But as $\quad\quad \phi_B = \int_S \vec{B} \cdot \vec{ds}$

So, $\quad\quad \oint_C \vec{E} \cdot \vec{dl} = \dfrac{d}{dt}\left(\int_S \vec{B} \cdot \vec{ds}\right)$

Transforming the line integral by Stokes' theorem into surface integral, we get

$$\int_S \text{curl } \vec{E} \cdot \vec{ds} = -\dfrac{d}{dt}\int (\vec{B} \cdot \vec{ds})$$

Assuming that surface S is fixed in space and only $\vec{B}$ changes with time, above eqn. yields

$$\int_S \left(\text{curl } \vec{E} + \frac{\partial \vec{B}}{\partial t} \right) \cdot \vec{ds} = 0$$

As the above integral is free for any arbitrary surface, the integral must vanish

$$\text{curl } \vec{E} = -\frac{\partial \vec{B}}{\partial t} \qquad \qquad \text{...(D)}$$

5.5 PARTICULAR CASES

(a) In a conducting medium of relative permittivity ϵ_r and permeability μ_r as,

$$\vec{D} = \epsilon \vec{E} = \epsilon_r \epsilon_0 \vec{E}$$

and

$$\vec{B} = \mu \vec{H} = \mu_r \mu_0 \vec{H}$$

Maxwell's eqn. reduces to

(i) $\nabla \cdot \vec{E} = \rho / \epsilon_r \epsilon_0$ (ii) $\nabla \cdot \vec{H} = 0$

(iii) $\nabla \times \vec{H} = \vec{J} + \epsilon_r \epsilon_0 \dfrac{\partial \vec{E}}{\partial t}$ (iv) $\nabla \times \vec{E} = -\mu_r \mu_0 \dfrac{\partial \vec{H}}{\partial t}$

(b) In a non-conducting medium of relative permittivity ϵ_r and permeability μ_r as,

$$\rho = \sigma = 0$$

and $$\vec{J} = \sigma \vec{E} = 0$$

and hence Maxwell's equation becomes

(i) $\nabla \cdot \vec{E} = 0$ (ii) $\nabla \cdot \vec{H} = 0$

(iii) $\nabla \times \vec{H} = \epsilon_r \epsilon_0 \dfrac{\partial \vec{E}}{\partial t}$ (iv) $\nabla \times \vec{E} = -\mu_r \mu_0 \dfrac{\partial \vec{H}}{\partial t}$

(c) In free space as,

$$\epsilon_r = \mu_r = 1$$

and $$\rho = \sigma = 0$$

Maxwell's equation becomes

(i) $\nabla \cdot \vec{E} = 0$ (ii) $\nabla \cdot \vec{H} = 0$

(iii) $\nabla \times \vec{H} = \epsilon_0 \dfrac{\partial \vec{E}}{\partial t}$ (iv) $\nabla \times \vec{E} = -\mu_0 \dfrac{\partial \vec{H}}{\partial t}$

5.6 ENERGY IN ELECTROMAGNETIC FIELDS AND POYNTING THEOREM

Electromagnetic potential energy $\quad U_e = \dfrac{1}{2} \displaystyle\int_V \vec{E} \cdot \vec{D} \, dv$ $\qquad$ ...(5.6)

and energy stored in a magnetic field $U_m = \dfrac{1}{2} \displaystyle\int_V \vec{H} \cdot \vec{B} \, dv$ $\qquad$ ...(5.7)

Now let us see whether these expressions apply to non-static situations.

Maxwell's equations in differential form are

$$\nabla \cdot \vec{D} = 0 \qquad \text{...(5.8)}$$

$$\nabla \cdot \vec{B} = 0 \qquad \text{...(5.9)}$$

$$\nabla \times \vec{E} = -\frac{\partial \vec{B}}{\partial t} \qquad \text{...(5.10)}$$

$$\nabla \times \vec{H} = \vec{J} + \frac{\partial \vec{D}}{\partial t} \qquad \text{...(5.11)}$$

Taking scalar product of equation (5.10) with $\vec{H}$ and equation (5.11) with $\vec{E}$, we get

$$\vec{H} \cdot \text{curl } \vec{E} = -\vec{H} \cdot \frac{\partial \vec{B}}{\partial t} \qquad \text{...(5.12)}$$

and $\qquad \vec{E} \cdot \text{curl } \vec{H} = \vec{E} \cdot \vec{J} + \vec{E} \cdot \dfrac{\partial \vec{D}}{\partial t}$ $\qquad$ ...(5.13)

Now subtracting eqn. (5.13) from eqn. (5.12), we get

$$\vec{H} \cdot \text{curl } \vec{E} - \vec{E} \cdot \text{curl } \vec{H}$$

$$= -\vec{H} \cdot \frac{\partial \vec{B}}{\partial t} - \vec{E} \cdot \frac{\partial \vec{D}}{\partial t} - \vec{E} \cdot \vec{J}$$

$$= -\left[\vec{H} \cdot \frac{\partial \vec{B}}{\partial t} + \vec{E} \cdot \frac{\partial \vec{D}}{\partial t} \right] - \vec{E} \cdot \vec{J} \qquad \text{...(5.14)}$$

Using vector identity.

$$\text{div.}(\vec{E} \times \vec{H}) = \vec{H} \cdot \text{curl } \vec{E} - \vec{E} \cdot \text{curl } \vec{H} \qquad \text{...(5.15)}$$

equation (5.14) may be expressed as,

$$\text{div}.(\vec{E} \times \vec{H}) = -\left(\vec{H} \cdot \frac{\partial \vec{B}}{\partial t} + \vec{E} \cdot \frac{\partial \vec{D}}{\partial t}\right) - \vec{E} \cdot \vec{J} \qquad \qquad ...(5.16)$$

Now, if the medium is linear so that the relations $\vec{B} = \mu \vec{H}$ and $\vec{D} = \epsilon \vec{E}$ ...(5.17) apply, then we may write

$$\vec{E} \cdot \frac{\partial \vec{D}}{\partial t} = \vec{E} \cdot \frac{\partial}{\partial t}(\epsilon \vec{E}) = \frac{1}{2} \epsilon \frac{\partial}{\partial t}(\vec{E})^2 = \frac{\partial}{\partial t}\left(\frac{1}{2}\vec{E} \cdot \vec{D}\right)$$

and

$$\vec{H} \cdot \frac{\partial \vec{B}}{\partial t} = \vec{H} \cdot \frac{\partial}{\partial t}(\mu \vec{H}) = \frac{1}{2} \mu \frac{\partial}{\partial t}(\vec{H})^2 = \frac{\partial}{\partial t}\left(\frac{1}{2}\vec{H} \cdot \vec{B}\right)$$

Using the relations, eqn. (5.16) takes the form

$$\text{div}(\vec{E} \times \vec{H}) = -\frac{\partial}{\partial t}\left(\frac{1}{2} \cdot (\vec{E} \cdot \vec{D} + \vec{H} \cdot \vec{B})\right) - \vec{J} \cdot \vec{E} \qquad \qquad ...(5.18)$$

Each term in above equation have certain physical significance which may be seen by integrating eqn. (5.18) over a volume V bounded by surface S. Thus,

$$\oint_V \text{div}(\vec{E} \times \vec{H})dv = -\int_V \left\{\frac{\partial}{\partial t}\frac{1}{2}(\vec{E} \cdot \vec{D} + \vec{H} \cdot \vec{B})\right\}dv - \int_V \vec{J} \cdot \vec{E} dv$$

Using Gauss divergence theorem to change volume integral on LHS of above equation into surface integral, we get

$$\int_S (\vec{E} \times \vec{H}) \cdot \vec{ds} = -\frac{d}{dt}\int_V \frac{1}{2}(\vec{E} \cdot \vec{D} + \vec{H} \cdot \vec{B})dv - \int_V \vec{J} \cdot \vec{E} \, dv$$

Rearranging this equation, we get,

$$-\int_V \vec{J} \cdot \vec{E} dv = \frac{d}{dt}\int_V \frac{1}{2}(\vec{E} \cdot \vec{D} + \vec{H} \cdot \vec{B})dv + \oint_S \vec{E} \times \vec{H} dv$$

5.7 INTERPRETATION OF $\int_V \vec{J} \cdot \vec{E} \, dv$

To understand the meaning of this term let us consider a charged particle 'q' moving with velocity v, under the combined effect of mechanical, electric and magnetic forces.

The electromagnetic force due to field vectors $\vec{E}$ and $\vec{B}$ acting on the charged particle is

$$\vec{F} = q(\vec{E} + \vec{V} \ddot{Y} \vec{B})$$

As the magnetic force $q(\vec{V} \times \vec{B})$ is always perpendicular to velocity, hence the magnetic field does not work.

Therefore, for a single charge q, the rate of doing work by electromagnetic field $\vec{E}$ and $\vec{B}$ is,

$$\frac{\partial W}{\partial t} = \vec{F} \cdot \vec{V} = q(\vec{E} + \vec{V} \ddot{Y} \vec{B}) \cdot \vec{V}$$

$$= q \vec{E} \cdot \vec{V}$$

If $\vec{F}_m$ is the mechanical force, then work done by mechanical force against electromagnetic field vectors per unit time, i.e. the rate at which mechanical work is done on the particle is

$$\frac{\partial W_m}{\partial t} = \vec{F}_m \cdot \vec{V} = -\vec{F} \cdot \vec{V} = q \vec{E} \cdot \vec{V} \qquad \qquad ...(5.19)$$

If the electromagnetic field consists of a group of charges moving with different velocities, e.g., n_i charge carries each of charge q_i moving with velocity V_i ($i = 1, 2, 3, ...$); then eqn. (5.19) may be written as,

$$\frac{\partial W_m}{\partial t} = -\sum_i n_i q_i \vec{V}_i \cdot \vec{E}_i \qquad \qquad ...(5.20)$$

In this case, current density $\vec{J} = \sum_i \vec{J}_i = \sum_i n_i q_i v_i$

Using this substitution, eqn. (5.20) becomes $\dfrac{\partial W_m}{\partial t} = -\sum_i \vec{J}_i \cdot \vec{E}_i = -\vec{J} \cdot \vec{E}$ ...(5.21)

This eqn. represents the power density that is transferred into electromagnetic field.

Therefore the expression $-\int \vec{E} \cdot \vec{J} \, dv$ represents rate of energy transferred into electromagnetic field through the motion of free charge in volume V.

If there are no sources of emf in volume V, then the term

$$-\int \vec{E} \cdot \vec{J} \, dv = -\int \frac{\vec{J}^2}{\sigma} dv \qquad \qquad (\because \ \vec{J} = \sigma \vec{E})$$

is negative and represents negative rate of heat energy produced.

5.8 INTERPRETATION OF $\dfrac{d}{dt} \int \dfrac{1}{2} (\vec{E} \cdot \vec{D} + \vec{H} \cdot \vec{B}) \, dv$

We know $\displaystyle\int_V \frac{1}{2} \vec{E} \cdot \vec{D} \, dv = U_e$ electrostatic potential energy in volume V

$$\int_V \frac{1}{2} \vec{H} \cdot \vec{B} \, dv = U_m$$

magnetic energy in volume V.

Obviously $\qquad U = \int \frac{1}{2} (\vec{E} \cdot \vec{D} + \vec{H} \cdot \vec{B}) \, dv$ ...(5.22)

represents some sort of potential energy of electromagnetic field. One need not ascribe this potential energy to the charged particle but consider this term as a field energy. This is known as electromagnetic field energy in volume V. A concept such as energy stored in the field itself rather than residing with the particles is a basic concept of the electromagnetic theory.

Obviously $\frac{1}{2} (\vec{E} \cdot \vec{D} + \vec{H} \cdot \vec{B})$ represents energy density of electromagnetic field, i.e.

$$V = \int \frac{1}{2} (\vec{E} \cdot \vec{D} + \vec{H} \cdot \vec{B})$$ ...(5.23)

consequently the term $\frac{d}{dt} \int \frac{1}{2} (\vec{E} \cdot \vec{D} + \vec{H} \cdot \vec{B}) \, dv$ represents the rate of change of electromagnetic energy stored in volume V.

5.9 INTERPRETATION OF $\oint_S (\vec{E} \times \vec{H}) \cdot \vec{ds}$

Since surface integral in this term involves only electric and magnetic fields, it is feasible to interpret this term as the rate of energy flow across the surface. It means that ($\vec{E} \times \vec{H}$) itself represents the energy flow per unit time per unit area. The latter interpretation, however, heads to certain difficulties the only interpretation which services is that the surface integral of ($\vec{E} \times \vec{H}$) over a closed surface represents the amount of electromagnetic energy crossing the closed surface per second. The vector $\vec{E} \times \vec{H}$ is known as the Poynting vector and is usually represented by the symbol $\vec{S}$, i.e.

$$\vec{S} = \vec{E} \times \vec{H}$$ ...(5.24)

5.10 INTERPRETATION OF ENERGY EQUATION

In view of above interpretation, eqn. (5.18) may be expressed as

$$-\vec{J} \cdot \vec{E} = \frac{\partial u}{\partial t} + \nabla \cdot \vec{S}$$ ...(5.25)

The physical meaning of this equation is that the time rate of change of electromagnetic energy within a certain volume, plus time rate of the energy flowing out through the boundary surfaces, is equal to the power transferred into the electromagnetic field.

This is the statement of conservation of energy in electromagnetism and is known as Poynting theorem.

Poynting Vector: In preceding section we have seen that $\vec{S} = \vec{E} \times \vec{H}$

...(5.26)

is known as Poynting vector and is interpreted as the power flux, i.e. amount of energy crossing unit area placed perpendicular to the vector per unit time. The conception of energy of the electromagnetic field as residing in the medium is very fundamental one and has great advantage in the development of the theory. Maxwell through the medium as resembling an elastic solid, the electrical energy representing the potential energy of strain of the medium, the magnetic energy the kinetic energy of the motion. Though such a mechanical view no longer exists, still the energy is regarded as being localised in space and as travelling in the manner indicated by Poynting vector. In a light wave, there is certain energy per unit volume, proportional to the square of the amplitude $\vec{E}$ or $\vec{H}$. This energy travels along and Poynting vector is the vector that measures the rate of flow or the intensity of the wave. In a plane electromagnetic wave, $\vec{E}$ or $\vec{H}$ are at right angles to each other and at right angles to the direction of flow; thus $\vec{E} \times \vec{H}$ must be along the direction of flow. In more complicated waves as well Poynting vector points along the direction of flow of radiation. For example, if we have a source of light and we wish to find at what rate it is emitting energy, we surround it by a closed surface and integrate the normal component at Poynting vector over the surface. The whole conception of energy being transported in the medium is fundamental to the electromagnetic theory of light.

In case of time varying fields, $\vec{S} = \vec{E} \times \vec{H}$ gives the instantaneous value of the Poynting vector. Let us find the form of Poynting vector for such cases.

Let the fields $\vec{E}$ and $\vec{H}$ be given by real parts of complex exponentials of the form:

$$\vec{E} = \vec{E}_0(\vec{r})e^{i\omega t}$$

At a given point of space let us assume that E is given by the real part of $E_0 e^{i\omega t}$ and H by real part of $H_0 e^{i\omega t}$ where E_0 and H_0 are complex vector functions of position. Let the real and imaginary parts of E_0 be denoted by E_r and E_m respectively. Similarly real and imaginary parts of H_0 are H_r and H_{im}.

Then, $\quad\quad$ E = Real part of $E_0\, e^{i\omega t}$

$$= R_e(E_0\, e^{i\omega t})$$

where 'R_e' denotes real part of $\vec{E} = R_e\left(E_r + iE_{im}\right)\left(\cos\omega t + i\sin\omega t\right)$

$$= E_r\cos\omega t - E_{im}\sin\omega t$$

Similarly, $\quad\quad$ $H = H_r\cos\omega t - H_{im}\sin\omega t$

Then Poynting vector is, $\vec{S} = \vec{E} \times \vec{H} = (E_r \times H_r)\cos^2 \omega t + (E_{im} \times H_{im})\sin^2 \omega t$

$$-\left[(E_r + H_{im}) + (E_{im} + H_r)\right]\sin \omega t \cos \omega t$$

we notice that there are two types of terms in the above expression; the first two whose time average is different from zero, since $\cos^2\omega t$ and $\sin^2\omega t$ average to $\frac{1}{2}$; and the last term whose time average is zero since $\sin\omega t.\cos\omega t$ averages to zero. Thus the time average of Poynting vector (average being taken over a complete cycle) is

$$\langle\vec{S}\rangle = \langle\vec{S}\times\vec{H}\rangle = \frac{1}{2}\left[(E_r \times H_r) + (E_{im} \times H_{im})\right] \qquad ...(5.27)$$

This eqn. can be written in a convenient way by using the rotation of complex conjugates, where the complex conjugate of a complex number is the number obtained from the original one by changing the sign of i, whenever it appears and is indicated by a '*' over the number. In terms of this notation.

$$\langle\vec{E}\times\vec{H}^*\rangle = \left(E_0\, e^{i\omega t}\right) \times \left(H_0^*\, e^{-i\omega t}\right)$$

$$= E_0 \times H_0^* = (E_r + i\, E_{im}) \times (H_r - i\, H_{im})$$

$$= (E_r \times H_r + E_{im} \times H_{im}) - i(E_{im} \times H_r - E_r \times H_{im}) \qquad ...(5.28)$$

Comparing (5.27) and (5.28), we note that except for the factor $\frac{1}{2}$, the real part of eqn. (5.28) is just the source as the quantity appearing in eqn. (5.27). That is we have,

$$\langle\vec{S}\rangle = \langle\vec{E} \times \vec{H}\rangle = \frac{1}{2}R_e(\vec{E} \times \vec{H}^*)$$

where $\vec{E}$ and $\vec{H}$ appearing on the RHS of above equation are the complex quantities whose real parts give the real $\vec{E}$ and $\vec{H}$ appearing on the LHS of the equation.

5.11 THE WAVE EQUATION

We shall now derive the equations for electromagnetic waves by the use of Maxwell's equation. This is one of the most important application of Maxwell's equation.

Let us consider a uniform linear medium having permittivity ϵ, permeability μ and conductivity σ, but not any charge or any current other than that determines by Ohm's law. Then

$$\vec{D} = \epsilon\vec{E},\ \vec{B} = \mu\vec{H},\ \vec{J} = \sigma\vec{E}\ \text{and}\ \rho = 0 \qquad ...(5.29)$$

So that Maxwell's equations,

$$\text{div } \vec{D} = \rho$$

$$\text{div } \vec{B} = 0$$

$$\text{curl } \vec{E} = -\frac{\partial \vec{B}}{\partial t} \qquad \qquad ...(5.30)$$

and

$$\text{curl } \vec{H} = \vec{J} + \frac{\partial \vec{D}}{\partial t}$$

In this case take the form $\text{div } \vec{E} = 0$ $\qquad \qquad ...(5.31)$

$$\text{div } \vec{H} = 0 \qquad \qquad ...(5.32)$$

$$\text{curl } \vec{E} = -\mu \frac{\partial H}{\partial t} \qquad \qquad ...(5.33)$$

$$\text{curl } \vec{H} = \sigma \vec{E} + \epsilon \frac{\partial \vec{E}}{\partial t} \qquad \qquad ...(5.34)$$

Taking curl of eqn. (5.33), we get $\text{curl curl } \vec{E} = -\mu \dfrac{\partial}{\partial t}(\text{curl } \vec{H})$

Substituting curl $\vec{H}$ from eqn. (5.34), we get

$$\text{curl curl } \vec{E} = -\mu \frac{\partial}{\partial t}\left(\sigma \vec{E} + \epsilon \frac{\partial \vec{E}}{\partial t} \right)$$

i.e., $\qquad \qquad \text{curl curl } \vec{E} = -\sigma\mu \dfrac{\partial \vec{E}}{\partial t} - \epsilon\mu \dfrac{\partial^2 \vec{E}}{\partial t^2} \qquad \qquad ...(5.35)$

Similarly if we take the curl of eqn. (5.34) and substitute curl $\vec{E}$ from eqn. (5.33),

we obtain $\text{curl curl } \vec{H} = -\sigma\mu \dfrac{\partial \vec{H}}{\partial t} - \epsilon\mu \dfrac{\partial^2 \vec{H}}{\partial t^2} \qquad \qquad ...(5.36)$

Now using vector identity, $\text{curl curl } \vec{A} = \text{grad div } \vec{A} - \nabla^2 \vec{A}$

and keeping in view equations (5.31) and (5.32); equations (5.35) and (5.36) take the form,

$$\nabla^2 \vec{E} = \sigma\mu \frac{\partial \vec{E}}{\partial t} + \epsilon\mu \frac{\partial^2 \vec{E}}{\partial t^2} \qquad \text{...(5.37)}$$

$$\nabla^2 \vec{H} = \sigma\mu \frac{\partial \vec{H}}{\partial t} + \epsilon\mu \frac{\partial^2 \vec{H}}{\partial t^2} \qquad \text{...(5.38)}$$

Equations (5.37) and (5.38) represent wave equations which govern the electromagnetic field in a homogeneous linear medium in which the charge density is zero; whether this medium is conducting or non-conducting. However, it is not enough that these equations be satisfied; Maxwell's equations must also be satisfied. It is clear that equations (5.37) and (5.38) are necessary consequence of Maxwell's equations; but the converse is not true. Now the problem is to solve wave equations (5.37) and (5.38) in such a manner that Maxwell's equations are also satisfied. One method that works very well for monochromatic waves is to obtain a solution

for $\vec{E}$. Then curl $\vec{E}$ will give time derivative of $\vec{B}$, $\left(\text{since curl } \vec{E} = -\frac{\partial \vec{B}}{\partial t} \right)$; so that $\vec{B}$ can be

computed.

It is more convenient to be the method of complex variable analysis for the solution of such wave equations. The time dependence of the field (for certainty we take vector $\vec{E}$) is taken to be $e^{-i\omega t}$, so that

$$\vec{E}(r,t) = \vec{E}_S(r) e^{-i\omega t} \qquad \text{...(5.39)}$$

It may be noted that the physical electric field is obtained by taking the real part of (5.39), further more $E_S(r)$ is in general complex so that the actual electric field is proportional to $\cos(\omega t + \phi)$ where ϕ is the phase of $E_S(r)$.

Using eqn. (5.39), eqn. (5.37) gives $\nabla^2 \vec{E}_S = \omega^2 \, \epsilon\mu \vec{E}_S + i\omega\sigma\mu \vec{E}_S = 0$ ...(5.40)

Here the spatial electric field depends on the space coordinate, i.e.

$$\vec{E}_S = \vec{E}_S(r)$$

for plane electromagnetic waves it is convenient to put

$$\vec{E}_S = \vec{E}_0 \, e^{i \cdot \vec{\kappa} \cdot \vec{r}}$$

where $\vec{\kappa}$ is the propagation wave vector defined as,

$$\vec{\kappa} = \left(\frac{2\pi}{\lambda} \right) \hat{n} = \frac{\omega}{V} \hat{n}, \ \hat{n} \text{ being unit vector along } \vec{\kappa}.$$

$\vec{V}$ is the phase velocity of the wave,

$\vec{r}$ is the position vector from origin.

With this in mind eqn. (5.39) may be written as,

$$\vec{E}(\vec{r},t) = \vec{E}_0 \, e^{i\vec{\kappa}\cdot\vec{r}-i\omega t} \qquad \qquad ...(5.41)$$

Here $\vec{E}_0$ is complex amplitude and is constant in space and time. It is important to note that when field vector is in form (5.41), i.e. operation of grad, div, and curl on field vector is equivalent to

$$\text{grad} \longrightarrow i\vec{\kappa}, \qquad \text{div} = \nabla. \longrightarrow i\vec{\kappa}.$$

$$\text{curl} = \nabla \times \longrightarrow i\vec{\kappa} \times \qquad \qquad ...(5.42)$$

Also $\qquad \qquad \dfrac{\partial}{\partial t} \to -i\omega$

5.12 PLANE ELECTROMAGNETIC WAVES IN FREE SPACE

Maxwell's equations are

$$\text{div } \vec{D} = \nabla \cdot \vec{D} = 0$$

$$\text{div } \vec{B} = \nabla \cdot \vec{B} = 0 \qquad \qquad ...(5.43)$$

$$\text{curl } \vec{E} = -\frac{\partial \vec{B}}{\partial t} \text{ and } \vec{B} = \mu \vec{H}$$

$$\text{curl } \vec{H} = \vec{J} + \frac{\partial \vec{D}}{\partial t} \text{ where } \vec{D} = \in \vec{E}$$

$$\vec{J} = \sigma \vec{E}$$

Free space is characterised by $\rho = 0$, $\sigma = 0$, $\mu = \mu_0$ and $\in = \in_0$ $\qquad ...(5.44)$
Therefore Maxwell's eqn. reduces to

$$\text{div } \vec{E} = 0 \qquad \qquad ...(5.45a)$$

$$\text{div } \vec{H} = 0 \qquad \qquad ...(5.45b)$$

$$\text{curl } \vec{E} = -\mu_0 \frac{\partial \vec{H}}{\partial t} \qquad \qquad ...(5.45c)$$

$$\text{curl } \vec{H} = - \epsilon_0 \frac{\partial \vec{E}}{\partial t} \qquad \qquad ...(5.45d)$$

Taking curl of eqn. (5.45c), we get

$$\text{curl curl } \vec{E} = -\mu_0 \frac{\partial}{\partial t}(\text{curl } \vec{H})$$

Subtracting curl $\vec{H}$ from (5.45d), we get

$$\text{curl curl } \vec{E} = -\mu_0 \frac{\partial}{\partial t}\left(\epsilon_0 \frac{\delta \vec{E}}{\partial t} \right) = -\mu_0 \epsilon_0 \frac{\partial^2 \vec{E}}{\partial t^2} \qquad ...(5.46)$$

Now using curl curl $\vec{E}$ = grade div $\vec{E} - \nabla^2 \vec{E}$

$$\text{curl curl } \vec{E} = - \nabla^2 \vec{E} \qquad \qquad (\because \nabla \cdot \vec{E} = 0)$$

Making this substitution eqn. (5.46) becomes

$$\nabla^2 \vec{E} - \mu_0 \epsilon_0 \frac{\partial^2 \vec{E}}{\partial t^2} = 0 \qquad \qquad ...(5.47)$$

Now taking curl of eqn. (5.45d), we get

$$\text{curl curl } \vec{H} = \epsilon_0 \frac{\partial}{\partial t}(\text{curl } \vec{E})$$

Substituting curl $\vec{E}$ from (5.45c), we get

$$\text{curl curl } \vec{H} = \epsilon_0 \frac{\partial}{\partial t}\left(-\mu_0 \frac{\partial \vec{H}}{\partial E} \right) = -\mu_0 \epsilon_0 \frac{\partial^2 \vec{H}}{\partial t^2} \qquad ...(5.48)$$

Again using identity curl curl $\vec{H}$ = grad div $\vec{H} - \nabla^2 \vec{H}$ and noting that div $\vec{H} = 0$ from (5.45b), we obtain

$$\text{curl curl } \vec{H} = - \nabla^2 \vec{H}$$

Making this substitution in eqn. (5.48), we get

$$\nabla^2 \vec{H} = \mu_0 \epsilon_0 \frac{\partial^2 \vec{H}}{\partial t^2} = 0 \qquad \qquad ...(5.49)$$

equations (5.47) and (5.49) represent wave equations governing electromagnetic field $\vec{E}$ and $\vec{H}$ in free space.

The field vectors $\vec{E}$ and $\vec{H}$ are propagated in free space as waves at a speed equal to 3×10^8 m/sec, the speed of light.

A plane wave is defined as a wave whose amplitude is the same at any point in a plane perpendicular to a specified direction. The electromagnetic field vectors $\vec{E}$ and $\vec{H}$ are both perpendicular to the direction of propagation vector $\vec{\kappa}$. This implies that electromagnetic waves are transverse in character. The flow of energy in a plane electromagnetic wave in free space is along the direction of propagation of wave. The electrostatic energy density is equal to magnetostatic energy density. The energy density associated with an electromagnetic wave in free space propagates with the speed of light with which the field vectors do.

5.13 PLANE ELECTROMAGNETIC WAVES IN A NON-CONDUCTING ISOTROPIC MEDIUM (i.e., ISOTROPIC DIELECTRIC)

A non-conducting medium which has same properties in all directions is called an isotropic dielectric.

Maxwell's equations are

$$\nabla \cdot \vec{D} = \rho$$
$$\nabla \cdot \vec{B} = 0 \qquad \qquad ...(5.50)$$

$$\nabla \times \vec{E} = -\frac{\partial \vec{B}}{\partial t}$$

$$\nabla \times \vec{H} = \vec{J} + \frac{\partial \vec{D}}{\partial t}$$

In an isotropic dielectric or non-conducting isotropic medium,

$$\vec{D} = \epsilon \vec{E}, \vec{B} = \mu \vec{H}, \ \vec{J} = \sigma \vec{E} = 0 \ \text{and} \ \rho = 0 \qquad ...(5.51)$$

Therefore, Maxwell's equations in this case take the form

$$\nabla \cdot \vec{E} = 0 \qquad \qquad ...(5.52a)$$

$$\nabla \cdot \vec{H} = 0 \qquad \qquad ...(5.52b)$$

$$\nabla \times \vec{E} = -\mu \frac{\partial \vec{H}}{\partial t} \qquad \qquad ...(5.52c)$$

$$\nabla \times \vec{H} = -\epsilon \frac{\partial \vec{E}}{\partial t} \qquad \qquad ...(5.52d)$$

Taking curl of eqn. (5.52c), we get

$$\text{curl curl } \vec{E} = -\mu \frac{\partial}{\partial t}(\text{curl } \vec{H})$$

Substituting curl $\vec{H}$ from (5.52d) in above eqn.

$$\text{curl curl } \vec{E} = -\mu \frac{\partial}{\partial t}\left(\in \frac{\partial \vec{E}}{\partial t}\right) = -\mu \in \frac{\partial^2 \vec{E}}{\partial t^2} \qquad ...(5.53)$$

Similarly if we take curl of (5.52d) and substitute curl $\vec{E}$ from (5.52c), we get

$$\text{curl curl } \vec{H} = -\mu \in \frac{\partial^2 \vec{H}}{\partial t^2} \qquad ...(5.54)$$

Using vector identity, curl curl $\vec{A} = \text{grad div } \vec{A} - \nabla^2 \vec{A}$

Equations (5.53) and (5.54) give $\nabla^2 \vec{E} - \mu \in \dfrac{\partial^2 \vec{E}}{\partial t^2} = 0 \qquad ...(5.55)$

$$\nabla^2 \vec{H} - \mu \in \frac{\partial^2 \vec{H}}{\partial t^2} = 0 \qquad ...(5.56)$$

There equations are vector equations of identical form which means that each of the fix components of $\vec{E}$ and $\vec{H}$ separately satisfies the same scalar wave eqn. of the form;

$$\nabla^2 u - \mu \in \frac{\partial^2 u}{\partial t^2} = 0 \qquad ...(5.57)$$

where 'u' is a scalar and can stand for any of components of $\vec{E}$ and $\vec{H}$. It is obvious that eqn. (5.57) resembles with the general wave equation,

$$\nabla^2 u - \frac{1}{V^2}\frac{\partial^2 u}{\partial t^2} = 0 \qquad ...(5.58)$$

when V is the speed of wave;

$$V = \frac{1}{\sqrt{\mu \in}} \qquad ...(5.59)$$

This means that the field vectors $\vec{E}$ and $\vec{H}$ are propagated in isotropic dielectric as waves with speed V given by,

$$V = \frac{1}{\sqrt{\mu \in}} = \frac{1}{\sqrt{\kappa_m \mu_0 \kappa_e \in_0}}$$

where κ_m is the relative permeability of medium and κ_e is relative permittivity or dielectric constant of the medium.

As $\dfrac{1}{\sqrt{\mu_0 \in_0}} = C$, speed of electromagnetic waves in free space.

$$\therefore \qquad V = \frac{C}{\sqrt{\kappa_m \kappa_0}} \qquad \qquad \text{...(5.60)}$$

Since $K_m > 1$ and $K_e > 1$; thereby indicating that the speed of electromagnetic waves in an isotropic dielectric is less than the speed of electromagnetic waves in free space.

As refractive index is defined as

$$n = \frac{C}{V} \ i.e., \ V = \frac{C}{n} \qquad \qquad \text{...(5.61)}$$

Hence comparing eqns. (5.60) and (5.61), we note that the refractive index in this particular case is

$$n = \sqrt{\kappa_e \kappa_m} \qquad \qquad \text{...(5.62)}$$

for a non-magnetic material $\kappa_m = 1$; therefore

$$n = \sqrt{\kappa_e} \ i.e., \ n^2 = \kappa_e \qquad \qquad \text{...(5.63)}$$

This relation is known as Maxwell's relation and has been verified by a number of experiments.

Replacing $\mu\in$ by $\dfrac{1}{V^2}$, wave equations (5.55) and (5.56) may be expressed as

$$\nabla^2 \vec{E} - \frac{1}{V^2} \frac{\partial^2 \vec{E}}{\partial t^2} = 0 \qquad \qquad \text{...(5.64)}$$

$$\nabla^2 \vec{H} - \frac{1}{V^2} \frac{\partial^2 \vec{H}}{\partial t^2} = 0 \qquad \qquad \text{...(5.65)}$$

The plane wave solutions of equations (5.64) and (5.65) in well-known forms may be written as

$$\vec{E}(\vec{r}, t) = \vec{E}_0 \, e^{i \, \vec{\kappa} \cdot \vec{r} - i\omega t} \qquad \qquad \text{...(5.66)}$$

$$\vec{H}(\vec{r}, t) = \vec{H}_0 \, e^{i \, \vec{\kappa} \cdot \vec{r} - i\omega t} \qquad \qquad \text{...(5.67)}$$

where $\vec{E}_0$ and $\vec{H}_0$ are complex amplitudes which are constant in space and time; while $\vec{\kappa}$ is wave propagation vector given by

$$\vec{\kappa} = \kappa \hat{n} = \frac{2\pi}{\lambda}\hat{n} = \frac{\omega}{V}\hat{n} \qquad \qquad ...(5.68)$$

Here $\hat{n}$ is a unit vector in the direction wave propagation vector.

Relative Direction of $\vec{E}$ and $\vec{H}$

The requirement $\nabla \cdot \vec{E} = 0$ and $\nabla \cdot \vec{H} = 0$ demand that $\vec{K} \cdot \vec{E} = 0$ and $\vec{K} \cdot \vec{H} = 0$ ...(5.69)

This means that the field vectors $\vec{E}$ and $\vec{H}$ are both perpendicular to the direction of propagation vector $\vec{\kappa}$. This implies that electromagnetic waves in isotropic dielectric are transverse in nature. Further restrictions are provided by curl of eqns. (5.52c) and (5.53d) viz.

$$\text{curl } \vec{E} = -\mu \frac{\partial \vec{H}}{\partial t}, \quad \text{curl } \vec{H} = \in \frac{\partial \vec{E}}{\partial t},$$

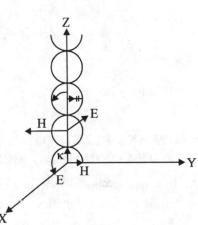

Fig. 5.5

Using (5.66) and (5.67), these equations yield

$$\vec{\kappa} \times \vec{E} = \mu\omega \vec{H} \qquad ...(5.70)$$

$$\vec{\kappa} \times \vec{H} = - \in\omega \vec{E} \qquad ...(5.71)$$

From these equations, it is obvious that field vectors $\vec{E}$ and $\vec{H}$ are mutually perpendicular to the direction of propagation vector $(\vec{E}, \vec{H}, \vec{\kappa})$ form a set of orthogonal vectors which form a right handed coordinate system in given order (Fig. 5.5).

Phase of $\vec{E}$ and $\vec{H}$ and Wave Impedance

From eqn. (5.70),
$$H = \frac{1}{\mu\omega}(\vec{\kappa} \times \vec{E}) = \frac{\vec{\kappa}}{\mu\omega}(\hat{n} \times \vec{E}) = \frac{1}{\mu V}(\hat{n} \times \vec{E})$$

$$= \frac{1}{\mu V}(\hat{n} \times \vec{E}) = \sqrt{\frac{\in}{\mu}}\,(\hat{n} \times \vec{E}) \qquad ...(5.72)$$

$$\left(\because \kappa = \frac{\omega}{V} \text{ and } V = \frac{1}{\sqrt{\mu \in}} \right)$$

Now the ratio of magnitude of $\vec{E}$ to the magnitude of $\vec{H}$ is symbolised by Z, i.e.

$$Z = \left|\frac{\vec{E}}{\vec{H}}\right| = \frac{E_0}{H_0} = \sqrt{\frac{\mu}{\epsilon}} = \sqrt{\frac{\kappa_m \mu_0}{\kappa_e \epsilon_0}} = \text{real quantity} \qquad ...(5.73)$$

This implies that the field vectors $\vec{E}$ and $\vec{H}$ are in the same phase, i.e. they have the same relative magnitude at all points at all time. The unit of 'Z' come out to be ohm, since

$$Z = \frac{\vec{E}}{\vec{H}} = \frac{\text{volt/m}}{\text{amp} - \text{turn/m}} = \frac{\text{volt}}{\text{amp.}} = \text{ohm}$$

hence the value of Z is referred to as wave impedance of isotropic dielectric medium. The wave impedance of medium is related to that of free space by relation

$$Z = \sqrt{\frac{\kappa_m \mu_0}{\kappa_e \epsilon_0}} = Z_0 \sqrt{\left(\frac{\kappa_m}{\kappa_e}\right)} \qquad ...(5.74)$$

where, $Z_0 = \sqrt{\dfrac{\mu_0}{\epsilon_0}}$ is called the wave impedance of free space.

5.14 PLANE ELECTROMAGNETIC WAVES IN AN ISOTROPIC NON-CONDUCTING MEDIUM (i.e., NON-CONDUCTING DIELECTRIC)

An isotropic medium is one in which electromagnetic field properties depend on direction. Let us consider a non-magnetic, non-conducting homogeneous anisotropic medium.

In such a medium, $\vec{J} = 0$; $\rho = 0$ and $\mu = \mu_0$ $\qquad ...(5.75)$

Moreover, the permittivity ϵ is no longer a scalar; but it is a tensor; so that the components of electric displacement $\vec{D}$ are in general related to components of $\vec{E}$ by the equations.

$$D_x = \epsilon_{xx} E_x + \epsilon_{xy} E_y + \epsilon_{xz} E_z$$

$$D_y = \epsilon_{yx} E_x + \epsilon_{yy} E_y + \epsilon_{yz} E_z \qquad ...(5.76)$$

$$D_z = \epsilon_{zx} E_x + \epsilon_{zy} E_y + \epsilon_{zz} E_z$$

where the coefficients are scalar constants for a homogeneous medium.

$$D_x = \epsilon_x E_x = \kappa_x \epsilon_0 E_x$$

$$D_y = \epsilon_y E_y = \kappa_y \epsilon_0 E_y \qquad ...(5.77)$$

$$D_z = \epsilon_z E_z = \kappa_z \epsilon_0 E_z$$

Maxwell's equation in an anisotropic dielectric medium take the form

$$\text{div } \vec{D} = 0 \qquad \qquad ...(5.78a)$$

$$\text{div } \vec{H} = 0 \qquad \qquad ...(5.78b)$$

$$\text{curl } \vec{E} = -\mu_0 \frac{\partial \vec{H}}{\partial t} \qquad \qquad ...(5.78c)$$

$$\text{curl } \vec{H} = \frac{\partial \vec{D}}{\partial t} \qquad \qquad ...(5.78d)$$

It may be noted that div $\vec{E}$ is not zero in this case by virtue of (5.78a) since $\vec{D}$ and $\vec{E}$ do not, in general possess the same direction.

Now, consider a plane wave advancing with angular frequency ω and phase velocity V along the direction of propagation vector $\vec{\kappa}$.

Then
$$\vec{E} = E_0 \, e^{i\vec{\kappa} \cdot \vec{r} - i\omega t} \qquad \qquad ...(5.79)$$

where $\vec{r}$ is the radius vector form origin and

$$\vec{\kappa} = \kappa \, \hat{n} = \frac{2\pi}{\lambda} \hat{n} = \frac{\omega}{V} \hat{n}$$

$\hat{n}$ being unit vector along $\vec{\kappa}$

From eqn. (5.78c) $\dfrac{\partial \vec{H}}{\partial t} = -\dfrac{1}{\mu_0} \text{curl } \vec{E} = -\dfrac{1}{\mu_0} \text{curl}(E_0 e^{i\vec{\kappa} \cdot \vec{r} - i\omega t}) \qquad ...(5.80)$

Using vector identity, $\text{curl } (\phi \vec{A}) = \phi \, \text{curl } \vec{A} - \vec{A} \times \text{grad } \phi \qquad ...(5.81)$

and keeping in view that curl $\vec{E}_0 = 0$ (since E_0 is spatially constant), eqn. (5.80) gives

$$\frac{\partial \vec{H}}{\partial t} = \frac{E_0}{\mu_0} \times \text{grad } (e^{i\vec{\kappa} \cdot \vec{r} - i\omega t})$$

Now, $\text{grad } (e^{i\vec{\kappa} \cdot \vec{r} - i\omega t}) = i\vec{\kappa} \, (e^{i\vec{\kappa} \cdot \vec{r} - i\omega t})$

$$\frac{\partial \vec{H}}{\partial t} = \frac{E_0}{\mu_0} \times (i\vec{\kappa} \, e^{i\vec{\kappa} \cdot \vec{r} - i\omega t}) = \frac{ie^{i\vec{\kappa} \cdot \vec{r} - i\omega t}}{\mu_0} \vec{E}_0 \times \vec{\kappa}$$

Integrating, $$\vec{H} = -\frac{i}{\mu_0} = \frac{e^{i\vec{\kappa}\cdot\vec{r}-i\omega t}}{i\omega}\vec{E}_0 \times \vec{\kappa}$$

$$= -\frac{i}{\mu_0\omega}e^{i\vec{\kappa}\cdot\vec{r}-i\omega t}\cdot\vec{E}_0 \times \vec{\kappa} = -\vec{H}_0 \, e^{i\vec{\kappa}\cdot\vec{r}-i\omega t}\,(\text{say}) \qquad ...(5.82)$$

i.e., $$\vec{H} = -\frac{i}{\mu_0\omega}\vec{E} \times \vec{\kappa} = \frac{i}{\mu_0\omega}\vec{\kappa} \times \vec{E} \,\,(\text{using eqn. 5.79}) \qquad ...(5.83)$$

This shows that $\vec{H}$ is normal to the plane of $\vec{E}$ and $\vec{K}$.

Now from (5.78d)

$$\frac{\partial \vec{D}}{\partial t} = \text{curl } \vec{H} = \text{curl}(\vec{H}_0 \, e^{i\vec{\kappa}\cdot\vec{r}-i\omega t})$$

Using vector identity (5.81) and noting that curl $\vec{H}_0 = 0$,

We obtain $$\frac{\partial \vec{D}}{\partial t} = -\vec{H}_0 \times \text{grad } (e^{i\vec{\kappa}\cdot\vec{r}-i\omega t}) = -\vec{H}_0 \times i\vec{\kappa}\,e^{i\vec{\kappa}\cdot\vec{r}-i\omega t}$$

Integrating $$\vec{D} = -\vec{H}_0 \times i\vec{\kappa} = \frac{1}{\omega}\vec{H} \times \vec{\kappa} \qquad (\text{using eqn. 5.82}) \,\,\,...(5.84)$$

This shows that $\vec{D}$ is normal to $\vec{\kappa}$ and $\vec{H}$. Thus eqns. (5.83) and (5.84) imply that both $\vec{H}$ and $\vec{D}$ are normal to the direction of propagation vector $\vec{\kappa}$. Therefore, the electromagnetic wave in anisotropic non-conducting medium is transverse with respect to $\vec{H}$ and $\vec{D}$. Substituting the value of $\vec{H}$ from eqn. (5.83) in eqn. (5.84) we get,

$$\vec{D} = \frac{1}{\omega}\left(\frac{1}{\mu_0\omega}\vec{\kappa} \times \vec{E}\right) \times \vec{\kappa}$$

$$= \frac{1}{\mu_0\omega^2}\vec{\kappa} \times (\vec{\kappa} \times \vec{E})$$

$$= \frac{1}{\mu_0\omega^2}\left[(\vec{\kappa}\cdot\vec{E})\vec{\kappa} - (\vec{\kappa}\cdot\vec{\kappa})\vec{E}\right]$$

$$= \frac{1}{\mu_0\omega^2}\left[\kappa^2\,\vec{E} - (\vec{\kappa}\cdot\vec{E})\vec{\kappa}\right]$$

$$= \frac{1}{\mu_0\omega^2}\left[(\vec{\kappa}\cdot\vec{E})\vec{\kappa} - (\vec{\kappa}\cdot\vec{\kappa})\vec{E}\right]$$

$$= \frac{1}{\mu_0\omega^2}\left[\kappa^2\,\vec{E} - (\vec{\kappa}\cdot\vec{E})\vec{\kappa}\right] \qquad ...(5.85)$$

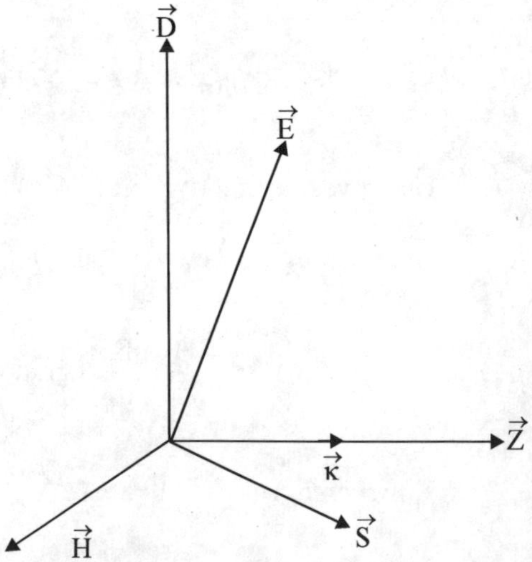

Fig. 5.6

This equation shows that vectors $\vec{D}$, $\vec{E}$ and $\hat{n}$ all lie in the same plane (Fig. 5.6).

5.15 POLARISATION OF ELECTROMAGNETIC WAVES

An electromagnetic wave in which the electric field vector $\vec{E}$ maintains a fixed direction relative to direction of propagation, is said to be linearly polarised with polarisation vector $\vec{E}$. In the foregoing sections we have seen that electric and magnetic vectors are generally perpendicular to each other and also perpendicular to direction of propagation of electromagnetic waves. Hence, such electromagnetic waves are plane polarised.

The plane containing the direction of propagation and field vector $\vec{H}$ is known as the plane of polarisation in optics; while the plane containing field vector $\vec{E}$ and direction of propagation is called the plane of vibration.

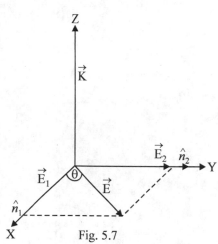
Fig. 5.7

Any plane polarised wave can be considered to be the sum of two plane-polarised components in the perpendicular directions, but in phase. For example, in Fig. 5.7, the field vector $\vec{E}$ can be resolved into two mutually perpendicular components E_1 and E_2. These two independent components which are in phase may be expressed as

$$\vec{E}_1 = \hat{n}_1\, E_1^0\ e^{i\vec{\kappa}\cdot\vec{r}-i\omega t} \qquad \qquad ...(5.86)$$

$$\vec{E}_2 = \hat{n}_2\, E_2^0\ e^{i\vec{\kappa}\cdot\vec{r}-i\omega t} \qquad \qquad ...(5.87)$$

The general solution of plane polarised wave may be expressed as,

$$\vec{E}(r,t) = \left(\hat{n}_1\, E_1^0 + \hat{n}_2\, E_2^0\right) e^{i\vec{\kappa}\cdot\vec{r}-i\omega t} \qquad \qquad ...(5.88)$$

Here E_1^0 and E_2^0 must have the same phase so that the wave might be plane polarised. The polarisation vector $\vec{E}$ has magnitude

$$E^0 = \sqrt{\left(E_1^0\right)^2 + \left(E_2^0\right)^2} \quad \text{and makes an angle}$$

$$\theta = \tan^{-1}\left(E_2^0 / E_1^0\right) \text{ with } \hat{n}_1.$$

If E_1^0 and E_2^0 have different phases, the wave (5.88) is elliptically polarised. To understand its meaning, let us consider the simplest case of circular polarisation. If E_1^0 and E_2^0 have the same real magnitude E_0 (say) but differ in phase by 90°, then eqn. (5.88) may be expressed as,

$$\vec{E}(r,t) = (\hat{n}_1 + \hat{n}_2)e^{i\vec{\kappa}\cdot\vec{r} - i\omega t} \qquad ...(5.89)$$

If we choose the axes so that the wave is propagating in the positive Z direction, while $\vec{E}_1$ and $\vec{E}_2$ are along X and Y direction respectively. Then the components of actual electric field, obtained by taking real part of (5.89) are

$$E_x(r,t) = E_0 \cos(\kappa z - \omega t) \qquad ...(5.90)$$

$$E_y(r,t) = E_0 \sin(\kappa z - \omega t)$$

Squaring eqn. (5.90) and adding, we obtain

$$E_x^2 + E_y^2 = E_0^2 \ \ i.e., \ \ \frac{E_x^2}{E_0^2} + \frac{E_y^2}{E_0^2} = 1 \qquad ...(5.91)$$

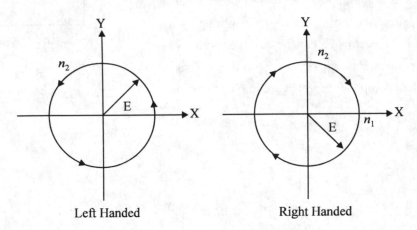

Left Handed   Right Handed

Fig. 5.8: Circularly Polarised Wave

which is the equation of a circle. Thus at a fixed point in space, the fields (5.90) are such that the electric field vector is constant in magnitude, but sweeps around in a circle at a frequency ω. For the upper sign $(\hat{n}_1 + i\hat{n}_2)$, the rotation is anti-clockwise when the observer is facing into the oncoming wave. This is called the left-handed circularly polarised light and sometimes this wave is said to have positive helicity (since the wave has a positive projection of angular momentum of Z-axis, the direction of propagation). For the lower sign $(\hat{n}_1 + i\hat{n}_2)$, the rotation of $\vec{E}$ is clockwise when looking into the wave; this wave is called right-handed circularly polarised light and is said to have negative helicity. If the amplitudes E_1^0 and E_2^0 are not equal in magnitude, then eqn. (5.90) may be written as

$$E_x(r,t) = E_1^0 \cos(\kappa z - \omega t)$$

$$E_y(r, t) = E_2^0 \sin(\kappa z - \omega t) \qquad ...(5.92)$$

Squaring and adding we obtain,

$$\frac{E_x^2}{\left(E_1^0\right)^2} + \frac{E_y^2}{\left(E_2^0\right)^2} = 1 \qquad ...(5.93)$$

which is the eqn. of ellipse. This indicates that the resultant wave is elliptically polarised as shown in Fig. 5.9 for a given point in space.

5.16 BOUNDARY CONDITION AT THE SURFACE OF DISCONTINUITY

In order to discuss the behaviour of electromagnetic waves at the boundary, we must find the boundary conditions holding at the surface of discontinuity between two media.

In each of the two media, let us assume that the solution of Maxwell's equations that we desire is a plane wave, just as in an infinite medium. At the boundary between the two media, however, certain boundary conditions are to be met and these demand that there be definite relations between the waves in the two media. In this section we shall investigate the boundary conditions which must be satisfied by field vectors $\vec{D}, \vec{B}, \vec{E}$ and $\vec{H}$ at the interface between media.

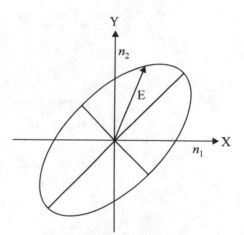

Elliptically Polarised Wave

Fig. 5.9

(i) Boundary Condition for Electric Displacement Vector $\vec{D}$

According to Maxwell's first equation div, $\vec{D} = \rho$ $\qquad ...(5.94)$

At any interface between two media, let us construct a pillbox like surface S composed of S_1, S_2, S_3 and S_4 as shown in Fig. 5.10. Integrating eqn. (5.94) over the pillbox shaped volume V, we obtain

$$\int_V \mathrm{div}\, \vec{D} \cdot dv = \int_V \rho dv \qquad ...(5.95)$$

Changing volume integral of LHS of above equation into surface integral by using Gauss divergence theorem, we get

$$\oint_S \vec{D} \cdot \vec{ds} = \int_V \rho \, dv$$

i.e.,
$$\int_{S_1} \vec{D}_1 \cdot \hat{n}_1 \, ds + \int_{S_2} \vec{D}_2 \cdot \hat{n}_2 \, ds + \int_{S_3} \vec{D}_1 \cdot \hat{n}_1' \, ds + \int_{S_4} \vec{D}_2' \cdot \hat{n}_2' \, ds = \int_V \rho \, dv \qquad ...(5.96)$$

where n's represent unit vector normal to concerned surface.

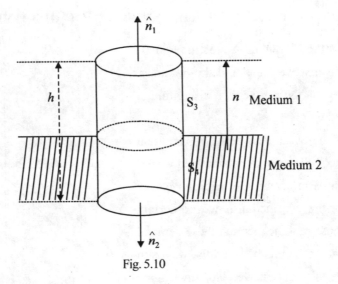

Fig. 5.10

In above eqn. the third and fourth terms give the contribution to the surface integral from the walls of the pillbox.

If $\vec{D}$ is bounded as $h \to 0$, then the third and the fourth terms vanish and S_1 approaches S_2 geometrically and the entire surface takes the form as shown in Fig. 5.10.

Thus in the limit $h \to 0$, equation (5.96) takes the form

$$\lim_{h \to 0} \left[\int_{S_1} \vec{D}_1 \cdot \hat{n}_1 \, ds + \int_{S_2} \vec{D}_2 \cdot \hat{n}_2 \, ds \right] = \lim_{h \to 0} \int_V \rho \, dv$$

If σ is the surface charge density of the interface, the above equation takes the form

$$\vec{D}_1 \cdot \hat{n}_1 \ S_1 + \vec{D}_2 \cdot \hat{n}_2 \ S_2 = \sigma A$$

i.e.,
$$(\vec{D}_1 \cdot \hat{n}_1 + \vec{D}_2 \cdot \hat{n}_2) A = \sigma A \qquad (\because S_1 = S_2 = A)$$

Since area is arbitrary and $\hat{n}_1 = \hat{n}; \ \hat{n}_2 = -\hat{n}$.

we get, $\vec{D}_1 \cdot \hat{n} - \vec{D}_2 \cdot \hat{n} = \sigma$

i.e., $\qquad D_{1n} - D_{2n} = \sigma$ ...(5.97)

where D_{1n} and D_{2n} are the normal components of electric displacement vector in the two media, the normal being chosen $\hat{n}$. Thus we conclude that the normal component of electric displacement is not continuous at the interface but charges by an amount equal to the free surface charge density of charge at the interface.

(ii) Boundary Condition for Magnetic Induction B

Maxwell's second eqn. is, $\nabla \cdot \vec{B} = 0$ ...(5.98)

Taking volume integral of the above eqn. over the entire volume of pillbox, we get

$$\int_V \nabla \cdot \vec{B} \, dv = 0$$

Using Gauss divergence theorem to change volume integral into surface integral

$$\int_S \vec{B} \cdot \vec{ds} = 0$$

i.e., $\qquad \int_{S_1} \vec{B}_1 \cdot \hat{n}_1 \, ds + \int_{S_2} \vec{B}_2 \cdot \hat{n}_2 \, ds + \int_{S_3} \vec{B}_1' \cdot \hat{n}_1' \, ds + \int_{S_4} \vec{B}_2' \cdot \hat{n}_2' \, ds = 0$

If $h \to 0$, third and fourth terms vanish while S_1 and S_2 approach each other; so that in the limit $h \to 0$, above eqn. takes the form

$$\int_A (\vec{B}_1 \cdot \hat{n}_1 + \vec{B}_2 \cdot \hat{n}_2) ds = 0 \qquad\qquad (\because S_1 = S_2 = A)$$

Since surface is arbitrary, we get

$$\vec{B}_1 \cdot \hat{n}_1 + \vec{B}_2 \cdot \hat{n}_2 = 0$$

Now taking into account the opposite direction of $\hat{n}_1$ and $\hat{n}_2$, viz. $\hat{n}_1 = \hat{n}$ and $\hat{n}_2 = (-\hat{n})$. We get,

$$\vec{B}_1 \cdot \hat{n} + \vec{B}_2 \cdot \hat{n} = 0$$

i.e., $\qquad B_{1n} = B_{2n}$ ...(5.99)

i.e., the normal component of magnetic induction $\vec{B}$ is continuous across the interface.

(iii) Boundary Condition for $\vec{E}$

Maxwell's third equation is curl $\vec{E} = -\dfrac{\partial \vec{B}}{\partial t}$ ...(5.100)

At the interface between two media, we now construct a rectangular loop *abcd* bounding a surface S as shown in Fig. 5.11.

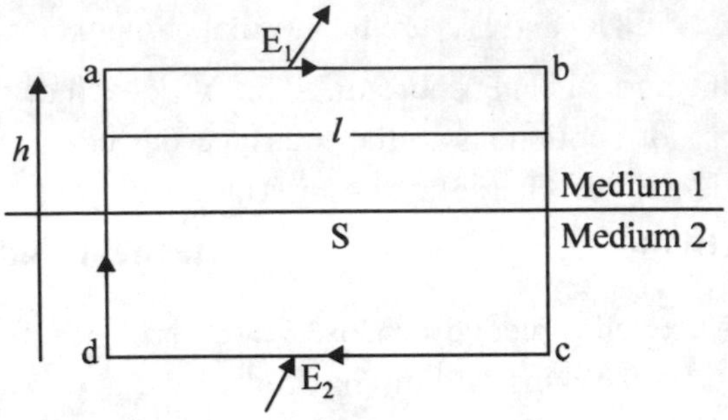

Integrating eqn. (5.100) over the loop *abcd*, we get,

$$\int_S \text{curl } \vec{E} \cdot \hat{n}\, ds = -\int_S \frac{\partial \vec{B}}{\partial t} \cdot \hat{n}\, ds$$

Fig. 5.11

Now transforming surface integral on LHS into line integral over the path *abcd* by using Stokes' theorem, we get

$$\oint_{abcd} \vec{E} \cdot \vec{dl} = -\int_S \frac{\partial \vec{B}}{\partial t} \cdot \hat{n}\, ds \qquad ...(5.101a)$$

or, $\displaystyle \int_{ab} \vec{E}_1 \cdot \vec{dl} + \int_{cd} \vec{E}_2 \cdot \vec{dl} + \int_{bc \text{ and } da} \vec{E} \cdot \vec{dl} = -\int_S \frac{\partial \vec{B}}{\partial t} \cdot \hat{n}\, ds$ ...(5.101b)

If $h \to 0$, then the contribution to the line integral from sides *bc* and *da* will vanish,

i.e., $\displaystyle \lim_{h \to 0} \int_{BC \text{ and } DA} \vec{E} \cdot \vec{dl} = 0$

and also the surface integral on RHS tends to zero provided $\partial \vec{B}/\partial t$ is everywhere

$$\left[\int_{ab} \vec{E}_1 \cdot \vec{dl} + \int_{cd} \vec{E}_2 \cdot \vec{dl} + \right] = 0$$

or, $E_{1t} l - E_{2t} l = 0$ *i.e.* $E_{1t} = E_{2t}$ ...(5.102)

where E_{1t} and E_{2t} are the tangential components of the electric field in the two media. Eqn. (5.102) represents that the tangential component of $\vec{E}$ must be continuous across the interface.

$$H_{1t} l - H_{2t} l = J_{S1} l$$

i.e., $H_{1t} - H_{2t} = J_{S1}$ ...(5.103)

Thus the tangential component of magnetic field intensity is not continuous at the interface, but changes by an amount equal to the component of surface current density perpendicular to tangential component of $\vec{H}$.

The surface current density is zero unless the conductivity is infinite; hence for finite conductivity $J_s = 0$; so

$$H_{1t} - H_{2t} = 0 \quad i.e., \quad H_{1t} = H_{2t}$$

That is, unless one medium has finite conductivity, the tangential component of magnetic field intensity is continuous.

5.17 EM VECTOR AND SCALAR POTENTIAL

Maxwell's first order differential equations relating the various components of electric field and magnetic field can be solved as they stand in certain simple cases. It is often convenient to introduce electromagnetic potentials, obtaining a small number of second order equations, while satisfying some of Maxwell's equations identically. Electromagnetic fields can be derived in terms of these potentials.

Maxwell's field equations are $\nabla \cdot \vec{D} = \rho$ ...(5.104a)

$$\nabla \cdot \vec{B} = 0 \qquad \qquad ...(5.104b)$$

$$\text{curl } \vec{E} = -\frac{\partial \vec{B}}{\partial t} \qquad \qquad (5.104c)$$

$$\text{curl } \vec{H} = \vec{J} + \frac{\partial \vec{D}}{\partial t} \qquad \qquad ...(5.104d)$$

According to eqn. (5.108b) $\nabla \cdot \vec{B} = 0$ (i.e., field vector $\vec{B}$ is solenoidal) and we know that div of curl of any vector is always zero; hence $\vec{B}$ can be expressed as the curl of a vector $\vec{A}$, i.e.

$$\vec{B} = \text{curl } \vec{A} = \nabla \times \vec{A} \qquad \qquad ...(5.105)$$

where $\vec{A}$ is known as electromagnetic vector potential. Substituting value of $\vec{B}$ from (5.105) in (5.104c), we get,

$$\text{curl } \vec{E} = -\frac{\partial}{\partial t} (\text{curl } \vec{A})$$

$$\Rightarrow \quad \text{curl } \left(\vec{E} + \frac{\partial \vec{A}}{\partial t} \right) = 0 \qquad \qquad ...(5.106)$$

(iv) Boundary Condition for $\vec{H}$

Maxwell's fourth eqn. is, $\text{curl } \vec{H} = \vec{J} + \dfrac{\partial \vec{D}}{\partial t}$ ...(5.107)

Again taking the surface integral over rectangular loop *abcd*, we get

$$\int_S \text{curl } \vec{H} \cdot \hat{n} \, ds = \int_S \left(\vec{J} + \frac{\partial \vec{D}}{\partial t} \right) \cdot \hat{n} \, ds$$ ...(5.108)

Using Stokes' theorem to change surface integral on LHS of (5.108) into line integral over the path *abcd*, we get

$$\oint_S \text{curl } \vec{H} \cdot \vec{dl} = \int_S \left(\vec{J} + \frac{\partial \vec{D}}{\partial t} \right) \cdot \hat{n} \, ds$$

or, $$\int_{ab} \vec{H}_1 \cdot \vec{dl} + \int_{cd} \vec{H}_2 \cdot \vec{dl} + \int_{bc+da} \vec{H}_1 \cdot \vec{dl} = \int_S \left(\vec{J} + \frac{\partial \vec{D}}{\partial t} \right) \cdot \hat{n} \, ds$$ ...(5.109)

If $h \to 0$, we note that the contribution to the line integral from sides *bc* and *da* vanishes

$$\lim_{h \to 0} \int_{bc+da} \vec{H} \cdot \vec{dl} \to 0$$ ...(5.110a)

$$\lim_{h \to 0} \int_S \frac{\partial \vec{D}}{\partial t} \cdot \hat{n} \, ds \to 0$$ ...(5.110b)

If $\dfrac{\partial \vec{D}}{\partial t}$ is bounded everywhere

and $$\lim_{h \to 0} \int_S \vec{J} \cdot \hat{n} \, ds \to J_{S\perp} l$$ ...(5.110c)

where $J_{S\perp}$ represents the component of surface current density perpendicular to the direction of $\vec{H}$ component which is being matched. The idea of a surface current density is closely analogous to that of a surface charge density, it represents a finite current in an infinitesimal layer.

Then in the limit $h \to 0$, eqn. (5.109) takes the form

$$\int_{ab} \vec{H}_1 \cdot \vec{ds} + \int_{cd} \vec{H}_2 \cdot \vec{dl} = J_{S\perp} l$$

This implies that the vector $\left(\vec{E} + \dfrac{\partial \vec{A}}{\partial t}\right)$ has zero curl, i.e. $\left(\vec{E} + \dfrac{\partial \vec{A}}{\partial t}\right)$ is irrotational and

we know that the curl of gradient of any scalar function is always zero; hence the vector

$\left(\vec{E} + \dfrac{\partial \vec{A}}{\partial t}\right)$, can be expressed as the gradient potential.

Equations (5.105) and (5.106) express electromagnetic field vectors $\vec{E}$ and $\vec{B}$ in terms of electromagnetic scalar and vector potentials ϕ and A. These electromagnetic potentials play a very important role in electrodynamics, particularly in relativistic electromagnetics (5.104d) is,

$$\text{curl } \vec{H} = \vec{J} + \frac{\partial \vec{D}}{\partial t}$$

In terms of $\vec{B}$ and $\vec{E}$ (using $\vec{B} = \mu \vec{H}$ and $\vec{D} = \epsilon \vec{E}$)

this eqn. takes the form, $\text{curl } \vec{B} = \mu \vec{J} + \mu \epsilon \dfrac{\partial \vec{E}}{\partial t}$ ...(5.111)

Using eqn. (5.109) and (5.110), we get,

$$\text{curl curl } \vec{A} = \mu \vec{J} + \mu \epsilon \frac{\partial}{\partial t}\left(-\frac{\partial A}{\partial t} - \text{grad } \phi\right)$$

i.e., $\qquad \text{grad div} \vec{A} - \nabla^2 \vec{A} = \mu \vec{J} - \mu \epsilon \dfrac{\partial^2 \vec{A}}{\partial t^2} - \mu \epsilon \dfrac{\partial}{\partial t}(\nabla \phi)$

Rearranging, we get, $\nabla^2 \vec{A} - \mu \epsilon \dfrac{\partial^2 \vec{A}}{\partial t^2} - \text{grad}\left(\nabla \cdot \vec{A} + \mu \epsilon \dfrac{\partial \phi}{\partial t}\right) = -\mu \vec{J}$ ...(5.112)

Similarly eqn. (5.104a) in terms of $\vec{E}$ takes the form $\nabla \cdot \vec{E} = \dfrac{\rho}{\epsilon}$

Using (5.112) we get, $\text{div}\left(-\dfrac{\partial \vec{A}}{\partial t} - \nabla \phi\right) = \dfrac{\rho}{\epsilon}$

$$\text{div grad } \phi + \frac{\partial}{\partial t}(\nabla \cdot \vec{A}) = -\frac{\rho}{\epsilon}$$

Adding and subtracting $\mu \in \dfrac{\partial^2 \phi}{\partial t^2}$, and rearranging

we get, $\qquad \nabla^2 \phi - \mu \in \dfrac{\partial^2 \phi}{\partial t^2} + \dfrac{\partial}{\partial t}\left(\nabla \cdot \vec{A} + \mu \in \dfrac{\partial \phi}{\partial t} \right) = -\dfrac{\rho}{\in}$ ...(5.113)

Equations (5.112) and (5.113) represent two inhomogeneous wave equations in terms of electromagnetic potentials $\vec{A}$ and ϕ in place of four Maxwell equations. These two equations contain $\vec{A}$ and ϕ coupled with each other. These inhomogeneous wave equations can be uncoupled by using the concept that $\vec{A}$ and ϕ are arbitrary.

Note: Generally $\vec{A}$ and ϕ are chosen subject to the condition

$$\text{div } \vec{A} + \mu \in \dfrac{\partial \phi}{\partial t} = 0 \qquad \qquad ...(5.114)$$

Since div $\vec{A}$ is so far arbitrary, we can do this. Such a condition is known as the Lorentz condition. Then the equations (5.112) and (5.113) for electromagnetic potentials become

$$\nabla^2 \vec{A} - \mu \in \dfrac{\partial^2 \vec{A}}{\partial t^2} = -\mu \vec{J} \qquad \qquad ...(5.115)$$

and $\qquad \nabla^2 \phi - \mu \in \dfrac{\partial^2 \phi}{\partial t^2} = -\dfrac{\rho}{\in}$ ...(5.116)

$\vec{A}$ and ϕ satisfy these equations then the fields determined by equations (5.105) and (5.106) satisfy Maxwell's equation. Equations (5.114), (5.115) and (5.116) form a set of equivalent in all respects to Maxwell's equation.

Equations (5.115) and (5.116) for $\vec{A}$ and ϕ take prescribed values of charge and current density.

Sometimes, however, we wish to assume that the current density obeys Ohm's law $(\vec{J} = \sigma \vec{E})$ and that the charge density is zero, as we should have in the interior of a conductor. In that case, proceeding in a similar manner, we find that in place of Lorentz condition (5.114), we should assume

$$\nabla \cdot \vec{A} + \mu \in \dfrac{\partial \phi}{\partial t} + \mu \sigma \phi = 0 \qquad \qquad ...(5.117)$$

and equations (5.115) and (5.116) become

$$\nabla^2 \vec{A} - \mu \sigma \dfrac{\partial \vec{A}}{\partial t} - \mu \in \dfrac{\partial^2 \vec{A}}{\partial t^2} = 0 \qquad \qquad ...(5.118)$$

SOLVED EXAMPLES

Q.1. **Calculate the speed of electromagnetic wave in free space.**

Sol.

For free space, $\epsilon_0 = 8.85 \times 10^{-12}$ coul^2/N–m^2

and $\qquad \mu_0 = 4\pi \times 10^{-7}$ weber/A–m

So, the speed of electromagnetic wave is

$$V = \frac{1}{\sqrt{\mu_0 \, \epsilon_0}} = \frac{1}{\sqrt{4\pi \times 10^{-7} \times 8.85 \times 10^{-12}}}$$

$$= \frac{1}{\sqrt{4 \times 3.14 \times 8.85 \times 10^{-19}}} = \frac{10^9}{\sqrt{4 \times 0.314 \times 8.85}}$$

$$= \frac{10^9}{2\sqrt{2.7789}} = \frac{10^9}{2 \times 1.667}$$

$$= \frac{10^9}{3.334} \times \frac{10 \times 10^8}{3.334} = 2.9994 \times 10^8 \text{ m/s}$$

$$\cong 3 \times 10^8 \text{ m/s}.$$

Q.2. **Find the speed of EM wave in a medium of relative permeability 2 and relative permittivity 8.**

Sol.

Here, $\quad \mu_r = 2, \ \mu_0 = 4\pi \times 10^{-7}$ wb/A–m

$\qquad \epsilon_r = 8, \ \epsilon_0 = 8.85 \times 10^{-12}$ coul^2/N–m^2

So, the speed of electromagnetic wave is

$$V = \frac{1}{\sqrt{\mu_r \, \mu_0 \, \epsilon_r \, \epsilon_0}} = \frac{1}{\sqrt{2 \times 4\pi \times 10^{-7} \times 8 \times 8.85 \times 10^{-12}}}$$

$$= \frac{10^9}{8\sqrt{3.14 \times 8.85 \times 10^{-1}}} = \frac{10^9}{8\sqrt{0.314 \times 8.85}} = \frac{10^9}{8 \times 1.667}$$

$$= \frac{10^9}{13.336} = \frac{1000 \times 10^9}{13.336} = 74.985 \times 10^6 \text{ m/sec}$$

$$\cong 75 \times 10^6 \text{ m/sec}.$$

Q.3. A medium has a dielectric constant 8. At a frequency of 1.5 MHz the displacement and conduction current are equal. What is conductivity of the medium?

Sol. The conduction and displacement currents are equal if $\sigma = \omega\in$

Here, $\qquad \in_r = 8, \ \in_0 = 8.85 \times 10^{-12} \ coul^2/N.m^2$

$\qquad\qquad f = 1.5 \ MHz = 1.5 \times 10^6 \ H_z$

Hence, $\qquad \sigma = \omega\in = 2\pi f \in_r \in_0$

$\qquad\qquad\qquad = 2 \times 3.14 \times 1.5 \times 10^6 \times 8 \times 8.85 \times 10^{-12}$

$\qquad\qquad\qquad = 6.67 \times 10^{-4}/ohm-m.$

Q.4. Calculate the wavelength of EM wave corresponding to frequency 8×10^{14} Hz travelling in glass ($\mu = 1.5$).

Sol. The velocity of wave in glass is $V = \dfrac{C}{\mu} = \dfrac{3 \times 10^8}{1.5} = 2 \times 10^8 \ m/sec$

we have $\qquad V = f\lambda$

or, $\qquad\qquad \lambda = \dfrac{V}{f} = \dfrac{2 \times 10^8}{8 \times 10^{14}} = \dfrac{1}{4} \times 10^{-6} = 0.25 \times 10^{-6} = 2500 \ Å.$

QUESTIONS

1. Derive Maxwell's electromagnetic field equations.

2. What is Poynting vector? Discuss its significance.

3. Obtain an expression for Poynting vector in electromagnetic field.

4. Discuss the propagation of electromagnetic waves in free space. Show that electric and magnetic fields are mutually perpendicular and perpendicular to direction of wave propagation.

5. Discuss the propagation of EM waves in a non-conducting isotropic medium.

6. Discuss on the polarisation of EM waves.

7. Derive boundary conditions for $\vec{E}, \vec{B}, \vec{D}$ and $\vec{H}$ at the surface of discontinuity.

8. Discuss on the electromagnetic vector and scalar potential. Find expressions for them.

9. Discuss on reflection and refraction of plane EM waves by a dielectric medium.

10. Discuss on reflection and transmission at a conducting surface for normal incidence.

11. Derive expressions for reflection and transmission coefficient of EM wave at normal incidence.

12. How should EM vector and scalar potential change simultaneously such that $\vec{E}$ and $\vec{B}$ remain the same?

Problems

1. A medium is characterised by relative permittivity $\epsilon_r = 45$ and relative permeability $\mu_r = 5$. Calculate the speed of electromagnetic wave in the medium and permeability of the medium.

   (**Ans.** $\mu = 20\pi \times 10^{-7}$, $V = 0.6 \times 10^{16}$ m/s)

2. The electromagnetic wave is propagating in free space with electric vector

   $$\vec{E} = 150 \cos(\omega t - KZ)\hat{x}$$

   How much average energy is passing through a rectangular hole of length 3 cm and width 1.5 cm on YZ or XZ plane in one minute time?

   (**Ans.** 0.806 J)

3. In free space, electric field intensity is given as $\vec{E} = 20 \cos(\omega t - 50x)\hat{y}$ volt/m. Calculate displacement current density.

   (**Ans.** $20\ \omega\epsilon_0 \sin(\omega t - 50x)$)

4. Calculate the speed of EM wave in vacuum using the following data:

   $$\epsilon_0 = 8.8547 \times 10^{-12}\ \text{Coul}^2/\text{N.m}^2$$

   and $\quad \mu_0 = 4\pi \times 10^{-7}$ weber/Amp.m

   (**Ans.** 3×10^8 m/s)

5. The electric component of a plane EM wave propagating in a non-magnetic medium is given by $\vec{E} = \hat{y}\ 40 \cos(10^8 t + 4Z)$ V/m. Find the direction of propagation.

   (**Ans.** The direction of propagation is along Z direction)

● ● ●

Chapter 6

Quantum Mechanics

INADEQUACY OF CLASSICAL MECHANICS

Sir Isaac Newton formulated three laws of motion, which form the footing of the most fundamental principles of classical mechanics. The equations of motion based on these laws are the simplest and they are suitable for solution of simple dynamical problems such as the motion of macroscopic bodies. Classical mechanics also includes Lagrange's equations, Hamilton's equations and Hamilton's principle, which form the fundamental principles of classical mechanics, because they are consistent with each other and with Newton's laws of motion. Lagrange's and Hamilton's equations are useful for solution of complicated dynamical problems.

As the bulk matter consists of electrons and atomic nuclei, its properties, can be deduced from the properties of electrons, protons and neutrons, etc. But, to our greater surprise, it is noticed that the observed properties of matter cannot be explained on the assumptions that the atomic particles obey the laws of classical mechanics. Such inadequacy of classical physics led to new concepts in order to explain the properties of bulk matter. Based on such new concepts, an advanced mechanics has been developed. The new mechanics is called the Quantum Mechanics.

The idea of Quantum sprang from Planck's theory of radiation. Max Planck, in 1900, gave his theory of radiation, according to which radiation is emitted or absorbed by matter in discrete packets or quanta of energy. Each packet of energy consists of 'hv' amount of energy, where 'v' is the frequency of radiation and 'h' is the Planck's constant ($h = 6.6262 \times 10^{-34}$ JS). The theory of Planck was comprised of a mixture of classical and non-classical concepts, and was not completely satisfactory.

In 1905, Einstein correlated the particle nature and wave nature of light. The former was given by Newton and the latter was given by Huygens. Einstein declared that the EM radiation travels in the form of wave which consists of energy packet, called photons (light particles), the energy in each photon is 'hv'. This led the conclusion that light or EM radiation is dualistic in nature.

Further, Quantum Mechanics embraces two new view points to its fold, viz. **Matrix Mechanics** and **Wave Mechanics**. First one was introduced by Werner Heisenberg in 1925. In this mechanics, unobserved quantities such as positions, velocities, etc. in electronic orbits are omitted and only observed quantities frequencies and intensities of spectral lines are taken into account. The other form of quantum mechanics (wave mechanics) was mathematically treated by Erwin Schrodinger in 1926. In this mechanics earlier concepts of classical wave theory are combined with Louis de Broglie's wave-particle relationship. The mathematical theories of Wave Mechanics and Matrix Mechanics appear to be different, but actually they are equivalent. Quantum Mechanics solves many unsolved problems of atomic physics satisfactorily. However, it has also some limitations. A more complete theory of particles called quantum field theory has been accepted since 1947.

In order to understand the development of Wave Mechanics, one must follow clearly the black body radiation proposed by Max Planck, and which could not be explained by Classical Mechanics. The wave theory of Huygens could not explain photoelectric effect of Einstein and the Compton effect. But, the Quantum theory of Planck could explain these two phenomena successfully.

Black Body Radiation and Energy Distribution in the Spectrum

An ideal black body is defined as one which absorbs all the radiation that falls upon it. No perfect black body is known, but a surface coated with lamp black is a good approximation. A small hole in the wall of a metal tube appears darker than the surrounding surface, because almost all the light entering the hole is absorbed. Thus a cavity having only a small aperture is almost a black body in the technical sense of absorbing all radiation incident upon it. Fig. 6.1(a) and (b) are the pictorial representations of the two black body structures. The form of black body used in radiation experiments is shown below (Fig. 6.2). It consists of a porcelain sphere 'S' having a small opening 'A'. The inner surface is coated with lamp black. Any radiation which enters the sphere through the opening suffers a few reflections. At each reflection, about 98% of the incident radiation is absorbed. Thus, after a few reflections at the inner surface, the radiation is completely absorbed. If the area of the opening is very small, the radiation cannot be reflected out of the sphere again. The sphere emits radiant energy through opening. To study the distribution, of radiant energy over different wavelengths in the black body radiation at a given temperature, the black body maintained at a constant temperature. By means of an infra-red spectrometer and a bolometer, the emissive powers of the black body for different wavelengths are measured.

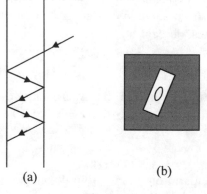

(a)   (b)

Fig. 6.1

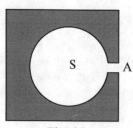

Fig. 6.2

The experimental results, as obtained by Lummer and Pringsheim (1899), are shown in Fig. 6.3.

The curves are plotted for temperature of the black body ranging from 700 to 1600 °C. The ordinates represent the emissive powers E(λ) and the abscissae represent wavelength λ between the beginning of the visible spectrum, and the value 6μ.

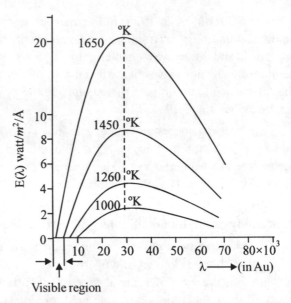

Fig. 6.3: Spectral distribution of black body radiation.

The emissive power E(λ) is defined in such a way that the quantity E(λ) dλ is the energy radiated in 1 sec per unit area for wavelengths in the range between λ and λ + dλ. Unit of E(λ) is watt/m² per unit range of wavelength, so that the corresponding unit of E(λ)dλ is watt/m².

From the curves of Fig. 6.3, two features are notable:

(i) The area under a curve, representing total power radiated, is proportional to the fourth power of the temperature of the source, i.e. $P \propto T^4$ or $P = rT^4$ which is called Stefan-Boltzmann law.

(ii) As the temperature is increased, the wavelength λ_{max} of maximum radiancy shifts towards higher frequency (or shorter wavelength) in accordance with the relation.

$$\lambda_{max} T = \text{constant} = 2897.2 \ \mu K° \qquad \text{...(6.1)}$$

which is known as Wien's displacement law.

Physicists tried to find an equal which would fit the experimental current of Fig. 6.3 and to devise a theory of black body radiation. In 1896, Wilhelm Wien suggested an empirical equation, still used in calculating temperatures of incandescent bodies,

$$R_x = c_1 \lambda^{-5} e^{-c_2/\lambda T} \qquad \text{...(6.2)}$$

where constants C_1 and C_2 were determined by fitting a curve to the experimental data. But, attempts to construct a satisfactory theory of radiation within the framework of 19th century physics failed.

Rayleigh-Jeans Law

In 1900, Lord Rayleigh applied the principle of equipartition of energy to the electromagnetic vibrations. Then, with a contribution from J.H. Jeans, this attempt led to the deduction of a formula for the energy per unit volume inside an enclosure with perfectly reflecting walls. This formula is called the Rayleigh-Jeans law. According to this law, the energy per unit volume with frequency in the range v and $v = dv$ is given by

$$U(v)dv = \frac{8\pi v^2 kT}{C^3}dv \qquad ...(6.3)$$

where $U(v)$ is the function of the frequency and it is the energy per unit volume per unit frequency range at frequency v, K is the Boltzmann's constant and C is the speed of light in free space.

Eqn. (6.3) can be transformed in terms of the wavelength λ by using the relation

$$v = \frac{c}{\lambda} \qquad ...(6.4)$$

and

$$dv = -\frac{c}{\lambda^2}d\lambda, \qquad ...(6.5)$$

The energy $U(v)\ dv = U(\lambda)d\lambda$. Hence, from eqn. (6.3), we have

$$U(\lambda)d\lambda = \frac{8\pi}{C^2}\left(\frac{C}{\lambda}\right)^2 KT\left(-\frac{C}{\lambda^2}d\lambda\right) = -\frac{8\pi\ KT}{\lambda^4}d\lambda \qquad ...(6.6)$$

This eqn. (6.6) is another form of Rayleigh-Jeans law.

The law explains the experimental results for long wavelengths at any given temperature, but not for shorter wavelengths (Fig. 6.3). According to eqn. (6.6), we have, as λ decreases, the energy density $U(\lambda)$ will continuously increase and as λ tends to become zero, $U(\lambda)$ approaches infinity. This is contrary to the experimental results.

According to the law, the total energy of radiation per unit volume of the enclosure for all wavelengths from 0 to ∞ is given by

$$U = \int_0^\infty U(\lambda)d\lambda = -\int_0^\infty \frac{8\pi\ kT}{\lambda^4}d\lambda = -8\pi\ kT\int_0^\infty \frac{1}{\lambda^4}d\lambda$$

$$= -8\pi\ kT\left[\frac{1}{3\lambda^3}\right]_0^\infty = \infty \qquad ...(6.7)$$

This result shows that for a given quantity of radiant energy, almost all the energy will finally be confined in vibrations of very small wavelengths. Thus, if the classical treatment is correct, on opening a shutter in the black body cavity, we would be bombarded with radiation of extremely short wavelengths. This is called ultraviolet catastrophe. This absurd result is because of the assumption that energy can be absorbed or emitted by the atomic oscillators continuously in any amount.

Planck's Quantum Hypothesis

In order to explain the distribution of energy in the spectrum of a black body, Max Planck in 1900, put forward the quantum theory. He assumed that the atoms in the walls of a black body behave like simple harmonic oscillators and each has a characteristic frequency of oscillation. In his theory he made the following two radical assumptions about the atomic oscillators:

(i) A simple harmonic oscillator cannot have any value of total energy. It can have only those values of the total energy E which satisfy the relation.

$$E = nhv,$$ ...(6.8)

where $n = 0, 1, 2, 3 \ldots$, and is called the quantum number, v is the frequency of oscillation and h is a universal constant called Planck's constant ($h = 6.625 \times 10^{-34}$ JS). Here 'hv' amount of energy is called Quantum of energy. Its value depends only on the frequency of radiation.

(ii) The oscillator does not emit or absorb energy continuously. The emission or absorption of energy occurs only when the oscillator jumps from one energy level to another. The energy lost or gained is in the form of discrete quanta of energy hv.

The quantum theory of Planck is applicable only to the process of emission and absorption of energy.

In 1905, Einstein extended Planck's quantum theory by assuming that a quantum of energy is the light particle 'photon'. Each photon has energy hv. The speed of light in space is the speed of its photon.

Planck's Law

According to Planck's quantum theory, the average energy of an oscillator is given by

$$E = \frac{hv}{e^{hv/KT} - 1}$$ ...(6.9)

It can be shown that the number of oscillations of degrees of freedom per unit volume in the frequency range v and $v + dv$ is given by

$$N(v)dv = \frac{8\pi v^2}{c^3} dv,$$ ...(6.10)

where c is the speed of light in vacuum. Then, following eqn. (6.3), Planck deduced the equation

$$U(v)dv = \frac{8\pi hv^3}{c^3} \cdot \frac{1}{e^{hv/KT} - 1} dv$$ ...(6.11)

where $U(v)\, dv$ is the energy per unit volume in the frequency range v and $v = dv$ and $U(v)$ is the energy per unit volume per unit frequency range at frequency v. In terms of nthe wavelength of the radiation, this equation (6.11) becomes

$$U(v)dv = +\frac{8\pi hc}{X^r} \cdot \frac{1}{e^{hc/\lambda KT} - 1} d\lambda \qquad \qquad ...(6.12)$$

Eqns. (6.11) and (6.12) are two forms of Planck's radiation law.

Plotting $U(\lambda)$ against λ, the curves of Fig. 6.4 are obtained. These curves agree very well with the experimental results over the whole range of wavelengths.

However, Planck's law of radiation is the generalisation of all the previous laws of electromagnetic radiation.

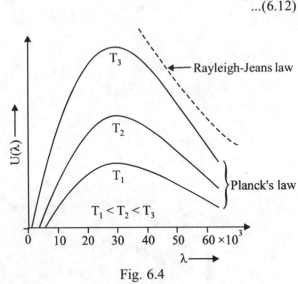

Fig. 6.4

Deductions from Planck's Law of EM Radiation

(i) Rayleigh-Jeans law: For small values of $hc/\lambda KT$, i.e. in the region of long wavelengths, the exponential term in Eqn. (6.5) can be expanded and retaining only the first term, we obtain

$$U(\lambda)d\lambda = +\frac{8\pi hc}{\lambda^5} \cdot \frac{1}{hc/\lambda KT} d\lambda = \frac{8\pi KT}{\lambda^4} d\lambda \qquad \qquad ...(6.13)$$

This is the Rayleigh-Jeans law.

(ii) Wien's law: In the region of low wavelengths, $\dfrac{hc}{\lambda KT}$ becomes large. Hence, 1 in the denominator can be neglected in comparison with the exponential term. Therefore, we get

$$U(\lambda)d\lambda = \frac{8\pi hc}{\lambda^5} \cdot e^{-hc/\lambda KT} d\lambda$$

$$= \frac{A}{\lambda^5} f(\lambda T) d\lambda$$

where, $\qquad A = 8\pi hc$ and $f(\lambda T) = e$

This is Wien's law.

(iii) Stefan-Boltzmann law: The energy density of the total radiation in a black body enclosure is given by

$$U = \int_0^\infty U(\lambda)d\lambda = \int_0^\infty \frac{8\lambda hc}{\lambda^5} \cdot \frac{1}{e^{hc/\lambda KT} - 1} d\lambda$$

Let $\qquad \qquad x = hc/\lambda KT$

$$\therefore \qquad \lambda = \frac{hc}{KTx}$$

or, $$d\lambda = -\frac{hc}{KTx^2}dx$$

when $\lambda = 0$, $x = \infty$ and when $\lambda = \infty$, $x = 0$.

Hence $$U = \int_{\infty}^{0} \frac{8\pi hc}{1} \times \left(\frac{KTx}{hc}\right)^5 \frac{1}{e^x - 1}\left(-\frac{hcd_x}{KTx^2}\right)$$

$$= -\int_{\infty}^{0}\left(\frac{8\pi K^4 T^4}{h^3 c^3}\right)\frac{x^3}{e^x - 1}dx$$

$$= \frac{8\pi K^4 T^4}{h^3 c^3}\int_{0}^{\infty}\frac{x^3}{e^x - 1}dx$$

The value of the integral is $\dfrac{\pi^4}{15}$.

$$\therefore \qquad U = \frac{8\pi K^4 T^4}{h^3 c^3}\cdot\frac{\pi^4}{15} = \frac{4}{c}\left(\frac{2}{15}\cdot\frac{\pi^5 K^4}{h^3 c^2}\right)T^4 \qquad \qquad ...(6.14)$$

It can be shown that the total radiation emitted per unit area in one sec by a black body at a given temperature is

$$R_B = \frac{c}{4}U \qquad \qquad ...(6.15)$$

Substituting the value of U from (6.14) in (6.15), we obtain

$$R_B = \frac{c}{4}\cdot\frac{4}{c}\left(\frac{2}{15}\cdot\frac{\pi^5 K^4}{h^3 c^2}\right)T^4 = \left(\frac{2}{15}\cdot\frac{\pi^5 K^4}{h^3 c^2}\right)T^4 \qquad \qquad ...(6.16)$$

$$= \sigma T^4 \qquad \qquad ...(6.17)$$

This is Stefan-Boltzmann law.

where $$\sigma = \frac{2}{15}\frac{\pi^5 K^4}{h^3 c^2} \qquad \qquad ...(6.18)$$

This constant is the Stefan's constant.

Substituting the values of σ, K and C in (6.18), h is calculated.

When $$\sigma = 5.67 \times 10^{-8} \text{ W/m}^2\text{K}^4$$

$$K = 1.346 \times 10^{-23} \, \text{J}/^\circ \text{K} \quad \text{(The value known at that time)}$$

$$C = 3 \times 10^8 \, m/\text{sec.}$$

The value of h comes as

$$h = 6.55 \times 10^{-34} \, \text{JS}.$$

This was the first calculated value of h. More recent and more precise experimental value of h is $6.6261965 \times 10^{-34}$ JS. When this value of h is substituted in eqn. (6.12) for obtaining the value of λ, it is found that the theoretical distribution curves agree excellently with the experimental curves over the whole range of wavelengths.

The success of Planck's hypothesis in explaining the distribution of energy in the spectrum of black body was the beginning of Quantum Mechanics. We now describe some more important phenomena which are explained by this hypothesis.

Photoelectric Effect

When a radiation of high frequency, such as ultraviolet light or X-rays is incident on a clean metal surface, electrons are emitted from the surface. Since the incident radiation causes the emission of electrons the phenomenon is called the photoelectric effect and the emitted electrons are called the photoelectrons. The whole range of electromagnetic radiation from gamma and X-rays to ultraviolet, the visible and the infrared rays produce this effect. The effect occurs also in solids, liquids and gases.

Thus, photoelectric effect is the phenomenon of emission of electrons from the surface of certain substance, mainly metals when light of shorter wavelength is incident upon them.

Laws of Photoelectric effect:

  (i) Photoelectric effect is an instantaneous process.

 (ii) Photoelectric current is directly proportional to the intensity of incident light and is independent of its frequency.

(iii) The stopping potential and hence the maximum velocity of the electrons depends upon the frequency of incident light and is independent of its intensity.

(iv) The emission of electrons stops below a certain minimum frequency known as threshold frequency.

Einstein's Theory of Photoelectric Effect

Einstein explained photoelectric effect on the basis of Planck's quantum theory of radiation.

According to Max Planck, radiation was composed of energy bundles only in the neighbourhood of the emitter. Einstein suggested that these energy packets are known as photons. Each packet carries energy

$$\text{E} = h\nu \qquad\qquad ...(6.19)$$

It is the matter of common observation that the electrons are ejected only when light is incident on a metal. This is due to the existence of a **'Potential barrier'** all round the surface of metal. The electron must possess certain minimum energy 'W_0' in order to cross the barrier. 'W_0' is known as work function. It is defined as the minimum energy required to pull an electron out from the surface of metal.

An incident photon supplies whole of its energy 'hv' to the electron, which consumes an energy 'W_0' to cross the potential barrier and comes out with the remaining energy as its Kinetic energy.

$$E_k = \frac{1}{2}mv^2$$

$$\frac{1}{2}m\,V_{max}^2 = hv - W_0 \qquad \qquad ...(6.20)$$

If v_0 is the threshold frequency, the energy 'hv_0' of the photon will be just sufficient to help the electron in overcoming the potential barrier.

$$W_0 = hv_0 \qquad \qquad ...(6.21)$$

So, $\qquad \frac{1}{2}m\,V_{max}^2 = hv - hv_0 = h(v - v_0) \qquad \qquad ...(6.22)$

Equation (6.22) is known as Einstein's eqn. of photoelectric effect. It is clear from this eqn. that:

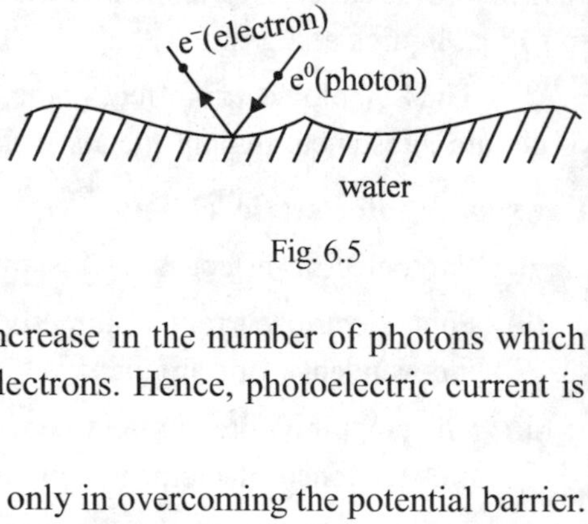

(i) Greater the frequency of incident radiation, greater is the kinetic energy of the electron and hence greater negative potential is required to stop it.

Fig. 6.5

(ii) An increase in intensity of light results in an increase in the number of photons which in turn results in ejection of more number of electrons. Hence, photoelectric current is proportional to the intensity of light.

(iii) Electrons coming from the surface spend energy only in overcoming the potential barrier. Therefore kinetic energy of these electrons in maximum. The incident radiation penetrates the metal to a thickness of about 10^{-6} cm. Thus, it will be able to eject the electrons from below the surface also. These electrons lose some of the energy as they rise to the surface and hence are emitted with a lesser kinetic energy. Different electrons lose different amounts of energy. This is the reason that electrons ejected with a variety of velocities.

(iv) If the frequency of incident radiation is less than v_0, it becomes unable to help the electrons in overcoming potential barrier. So, there exists a minimum frequency called threshold frequency.

Experimental Arrangement

The emitting surface A is mounted opposite a metal plate B in a highly evacuated glass bulb C. The plates A and B form two electrodes to which a variable PD can be applied. The glass-bulb is fitted with quartz window D. Ultraviolet light from the source S is transmitted by the quartz window and allowed to fall on the emitter surface.

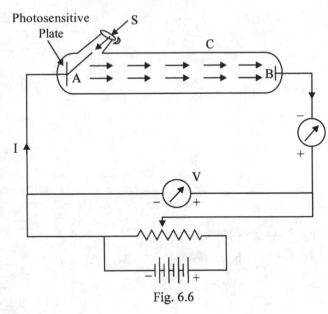

Fig. 6.6

When the applied PD between A and B is such that A is at negative potential with respect to B, the photoelectrons emitted from A are accelerated towards B. The resulting photoelectric current I flowing in the circuit is measured by the micro-ammeter μA and the accelerating PD is measured by the vacuum tube voltmeter 'V'.

If the PD is reversed so that now B is at negative potential with respect to A, the photo electrons are repelled towards A by the retarding PD. Consequently, the photoelectric current is reduced. The photoelectric current is found to depend on the following factors:

(i) the frequency of the incident radiation.

(ii) the intensity of the incident radiation.

(iii) the potential difference between the electrodes.

(iv) the nature of the emitting surface.

Stopping Potential:

If V_0 = stopping potential

$$\frac{1}{2}m\,V_{max}^2 = eV_0 \qquad \qquad ...(6.23)$$

From eqns. (2) and (5),

$$eV_0 = h\nu - h\nu_0 \qquad \left(\because \; V = \frac{\omega}{q} \right)$$

or, $\qquad V_0 = \dfrac{h}{e}\nu - \dfrac{\omega_0}{e}$

Let $\qquad \dfrac{h}{e} = A, \; \dfrac{\omega_0}{e} = B$

So, $V_0 = Av - B$ ...(6.24)

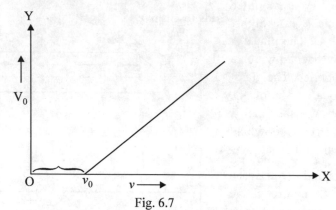

Fig. 6.7

Threshold Frequency:

If $V_0 = 0$, we have from eqn. (6.24)

$$O = Av = B$$

or, $Av = B$

or, $v = \dfrac{B}{A} = \dfrac{\omega_0}{e} \times \dfrac{l}{h} = \dfrac{hv_0}{e} \times \dfrac{e}{h} = v_0$

Thus, the intercept from v-axis is the value of threshold frequency.

Calculation of Work Function (ω_0):

If $v = 0$, $V_0 = V'$ (say)

Then, from eqn. (6.24), we obtain

$$V' = -B = -\dfrac{\omega_0}{e}$$

or, $\omega_0 = -eV' = eX(-V')$

$$= eX \text{ (intercept on } V_0\text{-axis)}$$

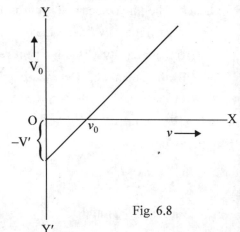

Fig. 6.8

Calculation of h:

Slope 'm' of the straight line given by eqn. (6.24) is $m = A = \dfrac{h}{e}$

$$rh = me$$

i.e., Planck's constant = eX slope of ($V_0 \sim v$) curve.

Experimental Observations

(i) Effect of Frequency on the Photoelectric Current

The collector plate B (Fig. 6.6) be made sufficiently positive with respect to the emitter plate A. The surface of A is illuminated with monochromatic light of different frequencies. It is found that photoelectric current is produced only when the frequency v of the incident light is greater than a certain minimum value v_0. This minimum frequency is called the **threshold frequency** for the surface.

(ii) Effect of the Intensity of the Incident Light

The collector plate B is made sufficiently positive with respect to the emitter plate A. By keeping the frequency v of the incident light and the PD (V) constant, the photoelectric current is measured for various intensities of the incident light.

The variation of the photoelectric current with intensity has been shown in the Fig. 6.9. It exhibits that **the photoelectric current is proportional to the intensity of the incident light**.

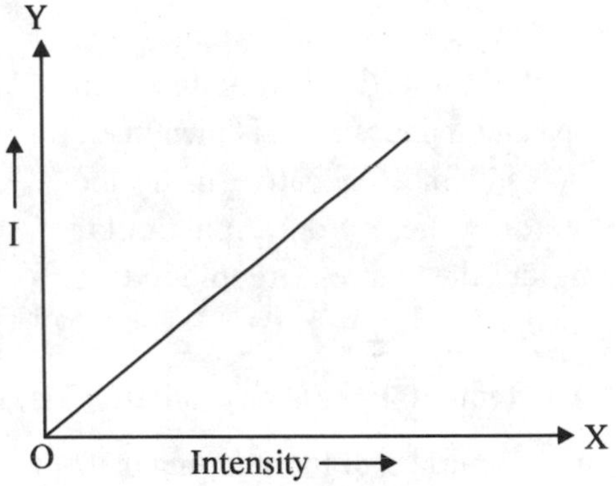

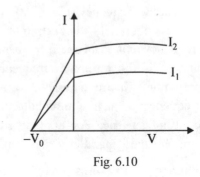

Fig. 6.9

(iii) Effect of the Potential Difference

Keeping the frequency and intensity I_1 of the incident light constant the potential difference (PD) between the electrodes is varied. Starting with a small positive potential, it is reduced to zero and then made negative, using the reversing key. The variation of photoelectric current with the PD has been shown in the Fig. 6.10. When the collector plate B is positive, the photoelectric current remains constant. When the collector plate B is made more and more negative, the photoelectric current goes on decreasing until it is stopped entirely. The retarding potential that stops the photocurrent is called the stopping potential. If the intensity of the incident light is increased to I_2, then initial photocurrent for positive potential V_0 is found to be the same for the light of the same frequency.

Fig. 6.10

The work done by the stopping potential V_0 is equal to the maximum kinetic energy of the photoelectron.

So, we have $eV_0 = \frac{1}{2}m V_{max}^2$

where m is the mass of the electron, e is the electronic charge and V_{max} is the maximum velocity of emission of the electrons.

(iv) Effect of Frequency on the Stopping Potential

The effect of frequency of the incident light on the stopping potential is shown in the Fig. 6.11.

The intensities of light of different frequencies are adjusted to produce the same maximum (I_m) value of the photoelectric current when B is the positive plate. The potential of B is then reduced in steps and made zero; it is then made more and more negative using the reversing key. For a particular emitter, the curves are plotted as above.

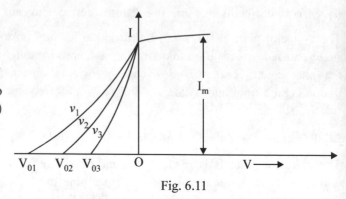

Fig. 6.11

(v) Effect of Frequency on the Maximum Kinetic Energy

The magnitudes of the stopping potentials at different frequencies, are converted into maximum kinetic energies. The maximum kinetic energy is plotted against the frequency (Fig. 6.12). Straight lines I and II are obtained. The straight line I meets the frequency axis at a certain frequency v_0. This shows that v_0 is the minimum frequency for which the maximum kinetic energy is zero. This minimum frequency v_0 is the threshold frequency for the emitting surface. For another emitting surface, a straight line II parallel to I is obtained.

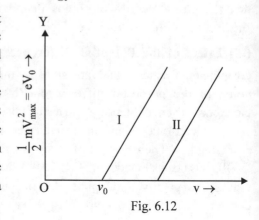

Fig. 6.12

Failure of Previous Theories on Light

The experimental observations on photoelectric effect cannot be explained by assuming light to be propagated as electromagnetic waves.

(i)   On the basis of this theory, if the intensity of illumination were increased, electrons would be emitted with greater energies. This is contrary to the experimental observation; on increasing the intensity, the electrons are ejected with the same kinetic energy.

(ii) Secondly, the existence of threshold frequency for a given material cannot be explained on the basis of the wave theory.

(iii) The fact that the electrons are ejected instantaneously cannot also be explained. If the illumination is very faint the classical theory indicates that time of several minutes would have to elapse before a single electron is released, which is contrary to the experimental observation.

Compton Effect

In 1922, A.H. Compton observed that when a monochromatic beam of X-rays of wavelength λ is scattered by a light element like carbon, it is observed that the scattered X-rays have maximum intensities at two wavelengths. One maximum occurs at the same wavelength λ as that of the incident beam and the other maximum occurs at a slightly longer wavelength X'. For his discovery, Compton was awarded the Nobel Prize in 1927 and the observed effect was named after Compton. The change in wavelength $\Delta\lambda = \lambda' - \lambda$, is called the Compton shift. It is found that $\Delta\lambda$ varies with the angle of scattering ϕ.

Experimental Arrangement of Compton:

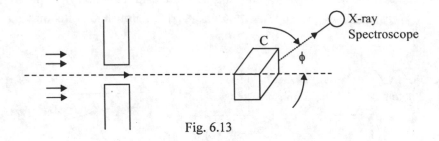

Fig. 6.13

Compton allowed a beam of monochromatic X-rays (Fig. 6.13) to fall on a scattering material C and then examined the scattered radiation with an X-ray spectroscope. Scattered radiation having the same wavelength as the incident beam was found, as expected from the electromagnetic theory. But, in addition, Compton observed a scattered wavelength λ' greater than that of the original beam (Fig. 6.14). The shift $\Delta\lambda = \lambda' - \lambda$, was found to become larger as the scattering angle ϕ was increased. This scattering with an increase in wavelength is called the Compton effect and cannot be explained by the classical wave theory, which suggests no way of introducing a frequency (or wavelength) change in scattering.

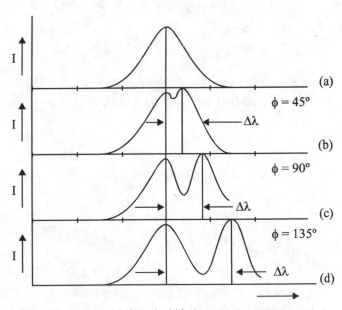

Fig. 6.14: Wavelength shift due to Compton scattering

Explanation of Compton Effect:

Compton effect could be explained on the basis of particle nature of electromagnetic radiation. We picture a photon as having zero rest mass; its total energy is its kinetic energy. Since the energy of a photon is E = hv, we can calculate its mass from E = mc^2.

$$mc^2 = hv, \text{ or } m = \frac{hv}{c^2}$$ ...(6.25)

The momentum of photon is $p = mc = \frac{hv}{c}$ ...(6.26)

We now interpret the Compton scattering as the elastic collision of a photon with an electron of the scattering material which is free and at rest before the collision (Fig. 6.15).

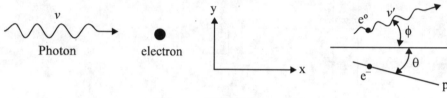

Fig. 6.15

Conservation of Momentum:

X-Components: $\dfrac{hv'}{c}\cos\phi + p\cos\theta = \dfrac{hv}{c}$

Y-Components: $\dfrac{hv'}{c}\sin\phi - p\sin\theta = 0$

where p is the momentum of the electron and v' is the frequency of the scattered photon. Since we assume an elastic collision, the final kinetic energy must be equal to the initial kinetic energy.

$$hv = hv' + E_K$$

where E_K is the final kinetic energy of the electron.

If we write the momentum equations as

$$pc\cos\theta = hv - hv'\cos\phi$$
$$pc\sin\theta = hv'\sin\phi$$

Squaring these equations and adding we get

$$p^2c^2 = (hv)^2 - 2hv \times hv'\cos\phi + (hv')^2$$ ...(6.27)

By combining two equations, we get a useful relationship between the total energy and the momentum of a particle,

$$E = m_0c^2 \Big/ \sqrt{1 - V^2/c^2}$$

and $\quad p = m_0 V \big/ \sqrt{1 - V^2/c^2}$

give $\quad E^2 = p^2 c^2 + m_0^2 c^4$ $\qquad\qquad$ ...(6.28)

Hence, for the recoil electron $p^2 c^2 = (E_K)^2 + 2(E_K) m_0 c^2$

From the energy equation $E_K = hv - hv'$

$\quad$ or, $\quad p^2 c^2 = (hv - hv')^2 + 2m_0 c^2 (hv - hv')$ $\qquad$ ...(6.29)

By substituting (6.29) in (6.27), we have obtain

$$(hv - hv')^2 + 2m_0 c^2 (hv - hv') = (hv)^2 + (hv')^2 - 2hv \times hv' - 2hv \times hv' \cos\phi$$

or, $\quad (hv)^2 + (hv')^2 - 2hv \times hv' + 2m_0 c^2 (hv - hv')$

$$= (hv)^2 + (hv')^2 - 2hv \times hv' \cos\phi$$

or, $\quad m_0 c^2 (hv - hv') = hv \times hv' - hv \times hv' \cos\phi$

$$= hv \times hv' (1 - \cos\phi)$$

on dividing by $h^2 c^2$ and substituting, $\dfrac{1}{\lambda} = \dfrac{v}{c}$

we have $\qquad \dfrac{m_0}{h}(v - v') = \dfrac{v \times v'}{c^2}(1 - \cos\phi)$

or, $\qquad v - v' = \dfrac{h}{m_0 c^2} vv' (1 - \cos\phi)$

or, $\qquad \dfrac{1}{\lambda} - \dfrac{1}{\lambda'} = \dfrac{h}{m_0 c} \times \dfrac{1}{\lambda\lambda'} (1 - \cos\phi)$

or, $\qquad \lambda' - \lambda = \dfrac{h}{m_0 c}(1 - \cos\phi)$ $\qquad\qquad$ ...(6.30)

which fits the experimentally measured wavelength shift as a function of scattering angle ϕ.

$\quad$ This part of the radiation which is scattered without wavelength change is accounted for by assuming that these photons were scattered by electrons strongly bound atoms. The recoiling mass is that of the whole atom, the photons experience negligible loss of energy in collision with the heavy atom, and hence the wavelength of these photons will be very close to that of the incident photons.

We can write eqn. (6.30) as

$$\Delta\lambda = \frac{2h}{m_0 c} \sin^2 \frac{\phi}{2} \qquad\qquad ...(6.31)$$

The quantity $\frac{h}{m_0 c}$ is known as the **Compton wavelength** of the electron and usually

denoted by the symbol λ_e. It corresponds the wavelength of a photon with energy equal to the rest energy $m_0 c^2$ of the electron. Its numerical value is given by

$$\lambda_e = \frac{h}{m_0 c} = \frac{6.6262 \times 10^{-34}}{9.11 \times 10^{-31} \times 2.998 \times 10^8} = \frac{6.6262 \times 10^{-11}}{9.11 \times 2.998} m$$

$$= 0.2426 \times 10^{-11} m = 0.02426 \,\text{Å}$$

Conclusions:

(i) When a beam of monochromatic X-rays of wavelength λ is scattered through an angle ϕ by a light element, the wavelength λ' of the scattered X-rays is greater than λ by the amount $\Delta\lambda = 0.2426(1-\cos\phi)\text{Å}$ when $\phi = 0°$, $\Delta\lambda = 0$ and when ϕ increases, $\Delta\lambda$ increases and it becomes maximum at $\phi = 180°$; the maximum value of $\Delta\lambda$ being $0.02426 \times 2 = 0.04852 \,\text{Å}$

At $\phi = 90°$, $\Delta\lambda = 0.02426 \,\text{Å}$.

Thus calculated value of $\Delta\lambda$ agrees very well with the experimental value $0.0236 \,\text{Å}$ at $\phi = 90°$ measured by Compton.

(ii) The Compton shift $\Delta\lambda$ is independent of the wavelength of the incident X-rays. It is also independent of the nature of substance scattering the X-rays.

Frequency of the scattered X-rays !

we have

$$\frac{1}{v'} = \frac{1}{v} + \frac{h}{m_0 c^2}(1-\cos\phi)$$

$$= \frac{1}{v}\left[1 + \frac{hv}{m_0 c^2}(1-\cos\phi)\right]$$

$$\therefore \qquad v' = \frac{v}{1 + \dfrac{hv}{m_0 c^2}(1-\cos\phi)}$$

or, $$v' = \frac{v}{1 + \propto (1-\cos\phi)} \qquad\qquad ...(6.32)$$

where, $$\propto = \frac{hv}{m_0 c^2}$$

Relationship Between θ and φ

Resolving the momenta along the direction of the incident photon

$$\frac{h\nu}{c} = \frac{h\nu'}{c}\cos\phi + p\cos\theta$$

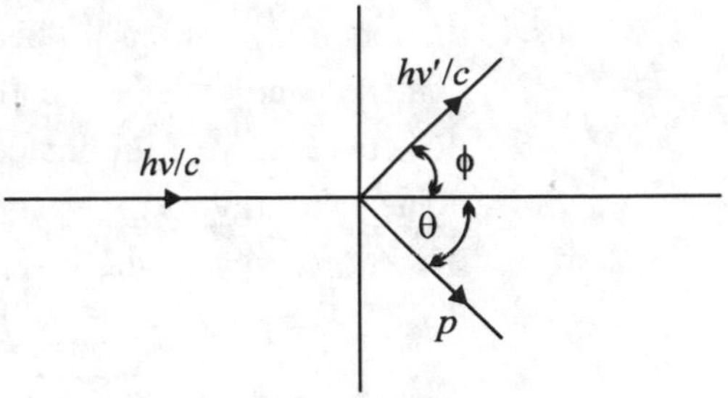

or, $\qquad p\cos\theta = \frac{h\nu}{c} - \frac{h\nu'}{c}\cos\phi$

or, $\qquad \frac{pc}{h}\cos\theta = \nu - \nu'\cos\phi \qquad ...(6.33)$

Fig. 6.16

Resolving the momenta perpendicular to the direction of the incident photon

$$p\sin\theta = \frac{h\nu'}{c}\sin\phi$$

or, $\qquad \frac{pc}{h}\sin\theta = \nu'\sin\phi \qquad\qquad\qquad\qquad ...(6.34)$

Dividing eqn. (6.33) by (6.34), we have

$$\cot\phi = \frac{\nu - \nu'\cos\phi}{\nu'\sin\phi} = \frac{1}{\sin\phi}\left[\frac{\nu}{\nu'} - \cos\phi\right] \qquad ...(6.35)$$

But, from eqn. (6.32), we have

$$\frac{\nu}{\nu'} = 1 + \propto (1 - \cos\phi)$$

$$\therefore \qquad \cot\phi = \frac{1}{\sin\phi}\left[1 + \propto(1 - \cos\phi) - \cos\phi\right]$$

$$= \frac{1}{\sin\phi}\left[(1 + \propto) - \cos\phi(1 + \propto)\right]$$

$$= (1 - \propto)\left(\frac{1 - \cos\phi}{\sin\phi}\right) = (1 + \propto)\frac{2\sin^2\frac{\phi}{2}}{2\sin\frac{\phi}{2}\cdot\cos\frac{\phi}{2}}$$

$$\therefore \qquad \cot\phi = (1 + \propto)\tan\frac{\phi}{2} \qquad\qquad\qquad ...(6.36)$$

or, $\qquad \dfrac{1}{\tan\dfrac{\phi}{2}} = \dfrac{1 + \propto}{\cot\theta}$

or, $\qquad \cot\dfrac{\phi}{2} = (1+\alpha)\tan\theta$ $\qquad\qquad$ ...(6.37)

This is the required relationship between θ and ϕ.

$\qquad$ Kinetic energy of the recoil electron:

$\qquad$ Kinetic energy of the incident photon $= hv$

$\qquad$ Kinetic energy of the scattered photon $= hv$

$\therefore \qquad$ Kinetic energy of the electron $= hv - hv'$

$$= hv - \frac{hv}{1+\alpha\left(1-\cos\phi\right)}$$

where, $\qquad \alpha = \dfrac{hv}{m_0 c^2}$

$$E_K = hv\left[1 - \frac{1}{1+\alpha(1-\cos\phi)}\right]$$

or, $\qquad E_K = hv\left[1 - \dfrac{\alpha(1-\cos\phi)}{1+\alpha(1-\cos\phi)}\right]$ $\qquad\qquad$ ...(6.38)

$$= hv\left[\frac{\alpha + 2\sin^2\dfrac{\phi}{2}}{1+\alpha\cdot 2\sin^2\dfrac{\phi}{2}}\right] = hv\left[\frac{2\alpha}{\operatorname{cosec}^2\dfrac{\phi}{2}+2\alpha}\right]$$

$$E_K = hv\,\frac{2\alpha}{1+2\cot^2\dfrac{\phi}{2}+2\alpha} = hv\,\frac{2\alpha}{1+2\cot^2\dfrac{\phi}{2}+2\alpha+\alpha^2-\alpha^2}$$

But $\qquad \cot\dfrac{\phi}{2} = (1+\alpha)\tan\theta$

So, $\qquad E_K = hv\,\dfrac{2\alpha}{1+2(1+\alpha)^2\tan^2\theta+2\alpha+\alpha^2-\alpha^2}$

$$= hv\,\frac{2\alpha}{2(1+\alpha)^2\tan^2\theta+\left(1+2\alpha+\alpha^2\right)-\alpha^2}$$

$$= hv\,\frac{2\alpha}{2(1+\alpha)^2\tan^2\theta+\left(1+\alpha^2\right)-\alpha^2}$$

$$= hv\,\frac{2\alpha}{2(1+\alpha)^2\left[2\tan^2\theta-\alpha^2+1\right]}$$

$$= hv \frac{2\alpha}{(1+\alpha)^2 \left[\sec^2\theta - \alpha^2\right]}$$

$$E_K = hv \frac{2\alpha\cos^2\theta}{(1+\alpha)^2 \left[1 - \alpha^2\cos^2\theta\right]} \qquad \qquad ...(6.39)$$

Limitation of Compton's Theory

Compton effect provides an explanation of the presence of X-rays of longer wavelength in the scattered beam, but it does not account for the presence of X-rays of the same wavelength as the incident X-rays. This unmodified radiation in the scattered beam can be explained in terms of a collision of a photon with a bound electron. Therefore, a certain amount of energy is necessary to detach an electron from an atom. If the energy imparted to the electron by a photon is much larger than the energy necessary to detach it from the atom, then the electron acts as a free electron, and eqn. (6) is applicable. If the collision is such that the electron is not detached from the atom, the rest mass m_0 of the electron must be replaced by the mass M of the atom, which is several thousand times greater than m_0. For example, the value of M is $12 \times 1840\ m_0$ for graphite scatterer. In this case, the maximum value of the Compton shift due to the collisions of photons with bounded electrons is

$$\frac{2h}{12 \times 1040\ m_0 c} = \frac{1}{12 \times 1840}\left(\frac{2h}{m_0 c}\right) = \frac{0.04852}{12 \times 1840} = 2.197 \times 10^{-6}\,\text{Å}$$

This value is too small to be detected. Thus, in this case when a photon is scattered from a bound electron, the wavelength of the scattered photon is not changed. This accounts for the presence of unmodified X-rays in the scattered beam. In case of light elements, such as Beryllium (Be), Carbon (C) and Aluminium (Al) the electrons are bound to the nuclei more loosely than in heavier elements. Therefore, in the scattered beam from such light elements, the radiation of the longer wavelength is more intense than the radiation of the original wavelength.

The study of the Compton effect leads to the conclusion that in its interaction with matter, radiant energy behaves as a stream of discrete particles (photons), each of energy hv and momentum hv/c. In other words, radiant energy is quantised. Consequently, the Compton effect is considered as a decisive phenomenon in support of quantisation of radiant energy.

Quantisation of Matter

Quantisation of matter proposed by Niels Bohr in 1913 was the first step to explain the stability properties of matter and the optical spectra of atoms. On the basis of Planck's quantum hypothesis, Bohr proposed a model of the hydrogen atom and he stated three quantisation rules to explain the stability and the optical spectrum of hydrogen atom. Bohr's theory explains quite successfully the spectrum of hydrogen and hydrogen like ions.

Drawbacks of the Old Quantum Theory

Planck's quantum hypothesis, Einstein's photoelectric effect, Compton effect and the variation of specific heat of solids with temperature and the hydrogen spectrum together form the old quantum theory. This theory has numerous drawbacks as noted below:

(i) Bohr's quantisation rules are arbitrary. The theory does not provide physical explanation for the assumptions.

(ii) The old quantum theory cannot be applied to explain the spectra of helium and of more complex atoms.

(iii) It can provide only a qualitative and incomplete explanations of the intensities of the spectral lines.

(iv) It cannot explain the dispersion of light.

(v) The theory cannot be applied to explain non-harmonic vibrations of systems.

Dual Nature of Radiation:

Newton's corpuscular theory of light could explain only two phenomena of light, i.e. reflection and refraction. Huygens wave theory of light could explain all the five phenomena of light, i.e. reflection, refraction, interference, diffraction and polarisation, but could not explain the phenomena like Einstein's photoelectric effect and Compton effect. Maxwell proposed his electromagnetic theory of light and which was confirmed by Hertz through experiments. Then Planck proposed his Quantum theory of radiation in 1900. He derived the law for the energy distribution of blackbody radiation and first showed the particle (quantum) aspect of electromagnetic radiation. In 1905, Einstein strikingly established this view point (quantum concept) with his explanation of photoelectric emission of electrons from the surface of matter (solids). In 1924, Compton explained that photons were endowed with momentum hv/c. He observed this by conducting scattering of X-rays on matter. Compton scattering is nothing but elastic collision of photon with electron. This phenomenon showed that there is increase in wavelength of the X-rays when these are scattered by matter.

Various characteristics of light are associated with the properties of waves. But, the phenomena like Einstein's photoelectric effect and Compton effect, etc. which involve transfer of energy, need particle properties. Both the particle and wave nature of radiation could be correlated as follows:

The electromagnetic radiation is considered as a stream of small packets of energy. These packets of energy are known as light quanta or photons. Each photon consists of hv amount of energy. In interaction of radiant energy with matter, any one of the photons can transfer all its energy to an electron of the matter. The frequency v is determined from the measurement of the wavelength λ of radiation, using the equation,

$$v = \frac{c}{\lambda}$$

Thus, the concept of frequency and wavelength is relevant to a wave, while the quantum having the isolated energy hv is the concept of a particle. Obviously, the electromagnetic radiation or light possesses dual character, i.e. particle nature and wave nature, but both the characteristics can never be exhibited by a single experiment.

SOLVED EXAMPLES

Q.1. **Two large closely spaced concentric spheres (both are black body radiators) are maintained at temperatures of 200 °K and 300 °K respectively. The space in-between the two spheres is evacuated. Calculate the net rate of energy transfer between the two spheres.**

(σ = 5.672 × 10⁻⁵ MKS units)

Sol.

Here, T_1 = 300 °K, $\qquad$ T_2 = 200 °K

$\qquad$ $\sigma = 5.672 \times 10^{-8}$ MKS units

$$R = \sigma\left(R_1^4 - T_2^4\right) = 5.672 \times 10^{-8}\left[(300)^4 - (200)^4\right] = 368.68 \text{ watts}/\text{m}^2.$$

Q.2. **Calculate the radiant emittance of a black body at a temperature of:**

(i) 400 °K (ii) 4000 °K.

(σ = 5.672 × 10⁻⁵ MKS units)

Sol.

Here, $R = \sigma T^4$

(i) $R = 5.672 \times 10^{-8}[400]^4 = 1452 \text{ watts}/\text{m}^2$

(ii) $R = \sigma T^4 = 5.672 \times 10^{-8}[4000]^4 = 1452 \times 10^4 \text{ watts}/\text{m}^2$

$\qquad\qquad = 14520 \text{ kilowatts}/\text{m}^2.$

Q.3. **Calculate the energy radiated per minute from the filament of an incandescent lamp at 2000 °K, if the surface area is 5 × 10⁻⁵ m² and its relative emittance is 0.85.**

Sol.

Here, $A = 5 \times 10^{-5} \text{ m}^2$, $\qquad\qquad$ $e = 0.85$

$\qquad$ $\sigma = 5.672 \times 10^{-8}$ MKS units, $\quad t = 60$ secs

$\qquad$ T = 2000 °K

So, $\quad E = A e \sigma t \, (T)^4$

$$= 5 \times 10^{-5} \times 0.85 \times 5.67 \Rightarrow 10^{-8} \times 60 \times (2000)^4 = 2315 \text{ Joules}.$$

Q.4. **An iron furnace radiates 1.53 × 10⁵ calories/hr through an opening of cross-section 10⁻⁴ m². If the relative emittance of the furnace is 0.80, calculate the temperature of the furnace. Given, σ = 1.36 × 10⁻⁸ cal/m² s–K⁴.**

Sol.

Here,　$E = 1.53 \times 10^5$ calories,　　　$A = 10^{-4}\ \mathrm{m}^2$

　　　$e = 0.80$,　　　$t = 3600$ secs,　$T = ?$

We have $E = Ae\sigma t(T)^4$

or,　　$T^4 = \dfrac{E}{Aeat}$

or,　　$T = \left(\dfrac{E}{Aeat}\right)^{\frac{1}{4}} = \left[\dfrac{153 \times 10^5}{10^{-4} \times 0.8 \times 1.36 \times 10^{-8} \times 3600}\right]^{\frac{1}{4}} = 2500\,^{\circ}\mathrm{K}$.

Q.5. **The wavelength of maximum emission for a certain star in our galaxy is 1.44×10^{-5} cm. Calculate the temperature of the star.**

Given, $b = 0.288$ cm degree.

Sol.

Here,　$\lambda_m = 1.44 \times 10^{-5}$ cm,　　　$b = 0.288$ cm degree

According to Wien's displacement law, we have $\lambda_m T = b$

or,　　$T = \dfrac{b}{\lambda_m} = \dfrac{0.288}{1.44 \times 10^{-5}} = 20{,}000\ ^{\circ}\mathrm{K}$.

Q.6. **Calculate the surface temperature of the sun and moon, given that $\lambda_m = 4753\ \text{Å}$ and 14 μm respectively, λ_m being the wavelength of maximum intensity of emission.**

Sol.

(i) Here,　　　$\lambda_m = 4753\ \text{Å}$ for the sun.

　　　According to Wien's displacement law, we have

$$\lambda_m T = \text{constant} = 0.2898 \times 10^{-2}\,\mathrm{mK}$$

　　　or,　　$\left(4753 \times 10^{-10}\,m\right) T = 0.2898 \times 10^{-2}\,\mathrm{mK}$

　　　or,　　$T = \dfrac{0.2898 \times 10^{-2}}{4753 \times 10^{-2}}\,\mathrm{K} = 6097\ \mathrm{K}$

(ii) For the moon, we have $\lambda_m T = 0.2898 \times 10^{-2}\,\mathrm{mK}$

　　　or,　　$T = \dfrac{0.2898 \times 10^{-2}\,\mathrm{mK}}{14 \times 10^{-6}\,\mathrm{m}} = 207\ \mathrm{K}$.

Q.7. **Calculate the electron velocity for an accelerating potential of 5000 V.**

Sol.

We have,　　$V_0 = 5000$ volts

　　　$\dfrac{e}{m} = 1.76 \times 10^{11}\ \mathrm{C/kg}$

　　　$V = ?$

Using the relation $\dfrac{1}{2}mV^2 = eV_0$, we have

or, $\qquad V = \sqrt{\dfrac{2eV_0}{m}}$

or, $\qquad = \sqrt{2 \times \left(\dfrac{e}{m}\right)V_0} = 2 \times 1.76 \times 10^{11} \times 5 \times 10^3 \text{ m/sec}$

$\qquad\qquad = 4.19 \times 10^7 \text{ m/sec}$

Q.8. **Calculate the number of photons from the green line of mercury (l = 4961 × 10⁻¹⁰ m) required to do one joule of work.**

Sol.

$\qquad \lambda = 4961 \times 10^{-10} \text{ m}, \qquad h = 6.6 \times 10^{-34} \text{ JS},$

$\qquad C = 3 \times 10^8 \text{ m/sec} \qquad n = ?$

According to Planck's law, we have $\quad E = hv = \dfrac{hc}{\lambda}$

If the required number of photons be n, then

$\qquad\qquad \dfrac{nhc}{\lambda} = 1 \text{ Joule}$

or $\qquad n = \dfrac{\lambda}{hc} = \dfrac{4961 \times 10^{-10}}{6.6 \times 10^{-34} \times 3 \times 10^8} = 2.505 \times 10^{18}.$

Q.9. **In photoelectric emission from a given target, the speed of emitted electron is 10⁶ m/sec. When light of wavelength 2.5 × 10⁻⁷m is used, calculate the work function of the target In electron volts.**

Sol.

$\qquad h = 6.6 \times 10^{-34} \text{ JS}, \qquad\qquad C = 3 \times 10^8 \text{ m/sec},$

$\qquad \lambda = 2.5 \times 10^{-7} \text{ m}, \qquad\qquad m = 9 \times 10^{-31} \text{ kg},$

$\qquad V = 10^6 \text{ m/sec}$

From Einstein's photoelectric equation, we have

$\qquad\qquad \dfrac{1}{2}mV^2 = hv - \omega_0$

or, $\qquad \omega_0 = hv - \dfrac{1}{2}mV^2 = \dfrac{hc}{\lambda} - \dfrac{1}{2}mV^2$

$\qquad\qquad = \dfrac{6.6 \times 10^{-34} \times 3 \times 10^8}{2.5 \times 10^{-7}} - \dfrac{1}{2} \times 9 \times 10^{-31} \times \left(10^6\right)^2$

$\qquad\qquad = 7.91 \times 10^{-19} - 4.55 \times 10^{-19}$

$\qquad\qquad = 3.36 \times 10^{-19} \text{ J} = \dfrac{3.36 \times 10^{-19}}{1.6 \times 10^{-19}} eV = 2.1 \, eV.$

Q.10. The work function of sodium is 2.3 eV. What is the maximum wavelength of light that will cause photoelectrons to be emitted from sodium? What will be the maximum kinetic energy of the photoelectrons if 2000 Å light falls on the sodium surface?

Sol.

Here, $\omega_0 = 2.3\,eV = 2.3 \times 1.6 \times 10^{-12}\,erg$ $\qquad h = 6.624 \times 10^{-27}\,ergs$

$C = 3 \times 10^{10}\,cm/sec$

We have, $\omega_0 = h\nu_0 = \dfrac{hc}{\lambda_0}$ $\qquad$ or, $\lambda_0 = \dfrac{hc}{\omega_0} = \dfrac{6.624 \times 10^{-27} \times 3 \times 10^{10}}{2.3 \times 1.6 \times 10^{-12}} = 5400\,\text{Å}$

The maximum kinetic energy of the photoelectrons is given by the relation,

$$\frac{1}{2}m\,V_{max}^2 = h\nu - \omega_0 = \frac{hc}{\lambda} - \omega_0, \quad (\text{where } \lambda = 2000 \text{ Å})$$

$$(E_K)_{max} = \left(\frac{6.624 \times 10^{-27} \times 3 \times 10^{10}}{2000 \times 10^{-8}} \right) - 2.3\,eV$$

$$= \left(9.936 \times 10^{-12}\,erg \right) - 2.3\,eV$$

$$= \frac{9.936 \times 10^{-12}}{1.6 \times 10^{-12}}\,eV - 2.3\,eV = 3.91\,eV.$$

Q.11. An X-ray photon of wavelength 0.3 Å is scattered through an angle 45° by a loosely bound electron. Find the wavelength of the scattered photon.

Sol.

Here, we have $\lambda = 0.3\,\text{Å}, \quad h = 6.63 \times 10^{-34}\,JS,$

$\phi = 45°, \quad m_0 = 9.11 \times 10^{-31}\,kg, \quad c = 3 \times 10^8\,m/sec$

The wavelength of the scattered photon is given by

$$\lambda' = \lambda + \frac{h}{m_0 C}(1 - \cos\phi) \qquad\qquad ...(1)$$

Now, $\qquad \dfrac{h}{m_0 C} = \dfrac{6.63 \times 10^{-34}}{9.11 \times 10^{-31} \times 3 \times 10^8}$

$$= 0.2426 \times 10^{-11}\,m = 0.2426 \times 10^{-10}\,m = 0.02426\,\text{Å}$$

From eqn. (1), we obtain

$$\lambda' = \lambda + 0.02426\,\text{Å}(1 - \cos 45°)$$

$$= 0.3 + 0.2426(1 - 0.7071) = 0.3071\,\text{Å}.$$

Q.12. A photon of frequency v is scattered by an electron of rest mass m_0. The scattered photon of frequency v' leaves in a direction at 90° to the incident photon. Show that the momentum acquired by the electron is given by

$$p = \frac{h}{c}\left(v^2 + v'^2\right)^{\frac{1}{2}}$$

Sol.

From the figure, we have

$$p^2 = \left(\frac{hv'}{c}\right)^2 + \left(\frac{hv}{c}\right)^2$$

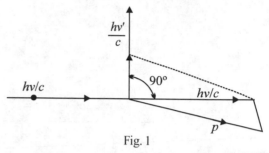

Fig. 1

or, $$p = \frac{h}{c}\left(v^2 + v'^2\right)^{\frac{1}{2}} \qquad \text{(Proved.)}$$

Q.13. An X-ray photon of frequency 3×10^{19} Hz collides with an electron and is scattered through 90°. Find the frequency of the scattered photon. Given: the Compton wavelength of the electron, $\lambda_e = 0.02426$ Å.

Sol.

Here, we have $\lambda_e = 0.02426 \times 10^{-10}$ m, $\phi = 90°$, $C = 10^8$ m/sec

$$v = 3 \times 10^{19} \text{ Hz}, \quad v' = ?$$

The frequency of the scattered photon is given by

$$v' = \frac{v}{1 + \dfrac{hv}{m_0 C^2}(1 - \cos\phi)} = \frac{v}{1 + \left(\dfrac{h}{m_0 C}\right)\dfrac{v}{C}(1 - \cos\phi)}$$

$$= \frac{v}{1 + \lambda_e\left(\dfrac{v}{C}\right)(1 - \cos 90°)} = \frac{3 \times 10^{19}}{1.2426} = 2.414 \times 10^{19} \text{ Hz}.$$

Q.14. Show that the maximum recoil energy of a free electron of rest mass m_0, when struck by a photon of frequency v, is given by

(i) $$K_{max} = \frac{(hv)}{hv + \dfrac{1}{2}m_0 C^2}$$

(ii) If λ is the wavelength of the photon and λ_e the Compton wavelength of the electron, then show that

$$K_{max} = \frac{2 m_0 C^2 \lambda_e^2}{\lambda^2 + 2\lambda_e \cdot \lambda}$$

Sol.

(i) The kinetic energy of the recoil electron is given by

$$K = hv \frac{\dfrac{hv}{m_0 C^2}(1 - \cos\phi)}{1 + \dfrac{hv}{m_0 C^2}(1 - \cos\phi)}$$

$$= \frac{(hv)^2 \times 2 \sin^2 \dfrac{\phi}{2}}{m_0 C^2 + hv \times 2 \sin^2 \dfrac{\phi}{2}} = \frac{2(hv)^2}{2 hv + m_0 C^2 \cosec^2 \dfrac{\phi}{2}}$$

K will be maximum when $\cosec \dfrac{\phi}{2}$ is minimum. The minimum value of $\cosec \dfrac{\phi}{2}$ is 1 and

this value occurs when $\dfrac{\phi}{2} = 90°$, i.e. when $\phi = 180°$

∴ The maximum value of K is given by $K_{max} = \dfrac{2(hv)^2}{2hv + m_0 C^2} = \dfrac{(hv)^2}{hv + \dfrac{1}{2}m_0 C^2}$ (Proved.)

(ii) From above, we have

$$K_{max} = \frac{2(m_0 c)^2 \times \left(\dfrac{h}{m_0 c}\right)^2 v^2}{2(m_0 c)\left(\dfrac{h}{m_0 c}\right)v + m_0 c^2}$$

Substituting $\dfrac{h}{m_0 C} = \lambda_e$ and $v = \dfrac{c}{\lambda}$, and simplifying we obtain

$$K_{max} = \frac{2(m_0 C)\lambda_e^2 \dfrac{C^2}{\lambda^2}}{2\lambda_e \dfrac{C}{\lambda} + C} = \frac{2m_0 \lambda_e^2 C^2}{2x \lambda_e + \lambda^2} \qquad \text{(Proved.)}$$

QUESTIONS

1. Explain inadequacy of classical mechanics.

2. On the basis of Planck's hypothesis, explain the energy distribution in the spectrum of black body radiation. Derive Planck's radiation law.

3. Write short-notes on:
   (i) Stefan's law of radiation,
   (ii) Planck's law of radiation,
   (iii) Rayleigh-Jeans law of radiation, and
   (iv) Wien's displacement law.

4. Explain photoelectric effect. Establish Einstein equation for the photoelectric effect.

5. State the laws of photoelectric effect. Explain Einstein's observations on photoelectric effect. Deduce Einstein's photoelectric equation.

6. Define the terms:
   (i) Threshold frequency,
   (ii) Photoelectric work function, and
   (iii) Stopping potential.

7. Discuss the inadequacy of the war theory of light to explain the photoelectric effect.

8. Show that Einstein's equation gives adequate explanation of all the facts about the photoelectric effect.

9. (i) Explain Compton effect.
   (ii) Why this effect cannot be explained on the basis of electromagnetic theory of radiation?
   (iii) Find an expression of the Compton shift $\Delta\lambda$.

10. What do you mean by Compton effect?
    Show that the change in the wavelength of X-ray photon on scattering from a free electron is independent of the wavelength of the incident radiation.

11. What is the Compton effect? Give its significance. Considering the Compton scattering of X-rays, derive expressions for: (i) the Compton shift, and (ii) the kinetic energy of the recoil electron.

12. Discuss dual nature of radiation.

Problems

1. What power is radiated from a tungsten filament 20 cm long and 0.01 mm in diameter when the filament is kept at 2500 °K in an evacuated bulb? The tungsten radiates at 30 percent of the rate of a black body at the same temperature. Neglect conduction losses.
   (P = 4.2 W)

2. Calculate the power in watts radiated from a filament of an incandescent lamp at 2000 °K if the surface area is 5×10^{-1} cm^2 and its emissivity is 0.85.

   (**Hint:** P = $e\sigma AT^4$)

3. At what rate does the sun lose energy by radiation? The temperature of the sun is about 6000 °K and its radius is 6.95×10^5 km. **(Hint: $P = \sigma AT^4$)**

4. How many watts will be radiated from a spherical black body 15 cm in diameter at a temperature of 800 °C? **(Ans. 5.4 KW)**

5. The threshold frequency for potassium is 3×10^{14} per second. What is the work function of potassium? **(Hint: $\omega_0 = h v_0$)**

6. How many photons of red light of wavelength 7×10^{-7}m constitute 1 J of energy? **(Ans. 3.6×10^{18} photons)**

7. Radiation of wavelength 3.75 Å falls upon a surface whose work function is 2.0 eV. Compute the maximum speed of the emerging photoelectrons. **(Ans. 6.8×10^5 m/sec)**

8. A photon of energy 12 eV falls on molybdenum whose work function is 4.15 eV. Find the stopping potential. **(Ans. 7.85 V)**

9. Find the maximum energy of the photoelectrons in electron-volts when light of frequency 1.5×10^9 MHz falls on a metal surface for which the threshold frequency is 1.2×10^9 MHz. **(Ans. 1.24 eV)**

10. A photon of energy 10 eV falls on molybdenum whose work function is 4.15 eV. Find the stopping potential. **(Ans. 5.85 volts)**

11. The threshold wavelength for a metal is 3800×10^{-10} m. Calculate the maximum kinetic energy of the photoelectrons ejected when ultraviolet light of wavelength 2500×10^{-10} m falls on it. **(Ans. 1.7 eV)**

12. Potassium has a work-function of 2 eV and is illuminated by monochromatic light of wavelength 3600 Å. Find:
    (i) the maximum kinetic energy of the ejected electrons,
    (ii) the maximum velocity of the ejected electrons, and
    (iii) the stopping potential in volts for the electrons.
    (Ans. (i) 2.325×10^{-19}J, (ii) 7.14×10^5 m/sec, and (iii) 1.45 volts.)

13. Prove that in the photoelectric effect from a metal surface, the maximum velocity of the photoelectrons is related to stopping potential by the equation $V_{max} = 5.923 \times 10^5 \sqrt{V_0}$.

14. The photoelectric threshold wavelength of silver is 2762 Å. Calculate:
    (i) the maximum kinetic energy of the ejected electrons,
    (ii) the maximum velocity of the electrons, and
    (iii) the stopping potential in volts for the electrons, when the silver surface is illuminated with ultraviolet light of wavelength 2000 Å.

15. A metallic surface, when illuminated with light of wavelength X_1, emits electrons with energies up to a maximum value E_1 and when illuminated with light of wavelength λ_2, where $\lambda_2 > \lambda_1$, it emits electrons with energies up to a maximum value E_2. Prove that Planck's constant h and the work function ω_0 of the metal are given by:

$$h = \frac{(E_2 - E_1)\lambda_1 \lambda_2}{c(\lambda_1 - \lambda_2)} \qquad \omega_0 = \frac{E_2\lambda_2 - E_1\lambda_1}{\lambda_1 - \lambda_2}$$

16. A metallic surface, when illuminated with light of wavelength 33.33 Å, emits electrons with energies up to 0.6 eV, and when illuminated with light of wavelength 2400 Å, it emits electrons with energies up to 2.04 eV. Calculate Planck's constant and the work function of the metal. **(Hint: Use the relations given in Q. no. 15)**

17. X-rays with $\lambda = 2.00$ Å are scattered from a carbon block and the scattered radiation is viewed at 90° to the incident beam. Find the Compton shift.

   (Ans. 0.02426 Å)

18. An X-ray photon of frequency 1.5×10^{19} Hz is scattered by a free electron and the frequency of the scattered photon is 1.2×10^{19} Hz. Find the kinetic energy imported to the electron. **(Hint: KE = $h\nu - h\nu'$)**

19. X-rays of wavelength 0.5×10^{-10} m are scattered by free electrons in a block of carbon through 90°. Find the momentum of:

   (i) incident photons,   (ii) scattered photons,
   (iii) recoil electrons, and   (iv) the energy of the recoil electrons.

   Ans.   (i) 1.326×10^{-23} kg m/sec,   (ii) 1.265×10^{-23} kg m/sec

   (iii) 1.832×10^{-23} kg m/sec, and   (iv) 1.79×10^{-16} J

20. Find the wavelength of X-radiation which is able to produce recoil electrons with maximum recoil energy of 5 KeV. (Given: the Compton wavelength of the electron is 0.02426 Å.)

$$\left[\textbf{Hint: } \lambda = \lambda_e \left(\sqrt{1 + \frac{2m_0 c^2}{K_{max}}} - 1 \right) \right]$$

   where K_{max} is the maximum kinetic energy.

21. A 250 KeV photon is scattered by a free electron (Compton effect). The kinetic energy of the recoil electron is 200 KeV. What is the wavelength of the scattered photon?

   (Ans. 2.484×10^{-11} m or 0.2484 Å)

22. In Compton scattering, the incident photons have wavelength 3×10^{-10} m. Calculate the wavelength of the scattered radiation, if they are viewed at an angle of 60° to the direction of incidence. **(Ans. 3.012 Å)**

● ● ●

Chapter 7

Wave Nature of Particles (Matter Waves)

INTRODUCTION

In 1925, Louis de Broglie gave the idea of matter waves. Phenomena like photoelectric effect and Compton effect show that light possess a dual nature, i.e. both wave and particle nature. A photon (light particle) has energy E = hv. It can be assigned a mass from the relation E = mc^2. Light being a wave motion, can also associated with material. Properties starting with a basic argument that nature loves symmetry, de Broglie postulated a new hypothesis called de Broglie's hypothesis. The mass of the photon is given by

$$mc^2 = hv$$

or,
$$m = \frac{hv}{c^2}$$
...(7.1)

Then the momentum of the photon is given by

$$p = mc = \frac{hv}{c}$$
...(7.2)

or,
$$p = h\left(\frac{v}{c}\right) = h/\lambda$$

or,
$$\chi = \frac{h}{p} = \frac{h}{mc}$$
...(7.3)

where χ is the wavelength of the wave, h is the Planck's constant and c is the speed of light. Thus, a material particle of mass 'm' moving with certain velocity 'V' can be treated as equivalent to a wave of wavelength λ, which is given by

$$\lambda = \frac{h}{m\text{V}} = \frac{h}{P}$$
...(7.4)

This relation (7.4) is called the de Broglie's relation. The wave associated with the particle is shown in the Fig. 7.1.

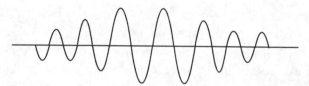

Fig. 7.1: Matter Wave

The amplitude of the wave, at any point, is directly proportional to the probability of finding the particle at that point.

This hypothesis was put to experimental confirmation by Davisson and Germer. They studied diffraction grating. The wavelength 'λ' obtained from experiment agreed with the value given by de Broglie's relation.

Superposition of Two Waves:

Wave properties of matter can be reconciled with particle properties by combining waves of different wavelengths to form a group of waves (wave packets). Since the effect of a particle in motion at any instant of time is confined to a small region in space, a wave packet can be used to represent a particle in motion.

Phase Velocity and Group Velocity:

The velocity of propagation of a wave is the velocity with which a displacement of a given phase moves forward. It is also called the phase velocity or wave velocity. When a number of plane waves of slightly different wavelengths and frequencies, travel in the same direction along a straight line, the resultant wave travels in the form of groups of waves or wavepackets (Fig. 7.2).

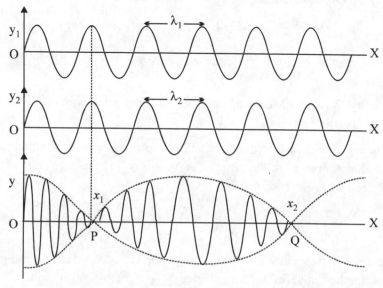

Fig. 7.2

Each group travels with a velocity, called the group velocity υ_g. If all the waves travel with the same phase velocity υ, then each group travels with this velocity, is $V_g = V_p$ and its form remains unchanged. If the phase velocity of a wave depends on its wavelength, then the group velocity V_g is different from V_p, and the group changes its form as it moves forward. If the phase velocity of a wave of longer wavelength is greater than that of a wave of shorter wavelength, the group velocity is less than the phase velocity.

Analytical Expression for the Displacement of a Group of Waves:

Stokes first treated the group of waves analytically. He assumed that a group of waves is formed by the superposition of two infinite trains of waves of the same amplitude but of slightly different wavelengths, which travel simultaneously in the same direction along the same straight line. On the basis of this assumption Lord Rayleigh treated the problem in the following way and he first pointed out the importance of a group of waves in the interpretation of optical phenomena and experiments.

A plane simple harmonic wave travelling in the opposite direction is represented by the well-known equation

$$Y = A \sin \omega \left(t - \frac{x}{\upsilon_p} \right),$$

where A is the amplitude,

$\frac{\omega}{2\pi} = v$, the frequency of the wave and V_p is the phase velocity of the wave.

The relation between V_p, v and λ is $V_p = v_p \lambda$

The above wave equation is usually written in the simpler form as

$$Y = A \sin(\omega t - Kx)$$

where, $\qquad K = \dfrac{\omega}{V_p} = \dfrac{2\pi v}{v\lambda} = \dfrac{2\pi}{\lambda}$

The quantity K is called the propagation constant for the wave of wavelength λ. K is also called the propagation number.

Let us consider two simple harmonic waves of the same amplitude travelling simultaneously in the positive x-direction to form the group. Let Y_1 and Y_2 be the displacements of a particle in the group formed of these two waves.

So, we have, for one wave

$$Y_1 = A \sin(\omega t - Kx) \qquad\qquad ...(7.5)$$

where, ω is the angular frequency, K is the propagation constant and V_g is the group velocity. The equation of another wave of angular frequency $(\omega + \delta\omega)$, propagation constant $(K = \delta K)$ and phase velocity $(V_p + \delta V_p)$ is

$$Y_2 = A \sin\left[(\omega t - \delta\omega)t - (K + \delta K)x\right] \qquad \qquad ...(7.6)$$

The equation of the resultant wave formed of these two waves is

$$Y = Y_1 + Y_2$$

$$= A \sin(\omega t - Kx) + A \sin\left[(\omega + \delta\omega)t - (K + \delta K)x\right]$$

$$= 2A\left[\cos\left(-\frac{\delta\omega}{2}t + \frac{\delta K}{2}x\right)\sin\left[\left(\frac{2\omega - \delta\omega}{2}\right)t - \left(\frac{2K + \delta K}{2}\right)x\right]\right]$$

$$= 2A \cos\frac{1}{2}(t\,\delta\omega - x\,\delta K)\sin\left[\left(\frac{2\omega + \delta\omega}{2}\right)t - \left(\frac{2K + \delta K}{2}\right)x\right] \qquad ...(7.7)$$

Since $\delta\omega$ and δK are small compared with ω and K respectively, we can neglect $\delta\omega$ and δK in the sine term eqn. above. Then, we have

$$Y = 2A \cos\frac{1}{2}(t\delta - x\delta K)\sin(\omega t - Kx) \qquad \qquad ...(7.8)$$

This is the analytical expression for the group of waves, i.e. the wave packet formed by the two waves. From this equation, we conclude the following:

(i) The sine factor represents a carrier wave which travels with the phase velocity

$$V_p = \frac{\omega}{K}$$

(ii) The amplitude of the resultant wave is given by

$$R = 2A \cos\frac{1}{2}(t\,\delta\omega - x\,\delta K) \qquad \qquad ...(7.9)$$

At a given time, the amplitude varies with x. The variation of the amplitude divides the resultant wave into groups as shown in the Fig. 7.2.

(iii) Length of a group of waves: In the Fig. 7.2, the dotted curves represent the variation of the amplitude with x at a given time t. At this time, let $P(x_1, 0)$ and $R(x_2, 0)$, where $x_2 > x_1$, be two successive points at which the amplitude is zero. Then at $P(x_1, 0)$, we have

$$\frac{1}{2}(t\,\delta\omega - x_1\,\delta K) = (2n+1)\pi/2 \qquad \qquad ...(7.10)$$

and at Q $(x_2, 0)$, we have $\dfrac{1}{2}(t\,\delta\omega - x_2\,\delta K) = (2n-1)\pi/2 \qquad \qquad ...(7.11)$

when $n = 0, 1, 2, ...$

The negative sign is used in the factor $(2n - 1)$ because, $x_2 > x_1$ and as x increases, the phase decreases. Subtracting (7.11) from (7.10) we get,

$$\frac{1}{2}(x_2 - x_1)\delta K = (2n+1)\pi/2 - (2n-1)\pi/2 = \pi$$

$$x_2 - x_1 = 2\pi/\delta K \qquad \qquad ...(7.12)$$

This equation gives the length of the group of waves, or the wave packet. In terms of the wavelength, the expression is

$$x_2 - x_1 = \frac{2\pi}{\delta\left(\dfrac{2\pi}{\lambda}\right)} = \frac{1}{\delta\left(\dfrac{1}{\lambda}\right)} = -\frac{\lambda^2}{\delta\lambda} \qquad \qquad ...(7.13)$$

(iv) Group velocity: The group velocity V_g is the velocity with which the maximum amplitude moves, at $x = 0$ and $t = 0$, the maximum amplitude as given by eqn. (7.9) is

$$R_{max} = 2A$$

Let us suppose that, at $x = x'$ and $t = t'$, the amplitude is again maximum. Then, from eqn. (7.9), we obtain

$$\frac{1}{2}(t'\,\delta\omega - x'\,\delta K) = 0$$

or, $\qquad x'\,\delta K = t'\,\delta\omega$

or, $\qquad \dfrac{x'}{t'} = \dfrac{\delta\omega}{\delta K}$

If $\qquad \delta K \to 0, \quad \dfrac{x'}{t'} = \dfrac{d\omega}{dK}$

or, $\qquad V_g = \dfrac{d\omega}{\delta K} \qquad \qquad ...(7.14)$

Relation Between the Group Velocity and the Phase Velocity

We have, $\qquad \qquad K = \dfrac{2\pi}{\lambda}$

$\therefore \qquad \qquad \dfrac{dK}{d\lambda} = -\dfrac{2\pi}{\lambda^2} \qquad \qquad ...(7.15)$

and $\qquad \qquad \omega = 2\pi v = 2\pi\dfrac{V_p}{\lambda}$

$\therefore \qquad \dfrac{d\omega}{d\lambda} = 2\pi\left(-\dfrac{V_p}{\lambda^2} + \dfrac{1}{\lambda}\dfrac{dV_p}{d\lambda}\right) = -\dfrac{2\pi}{\lambda^2}\left(V_p - \lambda\dfrac{dV_p}{d\lambda}\right) \qquad ...(7.16)$

Dividing (7.16) by (7.15), we get

$$\frac{d\omega}{d\lambda}\cdot\frac{d\lambda}{dK} = V_p - \lambda\frac{dV_p}{d\lambda}$$

But the group velocity

$$V_g = \frac{d\omega}{dK} = \frac{d\omega}{d\lambda} \cdot \frac{d\lambda}{dK}$$

So,

$$V_g = V_p - \lambda \frac{dV_p}{d\lambda} \qquad \qquad ...(7.17)$$

This equation shows that V_g is less than V_p when the medium is dispersive, i.e. when V_p is a function of λ. In a medium in which there is no dispersion, i.e. in which waves of all wavelengths travel with the same speed, $\frac{dV_p}{d\lambda} = 0$. Hence, is such a medium $V_g = V_p$. This result is true for electromagnetic waves in vacuum and elastic waves in a homogeneous medium.

The analysis given above is only an approximate one, because only two components have been considered for the formation of the group of waves.

De Broglie's Hypothesis

In 1925, Louis de Broglie extended the wave-particle relationship for photons to all particles in motion (i.e., electrons, protons, neutrons, atoms, molecules, etc.) and put forth the following hypothesis:

All particles in motion have properties similar to waves. The wavelength λ and the frequency v of the wave associated with a particle in motion, are given by

$$\lambda = \frac{h}{p} = \frac{h}{mV} \qquad \qquad ...(7.18)$$

and

$$v = \frac{E}{h} \qquad \qquad ...(7.19)$$

where, h is the Planck's constant whose value is 6.6262×10^{-34} JS,

v is the particle velocity,

E is the kinetic energy,

m is the relativistic mass,

given by the relation $m = \dfrac{m_0}{\sqrt{1 - V^2/c^2}}$ $\qquad \qquad ...(7.20)$

with c is the speed of light in vacuum, and m_0 is the rest mass of the particle. If V is small compared with c, m can be taken equal to m_0. Eqn. (7.18) is called the wave-particle de Broglie relation which has been discussed earlier.

We can write eqn. (1) as $p = \dfrac{h}{\lambda} = \left(\dfrac{h}{2\pi} \right) \dfrac{2\pi}{\lambda} = \hbar K$ $\qquad \qquad ...(7.20a)$

where, $\hbar = \dfrac{h}{2\pi}$ and $\dfrac{2\pi}{\lambda} = K$

From eqn. (2), we have $E = hv = \dfrac{h}{2\pi}(2\pi v) = \hbar\omega$ ...(7.21)

where, $\omega = 2\pi v$

Derivation of de Broglie Relation

Let a particle of mass 'm_0' at rest moves with a velocity V comparable to the speed of light and m be its relativistic mass.

In order to derive the relation between the momentum p of the particle and the wavelength λ of the associated wave, de Broglie made the following assumptions:

(i) The frequency v of the wave associated with the particle in motion and the total relativistic energy E are related by the equation

$$E = hv$$ ...(7.22)

(ii) The group velocity V_g of the waves is equal to the particle velocity υ, i.e.

$$V_g = V$$

Equation (7.22) can be written in the form

$$\dfrac{h}{2\pi} \times 2\pi\, v = E = mc^2$$

where, $mc^2 = \dfrac{m_0 c^2}{\sqrt{1 - V^2/c^2}}$

$$m^2 c^4 = \dfrac{m_0^2\, c^4}{1 - V^2/c^2}$$

or, $\quad m^2\, c^4 - \left(m^2\, c^2\right) V^2 = m_0^2\, c^4$

or, $\quad m^2\, c^4 = \left(m^2\, V^2\right) c^2 + m_0^2\, c^4$

or, $\quad mc^2 = \sqrt{P^2 c^2 + m_0^2\, c^4}$

$\therefore\quad \hbar \times 2\pi v = \sqrt{P^2 c^2 + m_0^2 c^4}$

or, $\quad \hbar\omega = \sqrt{P^2 c^2 + m_0^2 c^4}$ ...(7.23)

Since ω is a function of K, the momentum must be a function of K.

Differentiating eqn. (7.23) with respect to K, we have

$$\hbar \frac{d\omega}{dK} = \frac{1}{2} \frac{1}{\sqrt{P^2 c^2 + m_0^2 c^4}} = \left(2pc^2 \frac{dp}{dK} \right)$$

$$= \frac{pc^2}{mc^2} \frac{dp}{dK} = \frac{p}{m} \frac{dp}{dK} \qquad \qquad ...(7.24)$$

But,
$$V_g = \frac{d\omega}{dK}$$

$$\therefore \qquad V = V_g = \frac{d\omega}{dK}$$

and
$$\frac{p}{m} = \frac{mV}{m} = V$$

Hence, from eqn. (7.24), we obtain

$$\hbar V = V \frac{dp}{dK}$$

or,
$$\frac{dp}{dK} = \hbar$$

or,
$$dp = \hbar \, dK$$

Integrating this eqn., we get

$$p = K\hbar + c$$

Assuming the constant c to be zero, we get

$$p = K\hbar \qquad \qquad ...(7.25)$$

or,
$$p = \frac{2\pi}{\lambda} \times \frac{h}{2\pi} = \frac{h}{\lambda}$$

or,
$$\lambda = \frac{h}{p} \qquad \qquad ...(7.26)$$

This is the required de Broglie relation. De Broglie wavelength in terms of mass of the particle and kinetic energy:

Two cases are to be taken into consideration:

(i) When K is small compared with the rest energy $m_0 c^2$:

In this case, K is the classical kinetic energy, i.e.,

$$\frac{P^2}{2m} = K \quad \text{or} \quad P = \sqrt{2mK}$$

$$\therefore \quad \lambda = \frac{h}{P} = \frac{h}{\sqrt{2mK}} \qquad \qquad ...(7.27)$$

(ii) When K is not small compared with m_0c^2:

In this case, the total relativistic energy E is given by

$$E = \sqrt{p^2c^2 + m_0^2 c^4}$$

and the rest mass energy $= m_0c^2$

$$\therefore \qquad K = E - m_0c^2$$

or, $\qquad E = K + m_0c^2$

or, $\qquad \sqrt{p^2c^2 + m_0^2c^4} = K + m_0c^2$

Squaring both the sides, we have

$$p^2c^2 + m_0^2c^4 = K^2 + 2K\, m_0c^2 + m_0^2c^4$$

or, $\qquad p^2c^2 = K^2 + 2K\, m_0c^2$

or, $\qquad p = \sqrt{\dfrac{K^2}{c^2} + 2K\, m_0} = \sqrt{2K\, m_0\left(1 + \dfrac{K}{2\, m_0c^2}\right)}$

or, $\qquad \dfrac{h}{\lambda} = \sqrt{2m_0K\left(1 + \dfrac{K}{2m_0c^2}\right)}$

or, $\qquad \lambda = \dfrac{h}{\sqrt{2m_0K\left(1 + \dfrac{K}{2m_0c^2}\right)}}$ $\qquad$ ...(7.28)

or, $\qquad \lambda = \dfrac{hc}{\sqrt{K\left(K + 2m_0c^2\right)}}$ $\qquad$ ...(7.29)

Using eqns. (7.27), (7.28) or (7.29), the de Broglie wavelength can be calculated for any particle, such as an electron, a proton, a helium atom or even a large body.

Phase Velocity of de Broglie Waves

A particle in motion bears the characteristics of a wave. The phase velocity V_p is not equal to the particle velocity V of the de Broglie wave. If K is the propagation constant and ω is the angular frequency of the wave, then V_p is written as

$$V_p = \frac{\omega}{K}$$ $\qquad$ ...(7.30)

According to de Broglie's hypothesis, the momentum p and energy E of the particle are given by

$$p = \hbar K$$ $\qquad$ ...(7.31)

$$E = \hbar \omega \qquad \ldots(7.32)$$

Dividing (7.32) by (7.31), we have

$$\frac{E}{p} = \frac{\omega}{K}$$

$$\therefore \qquad V_p = \frac{E}{p} \qquad \ldots(7.33)$$

The following two cases may be considered:

(i) When the particle velocity is small compared with the velocity of light, the energy E is given by

$$E = \frac{p^2}{2m}$$

$$\therefore \qquad V_p = \frac{p^2}{2m} \bigg/ p = \frac{p}{2m} = \frac{mV}{2m} = \frac{V}{2} \qquad \ldots(7.34)$$

(ii) When the particle velocity is comparable with the velocity of light c, the V^2/c^2 is not negligible compared with 1. Hence, E represents the total relativistic energy mc^2. In this case,

$$V_p = \frac{mc^2}{p} = \frac{mc^2}{mV} = \frac{c^2}{V} \qquad \ldots(7.35)$$

Since V is always less than c, the phase velocity V_p is greater than the velocity of light c. But, a velocity larger than c cannot be measured. Therefore, the phase velocity has no physical meaning.

Relation between V_p and λ:

The relativistic relation between the total energy E and the momentum p of a free particle is

$$E = \sqrt{p^2 c^2 + m_0^2 c^4} = pc \sqrt{1 + \frac{m_0^2 c^2}{p^2}} \qquad \ldots(7.36)$$

According to de Broglie's hypothesis, we have

$$p = \hbar K$$

and $\qquad E = \hbar \omega$

Substituting these expressions for p and E in eqn. (7.36), we obtain

$$\hbar \omega = \hbar KC \sqrt{1 + \frac{m_0^2 c^2}{\hbar^2 K^2}}$$

or, $\qquad \dfrac{\omega}{K} = C \sqrt{1 + \dfrac{m_0^2 c^2}{\hbar^2 K^2}}$

$$\text{or,} \quad V_p = C\sqrt{1 + \frac{m_0^2\, c^2}{\hbar^2\, K^2}}$$

Since
$$\hbar K = \frac{h}{2\pi} \cdot \frac{2\pi}{\lambda} = \frac{h}{\lambda}$$

we have
$$V_p = C\sqrt{1 + \left(\frac{m_0^2\, c^2}{h^2}\right)\lambda^2} \qquad \qquad ...(7.37)$$

Thus, it is evident from the above eqn. (7.37) that for a particle in motion the phase velocity of the associated wave is always greater than C, and even in free space, the phase velocity increases with the wavelength.

Nature of de Broglie Waves:

Regarding the physical nature of wave to be associated with a particle in motion, no definite conclusion was given by de Broglie.

A monochromatic progressive wave is of infinite extent. The phase velocity of monochromatic de Broglie's wave is greater than the velocity of light as deduced earlier. This indicates that a particle in motion cannot be represented by a monochromatic wave. At any given instant of time, the effect of particle in motion is localised over a very small region. Therefore, the particle must be represented by a wave packet. The wave packet is supposed to be formed in the following way:

Let λ be the wavelength of the de Broglie wave associated with a particle moving with velocity V. Then the superposition of a number of plane waves of wavelengths slightly different from λ travelling along the same straight line gives rise to groups of waves of wave packets. The amplitude of a wave packet varies from zero to a certain maximum value over a certain small region of space having the dimensions of the particle. Such a wavepacket is associated with the particle.

Wave Packet

A wave packet can be constructed by superposing a large number of waves chosen so that they interfere constructively in a localized region of space and interfere destructively elsewhere. If the localization is constant in time, this can represent a particle.

Let a wave packet be formed by superposition of a number of waves propagating in the x-direction. We represent this by the wave function $\psi(x, t)$.

$$\psi(x, t) = \sum_{n} A_n\, e^{i(k_n x - \omega_n t)} \qquad \qquad ...(7.38)$$

where, A_n = Amplitude of the n-wave,

K_n = Magnitude of wave vector, and

ω_n = Angular frequency.

Taking a continuum of possible values of K, eqn. (7.38) can be written as

$$\psi(x,t) = \int_{-\infty}^{\infty} A(K)e^{i(kx-\omega t)}dle \qquad \qquad ...(7.39)$$

where A(K) is the amplitude of ψ_K corresponding to the momentum $p = \hbar K$. The angular frequency ω is a function of the wave vector K.

$$\omega = \omega(K) \qquad \qquad ...(7.40)$$

The value of $\psi(x, t)$ depends on the form of A(K). For simplicity, we assume that A(K) is constant in some interval of width ΔK around a given value K_0 and is zero outside of that interval. That is

$$A(K) = \begin{cases} O & \text{for} \quad K < K_0 - \Delta K \\ A(K_0) & \quad K_0 - \Delta K \le K \le K_0 + \Delta K \\ O & \quad K > K_0 + \Delta K \end{cases} \qquad ...(7.41)$$

This shows that K does not have a precisely determined value. It is distributed uniformly over a band of width $2\Delta K$. This is exhibited in the Fig. 7.3 below.

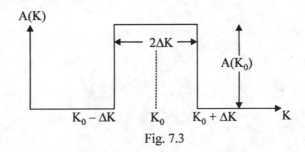

Fig. 7.3

In view of this, eqn. (7.39) becomes

$$\psi(x,t) = \int_{-\infty}^{K-\Delta K} O \times e^{i(kx-\omega t)}dK + \int_{K_0-\Delta K}^{K_0+\Delta K} A(K_0)e^{i(kx-\omega t)}dK + \int_{K_0+\Delta K}^{\infty} O \times e^{i(kx-\omega t)}dK$$

or,

$$\psi(x,t) = A(K_0) \int_{K_0-\Delta K}^{K_0+AK} e^{i(kx-\omega t)}dK \qquad \qquad ...(7.42)$$

We now assume that ω is a smooth function of K and expand it in Taylor series about K_0, i.e.

$$\omega(K) = \omega(K_0) + (K-K_0)\left(\frac{d\omega}{dK}\right)_{K=K_0} + \frac{1}{2}(K-K_0)\left(\frac{d^2\omega}{dK^2}\right)_{K=K_0}$$

Neglecting terms of the order $(K - K_0)^n$, $n \geq 2$, we obtain

$$\omega(K) \cong \omega(K_0) + (K - K_0)\left(\frac{d\omega}{dK}\right)_{K=K_0} \qquad ...(7.43)$$

or, $\qquad \omega(K) \cong \omega_0 + (K - K_0)\omega_0'$

where, $\quad \omega_0' = \dfrac{d\omega}{dK}\bigg|_{K=K_0}$

Substituting (7.43) in (7.42), we have $\psi(x,t) = A(K_0) \displaystyle\int\limits_{K_0 - \Delta K}^{K_0 + \Delta K} e^{i\left[Kx - \omega_0 t - (K - K_0)\omega_0' t\right]} dK$

$$= A(K_0) e^{i(K_0 x - \omega_0 t)} \int\limits_{K_0 - \Delta K}^{K_0 + \Delta K} e^{i(K - K_0)(x - \omega_0' t)} dK$$

$$= A(K_0) e^{i(K_0 x - \omega_0 t)} \int\limits_{-\Delta K}^{+\Delta K} e^{i(x - \omega_0' t)} d\xi \quad (K - K_0 = \xi)$$

$$= A(K_0) e^{i(K_0 x - \omega_0 t)} \left[\frac{e^{i(x - \omega_0' t)\xi}}{e(x - \omega_0' t)}\right]_{-\Delta K}^{+\Delta K}$$

$$= \frac{A(K_0) e^{i(K_0 x - \omega_0 t)}}{i(x - \omega_0' t)} \left[e^{i(x - \omega_0' t)\Delta K} e^{-i(x - \omega_0' t)\Delta K}\right]$$

$$= \frac{A(K_0) e^{i(K_0 x - \omega_0 t)}}{i(x - \omega_0' t)} \left[\begin{array}{l} \cos(x - \omega_0' t)\Delta K + i\sin(x - \omega_0' t)\Delta K - \cos(x - \omega_0' t)\Delta K \\ \hspace{5cm} + i\sin(x - \omega_0')\Delta K \end{array}\right]$$

$$= 2A(K_0) e^{i(K_0 x - \omega_0 t)} \frac{\sin\left[(x - \omega_0' t)\Delta K\right]}{x - \omega_0' t}$$

$$\psi(x,t) = 2A(K_0)\Delta K\, e^{i(K_0 x - \omega_0 t)} \times \frac{\sin\left[(x - \omega_0' t)\Delta K\right]}{\left[(x - \omega_0' t)\Delta K\right]}$$

or, $\qquad \psi(x,t) = A(x,t) e^{i(K_0 x - \omega_0 t)} \qquad\qquad ...(7.44)$

where, $\qquad A(x,t) = 2A(K_0)\Delta K \dfrac{\sin\eta}{\eta} \qquad\qquad ...(7.45)$

and $$\eta = (x - \omega_0' \, t)\Delta K \qquad \qquad ...(7.46)$$

Eqn. (7.44) represents the de Broglie wave whose phase is $(K_0 x - \omega_0 t)$ and momentum

is $\hbar K_0$. It is a plane wave modulated by the factor $\dfrac{\sin \eta}{\eta}$. The amplitude $A(x, t)$ of $\psi(x, t)$ does

not remain constant either in space or in time.

Group Velocity

It is noticed that $A(x, t)$ is maximum for maximum value of $\dfrac{\sin \eta}{\eta}$ which is equal to 1 corresponding

to $\eta = 0$. If we denote the coordinate of the centre of the packet as x_c, where $A(x, t)$ is

maximum, the condition $\dfrac{\sin \eta}{\eta} = 1$ corresponds to

$$\eta = 0 \quad \text{or} \quad (x_c - \omega_0' \, t)\Delta K = 0$$

Since $\quad \Delta K \neq O, \ x_c - \omega_0' \, t = 0 \quad \text{or,} \quad x_c = \omega_0$

The centre of the wave packet move with velocity

$$V_g = \frac{dx_c}{dt} = \omega_0'$$

or, $$V_g = \left(\frac{d\omega}{dK}\right) = \frac{dE_,}{dp} \qquad \qquad ...(7.47)$$

V_g is called the group velocity of the wave packet.

(i) Group velocity of photons:

For photons, $\quad m_0 = O, \ E = c_p,$

Hence, $$V_g = \frac{dE}{dp} = c$$

i.e., the group velocity of photons is equal to the velocity of light in vacuum.

(ii) Group velocity of electrons:

(a) *Classical:* $\quad E = \dfrac{p^2}{2m}, \ \dfrac{dE}{dp} = \dfrac{P}{m} = V$

or $\quad V_g = V \qquad \qquad ...(7.48)$

(b) *Relativistic:* The total energy E of an electron moving with relativistic speed is

$$E = \sqrt{p^2 c^2 + m_0^2 c^4}$$

or, $\dfrac{dE}{dp} = \dfrac{1}{2}\left[p^2c^2 + m_0^2c^4\right]^{-\frac{1}{2}} \times 2c^2 p$

$$= \dfrac{c^2 p}{\sqrt{p^2c^2 + m_0^2c^4}} = \dfrac{c^2 P}{E} = \dfrac{c^2 (mV)}{E}$$

$$= c^2 \dfrac{m_0 V}{\sqrt{1 + V^2/c^2}} \Big/ mc^2$$

$$= c^2 \dfrac{m_0 V}{\sqrt{1 + V^2/c^2}} \Big/ \dfrac{m_0 c^2}{\sqrt{1 + V^2/c^2}} = V$$

or, $V_g = V$ ...(7.49)

Thus, the wave packet moves with the same velocity as that of the particle.

Obviously, the group velocity is a meaningful entity but the phase velocity is not so. As the group velocity is associated with a wave packet, a particle has to be represented by a wave packet and not by a single monochromatic wave.

Spatial distribution of the wave packet:

From eqn. (7.45), the amplitude of the wave packet is

$$A(x,t) = 2A(K_0)\Delta K \dfrac{\sin\left[(x - \omega_0' \, t)\Delta K\right]}{(x - \omega_0' \, t)\Delta K}$$

At $t = 0$, we have

$$A(x,t) = 2A(K_0)\Delta K \dfrac{\sin(x\Delta K)}{(x\Delta K)}$$

$A(x, 0)$ will be maximum when $\dfrac{\sin(x\Delta K)}{x\Delta K}$ will be maximum. The maximum value of

$\dfrac{\sin(x\Delta K)}{x\Delta K}$ is 1. This corresponds to $x\Delta K = 0$.

Since, $\Delta K \neq 0, \quad x = 0.$

The form of A(x, 0) is schematically depicted in the Fig. 7.4 below.

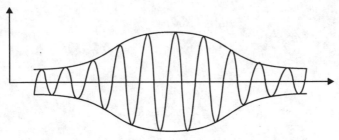

Fig. 7.4

A(x, 0) vanishes when $\dfrac{\sin(x\Delta K)}{x\Delta K} = 0$

or, $\sin(x\Delta K) = 0$

or, $x\Delta K = n\pi$ $(n = \pm 1, \pm 2, \pm 3, ...)$

or, $x = x_n = \dfrac{n\pi}{\Delta K}$

when $n = 1$, $x_1 = \dfrac{\pi}{\Delta K}$

when $n = -1$, $x_1 = \dfrac{-\pi}{\Delta K}$

or, $\Delta x = x_1 - x_1 = \dfrac{2\pi}{\Delta K}$

The value $\Delta x = \dfrac{2\pi}{\Delta K}$ can be considered as the spatial extension of the packet.

So, $\Delta x \cdot \Delta K = 2\pi$ ...(7.50)

Thus, the width of the packet in the x-space is related in the K-space by the eqn.

Since $p = \hbar K,\ \Delta P = \hbar \Delta K$ or, $\Delta K = \Delta P / p$

or, $\Delta x \cdot \Delta p = 2\pi \hbar = 2\pi \times \left(\dfrac{h}{2\pi}\right) = h$

$\therefore$ $\Delta x \cdot \Delta p = h$ ...(7.51)

Localization of wave packet in time:

The amplitude of $\psi(x, t)$ at $x = 0$ is given by

$$A(0, t) = 2A(K_0)\Delta K \frac{\sin(\omega_0' \ \Delta K)}{(\omega_0' \ \Delta K)} = 0$$

when $\sin(\omega_0' \ t \ \Delta K) = 0$

or, $\omega_0' \ t \ \Delta x = n\pi, \quad (n = \pm 1, \pm 2, \pm 3, ...)$

or, $\dfrac{\Delta\omega}{\Delta K} \ t_n \ \Delta K = n\pi$ $\left[\because \ \omega_0' = \left(\dfrac{d\omega}{dK}\right)_{K=K_0}\right]$

or, $t_n = \dfrac{n\pi}{\Delta\omega}$ or, $t_1 = \pm\dfrac{\pi}{\Delta\omega}$

So, the temporal extension Δt is given by

$$\Delta t = \frac{2\pi}{\Delta\omega}$$

or, $\Delta\omega \cdot \Delta t = 2\pi$ ...(7.52)

Multiplying both sides by $\hbar$, we obtain

$$\hbar \ \Delta\omega \cdot \Delta t = 2\pi\hbar$$

or, $\Delta E \cdot \Delta t \cong h$ ...(7.53)

The Gaussian wave packet:

The wave function representing a wave packet is given by

$$\psi(x, t) = \int\limits_{-\infty}^{+\infty} b(K)e^{i(Kx - \omega t)}dK$$ ...(7.54)

It represents a Gaussian wave packet of $b(K)$ of the form

$$b(K) = e^{-(K - K_0)^2 \alpha}$$ ...(7.55)

where $b(K)$ is a Gaussian function in the K-space peaked about the value K_0 and α is same constant. Then

$$\psi(x, t) = \int\limits_{-\infty}^{+\infty} b(K)e^{iK(x - ct)}dK$$

or, $\psi(x, t) = \psi_0(x - ct)$ ...(7.56)

It means that the wave packet simply travels along (+ve) x-direction with speed c, without any change in its space. But, this is not true in case of propagation with dispersion. In such a case, ω is a function of K.

$$\omega = \omega(K)$$

Then, $\quad \psi(x,t) = \int\limits_{-\infty}^{\infty} b(K) e^{i[Kx - \omega(K)t]} dK \qquad \qquad ...(7.57)$

Assuming the wave packet to be closely localised in the K-space about K_0, we can take

$$b(K) = e^{-\alpha(K - K_0)^2}$$

where α is a constant and it is large

Then, $\quad \tau(x,t) = \int\limits_{-\infty}^{+\infty} e^{-\alpha(K - K_0)^2} e^{i[Kx - \omega(K)t]} dK$

Assuming $\omega(K)$ as a smooth function of K, we can expand ω into a Taylor series, about K_0.

$$\omega(K) = \omega(K_0) + (K - K_0)\left(\frac{d\omega}{dK}\right)_{K=K_0} + \frac{1}{2}(K - K_0)^2 \left(\frac{d^2\omega}{dK^2}\right)_{K=K_0} + ...$$

Writing $\quad \omega(K_0) = \omega_0,$

and keeping terms only up to the second derivative of ω, we obtain

$$\omega(K) \cong \omega_0 + (K - K_0)\omega_0' + \frac{1}{2}(K - K_0)^2 \omega_0''$$

Taking $\quad \dfrac{\omega_0''}{2} = \beta$

$$\omega(K) \cong \omega_0 + (K - K_0)\omega_0' + (K - K_0)^2 \beta \qquad \qquad ...(7.58)$$

Substituting (7.58) in (7.57), we have

$$\psi(x,t) = \int\limits_{-\infty}^{+\infty} dK \, e^{-\alpha(K - K_0)^2} e^{i\left[Kx - \omega_0 t - (K - K_0)\omega_0' t - (K - K_0)^2 \beta t\right]}$$

$$= e^{i(K_0 x - \omega_0 t)} \int\limits_{-\infty}^{+\infty} dK \, e^{i(K - K_0)(x - \omega_0' t)} e^{-(\alpha + i\beta t)(K - K_0)^2}$$

$$= f(x,t) e^{i(K_0 x - \omega_0 t)}$$

where, $\quad f(x,t) = \int\limits_{-\infty}^{+\infty} dK' e^{iK'(x - \omega_0' t)} e^{-(\alpha + i\beta t)}$

and $\qquad K' = K - K_0$

The group velocity is $V_g = \omega_0' = \left(\dfrac{d\omega}{dK}\right)_{K=K_0}$

So, we can write $f(x,t) = \displaystyle\int_{-\infty}^{+\infty} dK' \, e^{iK'\xi} \, e^{-\alpha(K')^2}$ $\qquad$ ...(7.59)

where, $\qquad \xi = x - \omega_0' \, t = x - V_g t$ $\qquad$ ...(7.60)

and $\qquad \gamma = \alpha + i\beta t$ $\qquad$ ...(7.61)

Using the integral $\displaystyle\int_{-\infty}^{+\infty} e^{-\alpha S^2} e^{iSx} dx = \sqrt{\dfrac{\pi}{\alpha}} \, e^{-x^2/4\alpha}$

we have $\qquad \displaystyle\int_{-\infty}^{+\infty} dK' \, e^{iK'\xi} e^{-\gamma(K')^2} = \sqrt{\dfrac{\pi}{\gamma}} \, e^{-\xi/4\gamma}$

or, $\qquad f(x,t) = \sqrt{\dfrac{\pi}{\alpha + i\beta t}} \; e^{-\left[(x - V_g t)^2 \big/ 4(\alpha + i\beta t)\right]}$ $\qquad$ ...(7.62)

and $\qquad \psi(x,t) = \sqrt{\dfrac{\pi}{\alpha + i\beta t}} \; e^{-\left[(x - V_g t)^2 \big/ 4(\alpha + i\beta t)\right]} \times e^{i(K_0 x - \omega_0 t)}$ $\qquad$ ...(7.63)

So that $\qquad |\psi(x,t)|^2 = \dfrac{\pi}{\left(\alpha^2 + \beta^2 t^2\right)^{1/2}} \, e^{-\alpha(x - V_g t)^2 \big/ 2(\alpha^2 + \beta^2)}$ $\qquad$ ...(7.64)

This corresponds to a Gaussian distribution in the configuration space.

We·have $V_g = \dfrac{d\omega}{dt}$, the group velocity of this wave packet. If the wave packet is to be associated with a classical particle, the group velocity V_g must be given by the classical relation.

$$V_g = \frac{P}{m} = \frac{\hbar K}{m}$$

or, $\qquad \dfrac{d\omega}{dK} = \dfrac{\hbar K}{m}$

or, $\qquad \hbar \, d\omega = \dfrac{\hbar^2}{m} K \, dK$

Integrating this, we obtain

$$\hbar \omega = \frac{\hbar^2 K^2}{2m} + \text{constant} \qquad \qquad ...(7.65)$$

or, $\qquad \hbar \omega = \frac{P^2}{2m} + \text{constant} \qquad \qquad ...(7.66)$

The first term on the RHS of eqn. (7.66) is the KE of the particle. The second term is the constant of integration. It has the dimension of energy. We shall recognise this as the constant PE of the freely moving particle. Thus, if a free particle moves in a region where the PE is constant, it experiences no net force.

Experimental Confirmation of Matter Wave: Davisson-Germer Experiment

Theory: Wavelength of the de Broglie waves associated with a beam of electrons.

Let us consider a beam of electrons, each of rest mass m_0, accelerated by a potential difference V volts. Let υ be the velocity acquired by an electron in the beam. If υ is small in comparison with the velocity of light, then the KE of the electron is given by

$$\frac{1}{2} m_0 V^2 = Ve$$

or, $\qquad \upsilon = \sqrt{\frac{2\,Ve}{m_0}} \qquad \qquad ...(7.67)$

According to the de Broglie relation, the wavelength λ of the wave associated is given by

$$\lambda = \frac{h}{m_0\, \upsilon} = \frac{h}{m_0 \sqrt{\dfrac{2\,Ve}{m_0}}} = \frac{h}{2m_0\,Ve}$$

Substituting the numerical values of h, m_0 and e, we obtain

$$\lambda = \frac{12.28}{\sqrt{V}}\, \text{Å} \qquad \qquad ...(7.68)$$

If V = 100 volts, then λ = 1.228 Å.

This is the theoretical value of λ for the known PD in volts.

The wavelength of the waves associated with the beam of electrons calculated above is of the same order as that of X-rays. Therefore, if such a beam of electrons is reflected from a crystal, the reflected beam will show the same diffraction and interference phenomena as for X-rays of the same wavelength. This consideration was the basis of Davisson and Germer's experiment. In one set of experiments, the (1,1,1) face of the nickel crystal was arranged perpendicular to the incident beam of electrons.

Experimental Arrangement:

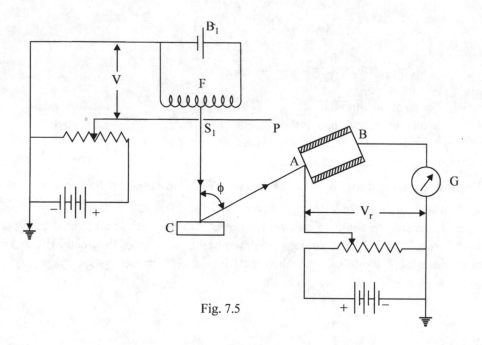

Fig. 7.5

The experimental arrangement of Davisson and Germer is shown in Fig. 7.5. F is a tungsten filament which emits electron when it is heated by passing a current from a low tension battery B_1. P is a metal plate having a narrow hole S_1. It is maintained at a positive voltage V with respect to F, so that the electrons are accelerated by the potential difference and emerge as a well collimated beam through the hole S_1. C is a nickel crystal with its (1,1,1) face normal to the beam of electrons. When the electron beam is incident on the crystal, electrons are scattered in all directions by the atoms in the crystal.

AB is the chamber in which the electron beam scattered in a given direction from the nickel crystal is received. The chamber can be rotated about an axis in the face of C passing through the point of incidence of the electron beam. The electron current is measured by means of a sensitive galvanometer G. The front and back walls of the chamber are insulated from one another and a retarding potential V_r is applied between them, so that only those electrons which have energy eV can enter the chamber. The apparatus is enclosed in an evacuated chamber.

Procedure:

A known potential difference is applied between the filament F and the metal plate P. The chamber AB is set at different angles, and for each setting of the chamber, the current is noted.

The current is directly proportional to the number of scattered electrons entering the chamber in 1 sec. Thus, the intensity of the scattered beam is measured as a function of the angle of scattering as a function of the angle of scattering. The crystal is held in a fixed position throughout the measurements. This observation is repeated for different known potential differences. For each potential difference, the intensity is plotted against the angle of scattering.

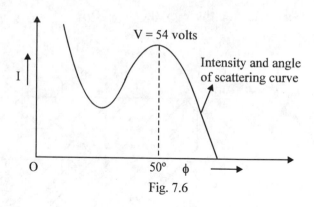

Fig. 7.6

Observations and Calculations:

It is seen that the intensity of the scattered beam is maximum at $\phi = 50°$ when the accelerating voltage V = 54 volts.

The theoretical value of λ is given by

$$\lambda = \frac{12.28}{\sqrt{V}} = \frac{12.2t}{\sqrt{54}} = 1.67 \text{ Å}$$

The maximum intensity at 50° is due to the first order constructive interference of the electron wave reflected from (3,3,1) planes in the nickel crystal. For the first order diffraction from the crystal, the wavelength of the waves is given by

$$\lambda = D \sin \phi \qquad \qquad ...(7.69)$$

where D is the distance between two adjacent lines of interaction of (3,3,1) planes with the (1,1,1) face of the nickel crystal. The value of D for nickel is 2.15 Å. Now, substituting D = 2.15 Å and $\phi = 50°$, we obtain

$$\lambda = 2.15 \sin 50°$$

$$= 2.15 \times 0.7660$$

$$= 1.65 \text{ Å}.$$

Thus, the experimental value of λ approaches theoretical value.

Derivation of the equation $\lambda = Dfi$

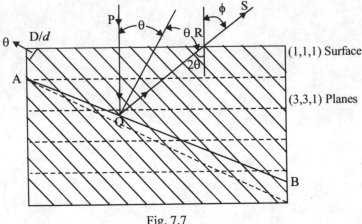

Fig. 7.7

The nickel crystal is face centered cubic and the length of the edge of the unit cube is 3.52 Å. In the crystal, there are regularly spaced (3,3,1) parallel planes (Fig. 7.7) which are rich in atoms. The interplanar spacing of these planes is

$$d = \frac{3.52}{\sqrt{3^3 + 3^2 + 1^2}} = \frac{352}{\sqrt{19}} = \frac{3.52}{4.359} = 0.807 \text{ Å}$$

The angle θ between (1,1,1) planes and (3,3,1) planes is 22°. Therefore, the distance D is given by

$$D \sin \theta = d$$

or, $$D = \frac{d}{\sin \theta} = \frac{0.807}{\sin 22} = 2.15 \text{ Å}$$

The electron beam when accelerated by the potential difference V in vacuum has kinetic energy K = eV. Therefore, the de Broglie wavelength of the beam, in vacuum, is given by

$$\lambda = \frac{h}{\sqrt{2m_0 K}} \qquad \qquad ...(7.70)$$

The beam falls normally on the (1,1,1) face of the crystal. Therefore, it enters the crystal without deviation. Due to the electrostatic fields of force between positive in cores, an electron entering the crystal with the initial kinetic energy K gains an amount of kinetic energy V_i, where V_i is the potential energy inside. Therefore inside the crystal, the de Broglie wavelength λ' of the electron beam is given by

$$\lambda' = \frac{h}{\sqrt{2m_0(K + V_i)}} \qquad \qquad ...(7.71)$$

from eqns. (1) and (2), it is clear that

$$\lambda = \lambda'$$

Inside the crystal, the beam of wavelength λ' is reflected from (3,3,1) planes and then it is refracted into vacuum as shown in Fig. 7.7.

Let QR be the reflected beam from the (3,3,1) plane AB,

RS be the refracted beam in vacuum,

θ be the angle of incidence at AB, and

ϕ be the angle of refraction in vacuum.

D be the separation between two adjacent lines of interaction of (3,3,1) planes with the (1,1,1) surface of the crystal.

α be the separation between two adjacent (3,3,1) reflecting planes.

The glancing angle of the beam

PQ at AB = $90° - Q$. Inside the crystal, the reflected beam, giving maximum intensity, satisfy Bragg's equation.

$$2d \sin(90° - \theta) = n\lambda'$$

where, $\qquad n = 1, 2, 3.$

or, $\qquad n\lambda' = 2d \cos\theta \qquad \qquad ...(7.72)$

The refractive index μ of the crystal for the electron beam is given by

$$\mu = \frac{\text{Phase Velocity in Vacuum}}{\text{Phase Velocity in the Crystal}} = \frac{\sin\phi}{\sin 2\theta}$$

or, $\qquad \mu = \dfrac{V}{V'} = \dfrac{\sin\phi}{\sin 2\theta}$

or, $\qquad \mu = \dfrac{v\lambda}{v\lambda'} = \dfrac{\sin\phi}{\sin 2\theta}$

or, $\qquad \mu = \dfrac{\lambda}{\lambda'} = \dfrac{\sin\phi}{\sin 2\theta}$

or, $\qquad \lambda' = \dfrac{\lambda \sin 2\theta}{\sin\phi} = \dfrac{2\lambda \sin\theta \cos\theta}{\sin\phi} \qquad \qquad ...(7.73)$

Substituting the value of λ' from eqn. (7.72) in eqn. (7.73), we obtain

$$\frac{2d\cos\theta}{n} = \frac{2\lambda\sin\theta\cos\theta}{\sin\phi}$$

or,
$$n\lambda = \frac{d\sin\phi}{\sin\theta}$$

But, we have $d = D\sin\theta$ from Fig. 7.7

Now,
$$n\lambda = \frac{D\sin\theta\sin\phi}{\sin\theta}$$

or,
$$n\lambda = D\sin\phi \qquad\qquad\qquad ...(7.74)$$

or,
$$1\lambda = D\sin\phi \qquad \text{when} \quad n = 1 \qquad\qquad ...(7.75)$$

This is for first order reflection.

Wave Particle Complementarity

It is noticed that the electromagnetic radiation has particle nature like other conventional waves. Similarly, the conventional particles (such as electron) have wave characteristics. Ostensibly both waves and particles exhibit dual nature. This is called wave-particle duality of course, this dual nature of a given entity is not manifested in the same situation. In a given observation, it behaves either as a wave or as a particle depending on the situation. So, the wave nature or particle nature is context sensitive and is not an intrinsic characteristic of a given system.

Wave nature and particle nature are complementary. Both the behaviour cannot be exhibited simultaneously. This is known as the complementarity principle as enunciated by Niels Bohr.

Bohr's Quantisation Condition

In the Bohr's theory of hydrogen atom, the quantisation condition

$$L = \frac{nh}{2\pi} \qquad\qquad\qquad ...(7.76)$$

for the angular momentum L of the electron, moving in a stationary circular orbit is only arbitrary. On the basis of de Broglie's hypothesis, this condition is easily obtained. For this purpose, following assumptions are made:

(i) The motion of the electron in a stationary circular orbit is represented by a standing matter wave (Fig. 7.8) of wavelength λ given by the de Broglie relation,

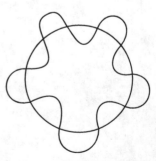

Fig. 7.8

$$\lambda = \frac{h}{m\text{V}} \qquad \qquad ...(7.77)$$

where m is the mass of the electron and V is velocity in the orbit.

(ii) The circular orbit contains an integral number of wavelengths, i.e.

$$\frac{2\pi\gamma_n}{\lambda} = n \qquad \qquad ...(7.78)$$

where, n = 1, 2, 3, ..., and γ_n is the radius of the orbit.

Substituting the value of λ in eqn. (7.78), we obtain

$$\frac{2\pi\gamma_n\, m\text{V}}{h} = n$$

or, $$m\text{V}\gamma_n = \frac{nh}{2\pi}$$

or, $$L = \frac{nh}{2\pi} \qquad \qquad ...(7.79)$$

which is Bohr's quantisation condition.

SOLVED EXAMPLES

Q.1. **Compute the de Broglie wavelength of a bike having mass 100 kg and moving with speed 100 km/hr.**

Sol. Given, m = 100 kg, V = 100 km/hr = $\dfrac{500}{9}$ m/sec.

De Broglie wavelength is given by $\lambda = \dfrac{h}{p} = \dfrac{h}{m V} = \dfrac{6.62 \times 10^{-34} \times 9}{10^2 \times 5 \times 10^2} m = 1.1916 \times 10^{-27}\,\text{Å}.$

Q.2. **Calculate the de Broglie wavelength of a particle of mass 10 gm moving with a speed of 310 m/s.**

Sol. Given, m = 10 gm = 10×10^{-3}kg,

$\qquad V$ = 310 m/s

De Broglie wavelength $\lambda = \dfrac{h}{m V}$

$$= \frac{6.6 \times 10^{-34}}{10 \times 10^{-3} \times 310} = 0.021 \times 10^{-32}\,\text{m}$$

$$= 0.021 \times 10^{-30}\,\text{cm} = 0.021 \times 10^{-22}\,\text{Å}.$$

Q.3. Calculate the de Broglie wavelength associated with a proton moving with a velocity equal to 1/20 th of the speed of light in vacuum.

Sol. Given, $h = 6.62 \times 10^{-34}$ JS,

$$m = 1.67 \times 10^{-27} \text{kg}$$

$$V = \frac{C}{20} = \frac{3 \times 10^8}{20} \text{m/s}$$

De Broglie wavelength is given by

$$\lambda = \frac{h}{mV} = \frac{6.62 \times 10^{-34} \times 20}{1.67 \times 10^{-27} \times 3 \times 10^8} = 2.643 \times 10^{-14} \text{m}.$$

Q.4. Calculate de Broglie wavelength of neutron of energy 28.8 eV. Given:

$$h = 6.62 \times 10^{-34} \text{JS}, \ m = 1.67 \times 10^{-27} \text{kg}$$

Sol. The rest mass energy of neutron is given by

$$E = mc^2 = 1.67 \times 10^{-27} \times \left(3 \times 10^8\right)^2 = 1.503 \times 10^{-10} \text{J} = 939.4 \text{ MeV}$$

As the KE of the neutron is 28.8 eV which is very small compared to its rest mass energy, so we may ignore the relativistic mass.

Hence, de Broglie wavelength for non-relativistic particle of mass m_0 may be related to the wavelength as

$$\lambda = \frac{h}{\sqrt{2m_0 E}} = \frac{6.62 \times 10^{-34}}{2 \times 1.67 \times 10^{-27} \times 28.8 \times 1.6 \times 10^{-19}}$$

$$= 5.336 \times 10^{-12} m = 5.336 \times 10^{-10} \text{cm} = 0.05336 \text{ Å}.$$

Q.5. Calculate the de Broglie wavelength of an electron which has been accelerated from rest through a potential difference of 100 V.

Sol. We know that $\lambda = \dfrac{12.25}{\sqrt{V}} \text{Å}$

Here, $V = 100$ Volts

So, $\lambda = \dfrac{12.25}{\sqrt{100}} = \dfrac{12.25}{10} = 1.225 \text{ Å}.$

Q.6. **Find the energy of the neutron in units of electron volt whose de Broglie wavelength is 1 Å. Given mass of the neutron = 1.674 × 10⁻²⁷ kg, Planck's constant** h **= 6.6×10⁻³⁴J.sec.**

Sol.

Given,
$$m = 1.674 \times 10^{-27}\,\text{kg}$$
$$\lambda = 1\text{Å} = 10^{-10}\,\text{m}$$
$$h = 6.6 \times 10^{-34}\,\text{J.sec.}$$

De Broglie relation is given by

$$\lambda = \frac{h}{mV} = \frac{h}{\sqrt{2mE}}$$

or,
$$E = \frac{h^2}{2m\lambda^2} = \frac{\left(6.6 \times 10^{-34}\right)^2}{2 \times 1.674 \times 10^{-27} \times \left(10^{-10}\right)^2}$$

$$= 13.01 \times 10^{-21}\,\text{J} = 8.13 \times 10^{-2}\,\text{eV}.$$

Q.7. **The de Broglie wavelength of a particle moving with velocity 200 m/s is 4.14 Å. What will be the de Broglie wavelength if its kinetic energy is doubled?**

Sol. De Broglie wavelength is given by

$$\lambda_1 = \frac{h}{mV_1} \qquad \qquad \text{...(1)}$$

Here,
$$\lambda_1 = 4.14 \text{ Å}, \ V_1 = 200 \text{ m/s} \ \text{ and}$$

$$\left(E_K\right)_2 = 2\left(F_K\right)_1 - 2 \times \frac{1}{2}mV_1^2 = mV_1^2$$

or,
$$\frac{1}{2}mV_2^2 = mV_1^2 \qquad \text{or,} \qquad V_2^2 = 2V_1^2 \qquad \text{or,} \ \ V_2 = \sqrt{2}\,V_1$$

From equation (1), we have $\dfrac{h}{m} = \lambda_1 V_1$

Again,
$$\lambda_2 = \frac{h}{mV_2} = \frac{h}{m\sqrt{2}\,V_1} = \frac{1}{V_2}\frac{h}{mV_1}$$

$$= \frac{1}{V_2}\lambda_1 = \frac{4.14 \text{ Å}}{\sqrt{2}}$$

$$= \frac{4.14\sqrt{2}}{2} = 2.07\sqrt{2} \text{ Å}$$

$$= \left(2.07 \times 1.414\right)\text{Å} = 2.92698 \text{ Å}.$$

QUESTIONS

1. What do you mean by matter waves? Discuss the dual nature of matter and waves.
2. Explain de Broglie's concept of matter waves. Obtain de Broglie relation.
3. Find the de Broglie wavelength of an electron of energy V electron volts.
4. A particle of charge q and mass m is accelerated from rest through a potential difference V volts. Find its de Broglie wavelength.
5. What do you mean by phase velocity and group velocity? Relate the two.
6. Explain de Broglie's hypothesis. Derive de Broglie relation for any particle in terms of its kinetic energy.
7. Obtain phase velocity of de Broglie waves.
8. Discuss on the nature of de Broglie waves.
9. What do you mean by wave packet? Show that the wave packet moves with the same velocity as that of the particle.
10. Describe Davisson-Germer Experiment to confirm the existence of matter waves.
11. What do you mean by wave particle complementarity?
12. Mention Bohr's quantisation condition with the derivation of angular momentum.

Problems

1. Calculate the value of de Broglie wavelength associated with 27 °C helium atom.
   (**Ans.** 1.26 Å)

2. Compute the de Broglie wavelength of an α-particle (mass = 6.576×10^{-27}kg and charge = 3.2×10^{-19} Coul) accelerated through 2000 volts.   (**Ans.** 2.36×10^{-13} m)

3. Calculate the de Broglie wavelength of an electron whose kinetic energy is 50 eV.
   $(h = 6.62 \times 10^{-34} \text{ JS}, m_0 = 9.1 \times 10^{-31} \text{kg})$ (**Ans.** 1.73 Å)

4. What is the de Broglie wavelength of an electron which has been accelerated from rest through a potential difference of 225 volts?   (**Ans.** 0.816 Å)

5. Calculate de Broglie wavelength associated with a proton moving with a velocity equal to 1/10th of the velocity of light in vacuum.   (**Ans.** 1.32×10^{-14}m)

●●●

Heisenberg's Uncertainty Principle

BACKGROUND

Diffraction of a beam of electrons leads to the conclusion that an electron is associated with a wave packet or a group of waves. Since a wave packet is of finite length, there will be an uncertainty in specifying the position of the electron. At the same time, the spectral distribution of the amplitude of a wave packet covers a range of wavelengths. According to the de Broglie relationship, this means that the momentum of the electron will also be uncertain.

Therefore, in general, it is not possible to determine precisely and simultaneously, the position and the momentum of the electron. If the momentum of the electron is accurately known, then by the de Broglie relation, the wavelength of the associated wave has a unique values, i.e. the associated wave is monochromatic. In such a case, the wave packet has infinite length and hence the position of the electron may be anywhere between minus infinity and plus infinity.

Based on this background, Werner Heisenberg, in 1927, enunciated the principle of indeterminacy. This term has been translated as uncertainty, indeterminacy or indefiniteness. The aim of this chapter is to focus on the principle of uncertainty and some applications of the principle.

Uncertainty Principle of Heisenberg

The principle states that for a particle of atomic magnitude, in motion, it is impossible to determine both the position and the momentum simultaneously with perfect accuracy. Quantitatively, the principle is given by the relation

$$\Delta x \cdot \Delta p \geq \hbar \qquad \qquad \qquad ...(8.1)$$

A rectangular coordinate of a particle and the corresponding component of the momentum are said to be canonically conjugate to each other. There are two more pairs of canonically conjugate variables. These are energy E of a particle and the time t at which it is measured, and a component L_Z of the angular momentum and its angular position ϕ in the perpendicular (XOY) plane. The uncertainty relations for these pairs of variables are:

$$\Delta E \cdot \Delta t \geq \hbar$$

and $\quad \Delta L_Z \cdot \Delta p \geq \hbar$

Elementary Proof of Uncertainty Relation

Suppose a particle is in motion along the x-axis. The de Broglie relation between the wavelength λ of the associated wave and the momentum P_x of the particle along the x-axis is

$$\lambda = \frac{h}{P_x}$$

$$\therefore \qquad P_x = \frac{h}{\lambda} = \frac{h}{2\pi} \cdot \frac{2\pi}{\lambda} = \hbar K \qquad \qquad \qquad ...(8.2)$$

The particle in motion is represented by a wave packet formed by the superposition of a number of plane waves of wavelengths different from λ. In our simple consideration, we consider the wave packet as the superposition of two simple harmonic plane waves of propagation constants K and K + δK. The length of such a wave packet is given by

$$\Delta x = \frac{2\pi}{\Delta K} \qquad \qquad \qquad ...(8.3)$$

Since the particle must be somewhere within the wave packet, Δx is the uncertainty in the position of the particle, and corresponding to the uncertainty in the position, the uncertainty in the propagation constant of the associated wave is ΔK.

Therefore, from eqn. (8.2), the uncertainty ΔP_x in the momentum in given by

$$\Delta P_x = \hbar\, \Delta K \qquad \qquad \qquad ...(8.4)$$

Multiplying eqns. (8.3) and (8.4), we obtain the product of the uncertainties as

$$\Delta x \cdot \Delta P_x = \left(\frac{2\pi}{\Delta K}\right) \hbar\, \Delta K = 2\pi\, \hbar$$

If wave packets have shapes different from that shown in Fig. 8.1 below, then the sign of equality is replaced by the sign $\geq$.

$$\therefore \qquad \Delta x \cdot \Delta P_x \geq \hbar \qquad \qquad \qquad ...(8.5)$$

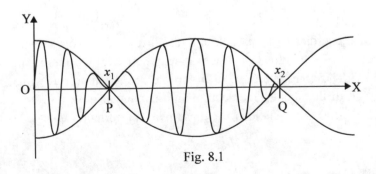

Fig. 8.1

In obtaining the uncertainty relation, we considered the wave packet formed by the superposition of two plane waves. But, superposition of two plane waves gives rise to series of wave packets instead of a single wave packet. A single wave packet is formed by superposition of an infinite number of plane waves of propagation constants slightly different from one another. By Fourier analysis of a single wave packet in one dimension, it can be shown that the width Δx of the wave packet and the range ΔK of the propagation constants of the waves, which give rise to the wave packet are given by

$$\Delta x \cdot \Delta K \approx 1$$

$$\therefore \qquad \Delta x \approx \frac{1}{\Delta K} \qquad\qquad ...(8.6)$$

Multiplying eqns. (8.4) and (8.6), we obtain

$$\Delta x \cdot \Delta P_x \approx \hbar$$

Since, this is the minimum limit of the product of the uncertainties we write,

$$\Delta x \cdot \Delta P_x \geq \hbar \qquad\qquad ...(8.7)$$

Illustration of Heisenberg's Uncertainty Principle by Thought-Experiments

A thought-experiment, or a Gedanken experiment is an imaginary experiment which does not violate any fundamental law of nature, but which cannot be performed in practice. For example, in a thought experiment, we can imagine a man to jump over the planet Jupiter to a height and calculate the required initial velocity.

We can illustrate the principle by considering the following thought-experiments.

(i) Gamma-ray microscope

We describe here a hypothetical experiment as proposed by Niels Bohr. He suggested the use of a microscope (Fig. 8.2) of high resolving power to determine simultaneously the x-components of position and linear momentum of an electron which is initially at rest. Monochromatic γ-rays are used to illuminate the electron.

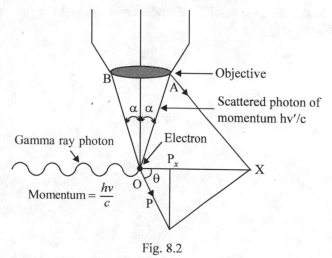

When a gamma-ray photon collides with the electron, one can view the electron through the microscope. Some momentum is transferred to the electron and the photon is scattered as noticed by the microscope.

Fig. 8.2

Let the electron be at 'O' initially having no velocity. In order to locate its position, let it be illuminated by a narrow beam of monochromatic γ-rays, proceeding in the x-direction.

Let v be the frequency and λ the wavelength of the incident γ-rays. Then the momentum of an incident γ-ray photon in the x-direction is

$$\frac{hv}{c} = \frac{h}{\lambda}$$

At least one photon should be scattered by the electron into the microscope, so that the electron is visible. In this process, the frequency and wavelength of the scattered photon is changed and the electron suffers a Compton recoil due to gain of momentum.

At the instant when the electron is observed in the microscope, let 2λ be the angle subtended at the electron by the diameter AB of the instrument's aperture.

The scattered photon may enter the microscope along the surface of the cone whose semi-vertical angle is α, or along any other direction within the cone. Suppose the photon enters the microscope along OA. Let v' be the frequency and λ' be the wavelength of the scattered photon. Then, the momentum of the scattered photon along OA is

$$\frac{hv'}{c} = \frac{h}{\lambda'}$$

The image of the electron formed by the microscope will be an interference pattern which consists of a central bright disc surrounded by alternate dark and bright rings. Since the position of the electron may be anywhere within the central bright disc, the uncertainty in the position of the electron is the diameter of the central disc. Let Δx be the diameter of the central disc. Then the uncertainty in the position is Δx.

According to Rayleigh's criterion in optics, the resolving power of an optical instrument is the distance between the peak intensity and the first minimum of the diffraction pattern, and its experession is

$$RP = \frac{\lambda'}{2\sin\alpha}$$

In this case, $$RP = \frac{\Delta x}{2}$$

$$\therefore \qquad \frac{\Delta x}{2} = \frac{\lambda'}{2\sin\alpha}$$

or, $$\Delta x = \frac{\lambda'}{\sin\alpha} \qquad\qquad ...(8.8)$$

This is the expression for the uncertainty in the position. Let p be the gain of momentum by the electron in the direction of recoil θ.

Resolving the momenta along OX, we obtain,

$$\frac{hv}{c} = \frac{hv'}{c}\cos(90°-\alpha) + p\cos\theta$$

or, $$\frac{hv}{c} = \frac{hv'}{c}\sin\alpha + p\cos\theta$$

or, $$\frac{h}{\lambda} = \frac{h}{\lambda'}\sin\alpha + p\cos\theta$$

or, $$p\cos\theta = \frac{h}{\lambda} - \frac{h}{\lambda'}\sin\alpha \qquad\qquad ...(8.9)$$

Here, $p\cos\theta$ is the x-component P_x of the momentum p.

Since the term h/λ on the right hand side of this equation is accurately known and since the scattered photon can enter the microscope along any other direction making angle less than α with the axis of the microscope, the second term on the right hand side of eqn. (8.9) represents maximum uncertainty in p_x. Let ΔP_x be this uncertainty in P_x. Then

$$\Delta P_x = \frac{h}{\lambda'}\sin\alpha \qquad\qquad ...(8.10)$$

Multiplying (8.8) and (8.10), we obtain

$$\Delta x \cdot \Delta P_x = h$$

Since, in this equation the value of ΔP_x is maximum, this equation is consistent with the uncertainty relation.

$$\Delta x \cdot \Delta P_x \geq \hbar$$

(ii) Diffraction of electrons through a slit

The diffraction of a monoenergetic beam of electrons through a single slit of width 'a' is illustrated in Fig. 8.3.

The incident beam is parallel to the x-axis and is represented by a plane wave of wavelength $\lambda = \dfrac{h}{p}$ where p is the magnitude of linear momentum $\vec{p}$ of the electron.

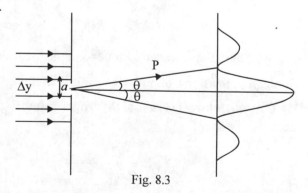

Fig. 8.3

$$\vec{p} = (p, 0, 0)$$

The beam, after being diffracted through the slit falls on a sensitive screen (photographic plate) which is kept behind the slit.

The incident electron passing through any point within the slit can be diffracted. So, the y-component of position of the electron is uncertain by an amount 'a'. This is

$$\Delta y = a \qquad \qquad ...(8.11)$$

Due to diffraction through the slit, a diffraction pattern is produced on the screen. If the position of the first minimum on the screen occurs at angle θ with the direction of the incident beam, then

$$a \sin\theta = \lambda \qquad \qquad ...(8.12)$$

The angle of diffraction lies between θ and $-\theta$.

The y-component of the linear momentum of the electron diffracted at an angle $(+\theta)$ is

$$P_y = p \sin \theta \qquad \qquad ...(8.13)$$

Similarly, the y-component of the momentum of the electron diffracted at an angle $(-\theta)$ is

$$p_y' = p\sin(-\theta) = -p\sin\theta \qquad \qquad ...(8.14)$$

So, the uncertainty in the y-component of the momentum of electron is

$$\Delta p_y = p_y - p_y = 2p\sin\theta \qquad \qquad ...(8.15)$$

Multiplying eqns. (8.11) and (8.15), we obtain

$$\Delta y \cdot \Delta P_y = a(2p\sin\theta) = 2p(a\sin\theta)$$

$$= 2p\lambda = 2\frac{h}{\lambda}\cdot\lambda$$

or, $\qquad \Delta y \cdot \Delta P_y = 2h \qquad$ or, $\qquad \Delta y \cdot \Delta P_y \geq \hbar \qquad \qquad ...(8.16)$

Thus, the uncertainty product is of the order of $\hbar$.

Some Applications of the Uncertainty Principle:

(i) Ground state energy of hydrogen atom: The energy of the electron in a hydrogen atom is

$$E = \frac{P^2}{2m} - \frac{e^2}{4\pi t_0 \gamma} \qquad \qquad ...(8.17)$$

where proton is at rest.

If the radius of the orbit of the electron is 'r' and its momentum is 'p', we have

$$r \cdot p \cong \hbar \qquad \text{or,} \qquad p \cong \frac{\hbar}{r} \qquad\qquad ...(8.18)$$

Substituting (8.18) in (8.17), we obtain

$$E \cong \frac{\hbar^2}{2mr^2} - \frac{e^2}{4\pi t_0 r} \qquad\qquad ...(8.19)$$

If the minimum energy E_0 (the ground state energy) corresponds to $r = r_0$, then

$$\left(\frac{\partial E}{\partial r}\right)_{r=0} = 0, \quad \text{which gives}$$

$$0 = \left(\frac{-\hbar^2}{mr^3} + \frac{e^2}{4\pi \in_0 r^2}\right)_{r=r_0}$$

$$\text{or,} \qquad \frac{\hbar^2}{mr_0^3} = \frac{e^2}{4\pi t_0 r_0^2} \qquad \text{or,} \qquad r_0 = \frac{4\pi t_0 \hbar^2}{me^2} \qquad\qquad ...(8.20)$$

From (8.19) and (8.20), we obtain

$$E_0 = -\frac{me^4}{32\pi^2 t_0^2 \hbar^2} \qquad\qquad ...(8.21)$$

which is the minimum energy of the hydrogen atom (−13.6 eV).

(ii) Ground state energy of one dimensional harmonic oscillator: The energy of one dimensional harmonic oscillator of mass m and angular frequency ω is

$$E = \frac{P^2}{2m} + \frac{1}{2}m\omega^2 x^2, \qquad\qquad ...(8.22)$$

We assume $\Delta p \sim p$ and $\Delta x \sim x$. According to uncertainty principle,

$$\Delta p \cdot \Delta x \cong \hbar \quad \text{gives} \quad p \cdot x \cong \hbar$$

$$\text{or,} \qquad P \cong \frac{\hbar}{x} \qquad\qquad ...(8.23)$$

From (8.22) and (8.23), we obtain

$$E = \frac{\hbar^2}{2mx^2} + \frac{1}{2}m\omega^2 x^2 \qquad\qquad ...(8.24)$$

In the ground state of harmonic oscillator, E has the minimum value for $x = x_0$ (say), i.e.

$$\left(\frac{\partial E}{\partial x}\right)_{x=x_0} = 0$$

Then, eqn. (8.24) gives

$$0 = \left(\frac{-\hbar^2}{mx^3} + m\omega^2 x \right)_{x=x_0} = \frac{-\hbar}{mx_0^3} + m\omega^2 x_0$$

or, $\dfrac{\hbar^2}{mx_0^3} = m\omega^2 x_0$ or, $x_0^4 = \dfrac{\hbar^2}{m^2\omega^2}$

or, $x_0^2 = \dfrac{\hbar}{m\,\omega}$ ...(8.25)

From (8.24) and (8.25), we obtain

$$E_0 = \frac{\hbar^2}{2m\left(\dfrac{\hbar}{m\omega}\right)} + \frac{1}{2}m\omega^2\left(\frac{\hbar}{m\omega}\right) = \frac{\hbar\omega}{2} + \frac{\hbar\omega}{2} = \hbar\omega \qquad ...(8.26)$$

This shows that the minimum energy of the harmonic oscillator is not zero but has a minimum non-zero value called the zero-point energy.

(iii) Non-existence of the electron inside the nucleus: The uncertainty Δx in the position of the electron (if it is assumed to be inside the nucleus) is of the order of

$$\Delta x \cong 10^{-14}\,\text{m}$$

From Heisenberg's uncertainty principle,

$$\Delta x \cdot \Delta p = \frac{\hbar}{2}$$

So, $\Delta p \cong \dfrac{\hbar}{2\Delta x} = 5.3 \times 10^{-21}\,\text{kg.m/sec}$

The momentum p of the electron must be of the order of Δp. The energy of the electron inside the nucleus must be

$$E = \sqrt{m_0^2 c^4 + p^2 c^2} = 10\,\text{MeV} \qquad ...(8.27)$$

But the energy of the emitted β-particle is much less than this. So, the electron cannot be contained within the nucleus.

Uncertainty and Complementarity:

The uncertainty principle is a direct consequence of the wave-particle duality. If the particle nature of a given entity (such as electron) is predominantly exhibited in a situation, its localization is better and hence the position uncertainty Δx is very small. But this leads to a large uncertainty Δp in the momentum. Hence, the de Broglie wavelength $\lambda = \dfrac{h}{P}$ becomes highly uncertain and a meaningful wave nature cannot be manifested in this situation. Both the wave nature and particle nature cannot be exhibited simultaneously.

SOLVED EXAMPLES

Q.1. **An electron has a speed of 500 m/s correct up to 0.01 %. With what minimum accuracy can you locate the position of this electron?**

Sol.

The momentum of electron, p = mV

$$= 9.11 \times 10^{-31} \times 500 = 4.55 \times 10^{-28} \, \text{kg ms}^{-1}$$

The uncertainty in momentum

$$\Delta p = m\Delta V = mV \frac{\Delta V}{V} = \frac{0.01}{100} mV$$

$$= 0.0001 \times 4.55 \times 10^{-28} = 4.55 \times 10^{-32} \, \text{kg ms}^{-1}$$

The minimum uncertainty in position is given by

$$\Delta x = \frac{h}{2\pi} \frac{1}{\Delta p} = \frac{6.625 \times 10^{-34}}{2\pi \times 4.55 \times 10^{-32}} = 2.318 \, \text{mm}.$$

Q.2. **An electron has a speed of 600 m/s with an accuracy of 0.005 %. Calculate the certainty with which we can locate the position of the electron. Given that:**

$$h = 6.6 \times 10^{-34} \, \text{JS}, \, m = 9.1 \times 10^{-31} \, \text{kg}, \, V = 600 \, \text{m/s}$$

Sol.

The momentum of electron = mV

$$= 9.1 \times 10^{-31} \times 600 \, \text{kg m/s}$$

$$\Delta p = \left(\frac{0.005}{100} \right) mV = 5 \times 10^{-5} \times 9.1 \times 10^{-31} \times 600$$

$$\Delta x = \frac{\hbar}{\Delta p} = \frac{h}{2\pi \Delta p} = \frac{6.6 \times 10^{-34} \times 10^{31}}{2 \times 3.14 \times 5 \times 10^{-5} \times 9.1 \times 600}$$

$$= \frac{6.6 \times 10^{-3}}{6.28 \times 5 \times 9.1 \times 6 \times 10^{-3}}$$

$$= \frac{6.6}{6.28 \times 30 \times 9.1} = \frac{6.6}{6.28 \times 3 \times 91} = \frac{2.2}{6.28 \times 91}$$

$$= \frac{1.1}{3.14 \times 91} = \frac{0.350318}{91} = 0.0038496 \, \text{m}.$$

Q.3. **If the uncertainty in position of an electron is 4×10^{-6} m, then compute the uncertainty in its momentum.**

Sol.

From the Heisenberg's uncertainty principle, we have

$$\Delta x \, \Delta p_x \approx h$$

or, $$\Delta p_x \approx \frac{h}{\Delta x} = \frac{6.6 \times 10^{-34}}{4 \times 10^{-10}} = 1.6 \times 10^{-24} \, \text{kg m/sec}.$$

QUESTIONS

1. State and explain Heisenberg's uncertainty principle. Using this principle show that an electron cannot reside in an atomic nucleus.

2. What is uncertainty principle? Give two examples to explain it.

3. State and prove the uncertainty relation.

4. Illustrate the uncertainty principle by two thought-experiments.

5. Explain the uncertainty principle by diffraction of electrons through a slit.

6. Using uncertainty relation, find the ground-state energy of hydrogen atom.

7. Calculate the ground-state energy of one dimensional harmonic oscillator by the help of uncertainty relation.

8. What do you mean by uncertainty and complementarity?

Problems

1. If the uncertainty in position of an electron is 6×10^{-10} m, then find the uncertainty in its momentum. **(Ans.** 1.1×10^{-24} kg m/sec)

2. An electron is confined to a box of length 10^{-9} m. Calculate the minimum uncertainty in its velocity. Given: $m_e = 9 \times 10^{-31}$ kg and $h = 6.6 \times 10^{-34}$ JS. **(Ans.** 7.3×10^5 m/s)

3. What is the uncertainty in the momentum of an electron which is restricted to a region of linear dimension equal to 1 Å? **(Ans.** 1.06×10^{-24} kg m/s)

4. What is the minimum uncertainty in the energy state of an atom if an electron remains in the state for 10^{-8} sec? (Given $h = 6.6 \times 10^{-34}$ JS and 1 eV $= 1.6 \times 10^{-19}$J)

   (Ans. 6.6×10^{-8} eV)

5. The accuracy with which the frequency of the atomic radiation can be determined by an amount $\Delta v \sim 1.6 \times 10^7$ Hz. What percentage of frequency of a 5000 Å photon is this? **(Ans.** 2.67×10^{-6}%)

●●●

Chapter 9
Basic Features of Quantum Mechanics

INTRODUCTION

In classical mechanics, a wave equation is a second order differential equation in space and time. Solutions of this equation represent wave disturbances in a medium. Therefore, a wave equation is the usual basis of mathematical theory of wave motion. For example, an electromagnetic wave travelling in the x-direction, is described by the wave equation

$$\frac{\partial^2 E_y}{\partial x^2} = \frac{1}{c^2} \frac{\partial^2 E_y}{\partial t^2}$$

where E_y is the y-component of the electric field intensity.

A differential equation for the wave associated with a particle in motion cannot be derived from first principles. The equation may be developed by any one of the following procedures:

(i) The equations of motion of classical mechanics are transformed into a wave equation in accordance with wave properties of matter based on de Broglie's hypothesis.

(ii) A complex variable quantity, called the wave function, is assumed to represent a plane simple harmonic wave associated with a free particle, and the classical expression for the total energy is used.

(iii) A particle at a given position and at a given time is represented by a wave packet which is obtained by superposition of a group of plane waves of nearly the same wavelength, which interfere destructively everywhere except at a packet, and the classical expression for the total energy is used.

(iv) In the classical expression for the total energy of a particle, the dynamical quantities are replaced by their corresponding operators. These operators are allowed to operate upon the wave function.

In this chapter, the Schrödinger's wave equation will be derived following the principle laid down in (b) above.

Wave Function for a Free Particle

Let a particle of mass 'm' be in motion along the positive x-direction with accurately known momentum p_x and total energy E. The position of the particle would be completely undetermined.

The wave associated with such a particle should be a plane, continuous harmonic wave travelling in the positive x-direction. The wavelength and frequency of the wave are given by

$$x = \frac{h}{p_x} \quad \text{and} \quad v = \frac{E}{h}$$

We rewrite these equations in terms of $\hbar = \dfrac{h}{2\pi}$ and the propagation constant $K = \dfrac{2\pi}{\lambda}$

$$p_x = \frac{h}{\lambda} = \frac{h}{2\pi} \cdot \frac{2\pi}{\lambda} = \hbar K \qquad \qquad ...(9.1)$$

$$E = hv = \frac{h}{2\pi} \cdot 2\pi v = \hbar\omega \qquad \qquad ...(9.2)$$

We assume that the plane wave is represented by a complex variable quantity $\psi(x, t)$. This quantity is called the wave function for the particle. We assume that the wave function has the following form:

$$\psi(x, t) = A\, e^{i\omega\left(t - \frac{x}{v}\right)} \qquad \qquad ...(9.3)$$

where A is a constant and $i\sqrt{-1}$. The phase velocity υ is given by

$$\upsilon = v\lambda = 2\pi v \cdot \frac{\lambda}{2\pi} = \frac{\omega}{K}$$

$$\therefore \qquad \frac{\omega}{V} = K$$

Now, from eqn. (3), we have $\psi(x, t) = A\, e^{-i(\omega t - Kx)}$

Substituting the values of K and ω as given by eqns. (1) and (2), we obtain,

$$\psi(x, t) = A\, e^{-i\left(\frac{Et}{\hbar} - \frac{P_x x}{\hbar}\right)}$$

$$= A\, e^{-i/\hbar(Et - P_x x)} = A\, e^{i/\hbar(P_x x - Et)} \qquad \qquad ...(9.4)$$

This equation represents a plane continuous simple harmonic wave associated with a particle of accurately known momentum p_x and total energy E moving in the positive x-direction. For a plane wave travelling in an arbitrary direction, the wave function is written as

$$\psi(\vec{r}, t) = A\, e^{i/\hbar(\vec{p}\,\vec{r} - Et)} \qquad \qquad ...(9.5)$$

where the vector $\vec{p}$ is the momentum and the vector $\vec{r}$ is the position vector of the particle.

One Dimensional Schrödinger's Wave Equation (Time Dependent)

Let us suppose that a particle of mass 'm' is in motion along the positive x-direction under the action of a force $F(x)$ which is a function of x. In such a field of force, the potential energy V of the particle does not explicitly depend on time; it is a function only of x. Such a force is called a conservative force.

Let the speed of the particle be small compared with that of light.

Let P_x be the momentum of the particle along the x-direction. Then the total energy E of the particle is given by

$$\frac{P_x^2}{2m} + V(x) = E \qquad \qquad ...(9.6)$$

The wave function $\psi(x, t)$ representing the plane wave associated with the particle is given by

$$\psi(x, t) = A\, e^{i/\hbar(P_x x - Et)} \qquad \qquad ...(9.7)$$

Differentiating eqn. (9.7) with respect to x, we have

$$\frac{\partial \psi}{\partial x} = A\left(\frac{i}{\hbar}\right)P_x\, e^{i/\hbar(P_x x - Et)} = \left(\frac{i}{\hbar}\right)P_x\, \psi \qquad \qquad ...(9.8)$$

and
$$\frac{\partial^2 \psi}{\partial x^2} = \left(\frac{i}{\hbar}\right)P_x\frac{\partial \psi}{\partial x} = \left(\frac{i}{\hbar}\right)P_x\left(\frac{i}{\hbar}\right)P_x\, \psi = -\frac{1}{\hbar^2}P_x^2\, \psi$$

or, $\qquad P_x^2\, \psi = -\hbar^2\frac{\partial^2 \psi}{\partial x^2} \qquad \qquad ...(9.9)$

Differentiating eqn. (9.7) with respect to t, we have

$$\frac{\partial \psi}{\partial t} = A\left(\frac{i}{\hbar}\right)(-E)\, e^{i/\hbar(P_x x - Et)} = -\frac{i}{\hbar}E\psi = \frac{E\psi}{i\hbar}$$

Therefore, $\qquad E\psi = i\hbar\frac{\partial \psi}{\partial t} \qquad \qquad ...(9.10)$

Now multiplying eqn. (9.6) by ψ, we have

$$\frac{P_x^2\, \psi}{2m} + V\psi = E\psi \qquad \qquad ...(9.11)$$

Substituting the values of $P_x^2\psi$ and $E\psi$ as given by eqns. (9.9) and (9.10) in eqn. (6), we obtain,

$$-\frac{\hbar^2}{2m} \cdot \frac{\partial^2 \psi}{\partial x^2} + V\psi = i\hbar\frac{\partial \psi}{\partial t} \qquad \qquad ...(9.12)$$

This is the time-dependent form of the Schrödinger wave equation for the motion of a particle in one dimension under the action of a conservative force. In this equation, ψ is a function of x and t, and V is a function of x.

One-Dimensional Wave Equation for a Free Particle

If there are no external forces, such as electrostatic, gravitation and nuclear forces acting on the particle, then the particle is said to be free and for the free particle V = 0. Therefore, the time-dependent Schrödinger's wave equation for the motion of a free particle in one dimension is

$$-\frac{\hbar^2}{2m} \cdot \frac{\partial^2 \psi}{\partial x^2} = i\hbar \frac{\partial \psi}{\partial t} \qquad \qquad ...(9.13)$$

Separation of the One-Dimensional Wave Equation into the Time-Dependent Part and Time-Independent Part

Eqn. (9.12) is a particle differential equation in two variables (x, t). The equation is separable into the time-independent part and time dependent part, if the potential energy V is not a function of time t.

Let $u(x)$ and $f(t)$ be the time-independent and time-dependent parts respectively of the wave function then

$$\psi(x, t) = u(x) f(t) \qquad \qquad ...(9.14)$$

Substituting this eqn. in eqn. (9.12), we obtain,

$$-\frac{\hbar^2}{2m} f \frac{d^2 u}{dx^2} + Vut = i\hbar u \frac{df}{dt}$$

We have used ordinary derivate in place of particle derivatives because each of the functions, u and f depends on only one variable.

Dividing both the sides by uf, we obtain,

$$-\frac{\hbar^2}{2m} \frac{1}{u} \frac{d^2 u}{dx^2} + V = i\hbar \frac{1}{f} \frac{df}{dt} \qquad \qquad ...(9.15)$$

(i) Time-dependent part

Now, we make use of eqn. (9.11).

$$i\hbar \frac{\partial \psi}{\partial t} = E\psi$$

Substituting eqn. (9) in this equation, we obtain

$$i\hbar u \frac{df}{dt} = E \qquad \qquad ...(9.16)$$

Thus, the right side of eqn. (9.15), which is a function only of time 't' is equal to the total energy E. This equation is the time-dependent part of the wave equation.

In order to solve eqn. (9.16), we separate the variables and obtain

$$\frac{df}{f} = \frac{E}{i\hbar} dt$$

or, $$\frac{df}{f} = -\frac{iE}{\hbar} dt$$

Integrating, we have $\quad f = C e^{-iEt/\hbar}$ ...(9.17)

where C is a constant.

(ii) Time-independent part

From eqns. (9.15) and (9.16), we have

$$-\frac{\hbar^2}{2m} \frac{1}{u} \frac{d^2u}{dx^2} + V = E$$

Multiplying both sides by u, we obtain

$$-\frac{\hbar^2}{2m} \frac{d^2u}{dx^2} + Vu = Eu$$ ...(9.18)

This is the one-dimensional time-independent Schrödinger's wave equation for a single particle. It is also called one-dimensional steady state Schrödinger equation, and its solution $u(x)$ is called time-independent wave function or steady state wave function. It is a linear equation because it contains no powers of u higher than the first; it is homogeneous in u because it contains no term independent of u or its derivatives; it is of the second order because its highest differential is of the second order. A convenient form of eqn. (9.18) is

$$\frac{d^2u}{dx^2} + \frac{2m}{\hbar^2}(E - V)u = 0$$ ...(9.19)

Substituting the solution of the time-dependent equation as given by eqn. (9.17) into (9.14) we have

$$\psi(x, t) = Cu(x) e^{-Et/\hbar}$$ ...(9.20)

Since $\quad E = h\nu = \dfrac{h}{2\pi} \times 2\pi\nu = \hbar\omega$

Eqn. (9.20) is also written in the following form

$$\psi(x, t) = Cu(x) e^{-i\omega t}$$ ...(9.21)

Eqn. (9.20) is a particular solution of eqn. (9.12), corresponding to a particular value of the constant energy E, and it satisfies the time-independent equation (9.18).

General Solution

Corresponding to various values of the constant energy E, there are various particular solutions of eqn. Therefore, the general solution of the equation is the sum of all particular solutions with arbitrary coefficients. Thus, the general solution is written as

$$\psi(x,t) = \sum_n C_n \, u_n \, (x) \, e^{-iE_n t/\hbar} \qquad \qquad ...(9.22)$$

where C_n are constants and the symbol $\sum_n$ represents the process of summation over discrete values of E_n. It should be noted that the general solution does not satisfy the time-independent equation (9.18). Complex conjugate wave function $\psi \times (x, t)$.

In the physical interpretation of the wave function $\psi(x, t)$, and solution of the wave equation, we will make use of the quantity $\psi(x, t)$ which is the complex conjugate of $\psi(x, t)$.

The one-dimensional time-dependent Schrödinger's wave equation for $\psi^*(x, t)$ is

$$-\frac{\hbar^2}{2m}\frac{\partial^2 \psi^*}{\partial x^2} + V\psi^* = -i\hbar\frac{\partial \psi^*}{\partial t} \qquad \qquad ...(9.23)$$

The general solution of this eqn. is

$$\psi^*(x,t) = \sum_n C_n^* \, u_n^* \, (x) \, e^{iE_n t/\hbar} \qquad \qquad ...(9.24)$$

Three-dimensional Schrödinger's Wave-Equation

Let us suppose a particle of mass m be in motion along any direction referred to three mutually perpendicular axes under the action of a force F (x, y, z) which is a function of the space coordinates (x, y, z). In such a field of force, the potential energy V of the particle does not explicitly depend on time; it is a function of the space co-ordinate. Such a force is called a conservative force.

Let $\vec{r}$ be the position vector of the particle at point (x, y, z) at time 't' and $\vec{p}$ be the momentum of the particle.

If the speed of the particle is small compared with that of light, the total energy E of the particle is given by

$$\frac{\vec{p}^2}{2m} + v(\vec{r}) = E$$

Since, $\qquad \vec{p}^2 = \vec{p} \cdot \vec{p} = p_x^2 + p_y^2 + p_z^2,$ we have

$$\frac{1}{2m}\left(p_x^2 + p_y^2 + p_z^2\right) + V(\vec{r}) = E \qquad \qquad ...(9.25)$$

For the motion of the particle in three dimensions, the wave function of the associated wave is given by

$$\psi(\vec{r}, t) = A e^{i/\hbar(\vec{p} \cdot \vec{r} - Et)}$$

Since, $\vec{p} \cdot \vec{r} = p_x x + p_y y + p_z z$, we have

$$\psi(\vec{r}, t) = A e^{i/\hbar(p_x x + p_y y + p_z z - Et)} \qquad \qquad ...(9.26)$$

Differentiating eqn. (9.26) w.r.t. x, we have

$$\frac{\partial \psi}{\partial x} = A \left(\frac{i}{\hbar} \right) p_x \, e^{i/\hbar\left(p_x x + p_y y + p_z z - Et \right)}$$

$$= \left(\frac{i}{\hbar} \right) p_x \, \psi \qquad \qquad ...(9.27)$$

and $$\frac{\partial^2 \psi}{\partial x^2} = \left(\frac{i}{\hbar} \right) p_x \frac{\partial \psi}{\partial x} = \left(\frac{i}{\hbar} \right) p_x \left(\frac{i}{\hbar} \right) p_x \psi = -\frac{1}{\hbar^2} p_x^2 \, \psi$$

From this eqn. we obtain $$p_x^2 \, \psi = -\hbar^2 \frac{\partial^2 \psi}{\partial x^2} \qquad \qquad ...(9.28)$$

Similarly, we have $$p_y^2 \, \psi = -\hbar^2 \frac{\partial^2 \psi}{\partial y^2} \qquad \qquad ...(9.29)$$

$$p_z^2 \, \psi = -\hbar^2 \frac{\partial^2 \psi}{\partial z^2} \qquad \qquad ...(9.30)$$

Differentiating eqn. (9.26) w.r.t. t, we obtain

$$\frac{\partial \psi}{\partial t} = A \left(\frac{i}{\hbar} \right) (-E) e^{i/\hbar\left(p_x x + p_y y + p_z z - Et \right)}$$

$$= -\frac{i}{\hbar} E \psi$$

or, $$E\psi = -\frac{\hbar}{i} \frac{\partial \psi}{\partial t} = i\hbar \frac{\partial \psi}{\partial t} \qquad \qquad ...(9.31)$$

Now multiplying eqn. (9.25) by ψ, we obtain

$$\frac{1}{2m} \left(p_x^2 + p_y^2 + p_z^2 \right) \psi + V\psi = E\psi \qquad \qquad ...(9.32)$$

Substituting the values of $p_x^2 \psi$, $p_y^2 \psi$, $p_z^2 \psi$ and $E\psi$ in eqn. (9.32), we have

$$-\frac{\hbar^2}{2m}\left(\frac{\partial^2\psi}{\partial x^2}+\frac{\partial^2\psi}{\partial y^2}+\frac{\partial^2\psi}{\partial z^2}\right)+V\psi = i\hbar\frac{\partial\psi}{\partial t} \qquad ...(9.33)$$

or, $\qquad -\frac{\hbar^2}{2m}\nabla^2\psi+V\psi = i\hbar\frac{\partial\psi}{\partial t} \qquad ...(9.34)$

This is the three-dimensional time dependent Schrödinger's wave equation for the motion of a single particle. The wave equation in the form of eqn. (9.34) was derived by Schrödinger in 1926.

Separation of the Three-Dimensional Wave Equation into Time-Dependent Part and Time-Independent Part

Equation (9.33) is a partial differential equation in four variables (x, y, z, t). The equation is separable into the time dependent part and time-independent part, if the potential energy V is not a function of time.

Let $u(\vec{r})$ and $f(t)$ be the time independent and time-dependent part respectively of the wave function.

Then $\qquad \psi(\vec{r},t)=u(\vec{\gamma})f(t) \qquad ...(9.35)$

Substituting this equation in eqn. (9.33), we obtain

$$=\frac{\hbar^2}{2m}f\left[\frac{\partial^2 u}{\partial x^2}+\frac{\partial^2 u}{\partial y^2}+\frac{\partial^2 u}{\partial z^2}\right]+Vuf = i\hbar u\frac{\partial f}{\partial t}$$

Dividing both the sides by uf, we have

$$-\frac{\hbar^2}{2m}\frac{1}{u}\left[\frac{\partial^2 u}{\partial x^2}+\frac{\partial^2 u}{\partial y^2}+\frac{\partial^2 u}{\partial z^2}\right]+V = i\hbar\frac{1}{f}\frac{\partial f}{\partial t} \qquad ...(9.36)$$

(i) Time-dependent part

The time-dependent part is obtained exactly in the same way as explained earlier. Thus the time-dependent part of the equation is

$$i\hbar\frac{1}{f}\frac{\partial f}{\partial t}=E \qquad ...(9.37)$$

The solution of this equation is

$$f=Ce^{-\frac{iE}{\hbar}t} \qquad ...(9.38)$$

where C is a constant.

(ii) Time-independent equation

From eqns. (9.36) and (9.37), we have

$$-\frac{\hbar^2}{2m}\frac{1}{u}\left[\frac{\partial^2 u}{\partial x^2}+\frac{\partial^2 u}{\partial y^2}+\frac{\partial^2 u}{\partial z^2}\right]+V = E$$

or,
$$-\frac{\hbar^2}{2m}\left[\frac{\partial^2 u}{\partial x^2}+\frac{\partial^2 u}{\partial y^2}+\frac{\partial^2 u}{\partial z^2}\right]+Vu=Eu \qquad \text{...(9.39)}$$

or,
$$-\frac{\hbar^2}{2m}\left[\frac{\partial^2}{\partial x^2}+\frac{\partial^2}{\partial y^2}+\frac{\partial^2}{\partial z^2}\right]u+Vu=Eu$$

or,
$$-\frac{\hbar^2}{2m}\,\nabla^2 u+Vu=Eu \qquad \text{...(9.40)}$$

Eqn. (9.39) can also be written as

$$\frac{\partial^2 u}{\partial x^2}+\frac{\partial^2 u}{\partial y^2}+\frac{\partial^2 u}{\partial z^2}+\frac{2m}{\hbar^2}\left(E-V\right)u=0 \qquad \text{...(9.41)}$$

Eqn. (9.39) or (9.40) or (9.41) is the three dimensional time-independent Schrödinger wave equation for a single particle. It is also called three dimensional steady state Schrödinger equation and its solutions are called time-independent or steady state wave-functions.

Three-Dimensional Time-Dependent Wave Function

Substituting the solution of the time-dependent equation, as given by eqn. (9.38) into eqn. (9.35), we obtain

$$\psi(\vec{r},t)=C\,u\,(\vec{r})\,e^{iEt/\hbar} \qquad \text{...(9.42)}$$

Since,
$$E=hv=\frac{h}{2\pi}\cdot 2\pi v=\hbar\omega$$

we have,
$$\psi\,(\vec{r},t)=C\,u\,(\vec{r})\,e^{-i\omega t} \qquad \text{...(9.43)}$$

The constant C is usually taken as 1. Eqn. (9.42) is a particular solution of eqn. (9.33), corresponding to a particular value of the constant energy E, and it satisfies the time independent eqn. (9.41). The wave function as given by this equation represents a monochromatic standing wave. The amplitude u ($\vec{r}$) of the standing wave is a function of $\vec{r}$.

Physical Interpretation of the Wave Function ψ

Solving the Schrödinger's wave equation, we obtain the wave function ψ. The quantum behaviour of a micro-particle of mass 'm' with potential energy V can be predicted from the wave function ψ. Hence, its behaviour is quite analogous to the classical trajectory $\vec{r}$ (t) is known, then velocity, acceleration, momentum, etc. of a particle can be determined exactly. Likewise, if ψ is known, the observable properties of a micro-particle can be understood fully. It is, therefore, important to have a precise understanding of the physical meaning of ψ.

The only thing known about ψ is that the wave function is large where the probability of finding the particle is large. It is small where the probability is less. This suggests that ψ should be interpreted in statistical terms. In this context, we follow the interpretation due to Max Born.

$\psi(\vec{r}, t)$ may represent the probability of finding a particle at time '*t*' at a position $\vec{r}$ with respect to the origin. But $\psi(\vec{r}, t)$ itself can *n* be a probability. This is because, the probability is a real and non-negative quantity whereas $\psi(\vec{r}, t)$ is, in general complex. We then assume that the product of $\psi(\vec{r}, t)$ and its complex conjugate $\psi^*(\vec{r}, t)$, i.e.,

$$\psi^*(\vec{r}, t)\, \psi(\vec{r}, t) = |\psi(\vec{r}, t)|^2$$

is the position probability density $\rho(\vec{r}, t)$, is

$$\rho(\vec{r}, t) = |\psi(\vec{r}, t)|^2 \qquad \qquad ...(9.44)$$

Thus, $\rho(\vec{r}, t)\, dV\ |\psi(\vec{r}, t)|^2\, dV$ is the probability of finding the particle at time '*t*' at the location. $\vec{r}$ in the volume element dV. The total probability of finding the particle somewhere in space is 1 or 100%. That is

$$\int \psi^*(\vec{r}, t)\, \psi(\vec{r}, t)\, dV = 1 \qquad \qquad ...(9.45)$$

If the motion of the particle is in one dimension (along *x*-axis say), then $\rho(x, t)|\psi(\vec{r}, t)|^2$ can be regarded as the probability per unit length of finding the particle at a given point *x* and a given time *t*.

$$\rho(x, t)dx = |\psi(x, t)|^2\, dx$$

is the probability of finding the particle in the element *dx* (i.e., between *x* and *x* + *dx*) about the point *x* at time *t*.

The probability of finding the particle between the points x_1 and x_2 is equal to the sum of probabilities $\rho(x)dx$ in the infinitesimal intervals between x_1 and x_2. It is thus represented by

$$\rho(x_1, x_2)t = \int_{x_1}^{x_2} \rho(x, t)dx = \int_{x_1}^{x_2} |\psi(x, t)|^2\, dx \qquad \qquad ...(9.46)$$

Since the particle is definitely present somewhere along the *x*-axis the probability of finding the particle somewhere along *x*-axis is 1 or 100%, i.e.

$$P = \int_{-\infty}^{+\infty} |\psi(x, t)|^2\, dx = 1 \qquad \qquad ...(9.47)$$

Normalization of ψ

The wave function ψ satisfying the equation $\int \psi^* \psi\, dV = 1$ is said to be normalized. Wave functions which are not normalized can be made so by the following method:

Let ψ_1 be a solution of the Schrödinger equation.

$$i\hbar \frac{\partial \psi_1}{\partial t} = \left[\frac{-\hbar^2}{2m} \nabla^2 + U \right] \psi_1 \qquad \qquad ...(9.48)$$

with $\qquad \int \psi_1^* \psi_1 \, dV = N \qquad \qquad ...(9.49)$

where N is finite, real and positive. ψ_1 is here not normalized to unity. In order to have

a normalized wave function, we now multiple, ψ_1 by the normalization constant $\frac{1}{\sqrt{N}}$ to obtain.

$$\psi = \frac{\psi_1}{\sqrt{N}} \qquad \qquad ...(9.50)$$

Substituting $\psi_1 = \sqrt{N} \, \psi$ in the Schrödinger's eqn. (9.48), we have

$$\sqrt{N} \left(i\hbar \frac{\partial \psi}{\partial t} \right) = \left(\frac{-\hbar^2}{2m} \nabla^2 + U \right) \sqrt{N} \, \psi$$

$$\Rightarrow \qquad i\hbar \frac{\partial \psi}{\partial t} = \left(\frac{-\hbar^2}{2m} \nabla^2 + U \right) \psi$$

Also eqn. (9.49) yields

$$\int \left(\sqrt{N} \, \psi \right)^* \left(\sqrt{N} \, \psi \right) dV = N$$

or, $\qquad \int \psi^* \psi \, dV = 1 \qquad \qquad ...(9.51)$

Thus, it is noticed that ψ_1 and $\frac{\psi_1}{\sqrt{N}}$ satisfy the Schrödinger's eqn. and $\psi = \frac{\psi_1}{\sqrt{N}}$ is

normalized. Since $\psi \left(= \frac{\psi_1}{\sqrt{N}} \right)$ and ψ_1 differ by a constant factor, they describe identical physical

systems.

N is called the norm of ψ_1. The ψ has unit norm. We conclude that normalizable wave functions have finite norm and wave functions with unit norm are normalized.

Norm is defined as the integral of $\psi^* \psi = |\psi|^2$ over all space. Its finiteness implies that $|\psi|^2$ must vanish at infinity. Hence,

$$\psi(\vec{r}, t) \to 0 \text{ as } r \to \infty \qquad \qquad ...(9.52)$$

This boundary condition must hold for all normalizable wave functions.

The wave function ψ for which the integral $\int \psi^* \psi dV$ exists and gives a finite value

is called square integrable.

It may be noted that if ψ in eqn. (4) is replaced by $e^{i/b}ψ$ where β is a real constant, this eqn. undergoes no change, i.e.,

$$\int \left(e^{iβ}ψ\right)^* \left(e^{iβ}ψ\right) dV = \int ψ^* ψ \, dV = 1$$

It implies that a normalized wave function ψ is determined only up to a phase factor of modulus one (i.e., $|e^{iβ}| = 1$). This lack of uniqueness is indicative of the fact that only $|ψ|^2$ has any physical relevance.

Limitations of ψ:

As $|ψ|^2$ represents a probability density, the solutions that can be allowed for ψ from the Schrödinger equation are subjected to certain limitations. The most important limitations are:

(i)   ψ must be finite for all values of x, y, z.

(ii) ψ must be single-valued, i.e., for each set of values of x, y, z, ψ must have one value only.

(iii) ψ must be continuous in all regions except in those regions where the potential energy $V(x, y, z) = ∞$.

Dimension of ψ:

For one-dimensional cos C, the dimension of $ψ(x)$ is $L^{-1/2}$ and for the three-dimensional case, the dimension of $ψ(x, y, z)$ is $L^{-3/2}$.

Probability Current Density

Let a particle of mass m be moving in the positive x-direction in region from x_1 to x_2 (Fig. 9.1).

For the one-dimensional motion of the particle, the wave function is ψ (x, t). Let dA be the area of cross-section of the region.

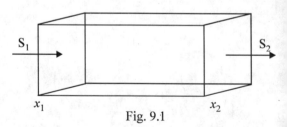

Fig. 9.1

The probability of finding the particle in the region is

$$\int_{x_1}^{x_2} ρ \, dx \, dA = \int_{x_1}^{x_2} ψ(x,t)ψ^*(x,t) \, dx \, dA \qquad ...(9.53)$$

and the probability density of finding the particle in the region is

$$P = ψ(x,t)ψ^*(x,t) \qquad ...(9.54)$$

If the probability of finding the particle in the region decreases with time, the rate of decrease of the probability that the particle is in the region from x_1 to x_2 per unit area is called the probability current density out of the region. Therefore, the probability current density $S_2 - S_1$ out of the region in the positive x-direction is given by

$$S_2 - S_1 = \frac{1}{dA}\left[-\frac{\partial}{\partial t}\int_{x_1}^{x_2} P\, dx\, dA\right] = -\frac{\partial}{\partial t}\int_{x_1}^{x_2}\rho\, dx$$

or, $$S_2 - S_1 = \frac{\partial}{\partial t}\int_{x_1}^{x_2}\psi\, \psi * dx$$

and the probability current density at position x_x is

$$S = -\frac{\partial}{\partial t}\int \psi\, \psi * dx \qquad \qquad ...(9.55)$$

Now, we will show that

$$S = -\frac{i\hbar}{2m}\left[\psi *\frac{\partial\psi}{\partial x} - \psi\frac{\partial\psi *}{\partial x}\right]$$

The Schrödinger equation for $\psi(x, t)$ and $\psi*(x, t)$ are

$$i\hbar\frac{\partial\psi}{\partial t} = -\frac{\hbar^2}{2m}\frac{\partial^2\psi}{\partial x^2} + V\psi \qquad \qquad ...(9.56)$$

$$-i\hbar\frac{\partial\psi *}{\partial t} = -\frac{\hbar}{2m}\frac{\partial^2\psi *}{\partial x^2} + V\psi * \qquad \qquad ...(9.57)$$

Multiplying eqn. (9.56) by $\psi*$ and eqn. (9.57) by ψ, we obtain

$$i\hbar\psi *\frac{\partial\psi}{\partial t} = -\frac{\hbar}{2m}\psi\frac{\partial^2\psi}{\partial x^2} + \psi * V\psi \qquad \qquad ...(9.58)$$

$$-i\hbar\psi\frac{\partial\psi *}{\partial t} = -\frac{\hbar^2}{2m}\psi\frac{\partial^2\psi *}{\partial x^2} + \psi\, V\psi * \qquad \qquad ...(9.59)$$

Subtracting (9.59) from (9.58), we have

$$i\hbar\left(\psi\frac{\partial\psi}{\partial t} + \psi\frac{\partial\psi *}{\partial t}\right) = -\frac{\hbar^2}{2m}\left[\psi *\frac{\partial^2\psi}{\partial x^2} - \psi\frac{\partial^2\psi *}{\partial x^2}\right]$$

or, $$i\hbar\frac{\partial}{\partial t}(\psi\, \psi *) = -\frac{\hbar^2}{2m}\frac{\partial}{\partial x}\left[\psi *\frac{\partial\psi}{\partial x^2} - \psi\frac{\partial\psi *}{\partial x}\right]$$

or, $$\frac{\partial}{\partial t}(\psi\, \psi *) = \frac{\hbar^2}{2m}\frac{\partial}{\partial x}\left[\psi *\frac{\partial\psi}{\partial x} - \psi\frac{\partial\psi *}{\partial x}\right] \qquad \qquad ...(9.60)$$

Substituting this eqn. into eqn. (9.55) we obtain,

$$S = -\frac{i\hbar}{2m} \int \frac{\partial}{\partial x}\left[\psi * \frac{\partial\psi}{\partial x} - \psi \frac{\partial\psi *}{\partial x}\right]dx$$

$$= -\frac{i\hbar}{2m}\left[\psi * \frac{\partial\psi}{\partial x} - \psi \frac{\partial\psi *}{\partial x}\right] \qquad ...(9.61)$$

To show that the probability current density for a free particle is equal to the product of its probability density and its speed.

For a free particle moving in the positive x-direction, the momentum P_x at position x is given by

$$\frac{\hbar}{i}\frac{\partial\psi}{\partial x} = P_x \psi$$

or, $$\frac{\partial\psi}{\partial x} = \frac{i}{\hbar}P_x \psi \qquad ...(9.62)$$

and $$-\frac{\hbar}{i}\frac{\partial\psi *}{\partial x} = P_x \psi *$$

or, $$\frac{\partial\psi *}{\partial x} = -\frac{i}{\hbar}P_x \psi * \qquad ...(9.63)$$

Substituting eqn. (9.62) and (9.63) in eqn. (9.61), we obtain

$$S = -\frac{i\hbar}{2m}\left[\psi * \frac{i}{\hbar}P_x \psi + \psi \frac{i}{\lambda}P_x \psi *\right]$$

$$S = \frac{1}{m}\left[\psi \psi P_x\right]$$

But, $$P_x = mV_x$$

So, $$S = \left(\psi \psi *\right)V_x$$

$$S = \left(\psi \psi *\right)V_x = \left|\psi(x,t)\right|^2 v_x \qquad ...(9.64)$$

Ehrenfest's Theorem

This theorem states that quantum mechanics gives the same results as classical mechanics for a particle for which the average or expectation values of dynamical quantities are involved.

We prove the theorem for one-dimensional motion of a particle by showing that:

(i) $$\frac{d\langle x\rangle}{dt} = \frac{\langle P_x\rangle}{m}$$

and   (ii) $\dfrac{d\langle P_x \rangle}{dt} = \langle F_x \rangle$

Let x be the position coordinate of a particle of mass m at time t.

The expectation value of x is given by $\langle x \rangle = \displaystyle\int\limits_{-\infty}^{+\infty} \psi*(x,t)\, x\, \psi(x,t)\, dx$ ...(9.65)

Differentiating this equation with respect to t, we have

$$\frac{d\langle x \rangle}{dt} = \int\limits_{-\infty}^{+\infty} x\, \frac{\partial(\psi\, \psi*)}{\partial t}\, dx,$$ ...(9.66)

But, $\dfrac{\partial(\psi\, \psi*)}{\partial t} = \dfrac{i\hbar}{2m} \dfrac{\partial}{\partial x}\left[\psi* \dfrac{\partial \psi}{\partial x} - \psi \dfrac{\partial \psi*}{\partial x} \right]$ ...(9.67)

Substituting this in eqn. (9.66), we obtain,

$$\frac{d\langle x \rangle}{dt} = \frac{i\hbar}{2m} \int\limits_{-\infty}^{+\infty} x\, \frac{\partial}{\partial x}\left[\psi* \frac{\partial \psi}{\partial x} - \psi \frac{\partial \psi*}{\partial x} \right] dx$$

Integrating the right hand side by parts, we obtain

$$\frac{d\langle x \rangle}{dt} = \frac{i\hbar}{2m}\left[x\left(\psi* \frac{\partial \psi}{\partial x} - \psi \frac{\partial \psi*}{\partial x} \right)\right]_{-\infty}^{+\infty} - \frac{i\hbar}{2m} \int\limits_{-\infty}^{+\infty}\left(\psi* \frac{\partial \psi}{\partial x} - \psi \frac{\partial \psi*}{\partial x} \right) dx$$

As x approaches either ∞ or $-\infty$, ψ and $\dfrac{\partial \psi}{\partial x}$ approaches zero and therefore the first term becomes zero.

Hence, we obtain $\dfrac{d\langle x \rangle}{dt} = -\dfrac{i\hbar}{2m}\displaystyle\int\left(\psi* \dfrac{\partial \psi}{\partial x} - \psi \dfrac{\partial \psi*}{\partial x} \right) dx$ ...(9.68)

The expectation value of P_x is given by $\langle P_x \rangle = \displaystyle\int\limits_{-\infty}^{+\infty} \psi* \dfrac{\hbar}{i} \dfrac{\partial \psi}{\partial x}\, dx$

$\therefore$ $\displaystyle\int\limits_{-\infty}^{+\infty} \psi* \dfrac{\hbar}{i} \dfrac{\partial \psi}{\partial x}\, dx = \dfrac{i}{\hbar}\langle P_x \rangle$ ...(9.69)

Similarly, $\displaystyle\int\limits_{-\infty}^{+\infty} \psi\, \dfrac{\partial \psi*}{\partial x}\, dx = -\dfrac{i}{\hbar}\langle P_x \rangle$ ...(9.70)

Substituting the values of these integrals in eqn. (4), we have

$$\frac{d\langle x \rangle}{dt} = -\frac{i\hbar}{2m}\left[\frac{i}{\hbar}\langle P_x \rangle + \frac{i}{\hbar}\langle P_x \rangle\right] = \frac{\langle P_x \rangle}{m} \qquad \text{...(9.71)}$$

(ii) To show that $\dfrac{d\langle P_x \rangle}{dt} = \langle F_x \rangle$

The expectation value of the momentum P_x is given by

$$\langle P_x \rangle = \int_{-\infty}^{+\infty} \psi *(x,t)\frac{\hbar}{i}\frac{\partial \psi}{\partial x}\psi(x,t)\,dx \qquad \text{...(9.72)}$$

$$= \frac{\hbar}{i}\int_{-\infty}^{+\infty}\psi *\frac{\partial \psi}{\partial x}\,dx \qquad \text{...(9.73)}$$

Differentiating eqn. (9.73) with respect to t, we have

$$\frac{d\langle P_x \rangle}{dt} = \frac{\hbar}{i}\int_{-\infty}^{+\infty}\left[\frac{\partial \psi *}{\partial t}\cdot\frac{\partial \psi}{\partial x} + \psi *\frac{\partial^2 \psi}{\partial x\,\partial t}\right]dx \qquad \text{...(9.74)}$$

Now, the time-dependent Schrödinger equations for ψ and $\psi*$ are

$$i\hbar\frac{\partial \psi}{\partial t} = -\frac{\hbar^2}{2m}\frac{\partial^2 \psi}{\partial x^2} + V\psi \qquad \text{...(9.75)}$$

$$-i\hbar\frac{\partial \psi *}{\partial t} = -\frac{\hbar^2}{2m}\frac{\partial^2 \psi *}{\partial x^2} + V\psi * \qquad \text{...(9.76)}$$

Differentiating eqn. (9.75) with respect to x, we have

$$i\hbar\frac{\partial^2 \psi}{\partial x\,\partial t} = -\frac{\hbar^2}{2m}\frac{\partial^3 \psi}{\partial x^3} + \frac{\partial(V\psi)}{\partial x} \qquad \text{...(9.77)}$$

Eqn. (9.74) can be written in the form,

$$\frac{d\langle P_x \rangle}{dt} = \int_{-\infty}^{+\infty}\left(-i\hbar\frac{\partial \psi *}{\partial t}\frac{\partial \psi}{\partial x} - \psi *i\hbar\frac{\partial^2 \psi}{\partial x\,\partial t}\right)dx$$

Substituting the expression for $-i\hbar\dfrac{\partial \psi *}{\partial t}$ and $i\hbar\dfrac{\partial^2 \psi}{\partial x\,\partial t}$ in this equation, we obtain

$$\frac{d\langle P_x \rangle}{dt} = \int\left[\left(-\frac{\hbar^2}{2m}\frac{\partial^2 \psi *}{\partial x^2} + V\psi *\right)\frac{\partial \psi}{\partial x} - \psi *\left(-\frac{\hbar^2}{2m}\frac{\partial^3 \psi}{\partial x^3} + \frac{\partial(V\psi)}{\partial x}\right)\right]dx$$

$$\frac{d\langle P_x \rangle}{dt} = -\frac{\hbar^2}{2m} \int\limits_{-\infty}^{+\infty} \left[\frac{\partial^2 \psi *}{\partial x^2} \frac{\partial \psi}{\partial x} - \psi * \frac{\partial^3 \psi}{\partial x^3} \right] dx + \int\limits_{-\infty}^{+\infty} \left[V\psi * \frac{\partial \psi}{\partial x} - \psi * \frac{\partial}{\partial x} (V\psi) \right] dx$$

$$= -\frac{\hbar^2}{2m} \int\limits_{-\infty}^{+\infty} \left[\frac{\partial}{\partial x} \left(\frac{\partial \psi}{\partial x} \cdot \frac{\partial \psi *}{\partial x} - \psi * \frac{\partial^2 \psi}{\partial x^2} \right) dx \right] + \int\limits_{-\infty}^{+\infty} \left[V\psi * \frac{\partial \psi}{\partial x} - \psi * \left(\psi \frac{\partial V}{\partial x} + V \frac{\partial \psi}{\partial x} \right) \right] dx$$

$$= -\frac{\hbar^2}{2m} \left[\frac{\partial \psi}{\partial x} \cdot \frac{\partial \psi *}{\partial x} - \psi * \frac{\partial^2 \psi}{\partial x^2} \right]_{-\infty}^{+\infty} + \int\limits_{-\infty}^{+\infty} \psi * \frac{\partial V}{\partial x} \psi \, dx \qquad \qquad ...(9.78)$$

As x approaches either ∞ or $-\infty$, ψ and $\dfrac{\partial \psi}{\partial x}$ approach zero. Therefore, the first term on the right hand side of cqn. (9.78) is zero. The second term represents the expectation value of the differential coefficient of the potential energy V with respect to x, i.e.,

$$\left\langle \frac{\partial V}{\partial x} \right\rangle = \int\limits_{-\infty}^{+\infty} \psi * \frac{\partial V}{\partial x} \psi \, dx$$

$$\therefore \qquad \frac{d\langle P_x \rangle}{dt} = -\left\langle \frac{\partial V}{\partial x} \right\rangle$$

But, $\quad -\dfrac{\partial V}{\partial x}$ is the classical force F_x.

$$\therefore \qquad \frac{d\langle P_x \rangle}{dt} = \langle F_x \rangle \qquad \qquad ...(9.79)$$

This equation represents Newton's second law of motion. Thus, if the expectation values of dynamical quantities for a particle are considered, quantum mechanics gives the equation of classical mechanics.

SOLVED EXAMPLES

Q.1. **For a free particle show that Schrödinger wave equation leads to the de Broglie relation** $\lambda = \dfrac{h}{p}$.

Sol. The time-independent solution of Schrödinger wave equation may have the form

$$E_n = \frac{n^2 h^2}{8ma^2}$$

Hence, $\quad E_1 = \dfrac{h^2}{8ma^2}$

In this case $a = \dfrac{\lambda}{2}$

$$\therefore \quad E_1 = \frac{h^2}{8ma^2} = \frac{4h^2}{8m\lambda^2} = \frac{h^2}{2m\lambda^2} \qquad \qquad ...(1)$$

Also $\qquad \qquad E_1 = \dfrac{p^2}{2m} \qquad \qquad \qquad \qquad ...(2)$

From equations (1) and (2), we obtain

$$\frac{h^2}{2m\lambda^2} = \frac{p^2}{2m}$$

or, $\qquad p^2 = \dfrac{h^2}{\lambda^2}$ or $p = \dfrac{h}{\lambda}$ or $\lambda = \dfrac{h}{p} \qquad \qquad ...(3)$

Equation (3) is the required de Broglie relation.

Q.2. **Calculate the value of lowest energy of an electron moving in a one-dimensional force-free region of length 4 Å.**

Sol. We have, $m = 9.1 \times 10^{-31}$ kg, $L = 4 \times 10^{-10}$ m. Since energy of electron in nth state is given by

$$E_n = \frac{n^2 \pi^2 \hbar^2}{2ma^2} = \frac{n^2 h^2}{8ma^2}$$

For ground state, $n = 1$

So, $\qquad E_1 = \dfrac{h^2}{8ma^2} = \dfrac{\left(6.625 \times 10^{-34}\right)^2}{5 \times 9.1 \times 10^{-31} \times \left(4 \times 10^{-10}\right)^2}$

$$= 0.376 \times 10^{-18} \, \text{J.}$$

Q.3. **A particle is moving in one-dimensional box and its wave function is given by**

$$\psi_n = A \sin \frac{n\pi x}{a}$$

Find the expression for the normalized wave function.

Sol. The probability that the particle in-between x and $x + dx$ for the nth state is given.

$$P_n \, dx = \left| \psi_n(x) \right|^2 dx$$

$$= A^2 \sin^2 \frac{n\pi x}{a} dx$$

Since the particle has to be in the region $x = 0$ and $x = a$ for all times, so we have

$$\int P_n \, dx = \int_0^a |\psi_n(x)|^2 \, dx = 1$$

or, $\qquad \int_0^a A^2 \sin^2 \dfrac{n\pi x}{a} \, dx = 1$

or, $\qquad A^2 \int_0^a \dfrac{1}{2}\left[1 - \cos\dfrac{2\pi n x}{a}\right] dx = 1$

or, $\qquad \dfrac{A^2}{2} \int_0^a \dfrac{1}{2}\left[x - \dfrac{a}{2\pi n x}\sin\dfrac{2\pi n x}{a}\right]_0^a = 1$

or, $\qquad \dfrac{A^2}{2} a = 1$

So, we have $\qquad A = \sqrt{2/a}$

$\therefore \qquad$ Normalized wave function is $\psi_n(x) = \sqrt{\dfrac{2}{a}} \sin\dfrac{n\pi x}{a}$.

Q.4. What are the SI units of wave function in one and two dimensions?

Sol. In one dimension, the probability density is given by

$$\int |\psi(x,t)|^2 \, dx = 1$$

Since probability is dimensionless. So, we have

$$|\psi|^2 \, [L] = [M^0 L^0 T^0]$$

or, $\qquad |\psi|^2 = [M^0 L^{-1} T^0]$

or, $\qquad |\psi| = [M^0 L^{-1/2} T^0]$

Thus, the dimension of wave function ψ is $[M^0 L^{-1/2} T^0]$.

Similarly, in two dimensions the probability density is given by

$$\int |\psi(x, y, t)|^2 \, dx \, dy = 1$$

As probability is dimensionless, we have

$$|\psi|^2 \, [L] \, [L] = [M^0 L^0 T^0]$$

or, $|\psi|^2 \left[L^2 \right] = \left[M^0 L^0 T^0 \right]$

or, $|\psi|^2 = \left[M^0 L^{-2} T^0 \right]$

or, $|\psi| = \left[M^0 L^{-1} T^0 \right]$

Thus, the dimension of a wave function ψ in two dimensions is $\left[M^0 L^{-1} T^0 \right]$

QUESTIONS

1. What is the meaning of wave function in quantum mechanics?

2. How is the wave function related to probability?

3. What do you mean by normalization of wave function?

4. Write the Schrödinger's time-independent equation for a free particle moving along Y-axis with low speed.

5. Write the Schrödinger's time-dependent non-relativistic equation in quantum physics.

6. Write the: (i) time-dependent, and (ii) time-independent Schrödinger's equation for a particle of mass m and energy E moving along the Z-axis, under a potential V.

7. Write the: (i) time-dependent, and (ii) time-independent Schrödinger's wave equation for a particle of mass m and energy E moving in the XY plane, under a potential V.

8. Derive the time-dependent and time-independent Schrödinger's wave equation.

9. Give the physical interpretation of the wave function ψ.

10. What do you mean by stationary state?

11. Solve time-dependent and time-independent Schrödinger's wave equation in one dimension.

12. What do you mean by probability current density? Show that the probability current density for a free particle is equal to the product of its probability density and its speed.

13. What do you mean by expectation value of a physical quantity?

14. State and prove Ehrenfest's theorem.

15. Considering the expectation values of dynamical quantities for a particle, show that quantum mechanics gives the equations of classical mechanics.

Problems

1. The wave function ψ of a system is a linear combination of eigenfunction θ_1, θ_2, θ_3, θ_4 and θ_5 as

    $$\psi = \left(\frac{1}{\sqrt{3}} \right)\theta_1 + \left(\frac{1}{\sqrt{3}} \right)\theta_2 + \left(\frac{1}{\sqrt{6}} \right)\theta_3 + \left(\frac{1}{\sqrt{24}} \right)\theta_4 + \left(\frac{1}{\sqrt{8}} \right)\theta_5$$

    What is the probability of the system being in the state given by θ_3?   **(Ans. 1/6)**

2. Show that $\psi(x) = e^{icx}$, where c is some finite constant in acceptable eigenfunction. Also normalize it over the region $-a \le x \le a$.

   (**Ans.** The normalized wave function $\psi(x) = \dfrac{x}{\sqrt{2a}} e^{icx}$)

3. Considering nucleus as a box with a size of 10^{-14} m across. Compute the lowest energy of a neutron confined to the nucleus. (**Ans.** 6.43 MeV)

4. The wave function ψ of a system is a linear combination (orthogonal set) of ϕ_1, ϕ_2 and ϕ_3 like

$$\psi = \left(\frac{1}{\sqrt{3}}\right)\theta_1 + \left(\frac{1}{\sqrt{2}}\right)\theta_2 + \left(\frac{1}{\sqrt{6}}\right)\theta_3$$

   What is the probability of that if the system is in the state ϕ_2? (**Ans.** 1/2)

5. Calculate the first two energy levels of an electron confined to a box 1Å wide.
   $(h = 6.254 \times 10^{-34} \text{ JS and } m_e = 9.1 \times 10^{-31} \text{ kg})$ (**Ans.** (i) 37.7 eV (ii) 156.8 eV)

●●●

Chapter 10

Applications of Quantum Mechanics

Time Dependent Schrödinger's Wave Equation (Operator Form)

The wave function may be written as

$$\psi = \psi_0 \, e^{-i\omega t}$$

Differentiating w.r.t. time, we get

$$\frac{\partial \psi}{\partial t} = -i\omega \psi_0 e^{-i\omega t} \qquad \qquad ...(10.1)$$

or, $$\frac{\partial \psi}{\partial t} = -i(2\pi v)\psi \qquad \qquad ...(10.2)$$

and $E = hv$ or, $v = E/h$

Then from (10.2), we have

$$\frac{\partial \psi}{\partial t} = -i\,(2\pi)\left(\frac{E}{h}\right)\psi$$

or, $$E\psi = -\frac{1}{i}\left(\frac{h}{2\pi}\right)\frac{\partial \psi}{\partial t}$$

or, $$E\psi = -\frac{i^2}{i}\left(\frac{h}{2\pi}\right)\frac{\partial \psi}{\partial t}$$

or, $$E\psi = i\hbar \, \frac{\partial \psi}{\partial t} \qquad \qquad ...(10.3)$$

Now Schrödinger's equation is given by

$$\nabla^2 \psi + \left(\frac{2m}{\hbar^2}\right)(E - V)\psi = 0$$

or, $$\nabla^2 \psi + \frac{2m}{\hbar^2} E\psi - \frac{2m}{\hbar^2} V\psi = 0 \qquad \qquad ...(10.4)$$

Substituting the value of $E\psi = i\hbar \dfrac{\partial \psi}{\partial t}$ from (10.3) in (10.4), we get

$$\nabla^2 \psi + \left(\frac{2m}{\hbar^2}\right) i\hbar \frac{\partial \psi}{\partial t} - \frac{2m}{\hbar^2} V\psi = 0$$

or,
$$\nabla^2 \psi - \frac{2m}{\hbar^2} V\psi = -\left(\frac{2m}{\hbar^2}\right) i\hbar \frac{\partial \psi}{\partial t}$$

or,
$$-\left(\frac{\hbar^2}{2m}\right)\nabla^2 \psi + V\psi = i\hbar \frac{\partial \psi}{\partial t} \qquad\qquad ...(10.5)$$

This equation represents Schrödinger's time dependent equation. Eqn. (10.5) can be written as

$$\left(-\frac{\hbar^2}{2m}\nabla^2 + V\right)\psi = i\hbar \frac{\partial \psi}{\partial t}$$

or,
$$H\psi = E\psi \qquad\qquad ...(10.6)$$

where,
$$h = \left[-\frac{\hbar^2}{2m}\nabla^2 + V\right]$$ is an operator called Hamiltonian. Eqn. (10.6)

represents the motion of a material particle in terms of Hamiltonian operator (non-relativistic particle).

Eigenvalues and Eigenfunctions

Any system is characterised by its position, momentum, energy, etc. In case of a wave function corresponding to a system is assigned, the state of the system can also be precisely known. In case the state of the system changes, its wave function also changes accordingly.

Schrödinger's wave equation has many solutions. Out of these, only those solutions which have meaningful results are to be selected and these wave functions are called acceptable wave functions.

Schrödinger's time independent equation actually belongs to a class of eigenvalue equations. Acceptable solutions of Schrödinger's wave equation exist only for certain discrete values which satisfy the relation

$$E_n = \frac{\eta^2 - h^2}{8\,m\,a^2}$$

The allowed values of energies of a particular system are called the eigenvalues [Eigen means proper wave functions are called eigenfunctions. The whole set of eigenvalues of energy is called the Energy Spectrum of the particle (or a system of particle)].

In general, if an operator $\hat{\alpha}$ operates on a function $\phi_n(x)$ and gives a set of wave functions $\phi_n(x)$ multiplied by a general set of eigenvalues a_n which may be defined by the equation

$$\hat{\alpha}\, \phi_n(x) = a_n\, \phi_n(x)$$

Expectation Values

Dynamic quantities, viz. position, potential energy, kinetic energy, momentum, etc. which from the basis of physical measurements are called observable in Physics. In quantum mechanics, each observable is represented by an operator which acts on a wave function ψ to give a new wave function.

We define the average or expectation value $< \alpha >$ of any observable α as

$$< \alpha >= \frac{\int\limits_{-\infty}^{+\infty} \psi^*(r,t)\, \alpha_{op}\, \psi(\vec{r},t)dV}{\int\limits_{-\infty}^{+\infty} \psi^*(\vec{r},t)\, \psi(r,t)\, dV}$$

From normalization condition

$$\int\limits_{-\infty}^{+\infty} \psi^*(\vec{r},t)\, \psi(\vec{r},t)\, dV = 1$$

$$< \alpha >= \int\limits_{-\infty}^{+\infty} \psi^*(\vec{r},t)\alpha_{op}\, \psi(\vec{r},t)\, dV \qquad \qquad ...(10.7)$$

We may give the expectation value of the vector $\vec{r}$

$$<r>= \int\limits_{-\infty}^{+\infty} \psi^*(r,t)\, \vec{r}\, \psi(\vec{r},t)\, dV \qquad \qquad ...(10.8)$$

Quantum Mechanics

The expectation value of position vector is also a vector but it is a function of time only since the space coordinate disappears on integration. Eqn. (10.8) may be written in terms of component equation as

$$<x>= \int\limits_{-\infty}^{+\infty} \psi^* x\, \psi\ dV \qquad \qquad ...(10.9)$$

$$<y>= \int\limits_{-\infty}^{+\infty} \psi^* y\, \psi\ dV$$

$$<z> = \int_{-\infty}^{+\infty} \psi * z \psi \ dV$$

In the same way, we may write the expectation value of potential energy V as

$$<V> = \int_{-\infty}^{+\infty} \psi * (r,t) V (\vec{r}, t) \psi (r,t) dV \qquad ...(10.10)$$

Some other examples value of momentum is given by the relation

$$<p> = \int_{-\infty}^{+\infty} \psi * (x,t) \, p \, \psi (r,t) dx \qquad ...(10.11)$$

To solve the integral (10.11) we have to express $\psi* (x, t) \, p \, \Psi (x, t)$ in terms of variable x and t.

Classically we may express p as a function of x and t. However, in quantum mechanics, due to uncertainty principle, p cannot be expressed in terms of x because p and x cannot be known simultaneously.

In the same way we cannot write p as a function of t. So we have to find some other way to solve the integral.

In order to evaluate $< p >$ and $< E >$, we can differentiate the free particle wave function $\psi = de^{-(Vk)(Et - px)}$ w.r.t x and t. Then we get,

$$\frac{\partial \psi}{\partial t} = i \hbar \, P \psi \quad \text{and} \quad \frac{\partial \psi}{\partial t} = -\frac{i}{\hbar} E \psi$$

which can be written as

$$P \psi = \frac{\hbar}{i} \frac{\partial}{\partial x} \psi \qquad ...(10.12)$$

and

$$E \psi = i \hbar \frac{\partial}{\partial t} \psi \qquad ...(10.13)$$

which shows that the dynamical quantity P in some sense corresponds to differential operator

$$\frac{\hbar}{i} \frac{\partial}{\partial x}$$

Similarly, dynamical quantity E corresponds to differential operator $t\hbar \frac{\partial}{\partial x}$, where an operator bills us about the operation to be carried out on a dynamical quantity.

For example, operator $t\hbar \frac{\partial}{\partial x}$ tells us to take the partial derivation of what comes after it and multiply the result by $t\hbar$.

Generally, the expectation value is written in **Bold face letters**, so that P and E correspond momentum and energy operators.

Momentum operator $\quad P = \dfrac{\hbar}{i} \dfrac{\partial}{\partial x}$ ...(10.14)

Energy operator $\quad E = i\hbar \dfrac{\partial}{\partial t}$ ...(10.15)

Derivation of Schrödinger Equation: Though we have only shown that the correspondences expressed in eqns. (10.14) and (10.15) hold for free particles, they are entirely general results whose validity is the same as that of Schrödinger's equation. To support this statement, we can replace the equation $E = T + V$ for the total energy of a particle with the operator equation.

$$E = T + V$$ ...(10.16)

Writing KE, T in terms of momentum P, we have

$$T = \dfrac{p^2}{2m}$$

which gives Kinetic energy operator

$$T = \dfrac{p^2}{2m} = \dfrac{1}{2m}\left(\dfrac{\hbar}{i}\dfrac{\partial}{\partial x}\right)^2 = \dfrac{-\hbar^2}{2m}\dfrac{\partial^2}{\partial x^2}$$ ...(10.17)

So eqn. (10.16) may be written as

$$i\hbar\dfrac{\partial}{\partial t} = -\dfrac{\hbar^2}{2m}\dfrac{\partial^2}{\partial x^2} + V$$ ...(10.18)

Multiplying the identity $\psi = \psi$ by eqn. (12), we get

$$i\hbar\dfrac{\partial \psi}{\partial t} = -\dfrac{\hbar^2}{2m}\dfrac{\partial^2 \psi}{\partial x^2} + V\psi$$

which is Schrödinger's equation. It shows that eqns. (10.14) and (10.15) are equivalent to postulating Schrödinger's equation.

Operator and Expectation Value: Since p and E can be replaced by their corresponding operators in an equation, we can use these operators to get expectation values for p and E. So, the expectation value for p is

$$<p> = \int\limits_{-\infty}^{+\infty} \psi * p\psi \, dx = \int\limits_{-\infty}^{+\infty} \psi * \left(\dfrac{\hbar}{i}\dfrac{\partial}{\partial x}\right)\psi \, dx = \dfrac{\hbar}{i}\int\limits_{-\infty}^{+\infty} \psi * \dfrac{\partial \psi}{\partial x} \, dx$$ ...(10.19)

and the expectation value of E is given by

$$<E> = \int\limits_{-\infty}^{+\infty} \psi * E\psi \, dx = \int\limits_{-\infty}^{+\infty} \psi * \left(i\hbar\dfrac{\partial}{\partial t}\right)\psi \, dx = i\hbar\int\limits_{-\infty}^{+\infty} \psi * \dfrac{\partial \psi}{\partial x} \, dx$$ ...(10.20)

Both eqns. (10.19) and (10.20) can be evaluated for any acceptable wave function $\psi\,(x,\,t)$.

So it is necessary to expose expectation value involving operators in the form

$$(p) = \int\limits_{-\infty}^{+\infty} \psi * p\,\psi\,dx$$

Every observable quantity G characteristic of a physical system may be represented by suitable quantum-mechanical operator G. To obtain this operator, it is only necessary to express in terms of x and p and then replace p by $(\hbar/i)\,\partial/\partial x$. If the wave function ψ of the system is known, the expectation value of $G\,(x,\,p)$ is

$$< G\,(x,\,p) > = \int\limits_{-\infty}^{+\infty} \psi * G\,\psi\,dx \qquad \qquad ...(10.21)$$

This result substantiates the statement made earlier that from ψ can be obtained all to information about a system that is permitted by the *uncertainty principle*.

Conditions for Quantization of a Variable: The condition that a certain dynamical variable G be restricted to the discrete values G_n – in other words, that G be quantized – is that the wave function ψ_n of the system be such that

$$G\Psi_n = G_n\Psi_n \qquad \qquad ...(10.22)$$

where G is the operator that corresponds to G and each G_n is a real number. When eqn. holds for the wave functions of a system, it is a fundamental postulate (in fact, the fundamental postulate) of quantum mechanics that any measurement of G can only yield one of the values of G_n. If measurements of G are made on a number of identical systems all in states described by the particular eigenfunction Ψ_k, each measurement will yield the single value G_k.

Hamiltonian Operator for Total Energy: The total-energy operator E of eqn. (10.16) is usually written in the form

$$H = -\frac{\hbar^2}{2m}\frac{\partial^2}{\partial x^2} + V \qquad \qquad ...(10.23)$$

and is called the Hamiltonian operator because it is reminiscent of the Hamiltonian function in advanced classical mechanics, which is an expression for the total energy of a system in terms of coordinates and momenta only. So the steady state Schrödinger equation can be written simply as

$$E_n\,\Psi_n = H\,\Psi_n \qquad \qquad ...(10.24)$$

so we can say that the various E_n are the eigenvalues of the Hamiltonian operator H. This kind of association between **eigenvalues and quantum-mechanical operators** is quite general.

Application of Schrödinger Equation: Particle in a Box (Motion in one dimension) Bound States

Let us consider a free particle in one-dimensional box. The particle has a mass m and is restricted to move in a straight line along the x-axis. The range of the particle is $0 < x < a$ and the particle is reflected back whenever it reaches the end of the range. We have the general equation

$$\frac{\partial^2 \psi}{\partial x^2} + \left(\frac{2m}{\hbar^2}\right)(E - V)\Psi = 0 \qquad \qquad ...(10.25)$$

For a free particle $V = 0$ for $0 < x < a$.

Also the wave function Ψ vanishes at $x = 0$ and $x = a$. For one dimension, the modified equation is given by

$$\frac{\partial^2 \psi}{\partial x^2} + \left(\frac{2m}{\hbar^2}\right) E\,\Psi = 0$$

$$\frac{\partial^2 \psi}{\partial x^2} + \left(\frac{8\pi^2\,mE}{\hbar^2}\right) E\,\Psi = 0 \qquad \left[\because \hbar = \frac{h}{2\pi}\right] \qquad ...(10.26)$$

Let us take $\qquad \dfrac{8\pi^2 mE}{h^2} = k^2$

$\therefore \qquad E = \dfrac{k^2 h^2}{8\pi^2 m}$

So eqn. (10.26) may be written as

$$\frac{\partial^2 \psi}{\partial x^2} + k^2 \psi = 0 \qquad \qquad ...(10.27)$$

The general solution for equation (10.27) is given by

$$\Psi(x) = A\sin kx + B\cos kx \qquad \qquad ...(10.28)$$

and from the boundary condition

$$\Psi = 0 \; at \; x = 0$$

So from equation (4), B must be zero

$$\Psi(a) = A\sin kx = 0$$

Since we cannot take A to be zero because it does not yield any solution, so we take

or $\qquad \sin ka = 0$

$$k = \frac{n\pi}{a} \qquad \qquad \text{where } n = 1, 2, 3, \text{ etc.}$$

Hence the only permissible solution of the wave equations are

$$\Psi_n(x) = A\sin\left(\frac{n\pi}{a}\right)x \qquad \qquad ...(10.29)$$

and the value of energy is given by $E = \dfrac{k^2 h^2}{8\pi^2 m}$ $\qquad \qquad$ ...(10.30)

Putting the value of $k = \dfrac{n\pi}{a}$ in equation (6), we get

$$E_n = \frac{n^2 h^2 \pi^2}{8\pi^2 ma^2} = \frac{n^2 h^2}{8 ma^2} \qquad \qquad ...(10.31)$$

It shows E_n depends upon the value of n. Each value E_n is called **Eigenvalue or proper value** and each of Ψ_n is called an **Eigenfunction or proper function**.

From normalization condition

$$\int_0^a |\Psi_n(x)|^2 \, dx = A^2 \int_0^a \sin^2\left(\frac{n\pi x}{a}\right) dx = 1 \qquad \qquad ...(10.32)$$

or, $\qquad \dfrac{A^2 a}{2} = 1$

$\therefore \qquad A - \sqrt{\dfrac{2}{a}}$

$\therefore \qquad \psi_n(x) = \sqrt{\dfrac{2}{a}} \sin\left(\dfrac{n\pi}{a}\right)x \qquad \qquad$ ...(10.33)

Now, for $n = 1$ $\qquad \psi_1(x) = \sqrt{\dfrac{2}{a}} \sin\left(\dfrac{n\pi}{a}\right)x$

for $n = 2$ $\qquad \psi_2(x) = \sqrt{\dfrac{2}{a}} \sin\left(\dfrac{2\pi}{a}\right)x$

for $n = 3$ $\qquad \psi_3(x) = \sqrt{\dfrac{2}{a}} \sin\left(\dfrac{3\pi}{a}\right)x \qquad \qquad$ ...(10.34)

Graphically, the first three functions are shown in Fig. 10.1.

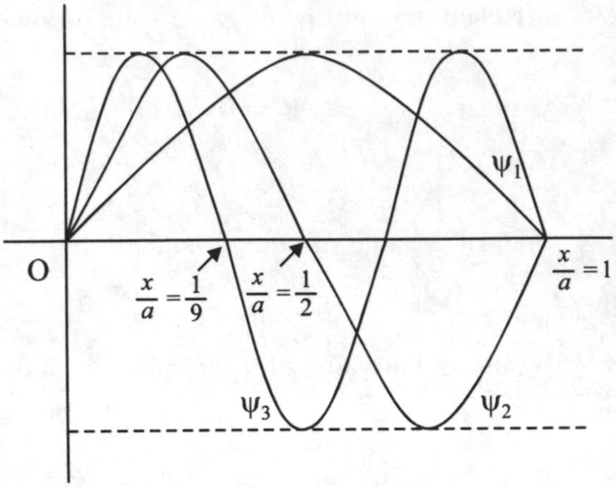

Equation (10.31) shows that energy E_n of the particle is quantized, we have

$$E_n = \frac{p_n^2}{2m}$$

or, $P_n = (2mE_n)^{1/2} = \left(\frac{2\,m\,n^2\,h^2}{8\,m\,a^2}\right)^{1/2}$

or, $P_n = \frac{nh}{2a}$ ...(10.35)

$\therefore \quad P_n \propto n$

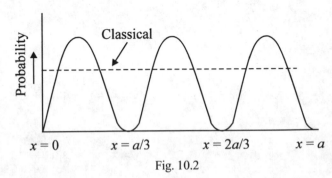

Fig. 10.1

which shows that the momentum of the particle is also quantized. Moreover, for each value of E_n, there is corresponding value of P_n.

From quantum mechanics, the probability is given by

$$|\psi_n(x)|^2\,dx = \frac{2}{a}\sin^2\left(\frac{n\pi x}{a}\right)dx \qquad \qquad ...(10.36)$$

This is wave like function and for $n = 3$ (say)

$$|\psi_n(x)|^2\,dx = \frac{2}{a}\sin^2\left(\frac{3\pi x}{a}\right)dx \qquad \qquad ...(10.37)$$

The graph between probability and x is given in Fig. 10.2.

In the graph, the dotted line represents the constant probability according to classical mechanics. But the curve represents the probability variation with variation of x according to quantum mechanics probability is zero for values of

Fig. 10.2

$$x = 0,\; x = \frac{a}{3},\; x = \frac{2a}{3} \text{ and } x = a \text{ and so on.}$$

Wave Function and Probability Density

The basic postulate of quantum mechanics is that the quantity $\Psi\Psi^*\ dxdydz$ is the probability that a particle represented by the wave function $\psi\ (x, y, z, t)$ is in the volume $dV = (dxdydz)$ at the time t.

$$P(t) = \psi\psi^* = |\psi(x,y,z,t)|^2$$

and the wave function y corresponds to the probability amplitued for the position of the particle.

Since the particle must be somewhere in the space, it is necessary that the probability density must satisfy the normalisation condition, i.e.

$$\int_{-\infty}^{+\infty}\int_{-\infty}^{+\infty}\int_{-\infty}^{+\infty} \psi\psi^*\ dxdydz = 1$$

In this case
$$\psi = \psi_0\ e^{-i\omega t} \qquad\qquad ...(10.38)$$

or,
$$\psi = \psi_0\ e^{-e2\pi Et}ih \qquad \left(\because E = hv = \frac{h\omega}{2\pi} \therefore \omega = \frac{2\pi E}{h} \right)$$

or,
$$\psi = \psi_0\ e^{iEt/\hbar} \qquad\qquad ...(10.39)$$

The complex conjugate is given by

$$\psi^* = \psi_0^*\ e^{iEt/\hbar} \qquad\qquad ...(10.40)$$

Substituting these values of ψ and ψ^* in equation (10.38)

$$\int_{-\infty}^{+\infty}\int_{-\infty}^{+\infty}\int_{-\infty}^{+\infty} \psi_0\ \psi_0^*\ dxdydz = 1 \qquad\qquad ...(10.41)$$

So far a wave function satisfying the time independent Schrödinger equation, the probability is also independent of time.

However, ψ has the following limitations:

(i) The wave function ψ partial being normalizable must be single valued as $P(t)$ can have only one value at a particular place at a particular time.

(ii) Wave function ψ and its partial derivation $\dfrac{d\psi}{dx}, \dfrac{d\psi}{dy}$ and $\dfrac{d\psi}{dz}$ must be continuous everywhere.

(iii) ψ must be finite for all values of x, y and z.

Probability Current Density

In order to find the quantum mechanical relation to express the particle flow, let us first begin with the classical relation.

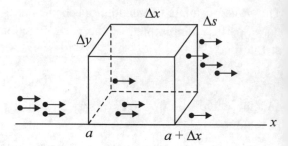

Let $\rho(x, y, z, t)$ be the charge density and let the charge be concentrated in a small volume element $\Delta x \Delta y \Delta z$ (Fig. 10.3). Let us first consider the case when all the velocities are along x-axis. Net charge per unit time $\Delta x \Delta y \Delta z$ at left is ρv evaluated at $x = a$ time $\Delta y \Delta z$ and that leaving right is

Fig. 10.3: Charge entering and leaving a small cubical volume element $\Delta x \Delta y \Delta z$ for a one-dimensional flow

$$(\rho v_x)_{a+\Delta x} \Delta y \Delta z = \left[\rho v_x + \partial(\rho v_x)/\partial \Delta x \right] \Delta y \Delta z \qquad \text{...(10.42)}$$

The charge in $\Delta x \Delta y \Delta z$ is of course, $\rho \, \Delta x \Delta y \Delta z$. If more charge leaves per second than arrives, ρ decreases in time and conservation of charge gives the relation.

$$(\rho v_x)_{a+\Delta x} \Delta y \Delta z = (\rho v_x)_a \Delta y \Delta z \frac{\partial(\rho v_x)}{\partial x} \Delta x \Delta y \Delta z = \frac{\partial \rho}{\partial t} \Delta x \Delta y \Delta z \qquad \text{...(10.43)}$$

which gives $\qquad \dfrac{\partial(\rho v_x)}{\partial x} = \dfrac{\partial p}{\partial t}$ $\qquad\qquad$ ...(10.44)

in one dimension. If we extended the case to three dimensions, we get

$$\nabla \cdot \rho v = -\frac{\partial \rho}{\partial t}$$

which is the equation of continuity, applicable to any flow process in which ρ represents the density of something that is conserved.

In order to find the quantum mechanical analogue of eqn. (10.43), we replace ρ with $q \, \psi^* \psi$ where q is the charge associated with the particle and $\psi^* \psi \, dxdydz$ is the probability at time t to locate the particle in volume element $dxdydz$ (where we have considered $\int \int \int \psi^* \psi \, dxdydz$ over all space to be normalized so as to give total number of particles enclosed). We assume that there is a quantity S which represents the probability stream density of the particles in such a way that qS plays the role of ρv in the equation of continuity. We seek S in terms of ψ by writing the equation of continuity in the form

$$\overline{\nabla} \cdot q\overline{S} = \partial(q\psi^* \psi)$$

or, $\nabla \cdot \overline{S} = \dfrac{\overline{\partial}}{\partial t}(\psi * \psi)$ ...(10.45)

using the eqn. $\dfrac{\hbar^2}{2m}\nabla^2\psi + v\psi = \dfrac{\hbar}{i}\dfrac{\partial\psi}{\partial t}$

$$\nabla \cdot \overline{S} = \dfrac{\partial\psi *}{\partial t}\psi - \psi *\dfrac{\partial\psi}{\partial t} = \dfrac{\hbar}{2mi}\left(\psi\nabla^2\psi * -\psi *\nabla^2\psi\right)$$ ...(10.46)

which gives $\overline{S} = \dfrac{\hbar}{2mi}\left(\psi *\nabla\psi - \psi\nabla\psi *\right)$ ...(10.47)

The quantity S is a vector probability current density for the flow of particles, if each particle bears charge q, the corresponding probability current density is qS. If A denotes any surface area enclosing a volume v, the integral of the normal component of S ever the surface is equal to the time rate of decrease of probability of finding a particle within the volume, so the left-hand integral represents the rate at which probability is streaming outward across the closed surface A, while the right-hand integral represents the rate at which the probability is decreasing within the volume.

When ψ has the form, $\psi = Ae^{i(p_x' - E)/\hbar}$ then $\psi* = A*e^{i(E_x - p_x')/\hbar}$

so we have, $|\overline{S}| = \dfrac{P}{m}|\vec{A}|^2$

where $|S|$ corresponds to the magnitude of S. So the significance of the beam represented by eqn. $\psi = Ae^{i(p_x' - E)/\hbar}$ can be based on the vectors $\vec{S}$ by assuming the number of particles crossing/unit area per second is $n = \dfrac{P|A|^2}{m}$ where $\dfrac{P}{m}$ corresponds to the speed of a particle.

The Free Particle

Suppose a particle of mass m is in motion along the x-axis. Suppose no force is acting on the particle, so that the potential energy of the particle is constant. For convenience, the constant potential energy is taken to be zero. Therefore, the time-independent Schrödinger equation for the free particle is

$$-\dfrac{\hbar^2}{2m}\dfrac{d^2\psi}{dx^2} = E\psi$$ ...(10.48)

Since the particle is moving freely with zero potential energy, its total energy E is the kinetic energy given by

$$E = \dfrac{p_x^2}{2m}$$ ...(10.49)

where p is the momentum of the particle.

From eqn. (1) we have $\dfrac{d^2\psi}{dx^2} + \dfrac{2mE}{\hbar^2}\phi = 0$

or, $\qquad \dfrac{d^2\psi}{dx^2} + k^2\psi = 0$ ...(10.50)

where $\qquad k = \sqrt{\dfrac{2mE}{\hbar^2}}$ ...(10.51)

So that $\qquad E = \dfrac{k^2\hbar^2}{2m},$ ...(10.52)

The solution of eqn. (3) is

$$\phi(x) = Ae^{ikx} + Be^{-ikx}$$ ...(10.53)

where A and B are constants.

Eqn. (6) gives the time-independent part of the wave function. The complete wave function for particle is given by

$$\phi(x, t) = \psi(x)\, e^{-i\omega t}$$

$$= \left(Ae^{ikx} + Be^{-ikx}\right) e^{-i\omega t}$$

$$= Ae^{-i(\omega t - kx)} + Be^{-i(\omega t + kx)}$$ ...(10.54)

where $\qquad \omega = \dfrac{E}{\hbar} = \dfrac{k^2\hbar^2}{2m}\dfrac{1}{\hbar} = \dfrac{\hbar k^2}{2m}$ ...(10.55)

Eqn. (10.54) represents a continuous plane simple harmonic wave. The first term on the right side of eqn. (10.54) represents the wave travelling in the positive x-direction and the second term represents the wave travelling in the negative x-direction. Therefore, for the motion of the particle in the positive x-direction, we have

$$\phi(x) = Ae^{ikx}$$ ...(10.56)

or, $\qquad \phi(x, t) = Ae^{-i(\omega t - kx)}$ ...(10.57)

The momentum operator, $\dfrac{\hbar}{i}\dfrac{\partial}{\partial x}$, operating on the wave function $\psi(x,\, t)$ gives

$$\frac{\hbar}{i}\frac{\partial\psi}{\partial x} = \frac{\hbar}{i}\left[A(ik)e^{-i(\omega t - kx)}\right] = \hbar k Ae^{-i(\omega t - k)} = \hbar k\,\psi$$

But from eqns. (10.49) and (10.52)

$$p_x = \hbar k$$ ...(10.58)

$$\therefore \qquad \frac{\hbar}{i} \frac{\partial \phi}{\partial x} = p_x \phi \qquad \qquad ...(10.59)$$

This equation shows that the wave function $\phi(x, t)$ for the particle is an eigenfunction of the linear momentum operator, and the momentum p_x is the eigenvalue of the operator. Hence, the momentum remains sharp with the value p_x.

The probability of finding the particle between x and $x + dx$ is given by

$$Pdx - \phi(x,t)\, \phi*(x,t)\, dx = A^2 dx \qquad \qquad ...(10.60)$$

$\therefore$ The probability density P for the position of the particle with the definite value of momentum is constant over the x-axis, i.e. all positions of the particle are equally probable. This conclusion is also obtained from the principle of uncertainty.

According to the interpretation of the wave function, the probability of finding the particle somewhere in space must be equal to 1, i.e.

$$\int_{-\infty}^{\infty} \phi(x,t)\, \phi*(x,t)\, dx = 1 \qquad \qquad ...(10.61)$$

In this case $\phi(x, t)\, \phi*(x, t) = $ constant $ = A^2$.

Therefore, the integral on the left side of eqn. (10.61) is infinite. Hence, the wave function for the particle cannot be normalised, A must remain arbitrary. The difficulty arises because we are derive with an ideal case. In particle, we cannot have an absolutely particle. The particle will always be confined within an enclose in the laboratory, and hence its position can be determined. It means that its momentum cannot be determined with absolutely accuracy.

The Rectangular Potential Barrier: The Tunnel Effect

A rectangular potential barrier of height V_0 and width a for a particle is shown in Fig. 10.4. It extends over the region (II) from $x = 0$ to $x = a$ in which the potential energy V of the particle will be constant equal to V_0. On both sides of the barrier, in regions (I) and (III) $V = 0$; this means that when the particle is in these regions no forces act on it.

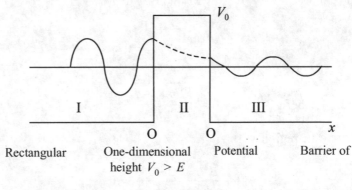

Rectangular   One-dimensional   Potential   Barrier of
height $V_0 > E$

Fig. 10.4

Suppose a beam of particles travelling parallel to the x-axis from left to right is incident on the potential barrier. In the regions (I) and (III), the energy E of a particle is wholly kinetic, and in the region (II) it is partly kinetic and partly potential. If $E = V_0$, then according to classical mechanics, the probability of any particle reaching the region (III) after crossing the region (II) is zero. However, according to quantum mechanics, the transmission probability has a small but definite value. This behaviour is called *tunnel effect*.

Let $\phi_1(x)$, $\phi_2(x)$ and $\phi_3(x)$ be the wave functions for the motion of particle in regions (I), (II) and (III) respectively. In these regions the time-independent Schrödinger wave equations are as follows:

In region (I): $-\infty < x < 0$

$$-\frac{\hbar^2}{2m}\frac{d^2\phi}{dx^2} = E\phi_1$$

or, $\qquad \frac{d^2\phi_1}{dx} + \frac{2mE}{\hbar^2}\phi_1 = 0$

or, $\qquad \frac{d^2\phi_1}{dx^2} + k^2\phi_1 = 0 \qquad\qquad\qquad ...(10.62)$

where $\qquad k = \sqrt{\frac{2mE}{\hbar^2}}$

In region (II): $0 < x < a$

If $E < V_0$, then equation is

$$\frac{d^2\phi_2}{dx^2} + \frac{2m}{\hbar^2}\left(E - V_0\right)\phi_2 = 0$$

or, $\qquad \frac{d^2\phi_2}{dx^2} - \frac{2m}{\hbar^2}\left(V_0 - E\right)\phi_2 = 0$

or, $\qquad \frac{d^2\phi_2}{dx^2} - \beta^2\,\phi_2 = 0 \qquad\qquad\qquad ...(10.63)$

where $\qquad \beta = \sqrt{\frac{2m(V_0 - E)}{\hbar^2}}$

In region (III): $a \leq x < \infty$

$$\frac{d^2\phi_3}{dx^2} + \frac{2mE}{\hbar^2}\,\phi_3 = 0$$

$$\frac{d^2\phi_3}{dx^2} + k^2\,\phi_3 = 0 \qquad\qquad\qquad ...(10.64)$$

The general solutions of eqns. (10.62), (10.63) and (10.64) are

$$\phi_1 = Ae^{ikx} + Be^{-ikx} \qquad \text{...(10.65)}$$

$$\phi_2 = Ce^{\beta x} + De^{-\beta x} \qquad \text{...(10.66)}$$

$$\phi_3 = Ge^{ikx} + He^{-ikx} \qquad \text{...(10.67)}$$

Since there is no particle coming from left in the region (III), we must have $H = 0$. Therefore,

$$\phi_3 = Ge^{ikx} \qquad \text{...(10.68)}$$

The interpretation of the terms in the above solutions is as follows:

In eqn. (10.65) the term Ae^{ikx} is a wave of amplitude A travelling in the positive x-direction and the term Be^{-ikx} is the wave of amplitude B reflected in the negative x-direction from the potential barrier when the incident wave falls on the barrier. In eqn. (10.66) the term $De^{-\beta x}$ is an exponentially decreasing wave function representing non-oscillatory disturbance which moves through the barrier in the positive x-direction, and the term $Ce^{\beta x}$ is the reflected disturbance within the barrier; it is an exponentially decreasing wave function.

Eqn. (10.68) represents the transmitted wave in the region (III). This wave travels in the positive x-direction.

Expression for the coefficients A and B in terms of the coefficient

(i) At $x = 0$, we have

$$\phi_1(0) = \phi_2(0)$$

$\therefore$ From eqns. (10.65) and (10.66), we get

$$A + B = C + D \qquad \text{...(i)}$$

we also have

$$\left(\frac{d\phi_1}{dx}\right)_{x=0} = \left(\frac{d\phi_2}{dx}\right)_{x=0}$$

$\therefore \qquad Aik - Bik = C\beta - D\beta$

or, $\qquad A - B = \dfrac{\beta}{ik}(C - D) \qquad \text{...(ii)}$

From eqns. (i) and (ii), we get

$$2A = \left(1 + \frac{\beta}{ik}\right)C + \left(1 - \frac{\beta}{ik}\right)D \qquad \text{...(10.69)}$$

and $\qquad 2B = \left(1 - \dfrac{\beta}{ik}\right)C + \left(1 + \dfrac{\beta}{ik}\right)D \qquad \text{...(10.70)}$

(ii) At $x = a$, we have

$$\phi_2(a) = \phi_3(a)$$

$\therefore$ From eqns. (10.66) and (10.67), we get

$$Ce^{\beta a} + De^{-\beta a} = Ge^{ika} \qquad \qquad ...(10.71)$$

We also have $\quad \left(\dfrac{d\phi_2}{dx}\right)_{x=a} = \left(\dfrac{d\phi_3}{dx}\right)_{x=a}$

$\therefore \qquad C\beta e^{\beta a} - D\beta e^{-\beta a} = Gike^{ika}$

or, $\qquad Ce^{\beta a} - De^{-\beta a} = \dfrac{ik}{\beta} Ge^{ika} \qquad \qquad ...(10.72)$

Now from eqns. (10.71) and (10.72), we get

$$C = \tfrac{1}{2}\left(1 + \dfrac{ik}{\beta}\right)e^{-\beta a} Ge^{ika} \qquad \qquad ...(10.73)$$

$$D = \tfrac{1}{2}\left(1 - \dfrac{ik}{\beta}\right)e^{\beta a} Ge^{ika} \qquad \qquad ...(10.74)$$

Substituting the values of C and D in eqn. (10.69), we get

$$2A = \left[\left(1 + \dfrac{\beta}{ik}\right)\tfrac{1}{2}\left(1 + \dfrac{ik}{\beta}\right)e^{-\beta a} + \left(1 - \dfrac{\beta}{ik}\right)\tfrac{1}{2}\left(1 + \dfrac{ik}{\beta}\right)e^{\beta a}\right] \times Ge^{ika}$$

Simplifying this equation, we get

$$A = \left[\left(\dfrac{e^{\beta a} + e^{-\beta a}}{2}\right) - \tfrac{1}{2}\left(\dfrac{\beta}{ik} + \dfrac{ik}{\beta}\right)\left(\dfrac{e^{\beta a} - e^{-\beta a}}{2}\right)\right]Ge^{ika}$$

Using the trigonometric relations:

$$\dfrac{e^{\beta a} + e^{-\beta a}}{2} = \cos h\beta a$$

and $\qquad \dfrac{e^{\beta a} - e^{-\beta a}}{2} = \sin h\beta a$

we obtain $\qquad A = \left[\cos h\beta a - \tfrac{1}{2}\left(\dfrac{\beta}{ik} + \dfrac{ik}{\beta}\right)\sin h\beta a\right]Ge^{ika}$

or $\qquad A = \left[\cos h\beta a + \dfrac{i}{2}\left(\dfrac{\beta}{k} - \dfrac{k}{\beta}\right)\sin h\beta a\right]Ge^{ika} \qquad \qquad ...(10.75)$

Similarly substituting the values of C and D in eqn. (10.70)

$$2B = \left[\left(1 - \frac{\beta}{ik}\right)\frac{1}{2}\left(1 + \frac{ik}{\beta}\right)e^{-\beta a} + \left(1 + \frac{\beta}{ik}\right)\frac{1}{2}\left(1 - \frac{ik}{\beta}\right)e^{\beta a}\right]Ge^{ika}$$

$$= \left[-\frac{1}{2}\left(\frac{\beta}{ik} - \frac{ik}{\beta}\right)e^{-\beta a} + \frac{1}{2}\left(\frac{\beta}{ik} - \frac{ik}{\beta}\right)e^{\beta a}\right]Ge^{ika}$$

$$= \left[\left(\frac{\beta}{ik} - \frac{ik}{\beta}\right)\left(\frac{e^{\beta a} - e^{-\beta a}}{2}\right)\right]Ge^{ika}$$

$$B = -\frac{i}{2}\left(\frac{\beta}{k} + \frac{k}{\beta}\right)\sin h\beta a\, Ge^{ika} \qquad \qquad ...(10.76)$$

Transmission Probability

The flux of particles in the incident beam is given by

$$S_1 = \left\{\begin{array}{l}\text{Probability density of particles in}\\ \text{the incident beam}\end{array}\right\} \times \text{Particle velocity}$$

$$= \left(Ae^{ikx}\right)\left(Ae^{ikx}\right)^* v_1 = Ae^{ikx}\, A^* e^{-ikx} v_1 = AA^* v_1$$

where A^* is the complex conjugate of A and v_1 is the particle velocity in the incident beam. Similarly the flux of particles in the transmitted beam is given by

$$S_3 = \left(Ge^{ikx}\right)\left(Ge^{ikx}\right)v_1 = GG^* v_1$$

The transmission probability T is defined by

$$T = \frac{GG^* v_1}{AA^* v_1} = \left(\frac{G}{A}\right)\left(\frac{G}{A}\right)^* = \left|\frac{G}{A}\right|^2 \qquad \qquad ...(10.77)$$

where $\left(\dfrac{G}{A}\right)^*$ is the complex conjugate of $\dfrac{G}{A}$. The quantity T is also called the transmission coefficient.

Now, from eqn. (15) we have

$$\frac{A}{G} = \left[\cos h\beta a + \frac{i}{2}\left(\frac{\beta}{k} - \frac{k}{\beta}\right)\sin h\beta a\right]e^{-ika}$$

$$\therefore \qquad \left(\frac{A}{G}\right)^* = \left[\cos h\beta a - \frac{i}{2}\left(\frac{\beta}{k} - \frac{k}{\beta}\right)\sin h\beta a\right]e^{-ika}$$

Hence
$$\left(\frac{A}{G}\right)\left(\frac{A}{G}\right)^* = \cos h^2\beta a + \frac{1}{4}\left(\frac{\beta}{k} - \frac{k}{\beta}\right)^2 \sin h^2\beta a$$

or,
$$\left|\frac{A}{G}\right|^2 = \cos h^2\beta a + \frac{1}{4}\left(\frac{\beta}{k} - \frac{k}{\beta}\right)^2 \sin h^2\beta a$$

Since
$$\cos h^2\beta a = 1 + \sin h^2\beta a$$

we get
$$\left|\frac{A}{G}\right|^2 = 1 + \left[\frac{1}{4}\left(\frac{\beta}{k} - \frac{k}{\beta}\right)^2 + 1\right]\sin h^2\beta a$$

$$= 1 + \frac{1}{4}\left[\left(\frac{\beta}{k} - \frac{k}{\beta}\right)^2 + 4\right]\sin h^2\beta a$$

or,
$$\frac{1}{T} = 1 + \frac{1}{4}\left(\frac{\beta}{k} + \frac{k}{\beta}\right)^2 \sin h^2\beta a \qquad \qquad ...(10.78)$$

Since
$$k = \sqrt{\frac{2mE}{\hbar^2}}$$

and
$$\beta = \sqrt{\frac{2m(V_0 - E)}{\hbar^2}}$$

$$\therefore \qquad \frac{\beta}{k} = \sqrt{\frac{V_0 - E}{E}} \quad \text{and} \quad \frac{k}{\beta} = \sqrt{-\frac{E}{V_0 - E}}$$

Substituting these values in eqn. (10.78) and simplifying we get

$$\frac{1}{T} = 1 + \frac{V_0^2}{4E(V_0 - E)}\sin h^2\beta a \qquad \qquad ...(10.79)$$

Thus in the case when $E < V_0$, the transmission probability is given by

$$T = \frac{1}{1 - \dfrac{V_0^2}{4E(V_0 - E)}\sin h^2\beta a} \qquad \qquad ...(10.80)$$

This is the expression for the transmission probability or the transmission coefficient in the case when $E < V_0$. The expression leads to the following conclusions:

Case (1): When $E < V_0$

1. There is a finite probability for particle to tunnel through a potential barrier of height V_0 even when the initial kinetic energy E of the particle is less than V_0.

2. When $\beta a > 1$, then $e^{\beta a}$ is large but $e^{-\beta a} = \dfrac{1}{e^{\beta a}}$ is very small compared with 1.

Hence, $\quad \sin h^{-2} \beta a = \left(\dfrac{e^{\beta a} - e^{-\beta a}}{2}\right)^2 \approx \dfrac{1}{4} e^{2\beta a}$

Then $\quad \dfrac{1}{T} \approx 1 + \dfrac{V_0^2}{4E(V_0 - E)} \dfrac{1}{4} e^{2\beta a}$

$\dfrac{1}{T} \approx 1 + \dfrac{V_0^2}{4E(V_0 - E)} \dfrac{1}{4} e^{2\beta a} \approx \dfrac{V_0^2 \, e^{2\beta a}}{16 \, E \, (V_0 - E)}$

$\therefore \quad T \approx \dfrac{16E}{V_0^2}(V_0 - E) e^{-2\beta a} \approx \dfrac{16E}{V_0}\left(1 - \dfrac{E}{V_0}\right) e^{-2\beta a}$ ...(10.81)

Eqn. (10.81) is true only when $E < V_0$.

Thus, when $\beta a \gg 1$, T becomes very small and it decreases exponentially with increase in the thickness of the barrier: Eqn. (10.81) also shows that if a is constant and if E/V_0 decreases, then T decreases exponentially. This variation is shown by the part AO of the curve for T (Fig. 10.5)

3. When the particle energy E approaches the potential energy V_0 of the top of the barrier, then

$\beta a = a\sqrt{\dfrac{2m(V_0 - E)}{\hbar^2}}$ will be $\ll 1$.

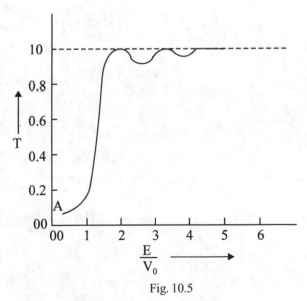

Fig. 10.5

$\therefore \quad \sin h\beta a \approx a\sqrt{\dfrac{2m(V_0 - E)}{\hbar^2}}$

and
$$\sin h^2 \beta a \approx \frac{2m(V_0 - E)a^2}{\hbar^2}$$

Hence,
$$\frac{1}{T} \approx 1 + \frac{V_0^2}{4E(V_0 - E)} \frac{2m(V_0 - E)a^2}{\hbar^2} \approx 1 + \frac{2m V_0^2 a^2}{4 E\hbar^2}$$

Since, $E \to V_0$, we get

$$\frac{1}{T} \approx 1 + \frac{mV_0^2 a^2}{2\hbar^2}$$

$$\therefore \qquad T \approx \frac{1}{1 + \dfrac{mV_0^2 a^2}{2\hbar^2}} \qquad \qquad ...(10.82)$$

Thus when E approaches V_0, T increases and when $E \approx V_0$, the transmission probability is given by eqn. (10.82) which shows that $T < 1$.

Case (2): When $E > V_0$

In this case,
$$\beta = \sqrt{\frac{2m(V_0 - E)}{\hbar^2}} = i \sqrt{\frac{2m(E - V_0)}{\hbar^2}} = i\alpha$$

where
$$\alpha = \sqrt{\frac{2m(E - V_0)}{\hbar^2}}$$

$$\therefore \qquad \sin h\beta a = \frac{e^{\beta a} - e^{-\beta a}}{2} = \frac{e^{i\alpha a} - e^{-i\alpha a}}{2} = i\left(\frac{e^{i\alpha a} - e^{-i\alpha a}}{2i}\right) = i \sin \alpha a$$

Squaring both the sides, we get

$$\sin \hbar^2 a = -\sin^2 \alpha a$$

Substituting this value in eqn. (10.79)

$$\frac{1}{T} = 1 - \frac{V_0^2}{4E(V_0 - E)} \sin^2 \alpha a = 1 + \frac{V_0^2}{4E(E - V_0)} \sin^2 \alpha a \qquad ...(10.83)$$

Hence,
$$T = \frac{1}{1 + \dfrac{V_0^2}{4E(E - V_0)} \sin^2 \alpha a} \qquad \qquad ...(10.84)$$

This is the expression for the transmission probability or the transmission coefficient in the case when $E > V_0$. The expression leads to the following conclusions:

(i) When E approaches V_0, then

$$\alpha a = a \sqrt{\frac{2m(E - V_0)}{\hbar^2}}, \quad \text{will be} \ll 1$$

$\therefore \qquad \sin \alpha a \approx a \sqrt{\frac{2m(E - V_0)}{\hbar^2}},$

and $\qquad \sin^2 \alpha a \approx \frac{2m(E - V_0)a^2}{\hbar^2}$

Hence $\qquad \frac{1}{T} \approx 1 + \frac{V_0^2}{4E(E - V_0)} \frac{2m(E - V_0)a^2}{\hbar^2} \approx 1 + \frac{2m V_0^2 a^2}{4E \hbar^2}$

Since $\qquad E - V_0$, we get

$$\frac{1}{T} \approx 1 + \frac{m V_0^2 a^2}{2\hbar^2}$$

$\therefore \qquad T \approx \dfrac{1}{1 + \dfrac{m V_0^2 a^2}{2\hbar^2}}$

This expression is the same as that obtained in the previous case ($E < V_0$).

(ii) Eqn. (10.84) shows that for $E > V_0$, $T = 1$, when
$$\alpha a = n\pi, \quad \text{where } n = 1, 2, 3, \ldots$$

or, $\qquad a = \dfrac{n\pi}{\alpha} = \dfrac{n\pi}{\sqrt{\dfrac{2m(E - V_0)}{\hbar^2}}} = \dfrac{n\pi\hbar}{\sqrt{2m(E - V_0)}}$ $\qquad$ ...(10.85)

or, $\qquad a = \dfrac{nh}{2\sqrt{2m(E - V_0)}}$ $\qquad$ ...(10.86)

But $\dfrac{h}{\sqrt{2m(E - V_0)}}$ is the de Broglie wavelength λ of the particle in the region of kinetic

energy $(E - V_0)$. Hence $T = 1$, when

$$a = n\left(\frac{\lambda}{2}\right)$$

This shows that if the width of the barrier is an integral multiple of half wavelengths, there is perfect transmission of the incident be of particles.

Eqn. (10.84) also shows that as E increases, T oscillates between the maximum value 1 and value less than 1 (Fig. 10.2).

Reflection Probability

The reflection probability is given by

$$R = \left(\frac{B}{A}\right)\left(\frac{B}{A}\right)^* = \left(\frac{B}{A}\right)^2 \qquad \qquad ...(10.87)$$

where $\left(\dfrac{B}{A}\right)^*$ is the complex conjugate of $\dfrac{B}{A}$. The quantity R is also called the reflection coefficient. Dividing eqn. (10.76) by eqn. (10.75).

$$\frac{B}{A} = \frac{-\dfrac{i}{2}\left(\dfrac{\beta}{k} + \dfrac{k}{\beta}\right)\sin h\beta a}{\cos h\beta a + \dfrac{i}{2}\left(\dfrac{\beta}{k} - \dfrac{k}{\beta}\right)\sin h\beta a}$$

$$\therefore \quad \left(\frac{B}{A}\right)^* = \frac{\dfrac{i}{2}\left(\dfrac{\beta}{k} + \dfrac{k}{\beta}\right)\sin h\beta a}{\cos h\beta a - \dfrac{i}{2}\left(\dfrac{\beta}{k} - \dfrac{k}{\beta}\right)\sin h\beta a}$$

Hence

$$R = \left(\frac{B}{A}\right)\left(\frac{B}{A}\right)^* = \frac{\dfrac{1}{4}\left(\dfrac{\beta}{k} + \dfrac{k}{\beta}\right)^2 \sin h^2\beta a}{\cos h^2\beta a + \dfrac{1}{4}\left(\dfrac{\beta}{k} - \dfrac{k}{\beta}\right)^2 \sin h^2\beta a}$$

$$= \frac{\dfrac{1}{4}\left(\dfrac{\beta}{k} + \dfrac{k}{\beta}\right)^2 \sin h^2\beta a}{1 + \dfrac{1}{4}\left(\dfrac{\beta}{k} - \dfrac{k}{\beta}\right)^2 \sin h^2\beta a} \qquad \qquad ...(10.88)$$

This is the expression for the reflection coefficient. Now from eqn. (10.78) we have the expression for the transmission coefficient.

$$T = \frac{1}{1 + \dfrac{1}{4}\left(\dfrac{\beta}{k} - \dfrac{k}{\beta}\right)^2 \sin h^2\beta a} \qquad \qquad ...(10.89)$$

Adding eqns. (10.88) and (10.89), we get

$$R + T = 1 \qquad\qquad ...(10.90)$$

The tunnel effect provides explanation for the following phenomena:

(i) the field emission of electrons from a cold metallic surface,

(ii) the electrical breakdown of insulators,

(iii) the reverse breakdown of semiconductor diodes,

(iv) the switching action of a tunnel diode, and

(v) the emission of α-particles from a radioactive element.

SOLVED EXAMPLES

Q.1. **The ground state energy of a particle confined to an infinite deep one dimensional potential well is 0.002 eV. Find the energy of the second excited state.**

Sol.

$$E_2 = n^2 E_0, \text{ here } E_0 = 0.002 \ eV$$

When is the ground state energy. E_2 is the energy of the second excited state, So $n = 3$.

$$\therefore \quad E_2 = (\delta)^2 \ E_0 = 9 E_0$$

$$= 9 \times 0.002 = 0.018 \ eV.$$

Q.2. **A particle trapped in an infinite deep potential well has de Broglie wavelength 'λ' in the ground state. What is the wavelength of the particle in the next excited state?**

Sol. The wavelength of an infinite deep potential well is given by

$$\lambda = \frac{2a}{n}, \text{ where } n = 1, 2, 3, ...$$

and a is the width of the potential well.

For the first excited state, $n = 2$.

So, $\qquad \lambda' = \dfrac{2a}{2} = \dfrac{\lambda}{2}.$

Q.3. **Prove that the momentum of a particle in one dimensional well of infinite height is quantized.**

Sol. In a well of infinite height, energy is given by

$$E_n = \frac{n^2 h^2}{8 m a^2}$$

or, $\dfrac{p^2}{2m} = \dfrac{n^2 h^2}{8\,ma^2}$ (where p is the momentum)

or, $p = \dfrac{nh}{2a}$

Thus, momentum is quantized. (Proved.)

Q.4. **12 million electrons with energy 3 eV are incident on a potential barrier of 9 eV and height of 1 nm and 0.5 nm width. Calculate how many electrons will tunnel through the barrier?**

Sol. Let x be the number of electrons passing through the 9 eV barrier of height 1 nm and width 0.5 nm.

Then the maximum value of x is given by

$$x \times 3 \text{ eV} = 9 \text{ eV}$$

or, $x = 3$.

Thus, 3 electrons can pass through the potential barrier.

Q.5. **A particle of mass 'm' is enclosed inside a potential well of infinite height and width 'a'. What is the maximum de Broglie wavelength of the particle in the ground state?**

Sol. In case of an infinite height box, the propagation vector K is given by

$$K = \frac{2\pi}{\lambda} \quad \Rightarrow \quad \frac{2\pi}{\lambda} = \frac{n\pi}{a}$$

$$\Rightarrow \quad \lambda = \frac{2a}{n}$$

For the ground state, $n = 1$

So, $\lambda = 2a$.

Thus, the maximum de Broglie wavelength is $\lambda = 2a$.

Q.6. **A particle is in one-dimensional infinitely deep potential well. What is the energy of second excited state if the ground state energy is 0.8 eV?**

Sol. The energy of the particle in one dimensional infinitely deep potential well is given by

$$E_n = \frac{n^2\,\pi^2\,h^2}{2ma^2}$$

For ground state, $n = 1$

So, $E_1 = \dfrac{\pi^2 h^2}{2ma^2} = 0.8\,\text{eV}$

For the second excited state, $n = 3$

$\therefore \quad E_3 = \dfrac{h^2 \pi^2}{2ma^2}(3)^2 = 9\,E_1$

$= 9 \times 0.8 = 7.2\,\text{eV}.$

Q.7. **A stream of electrons strike on a potential energy step of height 0.04 eV. Calculate the fraction of electrons reflected if energy of the incident electrons is 0.05 eV.**

Sol. The reflection coefficient is given by

$$R = \frac{\left(\sqrt{E} - \sqrt{E - V_0}\right)^2}{\left(\sqrt{E} + \sqrt{E - V_0}\right)^2}$$

Data given, $E = 0.05$ eV, $V_0 = 0.04$ eV.

$$\therefore \quad R = \frac{\left(\sqrt{0.05} - \sqrt{0.05 - 0.04}\right)^2}{\left(\sqrt{0.05} + \sqrt{0.05 - 0.04}\right)^2}$$

$$= \frac{\left(\sqrt{0.05} - \sqrt{0.01}\right)^2}{\left(\sqrt{0.05} + \sqrt{0.01}\right)^2} = 0.009959 \cong 0.01.$$

QUESTIONS

1. Obtain time-dependent Schrödinger's wave equation in operator form.

2. Find the expression for the energy state of a particle in one-dimensional box.

3. Solve the Schrödinger's equation for a particle enclosed in a one-dimensional rigid box of side L. Obtain its eigenvalues. Draw a graph of its first three eigenfunctions. Discuss the probability of finding the particle at different points in the box when it is in different states.

4. Obtain Schrödinger's wave equation for a particle in a square well potential and discuss energy levels when the well is infinitely deep.

5. Discuss quantum mechanically the problem of a particle in a finite square potential well. Draw diagrams showing the amplitude of the wave and probability density for the same. What will be the effect of increasing the width of the square well on energy levels?

6. Solve the problem of the leaking of a particle through a rectangular potential barrier of finite width.

7. Set up Schrödinger's wave equation for a particle of mass 'm' crossing a potential step

$$V(x) = 0 \text{ for } x < 0$$

$$= V_0 \text{ for } x > 0$$

8. Set up the time-independent Schrödinger's equation for one dimensional harmonic oscillator.

9. Write the time dependent Schrödinger's equation for a particle of mass 'm' moving along Y-axis under the potential $V = ay^2$.

10. A stream of independent particles of kinetic energy less than the potential barrier is incident on it. How classical physics differs from quantum physics in observing the phenomenon?

Problems

1. The width of an infinitely deep potential well is halved. What will happen to the energy of a particle trapped inside it?

(**Ans.** $E_2 = 4E_1$)

2. The particle trapped in a one-dimensional box of length 1 cm is described by the normalized wave function $\psi = x$. What is the expectation value of the particle's position (x)?

3. A particle is confined to move along a line of length L. Find the expectation value of its position $< x >$, if its normalized wave function is $\psi(x) = \sqrt{\dfrac{2}{L}} \sin \dfrac{n\pi x}{L}$.

$$\left(\textbf{Ans. } < x > = \frac{L}{2}\right)$$

4. A particle is in a one-dimensional infinitely deep potential well of width L and is in the first excited state. Evaluate the probability of finding the particle between $x = \dfrac{L}{4}$ to $x = \dfrac{L}{2}$.

$$\left(\textbf{Ans. } p = \frac{1}{4}\right)$$

5. The normalized wave function given by $\psi = \dfrac{1}{\sqrt{\pi a}}\exp\left[-\dfrac{x^2}{2a^2}+ikx\right]$ describes a free particle in a one dimensional region. In what region of space is the particle most likely found?

   (**Ans.** $x = 0$)

6. A particle of mass 1 gm moves with speed 1 cm/sec in a one-dimensional box of width 1 cm. Find the quantum number of the state.

   (**Ans.** $n = 3 \times 10^{26}$)

7. A particle of mass 9.1×10^{-31} kg is in a one-dimensional box. Radiation of wavelength 8.79×10^{-9} m is emitted when transition from $n = 5$ to $n = 2$ level takes place. Find the width of the one-dimensional box.

   (**Ans.** 1.8×10^{-9} m)

8. A stream of electrons strikes a potential energy step of height 0.24 eV. Compute the fraction of electrons reflected if energy of the incident electrons is 0.25 eV.

   (**Ans.** $R = 0.44$)

Chapter 11

Solid State Physics

Crystal Structures with Cubic Unit Cells

The matter is found to exist in three states, viz. solid, liquid and gaseous state. In solids and liquids the atoms are closely packed with interatomic distances of the order 10^{-10} m, while in gases the atoms are comparatively larger apart. The solid state of matter can be divided into two broad categories on the basis of structure:

  (i) Crystalline, and
  (ii) Amorphous (Non-crystalline)

Crystalline

Crystalline solids are a three dimensional collection of individual atoms, ions, or whole molecules organized in repeating patterns. These atoms, ions, or molecules are called **lattice points** and are typically visualized as round spheres. The two dimensional layers of a solid are created by packing the lattice point 'spheres' into square or closed packed arrays. These contain regular arrangement having short range as well as long range order. These have definite geometric shape, sharp melting point, definite heat of fusion, undergo clean cleavage and are also known as true solids. Metals and Non-metals belong to this category.

Amorphous

These are having irregular arrangement and short range order only. Irregular shape, melting over a range of temperature and do not have definite heat of fusion. These undergo an irregular cut and they are pseudo solids or supercooled solids, e.g. rubber, glass, plastic, etc.

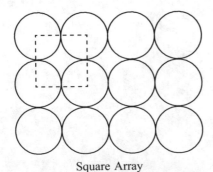

Square Array

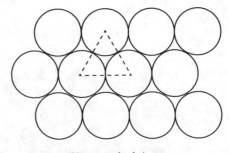

Close-packed Array

Fig. 11.1: Two possible arrangements for identical atoms in a 2-D structure

Stacking the two dimensional layers on top of each other creates a three dimensional lattice point arrangement represented by a unit cell.

Type	Structural Particles	Intermolecular Forces	Typical Properties	Examples
Metallic	Cations and delocalized electrons	Metallic bonds	Hardness varies from soft to very hard; melting point varies from low to very high; lustrous; ductile; good conductor of heat and electricity.	Na, Mg, Fe, Sn, Ag, etc.
Ionic	Cations and anions	Electrostatic attractions	Hard; moderate to very high melting point; non-conductor as a solid, but good electrical conductor as a liquid; soluble in polar solvent.	NaCl, MgO, KNO_3, etc.
Network covalent	Atoms	Covalent bond	Very hard and either sublime or melt at very high temperature; most are non-conductors of electricity.	Diamond, graphite, SiC, SiO_2, etc.
Molecular (i) Non-polar	Atoms on non-polar molecule	Dispersion forces	Soft; extremely low to moderate melting point; soluble in non-polar solvent.	He, Ar, H_2, I_2, CH_4, CCl_4, etc.
(ii) Polar	Polar molecule	Dispersion and dipole-dipole attraction	Low to moderate melting point; soluble in some polar and non-polar solvents.	$CHCl_3$, HCl, etc.
(iii) Hydrogen bonded	Molecule with H atom bonded to heteroatoms (O, S, F, N)	Hydrogen bonds	Low to moderate melting point; soluble in some hydrogen bonded solvent and some polar solvents.	H_2O, NH_3, etc.

The following are the characteristics of a crystal lattice:
 (i)   Each point in a lattice is called lattice point or lattice site.
 (ii) Each point in crystal lattice represents one constituent particle which may be an atom or molecule.
 (iii) Lattice point join straight line to bring out the geometry of lattice.

A unit cell is the smallest collection of lattice points that can be repeated to create the crystalline solid. The solid can be envisioned as the result of the stacking a great number of unit cells together. The unit cell of a solid is determined by the type of layer (square or close packed), the way each successive layer is placed on the layer below, and the coordination number for each lattice point (the number of 'spheres' touching the 'sphere' of interest.)

Parameters of a Unit Cell

Its dimensions along the three edges *a*, *b* and *c*, which may or may not be mutually perpendicular to each other.

The angle α between the edges (*b* and *c*), β (between *a* and *c*) and γ (between *a* and *b*) are denoted respectively. Therefore a unit cell is characterized by six parameters *a*, *b*, *c*, α, β and γ.

The Bravais lattice are the distinct lattice types which when repeated can fill the whole space. The lattice can therefore be generated by three unit vectors, a_1, a_2 and a_3 and a set of integers *k*, *l* and *m* so that each lattice point, identified by a vector *r*, can be obtained from:

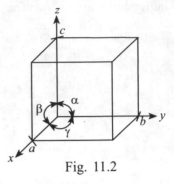

Fig. 11.2

$$r = ka_1 + la_2 + ma_3$$

In two dimensions there are five distinct Bravais lattices, while in three dimensions there are fourteen. These fourteen lattices are further classified as shown in the table below where a_1, a_2 and a_3 are the magnitudes of the unit vectors and β, α and γ are the angles between the unit vectors.

Primitive (Simple) Cubic Structure

Placing a second square array layer directly over a first square array layer forms a 'simple cubic' structure. The simple 'cube' appearance of the resulting unit cell (Fig. 11.2a) is the basis for the name of this three dimensional structure. This packing arrangement is often symbolized as 'AA...', the letters refer to the repeating order of the layers, starting with the bottom layer. The coordination number of each lattice point is six. This becomes apparent when inspecting part of an adjacent unit cell (Fig. 11.3b).

The unit cell in (Fig. 11.3a) appears to contain eight corner spheres, however, the total number of spheres within the unit cell is 1 (only 1/8th of each sphere is actually inside the unit cell). The remaining 7/8th of each corner sphere resides in 7 adjacent unit cells.

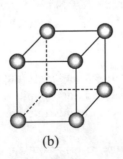

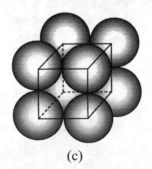

(a)   (b)   (c)

The considerable space shown between the spheres in (Fig. 11.2a) is misleading: lattice points in solids touch as shown in (Fig. 11.2b). For example, the distance between the centres of two adjacent metal atoms is equal to the sum of their radii. Referring again to (Fig. 11.2a) and imagine the adjacent atoms are touching. The edge of the unit cell is then equal to 2*r* (where *r* = radius of the atom or ion) and the value of the face diagonal as a function of

r can be found by applying Pythagorean's theorem ($a_2 + b_2 = c_2$) to the right triangle created by two edges and a face diagonal (Fig. 11.3a). Reapplication of the theorem to another right triangle created by an edge, a face diagonal, and the body diagonal allows for the determination of the body diagonal as a function of r (Fig. 11.3b).

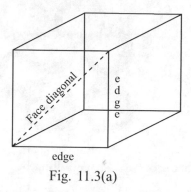

Fig. 11.3(a)

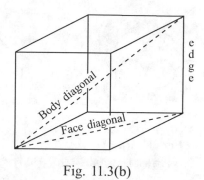

Fig. 11.3(b)

Few metals adopt the simple cubic structure because of inefficient use of space. The density of a crystalline solid is related to its 'percent packing efficiency'. The packing efficiency of a simple cubic structure is only about 52%. (48% is empty space!) How would you find the volume of a unit cell and the volume of atoms in a unit cell?

% packing efficiency = volume of atoms in a unit cell × 100

Body Centred Cubic Structure

A more efficiently packed cubic structure is the 'body-centred cubic' (bcc). The first layer of a square array is expanded slightly in all directions. Then, the second layer is shifted so its spheres nestle in the spaces of the first layer (Figs. 11.4a and b). This repeating order of the layers is often symbolized as 'ABA...'. Like previous figure, the considerable space shown between the spheres in Fig. 11.4c is misleading: spheres are closely packed in bcc solids and touch along the body diagonal. The packing efficiency of the bcc structure is about 68%. The coordination number for an atom in the bcc structure is eight. How many total atoms are there in the unit cell for a bcc structure? Draw a diagonal line connecting the three atoms marked with an 'x' in Fig. 11.4b.

S. No.	Crystal system	Axial length of unit cell	Interaxial angle	Number of lattice in the system
1.	Cubic	$a = b = c$	$\alpha = \beta = \gamma = 90°$	3
2.	Tetragonal	$a = b \neq c$	$\alpha = \beta = \gamma = 90°$	2
3.	Orthorhombic	$a \neq b \neq c$	$\alpha = \beta = \gamma = 90°$	4
4.	Rhombohedral	$a = b = c$	$\alpha = \beta = \gamma \neq 90°$	1
5.	Hexagonal	$a = b \neq c$	$\alpha = \beta = 90°\ \gamma = 120°$	1
6.	Monoclinic	$a \neq b \neq c$	$\alpha = \gamma = 90°, \beta \neq 90°$	2
7.	Triclinic	$a \neq b \neq c$	$\alpha \neq \beta \neq \gamma \neq 90°$	1
			Bravais lattice	14

Assuming the atoms marked 'x' are the same size, tightly packed and touching, what is the value of this body diagonal as a function of r, the radius? Find the edge and volume of the cell as a function of r.

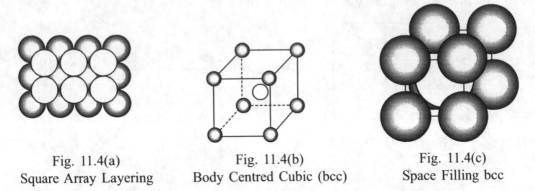

Fig. 11.4(a)	Fig. 11.4(b)	Fig. 11.4(c)
Square Array Layering	Body Centred Cubic (bcc)	Space Filling bcc

Cubic Closest Packed (ccp)

A cubic closest packed (ccp) structure is created by layering close packed arrays. The spheres of the second layer nestle in half of the spaces of the first layer. The spheres of the third layer directly overlay the other half of the first layer spaces while nestling in half the spaces of the second layer. The repeating order of the layers is 'ABC...' (Figs. 11.5 and 11.6). The coordination number of an atom in the ccp structure is twelve (six nearest neighbours plus three atoms in layers above and below) and the packing efficiency is 74%.

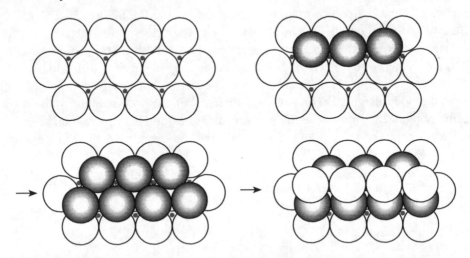

Fig. 11.5

Close packed array layering. The 1st and 3rd layers are represented by light spheres; the 2nd layer, dark spheres. The 2nd layer spheres nestle in the spaces of the 1st layer marked with an 'x'. The 3rd layer spheres nestle in the spaces of the 2nd layer that directly overlay the spaces marked with a '·' in the 1st layer.

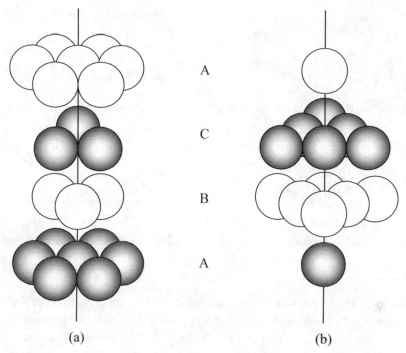

Figs. 11.6a and b: Two views of the Cubic Close Packed Structure

If the cubic close packed structure is rotated by 45° the face centred cube (fcc) unit cell can be viewed (Fig. 11.7). The fcc unit cell contains 8 corner atoms and an atom in each face. The face atoms are shared with an adjacent unit cell so each unit cell contains 1/2 a face atom. Atoms of the face centred cubic (fcc) unit cell touch across the face diagonal (Fig. 11.8). What is the edge, face diagonal, body diagonal and volume of a face centred cubic unit cell as a function of the radius?

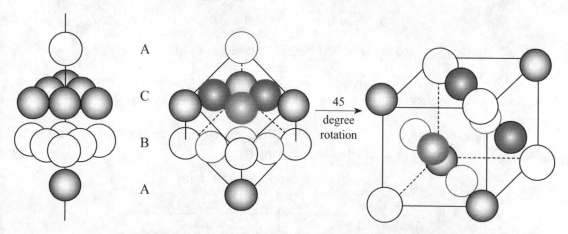

Fig. 11.7: The face centred cubic unit cell is drawn by cutting a diagonal plane through an ABCA packing arrangement of the ccp structure. The unit cell has 4 atoms (1/8 of each corner atom and 1/2 of each face atom)

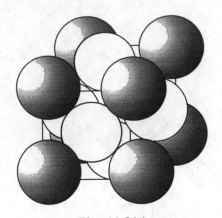

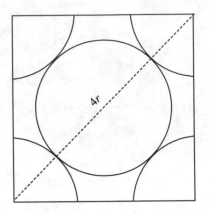

Fig. 11.8(a)
Space filling model of fcc

Fig. 11.8(b)
The face of fcc. Face diagonal = $4r$

Ionic Solids

In ionic compounds, the larger ions become the lattice point 'spheres' that are the framework of the unit cell. The smaller ions nestle into the depressions (the 'holes') between the larger ions. There are three types of holes: 'cubic', 'octahedral', and 'tetrahedral'. Cubic and octahedral holes occur in square array structures; tetrahedral and octahedral holes appear in close-packed array structures (Fig. 11.9). Which is usually the larger ion – the cation or the anion? How can the periodic table be used to prediction size?

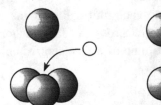
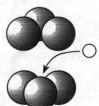
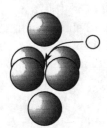
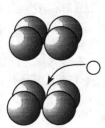

Fig. 11.9: Tetrahedral Octahedral Cubic

Holes in ionic crystals are more like 'dimples' or 'depressions' between the closely packed ions. Small ions can fit into these holes and are surrounded by larger ions of opposite charge.

The type of hole formed in an ionic solid largely depends on the ratio of the smaller ion's radius the larger ion's radius (r smaller/r larger).

Radius ratio and hole type

Hole Type	Radius Ratio
Tetrahedral	0.225 – 0.414
Octahedral	0.414 – 0.732
Cubic	0.732 – 1.000

Empirical Formula of an Ionic Solid

Two ways to determine the empirical formula of an ionic solid are: (i) From the number of each ion contained within 1 unit cell; and (ii) From the ratio of the coordination numbers of the cations and anions in the solid.

Position of atom and fraction contained in a single unit cell

Position of Atom in the Unit Cell	Adjacent Cells Sharing Atom	Fraction Contained Within a Single Unit Cell
Cube Corner	8	1/8
Edge	4	1/4
Face	2	1/2
Internal	1	1

Example: Find the empirical formula for the ionic compound shown in Figs. 11.10a and 11.10b.

First Method: When using the first method, remember most atoms in a unit cell are shared with other cells. Above table lists types of atoms and the fraction contained in the unit cell. The number of each ion in the unit cell is determined: 1/8 of each of the 8 corner X ions and 1/4 of each of the 12 edge Y ions are found within a single unit cell. Therefore, the cell contains 1 X ion (8/8 = 1) for every 3 Y ions (12/4 = 3) giving an empirical formula of XY3. Which is the cation? anion? When writing the formula of ionic solids, which comes first?

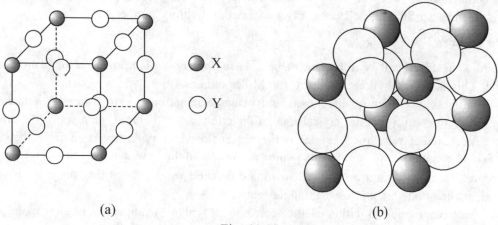

(a)   (b)

Fig. 11.10

Second Method: The second method is less reliable and requires the examination of the crystal structure to determine the number of cations surrounding an anion and vice-versa. The structure must be expanded to include more unit cells. Figure 11.11 shows the same solid in Figure 11.10 expanded to four adjacent unit cells. Examination of the structure shows that there are 2 X ions coordinated to every Y ion and 6 Y ions surrounding every X ion. (An additional unit cell must be projected in front of the page to see the sixth Y ion. A2 to 6 ratio gives the same empirical formula, XY3.)

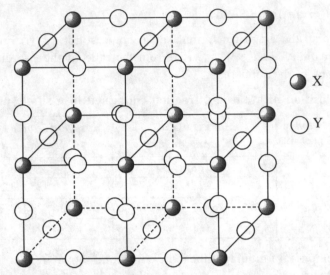

Fig. 11.11: Four unit cells of an ionic solid with formula XY3

Coordination number is the number of directly touching atoms around an atom in crystal lattice. In case of ionic crystals, the coordination number is the number of touching oppositely charged ions around a ion. For covalent ions, the coordination number is the number of atoms joined directly to a particular atom covalently. If a simple cubic cell is considered, the coordination number is 6 (4 atoms in the plane, 1 below the plane and 1 above the plane, in an octahedral way). There are two nearest neighbours on +X-axis, 2 on +Y-axis and 2 on Z-axis. Hence, an SC unit cell cubic crystal has coordination number 6.

Miller Indices

In Solid State Physics it is important to be able to specify a plane or a set of planes in the crystal. This is normally done by using the Miller indices.

Miller indices define directional and planar orientation within a crystal lattice. The indices may refer to a specific crystal face, a direction, a set of faces, or a set of directions.

Indices that refer to a crystal plane are enclosed in parentheses, indices that refer to a set of symmetrically equivalent planes are enclosed in braces (curly brackets), indices that represent a direction are enclosed in square brackets, and indices that represent a set of equivalent directions are enclosed in angle brackets.

Each plane oriented within a lattice corresponds with an arrangement of atoms; one plane might have a higher atomic density than another. It is apparent that Miller indices correspond with properties of the crystal which determine how the material responds to chemical and mechanical processes.

Miller indices are used to specify directions and planes.

- These directions and planes could be in lattices or in crystals.
- The number of indices will match with the dimension of the lattice or the crystal. Example: In 1D there will be 1 index and 2D there will be two indices, etc. (h,k,l) represents a point – note the exclusive use of commas.

- Negative numbers/directions are denoted with a bar on top of the number.
- [*hkl*] represents a direction.
- <*hkl*> represents a family of directions.
- (*hkl*) represents a plane.
- {*hkl*} represents a family of planes.

Miller Indices for Directions

- A vector *r* passing from the components by their common origin to a lattice point can be written as: $r = r_1 a + r_2 b + r_3 c$, where, *a*, *b*, *c* → basic vectors and miller indices → $(r_1\ r_2\ r_3)$
- Fractions in $(r_1\ r_2\ r_3)$ are eliminated by multiplying all denominators.
  [E.g., (1, ¾ ,½) will be expressed as (432)].

Let us understand why we need anything like Miller Indices. When we talk about different materials, each and every material will have a defined crystallographic plane and points. Now to know the structure of any material or compound (if that makes it easier for you), i.e. whether the material comes under BCC, FCC or SC. Miller indices define the crystal arrangement through the help of indexing points and these indexing points will let us know the kind of structure which that particular material has.

Elements have very different properties depending on how they are stacked together or what other elements they are stacked together with. We wouldn't put salt (composed of sodium and chlorine) on our food if it tasted like the chlorine that we put into our swimming pools.

A great example of the importance of crystal structure is the difference between two minerals; graphite and diamond. Graphite is the soft, dark coloured material that is found in pencil lead, while diamonds are very hard, often transparent and colourless, and very expensive gemstones. Both graphite and diamond are made out of only one element; carbon. The reason that graphite and diamond are so different from each other is because the carbon atoms are stacked together into two different crystal structures (See figure). Graphite is composed of carbon that forms loosely bonded sheets in their crystal structure. These sheets rub off easily to mark the paper when you write on it. Diamond is composed of carbon atoms stacked tightly together in a cubic crystal structure, making it a very strong material.

This shows us that it is not only important to know what elements are in the mineral, but it is also very important to know how those elements are stacked together.

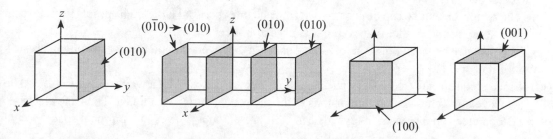

Electrical Conductivity (Band Theory)

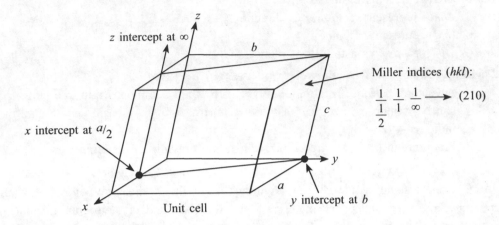

Fig. 11.12

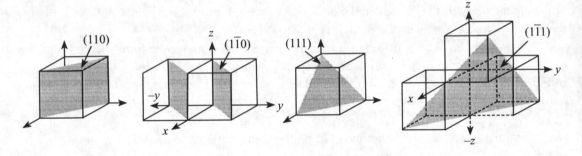

Reciprocal lattice

(i) Reciprocal lattice is an imaginary mathematical concept. It can also be called as Fourier space. The word lattice indicates a set of mathematical points in direct space which satisfy translational symmetry.

(ii) The word reciprocal lattice indicates a set of mathematical points in space where the distance between points has the dimensions of inverse length. The reciprocal lattice also satisfies the translational symmetry.

(iii) The vector that connects two points in reciprocal space is called as the reciprocal lattice vector 'G'.

(iv) The reciprocal lattice can be constructed for each direct crystal lattice. The points in reciprocal lattice represents Miller planes in direct crystal lattice.

(v) The distance between points in reciprocal space is the inverse of interplanar spacing in crystals in direct space.

(vi) Since electrons are treated as waves in quantum mechanics, with a wave vector $k = 2\pi\lambda$. The wave vector kk has similar dimensions as the reciprocal lattice vector GG. Thus, the electrons (Bloch electrons) present in crystals can be represented as waves in reciprocal space.

(vii) The regions in reciprocal space where the Bloch electron waves can travel are called the allowed energy regions. In contrast, where they can't travel are called as forbidden energy regions. These regions are formed due to the interaction of Bloch waves with the periodic potential of the crystal under investigation.

Moreover, this condition $(R_2 - R_1) \cdot \Delta k = 2\pi m \rightarrow T \cdot \Delta k = 2\pi m$, where $m =$ integer implies that the Δk's must be a discrete set of points.

To construct this (reciprocal) lattice we can start by the primitive translational vectors a, b, c of the (direct) lattice with $m = 1$ to find the primitive translation vectors a^*, b^*, c^* which generate all the k which verify the condition $\Delta k = G = ha^* + kb^* + lc^*$ (h, k, l) integers

$$a \cdot a^* = 2\pi \ b \cdot b^* = 2\pi$$

If we add the further conditions of perpendicularity $c \cdot c^* = 2\pi \ b \cdot a^* = 0$, $c \cdot a^* = 0$; $a \cdot b^* = 0$, $c \cdot b^* = 0$; $a \cdot c^* = 0$, $b \cdot c^* = 0$

$a^* = 2\pi b \times c/V \ b^* = 2\pi c \times a/V \ c^* = 2\pi a \times b/V$ and $V = |a \cdot b \times c|$ is the volume of the direct lattice.

Bragg's Law

Bragg's law means that the diffraction can occur only when the following equation is satisfied:

$$n\lambda = 2d\sin\theta \tag{1.1}$$

where,

n is a positive integer,
λ is the wavelength of the X-ray,
d is the distance between the lattice plane, and
θ is the incident glancing angle (supplement of the incident angle).

'Incident angle' is defined as the angle between the normal direction of a reflection plane and the incident ray (or more generally quantum beam), in a traditional theory of optics. The Bragg's law is often expressed in another way:

$$\lambda = 2d\sin\theta \tag{1.2}$$

and this expression may be more popular than that of eqn. (1.1). Be sure that the left side of eqn. (1.2) does not include 'n'.

The definition of d in eqn. (1.2) is different from the definition of d in eqn. (1.1), and it is equivalent with (d/n) for the definition of d in eqn. (1.2). In the following sections in this chapter, the expression by eqn. (1.1) will be used, because it looks a little easier to be understood the equation. It is recommended to draw illustrations several times, and to remember how the Bragg's law is derived.

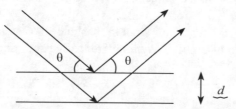

Fig. 11.13: Illustration to derive the Bragg's law difference in path length

The most significant point of the Bragg's law can be explained by the constructive interference, which occurs when the path difference of travelling waves matches with the integral multiplication of the wavelength. See the figure below.

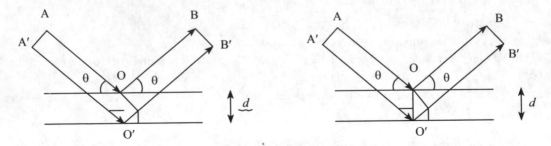

Fig. 11.14: It will take additional length when you move along the path A'O'B' as compared with the path AOB. How long is the additional length (path difference)?

The path difference corresponds to the length of the two segments drawn as heavier lines.

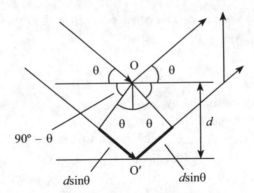

Fig. 11.15: Relations among path length, d and θ

The length of one segment should be '$d\sin\theta$', because the segment is the opposite side to 'the corner with the angle θ' in a 'right triangle with the oblique side of length d'.

Symmetric Reflection

The Bragg's law implies that the incident and reflected glancing angles are both equal to θ.

Why should the angles of incident and reflected beams be coincided?

Of course, the reflection by a mirror should be symmetric, but we are talking about the diffraction of X-ray by a lattice plane. The lattice plane should be something like planar arrangement of atoms.

Can you imagine existence of smooth surface like a mirror in the arrangement of atoms? What will occur on the asymmetric reflection shown in the figure below, for example?

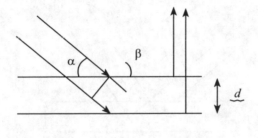

Fig. 11.16: What will occur on the reflection with the incident
(glancing) angle of α and reflected (glancing) angle of β?

The path difference should be '$d(\sin\alpha + \sin\beta)$', because it is also the sum of the two segments, the lengths of which are '$d\sin\alpha$' and '$d\sin\beta$', as shown in the figure below.

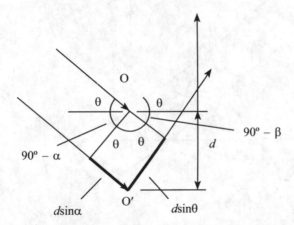

Fig. 11.17: Path difference for the incident (glancing) angle of α and reflected
(glancing) angle of β should be '$d(\sin\alpha + \sin\beta)$'

Don't you think diffraction may occur, if the value of '$d(\sin\alpha + \sin\beta)$' matches to the integral multiplication of the wavelength?

Actually, diffraction never occurs in such a case. Why the asymmetric reflection is forbidden? Now, the path difference in the asymmetric reflection appears to be because the reflection point O' on the second lattice plane is located right below the reflection point O of the first lattice plane, in Fig. 11.16 or Fig. 11.17. But the difference between the path reflected at O and O' should be '$d\sin(\alpha + \phi) + \sin(\beta + \phi)/\cos\phi$', if the reflection point O' is located at the point displaced by such distance given by '$d\tan\phi$', as shown in the Figs. 11.18 and 11.19 below.

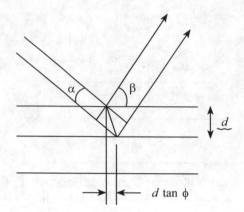

Fig. 11.18: How about displacing the reflection point for asymmetric reflection?

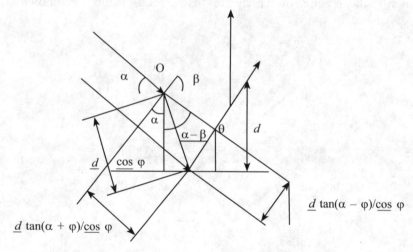

Fig. 11.19: Path difference can be 'd[sin $(\alpha + \varphi)$ + sin$(\beta - \varphi)$]/cos φ'
for asymmetric reflection

Both constructive and destructive interferences, depending on the location of the reflection point O' (different value of φ) can occur in the case of asymmetric reflection. There may be a special arrangement of atoms and angles α and β that lead only constructive interference, but it will be just a 'symmetric Bragg reflection' for different lattice planes.

On the other hand, the path difference should always be '2dsinθ' for symmetric reflection, no matter where the reflection point is located on a lattice plane. You can confirm this relation by the fact that the following equation holds for arbitrary φ,

$$\frac{\sin(\theta - \phi) + \sin(\theta + \phi)}{\cos \phi} = \sin \theta$$

which can easily be derived by applying trigonometric addition formulas. In conclusion, we can say that diffraction occurs only when the reflection is symmetric about the lattice plane.

Bragg's Condition

Furthermore, the Bragg's law does not only mean the constructive interference on eqn. (1.1), but it has a much more strict meaning to forbid any reflections not satisfying eqn. (1.1). How can this strict condition be derived?

So far, we have considered only two lattice planes, the top one and the 2nd one from the surface, but we must consider the 3rd, 4th, ..., and much more lattice planes to explain the strict condition.

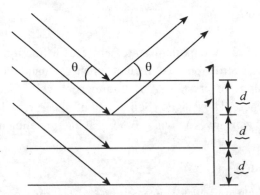

Fig. 11.20: We must take many lattice planes into consideration to understand the Bragg's law

The path difference is '$2d\sin\theta$' for the reflection from the 2nd lattice plane, as shown in Sec. 1–3, and it should be '$4d\sin\theta$' for the 3rd plane, '$6d\sin\theta$' for the 4th, ...,

and '$2(j-1)\,d\sin\theta$' for the j – th lattice plane.

All the reflections from any lattice planes (1st, 2nd, 3rd, ... planes) will always be interfered constructively, if the relation:

$$n\lambda = 2d\sin\theta$$

is satisfied. Then, how about the case:

$$(n+1/2)\lambda = 2d\sin\theta,$$

which means the path difference is given by half-integer multiplication of the wavelength? The reflection from the 1st and 2nd planes will be cancelled, and the reflection from the 3rd and 4th planes will also be cancelled, and so on. Most of the reflections will vanish except the last one for the odd number of planes.

Similarly, when the following relation:

$$(n+1/3)\lambda = 2d\sin\theta$$

is satisfied, the sum of the reflections from the 1st, 2nd and 3rd planes will be cancelled, the sum of the reflections from the 4th, 5th and 6th planes will be cancelled, and so on.

In general, when the following relation is satisfied,

$$(1/m)\lambda = 2d\sin\theta$$

where $(1/m)$ is an irreducible fraction,

the sum of the reflections from the 1st to the m-th planes are cancelled, and the sum from $(m + 1)$ to $(2m)$ will be cancelled, and so on. In conclusion, when the value of $(2d\sin \theta)/\lambda$ is not an integer, there will always be a combination of planes which cancels the amplitude of the reflected waves.

Therefore, diffraction will never occur, if eqn. (1.1) is not satisfied, in case of infinitely large number of lattice planes. In other words, diffraction may be observed, even if the situation is slightly deviated from the condition defined by eqn. (1.1), in case of finite number of lattice planes.

Brillouin Zones

For solid state physics the most important statement the diffraction condition was given by Brillouin. This is the only construction used in electron energy band theory and in the expression of the elementary excitations of crystals. A Brillouin zone cell is defined as a Wigner-Seitz cell in the reciprocal lattice. The Wigner-Seitz cell of the direct lattice was described by (Fig. 11.20a). The Brillouin zone gives a vivid geometrical interpretation of the diffraction condition $2k \cdot G = G^2$, or

$$k \cdot (1/2G) = (1/2G)^2 \tag{P}$$

We construct a plane normal to the vector G at the midpoint; any vector k from the origin to the plane will satisfy the diffraction condition. The plane thus described forms a part of the zone boundary, Fig. 11.20(a). An X-ray beam incident on the crystal will be diffracted if its wave vector has the magnitude and direction required by equation (P) and the diffracted beam will be in the direction of the vector $k - G$.

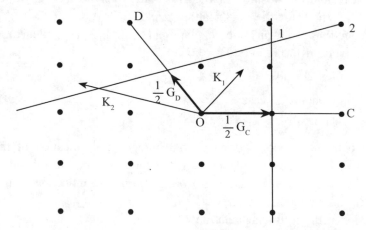

Fig. 11.20(a): Reciprocal lattice points near the point O at the origin of the reciprocal lattice. The reciprocal lattice vector G_C connects point OC; and G_D connects OD. Two planes 1 and 2 are drawn which are the perpendicular bisectors of G_C and G_D, respectively. Any vector from the origin to the plane 1, such as K_1, will satisfy the diffraction condition $K_1 \cdot (1/2\ G_C) = (1/2\ G_C)^2$. Any vector from the origin to the plane 2, such as K_2, will satisfy the diffraction condition $K_2 \cdot (1/2\ G_D) = (1/2\ G_D)^2$.

Fermions and Bosons

In a world where Einstein's relativity is true, space has three dimensions, and there is quantum mechanics, all particles must be either fermions (named after Italian physicist Enrico Fermi) or bosons (named after Indian physicist Satyendra Nath Bose). This statement is a mathematical theorem, not an observation from data. But data over the past 100 years seems to bear it out; every known particle in the Standard Model is either a fermion or a boson.

An example of a boson is a photon. Two or more bosons (if they are of the same particle type) are allowed to do the same exact thing. For example, a laser is a machine for making large numbers of photons do exactly the same thing, giving a very bright light with a very precise colour heading in a very definite direction. All the photons in that beam are in lockstep.

One can't make a laser out of fermions. An example of a fermion is an electron. Two fermions (of the same particle type) are forbidden from doing the same exact thing. Because an electron is a fermion, two electrons cannot orbit an atom in exactly the same way. This is the underlying reason for the Pauli exclusion principle that we learn in chemistry class, and has enormous consequences for the periodic table of the elements and for chemistry. The electrons in an atom occupy different orbits, in different shells around the atomic nucleus, because they cannot all drop down into the same orbit – they are forbidden from doing so because they are fermions. [More precisely, two electrons can occupy the same orbit as long as they spin around their own axes in opposite directions.] If electrons were bosons, chemistry would be unrecognizable!

The known elementary particles of our world include many fermions – the charged leptons, neutrinos and quarks are all fermions – and many bosons – all of the force carriers, and the Higgs particle(s).

Another thing boson fields can do is be substantially non-zero on average. Fermion fields cannot do this. The Higgs field, which is non-zero in our universe and gives mass thereby to the known elementary particles, is a boson field (and its particle is therefore a boson, hence the name Higgs boson that you will hear people use.)

Something else one can do with boson particles is form a Bose-Einstein condensate, a phenomenon predicted by Einstein back in the 1920.

What these experiments do in making this condensate is cause large numbers of identical boson atoms to all sit as still as a quantum mechanical object possibly can.

[This is all quantum mechanics, by the way. Einstein didn't like the implications of quantum mechanics, but you should not have the impression, despite some popular accounts, that he didn't understand it. In fact his work was crucial in the development of several aspects of quantum theory.]

All fundamental particles, such as electrons and photons, have an intrinsic degree of freedom called 'spin', which in many situations acts like a form of angular momentum.

A boson is a particle with a total spin that is equal to an integer (0, 1, 2, ...) times the Planck's constant ($\hbar\hbar$). For example, photons (the particles that compose light) are bosons having total spins of just $\hbar\hbar$. This is in contrast to electrons, which are fermions and have spins of $\hbar/2\hbar/2$. All fundamental force carrying particles (photon, gluon, W, Z) are bosons.

Bosons obey Bose-Einstein statistics, which means that any number of bosons of the same type can occupy the same physical state. This may seem of little practical consequence, but it implies that composite particles, including whole atoms, with integer spin numbers can condense into a special form of matter called a Bose-Einstein condensate.

It is also the reason why helium (He⁴ is a boson) becomes super fluid below 2.17 K, and photons in a region of space can have exactly the same wavelength and polarization (making lasers possible).

Bosons are thought to be particles which are responsible for all physical forces. Other known bosons are the photon, the W and Z bosons, and the gluon. Scientists do not yet know how to combine gravity with the Standard Model. It is the quantum excitation of the Higgs field – a fundamental field of crucial importance to particle physics theory.

If talking about Higgs, boson then it is, if nothing else, the most expensive particle of all time. It's a bit of an unfair comparison; discovering the electron, for instance, required little more than a vacuum tube and some genuine genius, while finding the Higgs boson required the creation of experimental energies rarely seen before on planet Earth. The Large Hadron Collider hardly needs any introduction, being one of the most famous and successful scientific experiments of all time, but the identity of its primary target particle is still shrouded in mystery for much of the public. It's been called the God Particle, but thanks to the efforts of literally thousands of scientists, we no longer have to take its existence on faith.

Maxwell-Boltzmann Distribution

A law describing the distribution of speeds among the molecules of a gas is known as Maxwell-Boltzmann distribution law. In a system consisting of N molecules that are independent of each other except that they exchange energy or collision, it is clearly impossible to say what velocity any particular molecule will have. However, statistical statements regarding certain functions of the molecules were worked out by James Clerk Maxwell and Ludwig Boltzmann. One form of their law states that

$$n = N \exp(-E/RT)$$

where n is the number of molecules with energy in excess of E, T is the thermodynamic temperature, and R is the gas constant.

Bose-Einstein Statistics

A statistical description of a system of particles that obey the rules of quantum mechanics rather than classical mechanics is known as quantum statistics. In this statistics, energy states are considered to be quantized. Bose-Einstein Statistics apply if any number of particles can occupy a given quantum state. Such particles are called bosons. Bosons have an angular momentum $nh/2\pi$ where n is zero or an integer and h is the Planck's constant. For identical bosons, the wave function is always symmetric. If only one particle may occupy each quantum state, Fermi-Dirac Statistics apply and the particles are called fermions. Fermions have a total angular momentum and any wave function that involves identical fermions is always asymmetric.

The relation between the spin and statistics of particles is given by the spin-statistics theorem. This is a fundamental theorem of relativistic quantum field theory that states that half integer spins can only be quantized consistently if they obey Fermi-Dirac statistics and even-integer spins can only be quantized consistently if they obey Bose-Einstein statistics. This theorem enables one to understand the result of quantum statistics that wave functions for bosons are symmetric and wave functions for fermions are asymmetric. It also provides foundation for the Pauli exclusion principle. It was first provided by Wolfgang Pauli in 1940.

In two-space dimensions, it is possible that there are particles (or quasiparticles) that have statistics intermediate between bosons and fermions. These particles are known as anions; for identical anions the wave function is not symmetric (a phase sign of +1) or symmetric (a phase sign of –1), but interpolates continuously between +1 and –1. Anions may be involved in the fractional quantum Hall effect.

Fermi-Dirac Distribution Function

Distribution functions are nothing but the probability density functions use to describe the probability with which a particular particle can occupy a particular energy level. When we speak of Fermi-Dirac distribution function, we are particularly interested in knowing the chance by which we can find a fermion in a particular energy state of an atom (more information on this can be found in the article 'Atomic Energy States'). Here, by fermions, we mean the electrons of an atom which are the particles with ½ spin, bound to Pauli's exclusion principle.

Necessity of Fermi-Dirac Distribution Function

In fields like electronics, one particular factor which is of prime importance is the conductivity of materials. This characteristic of the material is brought about the number of electrons which are free within the material to conduct electricity.

As per energy band theory(refer to the article 'Energy Bands in Crystals' for more information), these are the number of electrons which constitute the conduction band of the material considered. Thus in order to have an idea over the conduction mechanism, it is necessary to know the concentration of the carriers in the conduction band.

Fermi-Dirac Distribution Expression

Mathematically, the probability of finding an electron in the energy state E at the temperature T is expressed as

$$f(E) = \frac{1}{1 + e^{\left(\frac{E - E_f}{K_B T}\right)}}$$

where, $KR = 1.38 \times 10^{-23}\ JK^{-1}$ is the Boltzmann constant, T is the absolute temperature, E_f is the Fermi level or the Fermi energy.

Fermi Levels in Semiconductors

Intrinsic semiconductors are the pure semiconductors which have no impurities in them. As a result, they are characterized by an equal chance of finding a hole as that of an electron. This in turn implies that they have the Fermi-level exactly in between the conduction and the valence bands as shown by Fig. 11.21a.

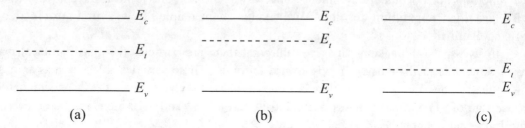

(a)   (b)   (c)

Fig. 11.21: Fermi levels of: (a) Intrinsic Semiconductor,
(b) *N*-type Semiconductor, and (c) *P*-type Semiconductor

Next, consider the case of an *N*-type semiconductor. Here, one can expect more number of electrons to be present in comparison to the holes. This means that there is a greater chance of finding an electron near to the conduction band than that of finding a hole in the valence band. Thus, these materials have their Fermi-level located nearer to conduction band as shown by Fig. 11.21b. Following on the same grounds, one can expect the Fermi-level in the case of *P*-type semiconductors to be present near the valence band (Fig. 11.21c). This is because, these materials lack electrons, i.e. they have more number of holes which makes the probability of finding a hole in the valence band more in comparison to that of finding an electron in the conduction band.

Effect of Temperature on Fermi-Dirac Distribution Function

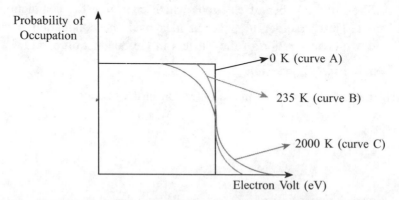

Fig. 11.22: Fermi-Dirac Distribution Function at Different Temperature

At $T = 0$ K, the electrons will have low energy and thus occupy lower energy states. The highest energy state among these occupied states is referred to as Fermi-level. This in turn means that no energy states which lie above the Fermi-level are occupied by electrons.

Thus we have a step function defining the Fermi-Dirac distribution function as shown by the curve A in Fig. 11.22. However, as the temperature increases, the electrons gain more and more energy due to which they can even rise to the conduction band. Thus at higher temperatures, one cannot clearly distinguish between the occupied and the unoccupied states as indicated by the curves B and C shown in Figure 11.22.

Formation of Energy Bands in Solids

- A single isolated atom has discrete energy levels. But if two atoms come closer to each other, they interact and significant changes in their outer energy levels can be observed.

- A solid, is an aggregate of atoms in very close proximity. To form a solid, a large number of atoms are to be brought very close to each other in the case of metals and atomic solids.

- Every atom is affected by the presence of the neighbouring atoms. During the formation of a solid, it is found that the energy levels of the inner shell electrons are not much affected by the presence of neighbouring atoms.

- However, the energy levels of outer shell electrons are changed considerably. These electrons are shared by more than one atoms in the crystal.

- As per Pauli's exclusion principle, not more than two interacting electrons can have the same energy level and therefore new energy levels must be established; which are discrete but infinitesimally different. This group of discrete but closely spaced energy levels is called an energy band.

- Thus in solids, the allowed energy levels of an atom are modified by the proximity of other atoms. Every discrete level of an individual atom gives rise to a band in solid. Each band contains as many discrete levels as the number of atoms in the solid, i.e. in a solid containing N atoms, there are N possible levels in each band and they can be occupied by 2N electrons.

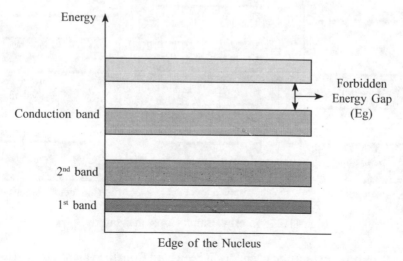

Fig. 11.23: Energy Band Diagram

- The splitting of energy levels is the greatest for the outermost electrons and least for inner electrons. The individual energies in the band are so close together that each band can be considered as continuous.

- In an atom, the electrons in the inner shells are tightly bound to the nucleus while the electrons in the outermost shell (i.e., the valence electrons) are loosely bound to the nucleus.
- During formation of a solid, a large number of atoms are brought very close together, the energy levels of these valence electrons are affected most.
- The energies of inner shell electrons are not affected much.
- The band formed by a series of energy levels containing the valence electrons called the Valence Band.
- It is the highest occupied energy band. It may be completely filled or partially filled with electrons.
- The next higher permitted energy band is called the Conduction Band.
- The electrons can move freely in the conduction band and hence the electrons in the conduction band are known as conduction electrons.
- The energy gap between the valence band and the conduction band is called Forbidden Energy Gap or the Forbidden Band.
- This band is formed by a series of non-permitted energy levels above the top of valence band and below the bottom of the conduction band.
- This energy gap is denoted by Eg and is the amount of energy to be supplied to the electron in valence band to get excited into the conduction band.

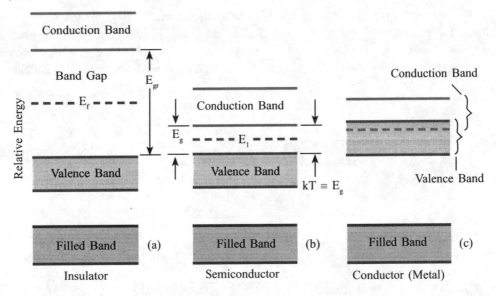

Fig. 11.24: Energy Band Gaps in Materials

- When an electron gains sufficient energy, it is ejected from the valence band. Because of this, a covalent bond is broken and a valence for electron, known as hole, is generated.
- It is supposed to behave as a positive charge.
- This hole can travel to the adjacent atom by acquiring an electron from an atom.

- When an electron is captured by a hole the covalent bond is again re-established.
- Thus conduction electrons are found in and move freely in the conduction band. The holes exist in and move in the valence band.

Classification of Solids

- Electrons in the conduction band are free and move easily under an electric field. The electrons in valence band are attached to the lattice and are not free to move. If they acquire sufficient energy to cross the forbidden gap Eg, they can occupy the conduction band states and are available for conduction.
- Hence Eg is the parameter which decides whether the material acts as conductor, insulator or semiconductor.
- On the basis of the band theory, solids are classified into three categories:

   (a) Insulators, (b) Semiconductors, and (c) Conductors.

(a) Insulators

- These are the materials in which the valence electrons are bound very tightly to their parent atoms.
- Hence very large electric field is required to remove them from attraction of nuclei.

The characteristic features of insulators are:

 (i) A full valence band.
(ii) An empty conduction band.
(iii) A large energy gap Eg (L5 to 10 ev). Example: Diamond, Silica.

(b) Semiconductors

- A semiconducting material has electrical properties between those of insulators and good conductors.
- The characteristic features of semiconductors are:

   (i) Almost empty conduction band and a filled valence band with a very narrow energy gap (Eg 1 ev).

   (ii) At 0K, the valence band is completely filled and the conduction band is empty. As temperature increases the electrons in valence band acquire enough energy to cross the small energy gap and move to conduction band. Thus the conductivity of semiconductors increases with temperature. The electrons moving to the conduction band leave behind positive holes in the valence band. Hence, semiconductor current is the sum of electron and hole currents flowing in opposite directions. Example: Silicon (Eg = 1.1 eV).

(c) Conductors

- The conductors have a large number of free electrons available for electrical conduction.
- Conductors may be defined as solids characterized by a single energy band called conduction band, which is partially filled at any temperature.
- Alternatively, we can also define a conductor as a solid in which the conduction and valence bands overlap and there is no energy gap between the two bands.

As there is no forbidden gap, there is no structure to establish holes. The total current is simply due to flow of electrons. Example: Metals, alloys, conductive polymers (doped polypyrrole, polythiophene, graphite, fullerene, tetrathiafulvalene, etc.)

QUESTIONS

Crystal Structure

1. For a cubic crystal lattice, what do the following represent?

   (i) [110]   (ii) [010]   (iii) (111)   (iv) {100}

2. Describe a primitive unit cell and a non-primitive rectangular cell.

3. What are typical crystal structures for metals and why?

4. Assuming iron (Fe) has a lattice parameter, a, of 0.287 nm. What is the atomic radius and atomic density of an iron crystal?

5. Calculate the atomic density of a simple orthorhombic crystal with $a = 1.046$ nm, $b = 1.288$ nm and $c = 2.448$ nm.

6. State the Bragg's law.

7. How can you determine the structure of a crystal?

8. What is the relationship between the real and reciprocal space lattices? Find the reciprocal lattice for the three 2D cases.

9. The intensity of a diffracted beam is much greater for crystal A than for crystal B. If you ignored any potential differences in intensity due to crystal structure, what could you deduce about the individual atoms that make up crystals A and B?

10. What types of radiation other than X-rays are commonly used to obtain diffraction patterns?

11. What are the energies needed to perform crystal diffraction?

●●●

Chapter 12

Lasers

Laser stands for light amplification by stimulated emission of radiation. An intense, concentrated, and highly parallel beam of coherent light can be produced by such a device. The Laser is the further developments of the maser which is a similar device using radio microwaves instead of visible light waves. Professor C.H. Townes and his associates at Columbia University had first built the maser successfully in between 1951 and 1954. Thereafter, maser technology was advanced a lot.

In the year 1958, A.H. Schawlow and C.H. Townes formulated the principles of the optical maser, or laser. In 1960, T.H. Maiman of the Hughes Aircraft Company Laboratories was able to produce laser, based on these principles. Due to scientific and technological applications of laser, extensive research on laser development and devices have been carried out with ten basic principles of operation.

Basic Principles of Operation

There are at least ten basic principles of operation of most lasers. These are:

(i)	Metastable states	(ii)	Optical pumping
(iii)	Fluorescence	(iv)	Population inversion
(v)	Resonance	(vi)	Stimulated emission
(vii)	Coherence	(viii)	Polarization
(ix)	Fabry-Perot interferometer	(x)	Cavity oscillation.

Out of these concepts, only the principle of coherence accompanying stimulated emission was the key to the realization of maser and laser operation.

Stimulated Emission

A gas is enclosed in a vessel containing free atoms having a number of energy levels, at least one of which is metastable. White light is passed into this gas. Consequently, many atoms are raised, through resonance, from the ground state to excited states. Many of the electrons will be trapped in the metastable state while they drop back. When the pumping light is intense enough, we expect a *population inversion*, i.e. more electrons in the metastable state than in the ground state.

A photon of energy $h\nu$ will be emitted if an electron in one of these metastable states spontaneously jumps to the ground state. This is usually a case. Such type of emission is called *fluorescent* or *phosphorescent radiation*. As the photon passes by another nearby atom in the same metastable state, it can, by the principle of resonance, immediately stimulate that atom to radiate a photon of the exact same frequency and return to its ground state. Surprisingly enough, this stimulated photon has exactly the same frequency, direction, and polarization as the primary photon (spatial coherence) and exactly the same phase and speed (temporal coherence). These two photons constitute primary wages, and upon passing close to other atoms in their metastable states, they stimulate them to emission in the same direction with the same phase. However, the stimulation of transitions from the ground state to the excited state can occur with the absorption of the primary wages. Therefore, a population inversion (more atoms in the metastable state than the ground state) is required by an excess of stimulate emission. Thus, a chain reaction can be developed with the right conditions in the gas which results in high-intensity coherent radiation.

Production of a Laser

A laser can be produced by collimating the stimulated emission, and this is done by properly designing a cavity in which the waves can be used over and over again. Here in optics, the principles of the Fabry-Perot interferometer are employed.

The space between the plates of a Fabry-Perot interferometer is filled with an active medium, which is a medium with population inversion density capable of amplifying the light flux passing through it. A light flux passing through the active medium will be amplified between successive reflections from the mirrors. This system forms an active optical resonator. The flux amplification during a passage through the active medium takes place in accordance with the formula

$$S(z) = S_o \, e^{\alpha z}$$

where, $S_o = S(0)$, $S(z)$ is the energy flux density in the direction of the Z-axis.

The radiation is partially attenuated as a result of reflection at the mirrors. One of the mirrors has the highest possible reflectivity, while the other mirror transmits a certain portion of light forming the radiation called *Laser* radiation.

Lasing begins only when the energy transferred per cycle from the active medium in the resonator to the light flux becomes higher than the total losses of the light flux in the resonator, including the energy carried away by the *laser* radiation. The beginning of lasing is characterized quantitatively by the lasing threshold condition.

Properties of Lasers

The following properties of lasers are discussed here:
   (i) Lasing threshold condition,
   (ii) Steady-state lasing conditions,
   (iii) Q-factor,
   (iv) Continuous-wave (CW) and pulsed lasers,
   (v) Enhancement of the emission power,
   (vi) Q-switching method,
   (vii) Radiation modes,

(viii) Mode synchronization,
 (ix) Pulse duration,
  (x) Attainment of mode synchronization, and
 (xi) Laser speckles.

(i) Lasing threshold condition: Lasing starts when the energy acquired by the high flux in the active medium per cycle exceeds the energy losses, including the energy of the laser radiation leaving the system. The lasing threshold condition is written in the form

$$\alpha_o L = f$$

where, α_o is the gain factor in the absence of a light flux,
L is the distance travelled by light in the active medium during half-cycle,

$$f = -\ln(\rho_1 \rho_2)$$

ρ_1 and ρ_2 are the effective reflectivities of the two mirrors of Fabry-Perot interferometer.

(ii) Steady-state lasing conditions: In this condition, the light beam receives energy from the active medium in order to compensate the losses. The steady-state lasing condition can be expressed as

$$\alpha L = f$$

where, α is the gain-factor per cycle.

(iii) Q-factor: The Q-factor is the ratio of the energy W stored in the system to the energy losses W per cycle.

$$Q = W / W$$

The necessary condition for lasing is that the amplification over a half wavelength be equal to (or larger than) the reciprocal of the Q-factor.

(iv) Continuous-wave (CW) and pulsed lasers: The creation of population inversion of levels is called pumping. Depending upon the nature of its time dependence, pumping may be continuous or pulsed. If pumping is carried out by pulses, the laser-radiation also will be pulsed. When the steady-state lasing condition is satisfied at all instants, the laser radiation is continuous.

(v) Enhancement of the emission power: The power of laser radiation can be increased by increasing the number of atoms participating in the amplification of the light flux in the laser resonator through induced emission and decreasing the pulse duration.

(vi) Q-switching method: The method of controlling the laser output power is called the Q-switching method. The duration of a laser pulse depends on the time during which the population inversion changes due to the emission of a pulse to such an extent that the system no longer satisfies the lasing condition. Pulse duration depends on many factors but is usually of the order of 10^{-7} to 10^{-8} second.

The switching of Q-factor is achieved with the help of a rotating prism. A higher pulse repetition frequency can be attained by switching the Q-factor with the help of a Kerr cell. Q-switching can also be attained by other methods.

(vii) **Radiation modes:** The steady-state laser radiation forms standing waves in optical resonator. In the most general form, a standing wave can be represented as a superposition of elementary standing waves called the oscillation modes. Laser radiation corresponding to these oscillation modes are called the radiation modes of a *laser*.

An oscillation mode depends on the geometrical properties of the resonator, the refractive index of the active medium, and generally on the boundary conditions of the resonator. There are two types of resonators, namely, a resonator with plane rectangular minors and a cylindrical resonator with spherical mirrors. The separation between the sided mode frequencies is much smaller than that between the axial mode frequencies.

(viii) **Mode synchronization:** The concentration of laser radiation in one direction in a resonator with plane mirrors is due to the fact that the Q-factor is maximum in this direction, while the lasing threshold is minimum. The divergence of a laser beam is mainly caused by diffraction phenomena.

The matching of phase of different modes is called mode synchronization. The instant when the amplitudes of modes are added in the same phase corresponds to a sharp increase in the intensity of laser radiation and a pulse is observed at this moment of time.

(ix) **Pulse duration:** Pulse duration is the time for which the resultant amplitude persists. The large the number of synchronized modes, the shorter the pulse duration.

(x) **Attainment of mode synchronization:** Mode synchronization may occur spontaneously. In this case, a small number of modes is usually synchronized and hence the pulse duration does not differ much from the period of their repetition (or the cycle duration). Various methods are applied for reducing the pulse duration. Involvement of nonlinear effects plays a very significant role in laser physics. Different modes are interconnected taking non-linearity into consideration. Mode synchronization is not a completely random process anymore, and it becomes possible to exert influence on it. Pulses with a duration much smaller than 10^{-12} s have been obtained at present.

(xi) **Laser speckles:** The surface of most objects illuminated by a laser beam appears to be granular. The random distribution of the granules over the surface can be photographed. If the camera is focused at a point above the surface or below it, the image of the granules is still formed on the film. The granules appear twinkling and move relative to the surface as the observer moves while looking at the surface. These granules are called speckles. The speckles emerge only due to a high degree of coherence of laser radiation. From the physical point of view, the origin of speckles is associated with the noise effect. This effect has a number of useful applications.

Speckles are formed due to diffuse scattering by a surface as well as by passing light through a scattering object because the phase difference between the waves arriving at a given point from two different scatterers is always the same. Such speckles have also wide applications.

Types of Lasers and Their Characteristics

Various type of lasers are found depending upon the choice of active medium, output power, operating conditions and other characteristics. Basic types of lasers are:

(i) Ruby laser

(ii) Helium-Neon gas laser

(iii) Carbon-dioxide laser

(iv) CW-mode CO_2–laser

(v) T-laser

(vi) Gas dynamic lasers

(vii) Dye lasers

(viii) Other lasers

(i) Ruby laser

The first successful laser developed by Maiman in 1960 was the Ruby laser. A single crystal of synthetic pink ruby was used as resonating cavity. The ruby is primarily a transparent crystal of Corundum (Al_2O_3) doped with approximately 0.05 per cent of trivalent chromium ions in the form of Cr_2O_3, the latter providing its pink colour. The aluminium and oxygen atoms of the Corundum are inert; the Chromium ions are the active ingredients.

The ruby crystal is grown in cylindrical form. Crystals having a length of about 5 cm and a diameter of about 1 cm are usually employed in lasers. The ends of the ruby crystal are polished flat and parallel. In a typical ruby laser, one end is highly reflective (about 96 per cent), and the other end is close to half-silvered (about 50 per cent).

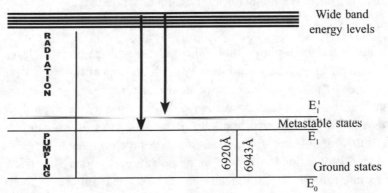

Fig. 12.1

When white light falls on ruby, the blue and the green regions of the spectrum are absorbed, while the red region is scattered. Optical pumping is carried out with the help of a xenon lamp emitting high-intensity pulses of light when a current pulse passing through it heats the gas to a temperature of several thousand kelvins. A continuous pumping is not possible because the lamp cannot operate continuously at such a high temperature. The radiation emitted by the lamp is similar is nature to the black-body radiation, and is absorbed by Cr^{3+} ions which consequently go over to the energy levels in the absorption bands. Because of non-radioactive transition, Cr^{3+} ions rapidly go over to levels E_1 and $E_1'P$ (Fig. 12.1). The excess energy is transferred to the lattice, i.e. transformed into the lattice vibration energy or the phonon energy.

Levels E_1 and E_1' P are metastable. The lifetime corresponding to level E_1 is equal to 4.3 ms. Pulsed pumping leads to an accumulation of excited atoms on levels E_1 and E_1' resulting in a considerable population inversion with respect to level E_o.

The two faces of the ruby crystal are cut in such a way that the radiation emitted by a ruby laser is linearly polarized.

Many other pumping light sources of energy like exploding wires, chemical reactions and concentrated sunlight have been designed and used successfully.

(ii) Helium-Neon gas laser

In 1961, Javan, Bennett and Harriott introduced the operation of gas laser for the first time. Since then various types of gas lasers, using gases of many kinds and mixtures, have been in use. As the He-Ne laser is inexpensive, unusually stable, and continuous in emission, it is widely used in optics and Physics laboratories of the world.

The active medium in He-Ne laser is a gaseous mixture of helium and neon. Lasing is caused by transitions between neon energy levels and helium plays the role of an intermediary through which energy is transferred to neon atoms for creating a population inversion.

In principle, neon can generate laser radiation as a result of more than 130 different transitions. However, the lines having wavelengths 6328 Å, 11,523 Å and 33,900 Å are the ones with highest intensity. The wavelength 6328 Å is in the visible part of the spectrum, while other wavelengths correspond to the infrared region.

The radiation from a helium-neon laser is linearly polarized.

(iii) Carbon-dioxide laser

Carbon-dioxide laser is a high-power molecular-gas laser. A continuous laser beam with a power output of several kilowatts is produced by such an optical device. At the same time, a relatively high degree of purity and coherence is also maintained.

The significance of such laser power is well understood by cutting a diamond and thick steel plates within a few seconds with the help of focused laser beam. Moreover, such lasers produce a wide range of infrared frequencies and are tunable over a range of wavelengths. The beams also find applications in optical communications systems, in optical radar, and in terrestrial and extraterrestrial system, because of the fact that infrared light is only slightly scattered or absorbed by the atmosphere.

The spectra of molecular gases are considerably more complicated than those of many atomic gases. In addition to the electronic energy levels of a free atom, a molecule can have levels arising from quantized vibrations and rotations of the atoms themselves. Consequently, there are a number of almost equally spaced vibration levels corresponding to a given electronic configuration in a molecule, and for each vibrational level there are a number of rotational levels.

Like other molecules, the CO_2 molecules have a band spectrum formed by rotational and vibrational energy levels. The CO_2 molecule is a linear one and has a centre of symmetry. It has three fundamental vibrational modes. The energy of the quanta of fundamental vibrational modes is given by: (i) $1/\lambda_1 = 1337$ cm^{-1}, (ii) $1/\lambda_2 = 667$ cm^{-1}, (iii) $1/\lambda_3 = 2349$ cm^{-1}. There may be one or more quanta corresponding to each vibrational mode. The vibrational states of a molecule are denoted by the number of quanta in the corresponding fundamental vibrational mode.

The addition of N_2 gas to the laser cavity results in the selective raising of CO_2 molecules to the desired laser levels. The efficiency of CO_2 laser is high, because little energy is required for excitation and a major part of this energy is transferred to the laser beam.

The transition used in a CO_2–laser gives radiation with a wavelength 10.6 µm, which lies in the infrared region of the spectrum. With the help of the vibrational levels, the emission frequency can be varied over an interval between 9.2 and 10.8 µm. When a current is passed through the mixture of CO_2 and N_2 gases, energy is transferred to CO_2 molecules by N_2 molecules which are themselves excited as a result of collisions with electrons.

(iv) CW-mode CO_2–laser

The CW-mode laser is the modified form of CO_2–laser. In this, a mixture of CO_2, N_2 and He gases is continuously pumped through the laser in the axial direction. Such a laser is in a position to continuously generate coherent radiation whose power is more than 50 W per metre length of the active medium.

(v) T-laser

When the working mixture of the gases in CO_2–laser is excited by a transverse electric field under atmospheric pressure, T-laser is obtained. Since the electrodes in such a laser are parallel to the axis of the resonator, a relatively small potential difference is required to generate a strong electric field in the resonator. Thus, the laser is operated in a pulsed mode under atmospheric pressure, when the concentration of CO_2 in the resonator is quite high. Consequently, high-power laser radiation is developed.

(vi) Gas dynamic lasers

An intensive cooling takes place by passing a strongly heated mixture of CO_2 and N_2 (at temperatures between 1000 and 2000 K) at a high speed through an expansion nozzle. In this process, the upper and lower energy levels are thermolyzed at different rates, and hence a population inversion is observed. Using this population inversion, laser radiation can be produced by mounting an optical resonator at the outlet from the nozzle. Such type of lasers is known as gas dynamic lasers, which are capable of generating very high power of radiation in the CW-mode.

(vii) Dye lasers

Dyes have very complex molecules with highly pronounced vibrational energy levels. The energy levels in a spectral band are almost continuous. The intermolecular interaction results rapid non-radiative transitions of the molecules to the lowest energy level in each band. Hence, in a very short time following molecular excitation, all excited molecules are concentrated on the lowest level of the E_1–band. These molecules can then undergo a radioactive transition to any energy level of the lower band. As a result of which, radiation of practically and frequency can be emitted in the interval corresponding to the width of the lowest energy band.

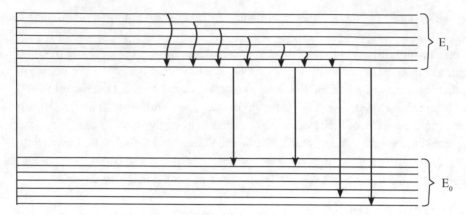

Fig. 12.2: Energy level diagram of a dye molecule

In this way, by using the dye molecules as the active medium for generation, laser radiation of any frequency can be obtained by an appropriate tuning of the resonator. Thus, tunable lasers are produced by employing dyes. Pumping in dye lasers is accomplished with the help of gas-discharge lamps or the radiation from some other type of lasers. As various types of materials are used in dye lasers, it becomes possible to obtain laser radiation in the entire visible range of the spectrum as well as over a considerable part of the infrared and ultraviolet region of the spectrum.

(viii) Other lasers

There are hundreds of different kinds of lasers which are produced by using many different materials. These lasers emit radiation over a wide range of wavelength, from the ultraviolet at one end of the spectrum to microwaves at the other. Lasing is also possible with many gas elements and the same is true of many diatomic and triatomic molecules and many metals.

One type of chemical laser derives energy from the dissociation by light of trifluoroiodomethane (CF_3I). As this complex molecule dissociates, the carbon-iodine bond is broken and an excited iodine atom is released. Coming back to ground state, the iodine atom gives off a photon with a wavelength of 13,150 Å.

Lasers using semiconductors in the form of *pn* junctions are very small. Such type of laser requires only low voltages and is modulated easily. The most commonly used material is gallium arsenide (GaAs) impregnated with zinc.

Laser Safety

He-Ne laser has an intensity of a fraction of a milliwatt, whereas CO_2 laser has an intensity of many kilowatts. Laser injuries have been few and their dangers are highly debatable. However, the greatest danger is the inadvertent direction of an undiverged laser beam directly into the eye.

Filtering glasses and shields are used to save eyes from the high-powered lasers. One should be aware of the fact that a laser beam incident upon a specular reflecting surface can redirect the beam undiminished in intensity.

Applications of Laser

Modulated laser beams are used in communication. Lasers have been employed by the medical professionals in surgery, where retinal tissue is cauterized to weld detached retinas. They have been used by surveyors and engineers for critical alignment, as well as for ranging in metrology and determining the distance to the moon. Attenuation and scattering of laser beams have been used in atmospheric research. Lasers are also utilized in the production and research with holograms.

New findings in the moon and the earth studies can be opened up from the changing distance between these two astronomical bodies. Such investigations are made possible by the application of laser technology. Lasers have been used, like radar, to determine large and small distances. Lasers have already had extensive military use of a different sort. Laser scanner deters terrorists.

QUESTIONS

1. What do you understand by the term Laser?
2. What is the requirement to produce laser beam?
3. What do you understand by population inversion?
4. What is pumping?
5. What are the various techniques of pumping?
6. What are the main properties of laser beam?
7. Differentiate between laser beam and ordinary light beam. Write some uses of laser.
8. Distinguish between spontaneous and stimulated emissions.
9. What are transition probabilities?

●●●

Chapter 13

Fibre Optics

Prolegomenon

Light passing from denser to rarer medium suffers total internal reflection at the boundary of the two media if the angle of incidence becomes greater than the critical angle. This idea had been used by the British physicist John Tyndall to demonstrate that light rays in a tank of water shining through a hole in the side follow the stream of water emerging from the orifice. This effect is commonly observed today in fountains illuminated by lights from under the water. Bundles of tiny rods or fibres of clear glass or plastic provide the basis for the sizeable industry of *fibre optics*. It is known from the observation on individual fibres over 50 m long that there are essentially no losses due to reflection on the sides. All attenuation of an incident beam is attributable to reflection from the two ends and absorption by the fibre material.

Soon after the discovery of the laser, some preliminary tests on the propagation of information-carrying light waves through the open atmosphere were conducted, but it was felt that the presence of rain in the terrestrial atmosphere obstruct the communication system. In order to have an efficient and dependable communication system, a guiding medium would be necessary in which the information-carrying light waves could be transmitted. This guiding medium is the optical fibre which is hair thin and guides the light beam from one place to another.

Glass Fibre

A glass fibre shown in Fig. 13.1 consists of a cylindrical central core cladded by a material of slightly lower refractive index. Light rays impinging on the core-cladding interfaced at an angle greater than the critical angle undergo total internal reflection at that interface and are trapped inside the core of the waveguide.

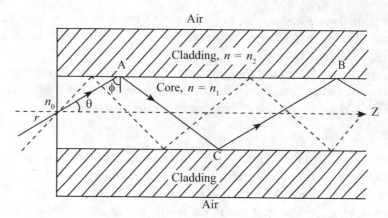

Fig. 13.1: Total internal reflection of light rays at the core-cladding interface

Rays making larger angles with the axis take a longer time to traverse the length of the fibre. Fig. 13.2 shows the transmission of light from a flashlight through a bent glass or plastic rod by total internal reflection.

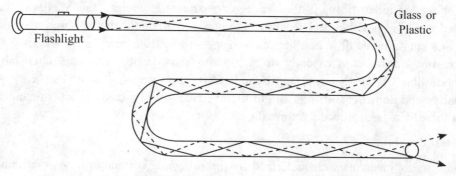

Fig. 13.2: Light from a flashlight follows a bent transparent rod
by total internal reflection

Fibres are usually coated with a thin transparent layer of glass or other material of lower refractive index. Total reflection still occurs between the two. A cladded fibre has advantage over a bare fibre in respect of transmission of light from one place to another. Because, when light is transmitted from one end to another, it is required to support the fibre. These supporting structures may considerably distort the fibre, thereby affecting the guidance of the light wave. This can be eliminated by thickly coating the fibre. The other advantage of coating the fibres is that the fibres of a bundle become separated from one another and this prevents light leakage between touching fibres and at the same time protects the fire-polished reflecting surfaces.

Production of Coated Fibres

Coated fibres are prepared by inserting a thick, high-refractive index glass rod in tubing of lower index. In a special furnace, the two are then drawn down to 1/1000 inch diameter, and the thickness is controlled within narrow limits. A bundle of these fibres can then be fused together to form a solid mass and drawn down a second time so that individual fibres are about 2 μm in diameter. This is about two wavelengths of visible light. Such bundles can resolve nearly 250 lines per millimetre.

If fibres are drawn down until their diameters are close to the wavelength of light, they cease to act like pipes and behave more like waveguides used in conducting microwaves. Two wavelengths of light is an approximate limit for image transmission.

Step-Index Fibre

As already described, the simplest type of optical fibre consists of a thin cylindrical structure of transparent glassy material of uniform refractive index n_1 surrounded by a cladding of another material of uniform but slightly lower refractive index n_2. These fibres are called step-index fibres due to the step discontinuity of the index profile at the core cladding interface.

A pulse of light sent into a fibre broadens in time as it propagates through the fibre. Such phenomenon is known as pulse dispersion. This happens because of the different times taken by different rays to propagate through the fibre. The smaller the pulse dispersion, the greater will be the information-carrying capacity of the system.

Fibres which permit a large number of guided optical ray paths are known as multimode fibres. For a step-index fibre, if the waveguide parameter is less than 2.4048, then only one guided mode is possible and the fibre is known as a single mode fibre. A single ray is possible in a single mode fibre and hence the dispersion will be completely absent. Therefore, the information transmission capacity of such a single mode fibre must be much larger than a multimode fibre. Typically, in a single mode fibre, the dispersion is of the order of a few picoseconds per kilometre and thus single mode fibres are obviously more suitable for uses in which there is a large flow of information.

Parabolic Index Fibre

A parabolic index medium is characterized by the following refractive index distribution

$$n^2 = n_1^2 \, [1 - 2\Delta \, (r^2/a^2)]$$
$$= n_1^2 \, [1 - 2\Delta/a^2 \, (x^2 + y^2)] \qquad \qquad ...(1)$$

Where Δ and a are constants and n_1 represents the axial refractive index. A medium characterized by equation (1) is usually called an infinitely extended square law medium. It should be noted here that the refractive index variation given by equation (1) cannot be extended to infinity and in an actual fibre, there is a boundary at $r = a$ beyond which the refractive index is constant. Here, 'a' is the core radius.

The refractive index 'n' is a function of 'r'

where, $r^2 = x^2 + y^2$ $\qquad \qquad$ ...(2)

Exact solutions of the scalar wave equation for an infinitely extended square law medium give a deep physical insight and also reveal many salient aspects of optical waveguides. Different modes in a parabolic index fibre travel with almost the same group velocity. This means all the rays take approximately the same amount of time to propagate through the fibre.

Uses

Light images can be transmitted around corners and over long distances by using an ordered array or bundle of tiny transparent fibres. If the individual fibres in a bundle are not arranged in an orderly array, then the emerging image is scrambled and hence meaningless.

Optical fibres have capability of carrying a huge amount of information from one place to the other. Recently developed fibres are characterized by extremely low losses (~0.2 dB/km). As a result of which, the distance between two consecutive repeaters can be as large as 250 kms. In a more advance fibre optic system, it has been possible to send 140 Mbit/S information through a 220 km link of one optical fibre optic systems have merits over copper cables, the people of the world have seriously started replacing the copper cables by the fibre optic systems. In addition to the long distance communication systems, optical fibres are also being extensively used for local area networks (LANS). Such networks wire up telephones, televisions, computers or robots in offices and cities.

One of the most important applications of fibre optics is in the field of medicine. A cystoscope, or catheter-type instrument, enables the surgeon to observe and operate by remote control on tiny area deep within the body.

Conclusion

Use of graded core fibres reduces significantly the large pulse dispersion in multimode step index fibres. Indeed, the first generation optical communication systems used multimode graded index fibres at 0.85 µm. It is evident that the information carrying capacity was limited by material dispersion. Shifting the wavelength to 1.3 µm, the material dispersion was decreased. In the second generation optical communication systems, the material dispersion is negligibly small and the information-capacity is limited by intermodal dispersion. Now, the intermodal dispersion can be eliminated by using single mode fibres. This is really the case with third generation optical communication systems which operate at 1.3 µm. Fourth generation optical communication systems use single mode fibres at 1.55 µm. The loss in typical single mode fibres is the lowest around 1.55 µm.

Optical frequencies (~10^{15} Hz) are larger than those of conventional radio waves (~10^{6} Hz) and microwaves (~10^{10} Hz). As a result of which, a light beam acting as a carrier wave is capable of carrying far more information in comparison to radio waves and microwaves. In the very near future, the demand for flow of information traffic will be so high that only a light wave will be able to cope with it. The evolution of different types of optical fibres fabricated with recently developed technology comes to the rescued of the long distance communication systems due to manifestation of high information-carrying ability.

QUESTIONS

1. Explain laser oscillation.
2. What are the applications of optical fibre communication system?
3. Write a short note on optical switches.
4. What is the role of photodiode in OFC?

5. Explain in brief about bending losses.

6. Describe APD in detail.

7. Write a note on fusion splicing method.

8. Explain optical digital link.

9. What are the properties of a good connector?

10. Explain the block diagram of optical fibre communication.

11. What is the principle of operation of distributed feedback laser?

12. What are the disadvantages of OFC system?

13. Explain in brief the performance characteristics in photodiode.

14. Write a short note on population inversion.

15. Write a short note on microbending.

●●●